Advancing an Ecosystem Approach in the Gulf of Maine

Advancing an Ecosystem Approach in the Gulf of Maine

Edited by

Robert L. Stephenson

St. Andrews Biological Station
531 Brandy Cove Road, St. Andrews, New Brunswick E5B 2L9, Canada

John H. Annala

Gulf of Maine Research Institute
350 Commercial Street, Portland, Maine 04101, USA

Jeffrey A. Runge

School of Marine Sciences, University of Maine, Gulf of Maine Research Institute
350 Commercial Street, Portland, Maine 04101, USA

Madeleine Hall-Arber

MIT Sea Grant College Program
77 Massachusetts Avenue E38-300, Cambridge, Massachusetts 02139, USA

American Fisheries Society Symposium 79

Proceedings of the Symposium
"Gulf of Maine Symposium: Advancing Ecosystem Research for the Future of the Gulf"
Held in St. Andrews, New Brunswick, Canada
October 4–9, 2009

American Fisheries Society
Bethesda, Maryland, USA
2012

Suggested citation formats follow.

Entire book

Stephenson, R. L., J. H. Annala, J. A. Runge, and M. Hall-Arber, editors. 2012. Advancing an ecosystem approach in the Gulf of Maine. American Fisheries Society, Symposium 79, Bethesda, Maryland.

Chapter in book

Friedland, K. D., J. P. Manning, and J. S. Link. 2012. Thermal phenological factors affecting the survival of Atlantic salmon in the Gulf of Maine. Pages 393–407 *in* R. L. Stephenson, J. H. Annala, J. A. Runge, and M. Hall-Arber, editors. Advancing an ecosystem approach in the Gulf of Maine. American Fisheries Society, Symposium 79, Bethesda, Maryland.

Printed in the United States of America on acid-free paper.

Library of Congress Control Number 2012945155

ISBN 978-1-934874-30-1
ISSN 0892-2284

American Fisheries Society Web site: www.fisheries.org

American Fisheries Society
5410 Grosvenor Lane, Suite 110
Bethesda, Maryland 20814
USA

Contents

Introduction vii
Symbols and Abbreviations xix

Ecosystem Services in the Gulf of Maine 1
Stephen S. Hale and Maxine Westhead

Linking Ecosystem Health and Services to Inform Marine Ecosystem-Based Management. 9
Katie K. Arkema, and Jameal F. Samhouri

Peopling the Marine Ecosystem 27
Bonnie J. McCay

We Really Can Get There from Here: Important Steps toward Ecosystem-Based Fishery Management 35
Sally McGee

Fishermen without Borders 41
Jean Guy d'Entrement

Tools for Integrated Policy and Management in the Gulf of Maine: A Summary of Major Advancements Since 1996 47
Maxine Westhead

State-of-the-Environment Reporting for the Gulf of Maine 61
Jay Walmsley

Collaborative Learning Strategies to Overcome Barriers to Ecosystem Management in Coastal Watersheds of the Gulf of Maine 73
Christine Baumann Feurt

Linking Science to Management and Policy through Strategic Communication 89
Verna DeLauer, Susan Ryan, Ivar Babb, Peter Taylor, and Pam DiBona

Observations and Analysis of Change in the Gulf of Maine: Present Status and Future Directions 103
Jeffrey A. Runge

Zooplankton of the Gulf of Maine—A Changing Perspective 115
Jonathan A. Hare and Joseph Kane

Status of the Northeast U.S. Continental Shelf Large Marine Ecosystem: An Indicator-Based Approach 139
Michael J. Fogarty, Kevin D. Friedland, Laurel Col, Robert Gamble, Jonathan Hare, Kimberly Hyde, Jason S. Link, Sean Lucey, Hui Liu, Janet Nye, William J. Overholtz, David Richardson, B. Rountree, and Maureen Taylor

Spatial Patterns of Subtidal Benthic Invertebrates and Environmental Factors in the Nearshore Gulf of Maine 167
Stephen S. Hale

Regime Shift in the Gulf of Maine 185
Peter C. Smith, Neal R. Pettigrew, Philip Yeats, David W. Townsend, and Guoqi Han

Zooplankton Monitoring in the Gulf of Maine: Past, Present, and Future 205
Catherine L. Johnson and Jonathan A. Hare

Quantifying Watershed-Based Influences on the Gulf of Maine Ecosystem: The Saint John River Basin 219
Glenn Benoy, Eric Luiker, and Joseph Culp

Anthropogenic and External Influences on the Gulf of Maine: Workshop Summary 235
Madeleine Hall-Arber, Judith Pederson, and Peter Wells

Useable Pasts of Harvested Waters: The Working Role of American Fisheries History 243
Michael J. Chiarappa

Advances in Understanding Ecosystem Structure and Function in the Gulf of Maine 261
Michael J. Fogarty, David W. Townsend, and Emily Klein

Seafloor Mapping for Ecosystem Management in the Gulf of Maine 273
Jonathan H. Grabowski and Tracy Hart

Ecosystem Modeling in the Gulf of Maine Region: Towards an Ecosystem Approach to Fisheries 281
Jason S. Link and Alida Bundy

Managing for Migrants: The Gulf of Maine as a Global "Hotspot" for Long-Distance Migrants 311
Antony W. Diamond

Stalk-Eyed Views of the Gulf of Maine—Through a Nepheloid Layer Dimly 321
Peter A. Jumars

Results of a Collaborative Project to Observe Coastal Zooplankton and Ichthyoplankton Abundance and Diversity in the Western Gulf of Maine: 2003–2008 345
Jeffrey A. Runge and Rebecca J. Jones

Organismal Biology in the Age of Ecosystem-Based Management 361
Jacob P. Kritzer and Jamie M. Cournane

Representation of Biodiversity Knowledge within Ecosystem-Based Management Approaches for the Gulf of Maine 375
Peter Lawton, Lewis S. Incze, and Sara L. Ellis

Thermal Phenological Factors Affecting the Survival of Atlantic Salmon in the Gulf of Maine 393
Kevin D. Friedland, James P. Manning, and Jason S. Link

A Perspective on Advancing Ecosystem Research for the Gulf of Maine 409
Robert L. Stephenson

Introduction

Among the greatest practical challenges facing research and management of the coastal zone are the development and implementation of more integrated, holistic, and sustainable policies and strategies for an ecosystem approach to management (EAM), also known as ecosystem-based management (EBM). Nowhere is this more apparent than in the Gulf of Maine—a region that is both well studied and highly managed by two nations committed to implementing an ecosystem approach to management.

This volume contains a collection of papers, perspectives, and syntheses arising from an international symposium held in October 2009. The objectives of the symposium were to review progress, share knowledge, and identify challenges in the development of an EAM for the Gulf of Maine. Specifically, the symposium sought to (1) examine what progress had been made over the past decade in developing and implementing an ecosystem approach, (2) evaluate how well positioned we are to move forward with an EAM in the Gulf of Maine, and (3) propose research priorities (Cooper et al. 2010; Stephenson and Annala 2012).

It is hoped that the papers collected in this volume provide insight into the evolution of diverse aspects of an ecosystem-based approach to management and that the volume is of interest and relevance to those interested in implementation of an EAM both in the Gulf of Maine and elsewhere.

Context for the 2009 Gulf of Maine Symposium

The 2009 Gulf of Maine Symposium was convened by the Regional Association for Research on the Gulf of Maine (RARGOM) and collaborating organizations for several compelling reasons.

First, it was perceived that there had been some substantial changes since a previous RARGOM Gulf of Maine symposium was held (also in St. Andrews) in 1996 (Wallace and Braasch 1997). Underlying themes of that symposium were ecosystem modeling and recognition of the importance of transboundary linkages among terrestrial (air and watershed) and adjacent marine ecosystems. Since that time, there had been significant advances in our understanding of the Gulf of Maine. Federal, provincial, and state jurisdictions in Canada and the United States had moved forward with ecosystem approaches to management. Both Canada and the United States had made, or had proposed, legislative changes referring specifically to the need for EBM. In Canada, these included the Oceans Act (proclaimed in 1997), Canada's Oceans Strategy (2002) and the Sustainable Fisheries Framework (2009). In the United States, the 1996 Sustainable Fisheries Act amended the Magnuson-Stevens Act with a requirement to identify essential fish habitat and to make progress towards reducing, avoiding, or mitigating both fishing and nonfishing impacts to these habitats. More recently, the U.S. National Oceans Policy (July 2010) established nine objectives, including marine spatial planning (See Westhead 2012, this volume).

Second, there had been considerable evolution, over more than a decade, in the international context related to the ecosystem approach to management, with the result that EAM is a principle ascribed to, and adopted, by many governments and international organizations and agreements. The World Summit on Sustainable Development (Johannesburg 2002) called for application of an ecosystem approach to management by 2010, the United Nations (UN) Convention on Biological Diversity has EAM as a core element, and the UN Food and Agriculture Organization has developed guidelines for application of the ecosystem approach to fisheries management (see, for example, Sinclair and Valdimarsson 2003; Bianchi and Skjoldal 2008).

Third, there had been evolution of the public perspective with respect to the need for enhanced stewardship in the face of ecological change. The public increasingly expressed, on the one hand, concern about current approaches to management and, on the other, an interest in stewardship and sustainability (including marketplace certification schemes). In addition, there was recognition that the ecosystem was changing and was likely to change further as the result of climate change and continued human activity.

The Gulf of Maine (GoM) serves as an excellent case study in the development of an ecosystem approach. The GoM has a relatively long history of research both in the United States and Canada from a number of universities, marine laboratories, and research centers, as well as regional networks (i.e., RARGOM, Gulf of Maine Council, and the Northeast Regional Association of Coastal Ocean Observing Systems [NERACOOS]), making it among the best studied ecosystems in the world. It is managed, with elaborate systems, by two advanced nations with an explicit interest in implementing an ecosystem-based approach. Both nations have implemented regulations, policies, and laws to move management toward a more holistic, ecosystem-based approach. Recognition of the increase in stressors on the marine/coastal system, the benefits of healthy ecosystems, and the need for sustainability (defined broadly) to secure resources for the future was widespread. Because of the desire to implement EAM, the need to share information and review progress from several recent workshops and research initiatives related to aspects of an ecosystem approach in the Gulf of Maine became obvious (King and MacKenzie 2004; Runge and Braasch 2005; Gulf of Maine Council on the Marine Environment 2007).

There was (and remains) increasing awareness of the need for an integrated approach to the management of multiple human activities in relation to a more diverse set of objectives that include a higher standard of ecological integrity and diverse aspects of sustainability (see Stephenson 2012, this volume; Stephenson and Annala 2012). The symposium was seen as a timely opportunity to review the evolving landscape of management, to update the past decade of scientific research, and to shape priorities needed to move forward with integrated management, or ecosystem approach to management, in the Gulf of Maine. Symposium participants were challenged to address three questions related to the implementation of EAM in the Gulf of Maine: What progress has been made in recent years? How well placed are we to implement EAM? What are the priorities going forward? (Cooper et al. 2010).

This Volume

As befitting the subject, this volume contains a diverse set of papers. Some are perspectives, some are research papers, and others are syntheses of theme sessions or technical sessions held at the symposium. The contributions vary in style and span a considerable suite of topics, from the scientific understanding of the biophysical system to the social, management/policy aspects of management. They offer a perspective on the current status of knowledge of the Gulf of Maine as it relates to the emerging ecosystem approach. Papers in the volume illustrate the evolution of ocean science and shift to a more holistic approach in management since the last symposium in 1996 (i.e., technology, monitoring programs, focus on climate change) and provide suggestions for addressing the challenges that arise from attempts to implement EAM.

All the presentations from the 2009 symposium are not represented here, however. Further symposium details, including a complete list of presentations and set of abstracts, are available in a technical report (Cooper et al. 2010). Some papers presented at the symposium have been published elsewhere (referenced in the syntheses), and a summary appears in the proceedings of the Lowell Wakefield "Ecosystems2010" symposium (Stephenson and Annala 2012).

While the ecosystem approach to management is the overarching theme to the volume, no single definition is used throughout. The number of definitions and approaches represented is a product of the evolving state of development of EAM in the Gulf of Maine area. While there is no single definition, the common feature is an overall need "for an integrated approach to the management of multiple activities, in relation to a more diverse set of objectives with a higher standard of ecological integrity, in a changing environment" (Stephenson and Annala 2012). Key attributes include the following: (1) it is place-based and entails the development of integrated management plans for defined social/ecological regions or systems, (2) it considers humans as integral components of the ecosystem, and (3) it requires an understanding of the interrelationships among the components of the system and the environment. Authors were encouraged to define the aspects of EAM relevant to their topic and contribution.

How Far Have We Come? What Have We Learned?

Development and Implementation of the Ecosystem Approach

Very apparent since the 1996 symposium is the rapid move towards implementation of an EAM. Several papers in this volume discuss the development and implementation of this approach. The concept of EAM has found its place in the legislation of both the United States and Canada and in several regional initiatives (DeLauer et al. 2012; Fogarty et al. 2012a; Hall-Arber et al. 2012; McGee 2012; Runge 2012; Stephenson 2012; Westhead 2012; all this volume). The papers in this volume demonstrate both the widespread acceptance of the general concepts and needs, as well as the current diversity of views regarding implementation. Some perspectives emphasize the importance of continued research into the physical and biological basis of ecosystem structure and function (e.g., Fogarty et al. 2012b; Hale 2012; Hare and Kane 2012; Jumars 2012; all this volume). Other perspectives propose a more robust framework of management that accommodates the major uncertainties associated with the difficulty in achieving comprehensive ecosystem understanding (e.g., Hale and Westhead 2012; Walmsley 2012; Westhead 2012; all this volume). Related to this is the recent emergence of marine spatial planning as a tool for implementation.

Context of Climate Change

Recognition of global climate change and its regional impact is a major development since the previous symposium. Evidence exists for at least decadal scale change in the physical properties in the Bay of Fundy and elsewhere in the Gulf of Maine (e.g., Hare and Kane 2012; Smith et al. 2012, this volume). A number of marine species are near the edge of their range (e.g., *Calanus finmarchicus*). In addition, some researchers express concern about the potential for major ecosystem "flips" (Fogarty et al. 2012b; Hare and Kane 2012; Runge 2012). Awareness of the importance of considering cumulative impacts (e.g., lower pH and sea surface temperature warming; acidity, salinity, temperature, and distribution of productivity) and of the uncertainty and variability caused by climate change is more common. One of the goals of recent research in modeling is to help constrain this uncertainty, including the development of scenarios based on advances in biological-physical models (e.g., C. Davis and colleagues, Woods Hole Oceanographic Institution, presentation at the 2009 Gulf of Maine Symposium) and the trophic centric ecosystem models of Link and Bundy (2012, this volume).

Impact of Large-Scale Perturbations

The GoM has experienced large-scale perturbations over the past several decades due to human activities, including top-down effects from overfishing as well as changes to habitat and environmental change. For example, there is evidence for a major shift in transport of water into the GoM—the inflow recently has been fresher with a greater contribution from the Scotian Shelf and less deep flow of nutrient-rich slope water (Smith et al. 2012). Plankton composition, size, and abundance have changed on interannual and interdecadal time scales (Hare and Kane 2012; Johnson and Hare 2012; Runge and Jones 2012; all this volume). The relative dominance of groundfish, pelagic fishes (herring), and elasmobranches has shifted, and there is a general increase in mollusk and crustacean biomass and a large increase in lobster (Fogarty et al. 2012a, 2012b; Runge 2012). These changes are interrelated, and their effects are complex as well as difficult to predict (see, for example, Benoy et al. 2012; Fogarty et al. 2012; Smith et al. 2012; all this volume).

Advances in Knowledge of Gulf of Maine Structure and Function

Over the past decade, there have been significant advances in the Gulf of Maine in a number of areas, including the development of monitoring and observation systems, ecosystem modeling, appreciation of ecology and life history of key species (especially in light of ecosystem change), biodiversity, and seafloor mapping (Stephenson and Annala 2012).

Recent international programs, including the GoMA (Gulf of Maine Area), NaGISA (Natural Geography in Shore Areas), and HMAP (History of Marine Animal Populations) projects of the Census of Marine Life, have documented biodiversity and developed a species registry (Fogarty et al. 2012b; Lawton et al. 2012, this volume). Research programs have investigated the linkage between physical processes and productivity (the U.S. GLOBEC [Global Ocean Ecosystem Dynamics] and the National Oceanic and Atmospheric Administration's ECOHAB [Ecology and Oceanography of Harmful Algal Blooms] programs) (Fogarty et al. 2012b). In addition, knowledge of oceanographic circulation has evolved considerably (Runge 2012; Smith et al. 2012). While there has been progress in physical modeling, including better understanding of climate forcing, we currently do not have the resources or infrastructure to integrate the full breadth of information being collected by major regional data collection schemes (e.g., NERACOOS [Northeastern Regional Association of Coastal and Ocean Observing Systems], the Atlantic Zone Monitoring Program, and the National Oceanic and Atmospheric Administration's ECOMON [Ecosystem Monitoring Program]).

There have certainly been advances in technology/data analyses/techniques with such modeling tools as EcoPath with EcoSim and InVEST, Atlantis, (Arkema and Samhouri 2012; Kritzer and Cournane 2012; Link and Bundy 2012; all this volume); mapping of the seafloor with acoustic imaging (Grabowski and Hart 2012 this volume); habitat mapping as predictive tool (Lawton et al. 2012); use of genomics; greater ability to identify microbes; isotope studies for diet studies; and the use of satellites to document temporal and spatial patterns in surface chlorophyll over time (Fogarty et al. 2012b). Coupled biological-physical models have improved capacity to understand the influence of environmental change on lower ecosystem processes. We have seen an increase in the availability of data online (i.e., OBIS [Ocean Biogeographic Information System], ocean observing data, and satellite imagery). In spite of these advances in ecosystem understanding, the focus remains on the large, higher trophic level organisms, while the representation of the ecosystem below fish in most models currently used in management is rudimentary.

With the recognition that ecosystems can rapidly undergo a change to an alternate state, an emphasis on the need to understand current status and historical responses to forcing factors has increased. The 2009 symposium highlighted research illustrating changes in marine communities, noting that the causes are often interrelated and difficult to unravel. Smith et al. (2012) describe a possible oceanographic regime shift that has only recently been recognized through analysis of time-series data, prompting questions related to change in the productivity and distribution of organisms. Populations of the two most iconic species, Atlantic cod *Gadus morhua* and American lobster *Homarus americanus*, have undergone dramatic changes in abundance. The zooplankton papers in the volume (e.g., Hare and Kane 2012; Runge and Jones 2012) call for sustained monitoring of coastal and offshore waters in order to detect changes over time. Friedland et al. (2012, this volume) discuss the interplay of environmental factors and predation on survival rates of Atlantic salmon *Salmo salar* smolts as they migrate from rivers to oceans. Diamond (2012, this volume) points out that migratory species (whales, shorebirds, seabirds, and tunas) constitute a significant portion of the GoM biomass and must be accounted for in EBM plans. Grabowski and Hart (2012) discuss developments in acoustic imaging techniques (side-scan sonar, multibeam echo sounders, and LIDAR) that are allowing scientists to rapidly examine small seafloor features over large areas, map variation in habitat, and elucidate species-bottom associations.

Progress in the knowledge about the natural GoM ecosystem and the key components of EBM has been made, but there remains a lack of integration of social sciences in the understanding of ecosystem structure and function. The papers by social scientists (DeLauer 2012; Feurt 2012; McCay 2012; all this volume) highlight the need for effective communication between science and decision makers, as well as other stakeholders, and provide best practices and recommendations for techniques to speak to people of all backgrounds (language, cultural lens, knowledge base, and assumptions). Chiarappa (2012, this volume) discusses the history of fishing in the United States and the changing perception/culture surrounding fishing in the eyes of the public. Synthesizing the diverse natural and social elements and applying them in EAM remains a challenge.

Looking Forward—What Is Needed to Implement Ecosystem-Based Management?

Need for an Overall Vision for Regional Ecosystem-Based Management and a Framework for Implementation

While the notion of ecosystem-based management has made it into legislation, there is no single vision or framework for implementation. The complexity of EBM requires a participatory framework for effective implementation. Rather than just having conservation and management goals, EBM must consider ecological, social, economic, and institutional aspects as well as the trade-offs among diverse objectives. Because several governments and agencies have jurisdiction over marine issues, political challenges abound. Several contributions in this volume suggest moving to more holistic, participatory management structures in order to manage with diverse sets of objectives (including conservation, social, economic, and institutional goals). The perspective offered by fisherman/processing plant owner J. G. d'Entremont (2012, this volume) highlighted the need for securing fish resources (such as by use of individual transferable quotas, assess risk to sustainability of changes to policy before implementing), securing year-round jobs in the fishing industry, encouraging value-added products to draw more value from the resource, competitive prices, and adequate control and monitoring to achieve sustainable certifications. He suggests that to date the ecosystem approach has been ad hoc. Fogarty et al. (2012b) also addressed the need for specific sustainable harvest levels, allocation strategies, and the need to define spatial units and determine the value of fisheries for each unit. Governance remains an outstanding issue as no governmental agency around the Gulf of Maine deals sufficiently with socio/economic aspects of management. A more holistic approach requires a more participatory governance structure with a collaborative management approach (McCay 2012; McGee 2012). Stephenson (2012) suggests the need for attention to governance that would allow implementation of a common vision for EAM that could be applied to all activities in a consistent way. It is unlikely that a single approach will emerge in the near future, so there is need to embrace the diversity of perspectives and watch (compare) how they perform.

A Need to Link Ecosystem Function and Structure (Health) with Ecosystem Services

The notion of ecosystem services is an emerging concept. As the main aims of EAM are to maintain ecosystem health and sustain the ecosystem services provided to humans, there is a need to incorporate ecosystem services into both models and management in order to quantify and value the services provided. Once services have an articulated value, it is possible to perform cost–benefit analysis and to evaluate trade-offs among ecosystem services and human activities (Arkema and Sambouri 2012; Hale and Westhead 2012). Several papers emphasize the need to maintain healthy, productive, and resilient systems (Fogarty et al. 2012b; Hale 2012; Johnson and Hare 2012; Lawton et al. 2012; Walmsley 2012). This requires information about abundance, diversity, and distribution of key species, as well as relationships, patterns of energy flow, functional roles of taxa, and the role of external stressors that will allow prediction of likely responses to natural and anthropogenic forcing. Consideration of the time scale of management is relevant. It is anticipated that climate forcing will shift biodiversity and ecosystem services and that adaptation to climate change will become an increasingly important aspect of management. Papers at the symposium suggested looking longer term (out to a decade or more) and adding greater historical perspective. What was the system like 200 years ago (e.g., HMAP [History of Marine Populations]). What will it be like in 10–20 years? While several papers discuss the need for an historical perspective, we lack an accurate baseline. By the time data began to be collected in late 1800s, the system was already greatly altered (i.e., low abundances of whales, seabirds, anadromous fishes; dams; contaminants; Benoy et al. 2012; Fogarty et al. 2012b).

Need for an Interdisciplinary or Multidisciplinary Approach

The diverse objectives of the ecosystem-based approach require integration of several disciplines. The syntheses from the symposium theme sessions, including Westhead (2012; tools), Fogarty et al. (2012b; structure and function of GoM), Hall-Arber et al. (2012; anthropogenic and external influences), Runge (2012; monitoring, observation, and data collection), and Hale and Westhead (2012; ecosystem services workshop), show the current range of perspectives. More attention has been paid to biophysical information than to social aspects, and the symposium did not fully develop the human dimensions (McCay 2012). Social scientists and natural scientists are not really fully engaged with each other. Members of different disciplines and groups employ different methodologies, read different literatures, and have different languages and different traditions. Not only are managers and scientists not collaborating, but natural scientists and social scientists are also not integrated. Agreement on a common list of objectives, including social, economic, and cultural objectives, as well as relevant performance indicators and reference points, is needed.

Need for Collection of New and Different Information

The evolution of a more holistic approach to management has created a challenge with respect to information. Understanding of the ecosystem is incomplete—our understanding is biased toward large organisms and processes that have been important to past activities—but we can predict that the EAM will require new and different information (Stephenson 2012). The driver-pressure-state-impact-response (DPSIR) framework and a management by objective structure provide a framework for assessing more diverse aspects of a system. In order to assess the status of system and changes over time, useful indicators must be identified. Fogarty et al. (2012a) used a time series of North Atlantic oscillation and Atlantic multidecadal oscillation data, human drivers (population, per capita income), physical factors (temperature, salinity, stratification), fishing effort and catch, biomass and distribution of taxa (phytoplankton, zooplankton, inverts, fish), primary production, and fishing revenue to assess state of the ecosystem. Walmsley (2012) discussed how this DPSIR approach can be used in state of environment reports such as the 2010 report by Gulf of Maine Council. The challenge of supporting long-term, comprehensive, consistent, and relevant data sets remains.

The more data and longer the time series, the more useful are models in interpreting change over time and for informing management. This requires funding and commitment to monitoring, as well as to availability of data and infrastructure committed to its analysis. Hare and Kane (2012) note that "complex adaptive systems present major challenges to EBM because regime shifts and phase shifts cannot be foreseen," and that monitoring programs can help to detect changes over time (i.e., of zooplankton), even if they cannot be predicted. Monitoring programs should occur in nearshore and offshore areas, in both the United States and Canada, and include collection of data on a variety of taxa and environmental indicators. New methods will be needed for the collection of some data (e.g., nets do not work for sampling gelatinous zooplankton or micronekton taxa such as euphausids, requiring instead acoustic or video sampling). To date, the focus has been on single species that are commercially valuable and/or large and resident. The importance of incorporating knowledge of a broad range of taxa (i.e., microbes, meiofauna, long-distance migrants such as migratory birds, whales, tunas, and basking shark *Cetorhinus maximus*) and habitats (Diamond 2012; Lawton et al. 2012), including the identification of biologically diverse sites (hotpots) and protection of these sites as well as representative habitats, is increasingly acknowledged. Furthermore, spatial units must be better delineated. Evidence suggests that the near shore requires special considerations, that the western and central Gulf of Maine are distinct from the eastern GoM and Scotian Shelf (Fogarty et al. 2012b), and that there are known biogeographic breaks (Hale 2012). Papers in this volume point to the importance of incorporating diverse sources of information and knowledge, including local ecological knowledge (e.g., Walmsley 2012). This raises the important issue of our institutional capacity to implement EAM.

Need for Enhanced Communication among Participants

Several authors (DeLauer et al. 2012; Feurt 2012; Lawton et al. 2012; Runge 2012) reference a critical need for strategic and effective communication among disciplines and stakeholders. Interactions between science and decision makers in particular need to be improved (transfer of knowledge) and framed in the context of feedback loops rather than a one-way transfer of knowledge. To make communication effective, there is a need to present scientific information so it can be easily accessed and understood by multiple stakeholders. Feurt (2012) points out that one of the biggest challenges to implementing EBM is social—while ecosystem management theory provides a conceptual framework, putting it in practice has been difficult. She evaluates the application of social science methods to create a program based on collaborative learning principles (cross-disciplinary dialogue, learning through adaptive management, involvement of all stakeholders, avoiding discipline specific language, cultural models, and acknowledgment of the value of all types of expertise—ecological, technical, local, land use, educational, governance, and science). The symposium pointed to the importance of enhanced ocean literacy and public understanding, appreciation of the suite of "pressures" (e.g., fishing and agriculture) and continued development of enhanced dialogue among managers, scientists, and stakeholders.

While not yet fully articulated or operational, an EAM will inevitably include more diverse objectives with respect to the impacts of fishing and other activities on the ecosystem, as well as increased awareness of the impact of the ecosystem (including ecosystem change) on management. It will demand integration of conservation (including maintenance of productivity at the organism and trophic levels, conservation of biodiversity of populations and communities, and protection of habitat) as well as social, economic, and institutional aspects of management. There is increasing awareness of the need for development of an integrated approach to the management of multiple human activities in relation to a more diverse set of objectives, a higher standard of ecological integrity, more diverse aspects of sustainability, and environmental change.

Impediments to implementation of an EAM in the Gulf of Maine include diversity in definition or approach, complexity in jurisdiction, ecosystem complexity, the need for enhanced monitoring and information to support evolving management landscape, and a need for institutional (governance) to support cross-disciplinary and interjurisdictional considerations.

EAM requires an integrated, interdisciplinary, participatory, and highly communicative environment. A major question, unresolved by the symposium, is whether we now have the infrastructure to promote this. EAM requires more (and different) information and increased participation. The move to EAM demands institutional capacity that will link the various elements and allow effective communication among diverse participants.

Acknowledgments

This volume draws heavily upon a technical report of the symposium (Cooper et al. 2010), and we thank the contributors to that report, especially the session chairs and rapporteurs: Lara Cooper, Heather Breeze, Marc Craig, Maxine Westhead, Leslie-Ann McGee, Arran McPherson, Tim Hall, Emily Klein, Michael Fogarty, David Townsend, Peter Wells, Madeleine Hall-Arber, Judy Pederson, Jim Abraham, Bob O'Boyle, Catherine Johnson, Ru Morrison, Jeffrey Runge, Mike Sinclair, Stephen Hale, Lew Incze, Peter Lawton, Sara Ellis, Jonathan Grabowski, Brian Todd, Page Valentine, Tracy Hart, Jake Kritzer, Jamie Cournane, Verna DeLauer, and David Keeley.

We are grateful to many people who contributed to reviews and editing associated with this volume, including Cameron Ainsworth, Chelsie Archibald, Katie Arkema, Brenda Best, Ben Blount, Harmon Brown, Jim Churchill, Patricia Clay, Flaxen Conway, Chad Demarest, Dorothy Diamond, Syma Ebbin, Julie Ekstrom, Gordon Fairchild, Shirley Fiske, Peter Fricke, Kevin Friedland, Beth Fulton, Walt Golet, David Craig Griffin, Vincent Guida, Brad Harris, Justin Huston, Pete Jumars, Vladimir Kostylev, Dan Lane, Kathleen Leyden, Steve McCormick, Ru Morrison, Dave Mountain, Barbara Neis, Bob O'Boyle, Fred Page, Andrew Pershing, Bill Peterson, Stephane Plourde, Howard Powles, Nicolas Record, Winnie Ryan, Nancy Shackell, Graham Sherwood, Rabindra Singh, Andre St. Hilaire, Jason Stockwell, Brian Todd, Howard Townsend, Thomas Trott, Gordon Waring, Jeff Watson, Priscilla Weeks, and Maxine Westhead.

We acknowledge financial contributions of the National Oceanic and Atmospheric Administration and Fisheries and Oceans Canada. We thank Aaron Lerner and Deborah Lehman of the American Fisheries Society for assistance with publication.

References

Arkema, K. K., and J. F. Samhouri. 2012. Linking ecosystem health and services to inform marine ecosystem-based management. Pages 9–25 *in* R. L. Stephenson, J. H. Annala, J. A. Runge, and M. Hall-Arber, editors. Advancing an ecosystem approach in the Gulf of Maine. American Fisheries Society, Symposium 79, Bethesda, Maryland.

Benoy, G., E. Luiker, and J. Culp. 2012. Quantifying watershed-based influences on the Gulf of Maine ecosystem: the Saint John River basin. Pages 219–234 *in* R. L. Stephenson, J. H. Annala, J. A. Runge, and M. Hall-Arber, editors. Advancing an ecosystem approach in the Gulf of Maine. American Fisheries Society, Symposium 79, Bethesda, Maryland.

Bianchi, G., and H. R. Skjoldal, editors. 2008. The ecosystem approach to fisheries. Food and Agriculturel Organization of the United Nations, Rome.

Chiarappa, M. J. 2012. Usable pasts of harvested waters: the working role of American Fisheries history. Pages 243–259 *in* R. L. Stephenson, J. H. Annala, J. A. Runge, and M. Hall-Arber, editors. Advancing an ecosystem approach in the Gulf of Maine. American Fisheries Society, Symposium 79, Bethesda, Maryland.

Cooper, L. L., R. L. Stephenson, and J. H. Annala. 2010. Gulf of Maine symposium—advancing ecosystem research for the future of the gulf: proceedings of a symposium held at St. Andrews, New Brunswick, October 5–9, 2009. Canadian Technical Report of Fisheries and Aquatic Sciences 2904.

DeLauer, V., S. Ryan, I. Babb, P. Taylor, and P. DiBona. 2012. Linking science to management and policy through strategic communication. Pages 89–101 *in* R. L. Stephenson, J. H. Annala, J. A. Runge, and M. Hall-Arber, editors. Advancing an ecosystem approach in the Gulf of Maine. American Fisheries Society, Symposium 79, Bethesda, Maryland.

d'Entrement, J. G. 2012. Fishermen without borders. Pages 41–45 *in* R. L. Stephenson, J. H. Annala, J. A. Runge, and M. Hall-Arber, editors. Advancing an ecosystem approach in the Gulf of Maine. American Fisheries Society, Symposium 79, Bethesda, Maryland.

Diamond, A. W. 2012. Managing for migrants: the Gulf of Maine as a global "hotspot" for long-distance migrants. Pages 311–320 *in* R. L. Stephenson, J. H. Annala, J. A. Runge, and M. Hall-Arber, editors. Advancing an ecosystem approach in the Gulf of Maine. American Fisheries Society, Symposium 79, Bethesda, Maryland.

Feurt, C. B. 2012. Collaborative learning strategies to overcome barriers to ecosystem management in coastal watersheds of the Gulf of Maine. Pages 73–88 *in* R. L. Stephenson, J. H. Annala, J. A. Runge, and M. Hall-Arber, editors. Advancing an ecosystem approach in the Gulf of Maine. American Fisheries Society, Symposium 79, Bethesda, Maryland.

Fogarty, M. J., K. D. Friedland, L. Col, R. Gamble, J. Hare, K. Hyde, J. S. Link, S. Lucey, H. Liu, J. Nye, W. J. Overholtz, D. Richardson, B. Rountree, and M. Taylor. 2012. Status of the northeast U.S. Continental Shelf large marine ecosystem: an indicator-based approach. Pages 139–165 *in* R. L. Stephenson, J. H. Annala, J. A. Runge, and M. Hall-Arber, editors. Advancing an ecosystem approach in the Gulf of Maine. American Fisheries Society, Symposium 79, Bethesda, Maryland.

Fogarty, M. J., D. Townsend, and E. Klein. 2012. Advances in understanding ecosystem structure and function in the Gulf of Maine. Pages 261–272 *in* R. L. Stephenson, J. H. Annala, J. A. Runge, and M. Hall-Arber, editors. Advancing an ecosystem approach in the Gulf of Maine. American Fisheries Society, Symposium 79, Bethesda, Maryland.

Friedland, K. D., J. P. Manning, and J. S. Link. 2012. Thermal phenological factors affecting the survival of Atlantic salmon in the Gulf of Maine. Pages 393–407 *in* R. L. Stephenson, J. H. Annala, J. A. Runge, and M. Hall-Arber, editors. Advancing an ecosystem approach in the Gulf of Maine. American Fisheries Society, Symposium 79, Bethesda, Maryland.

Grabowski, J. H., and T. Hart. 2012. Seafloor mapping for ecosystem management in the Gulf of Maine. Pages 273–280 *in* R. L. Stephenson, J. H. Annala, J. A. Runge, and M. Hall-Arber, editors. Advancing an ecosystem approach in the Gulf of Maine. American Fisheries Society, Symposium 79, Bethesda, Maryland.

Gulf of Maine Council on the Marine Environment. 2007. Gulf of Maine Council on the Marine Environment Action Plan 2007-2012. Available: www.gulfofmaine.org/actionplan/GOMC%20Action%20Plan%202007-2012.pdf (June 2012).

Hale, S. S. 2012. Spatial patterns of subtidal benthic invertebrates and environmental factors in the nearshore Gulf of Maine. Pages 167–183 *in* R. L. Stephenson, J. H. Annala, J. A. Runge, and M. Hall-Arber, editors. Advancing an ecosystem approach in the Gulf of Maine. American Fisheries Society, Symposium 79, Bethesda, Maryland.

Hale, S. S., and M. Westhead. 2012. Ecosystem services in the Gulf of Maine. Pages 1–8 *in* R. L. Stephenson, J. H. Annala, J. A. Runge, and M. Hall-Arber, editors. Advancing an ecosystem approach in the Gulf of Maine. American Fisheries Society, Symposium 79, Bethesda, Maryland.

Hall-Arber, M., J. Pederson, and P. Wells. 2012. Anthropogenic and external influences on the Gulf of Maine: workshop summary. Pages 235–241 *in* R. L. Stephenson, J. H. Annala, J. A. Runge, and M.

Hall-Arber, editors. Advancing an ecosystem approach in the Gulf of Maine. American Fisheries Society, Symposium 79, Bethesda, Maryland.

Hare, J. A., and J. Kane. 2012. Zooplankton of the Gulf of Maine—a changing perspective. Pages 115–137 *in* R. L. Stephenson, J. H. Annala, J. A. Runge, and M. Hall-Arber, editors. Advancing an ecosystem approach in the Gulf of Maine. American Fisheries Society, Symposium 79, Bethesda, Maryland.

Johnson, C. L., and J. A. Hare. 2012. Zooplankton monitoring in the Gulf of Maine: past, present, and future. Pages 205–218 *in* R. L. Stephenson, J. H. Annala, J. A. Runge, and M. Hall-Arber, editors. Advancing an ecosystem approach in the Gulf of Maine. American Fisheries Society, Symposium 79, Bethesda, Maryland.

Jumars, P. A. 2012. Stalk-eyed views of the Gulf of Maine—through a nepheloid layer dimly. Pages 321–343 *in* R. L. Stephenson, J. H. Annala, J. A. Runge, and M. Hall-Arber, editors. Advancing an ecosystem approach in the Gulf of Maine. American Fisheries Society, Symposium 79, Bethesda, Maryland.

King, P., and C. MacKenzie. editors. 2004. Gulf of Maine summit: committing to change. St. Andrews, New Brunswick, October 26–29, 2004. Summit Report. Gulf of Maine Council on the Marine Environment & Global Program of Action Coalition for the Gulf of Maine. Available: www2.gulfofmaine.org/gulfofmainesummit-org/report.html (July 2012).

Kritzer, J. P., and J. M. Cournane. 2012. Organismal biology in the age of ecosystem-based management. Pages 361–373 *in* R. L. Stephenson, J. H. Annala, J. A. Runge, and M. Hall-Arber, editors. Advancing an ecosystem approach in the Gulf of Maine. American Fisheries Society, Symposium 79, Bethesda, Maryland.

Lawton, P., L. S. Incze, and S. L. Ellis. 2012. Representation of biodiversity knowledge within ecosystem-based management approaches for the Gulf of Maine. Pages 375–391 *in* R. L. Stephenson, J. H. Annala, J. A. Runge, and M. Hall-Arber, editors. Advancing an ecosystem approach in the Gulf of Maine. American Fisheries Society, Symposium 79, Bethesda, Maryland.

Link, J. S., and A. Bundy. 2012. Ecosystem modeling in the Gulf of Maine region: towards an ecosystem approach to fisheries. Pages 281–310 *in* R. L. Stephenson, J. H. Annala, J. A. Runge, and M. Hall-Arber, editors. Advancing an ecosystem approach in the Gulf of Maine. American Fisheries Society, Symposium 79, Bethesda, Maryland.

McCay, B. J. 2012. Peopling the marine ecosystem. Pages 27–33 *in* R. L. Stephenson, J. H. Annala, J. A. Runge, and M. Hall-Arber, editors. Advancing an ecosystem approach in the Gulf of Maine. American Fisheries Society, Symposium 79, Bethesda, Maryland.

McGee, S. 2012. We really can get there from here: important steps toward ecosystem-based fishery management. Pages 35–40 *in* R. L. Stephenson, J. H. Annala, J. A. Runge, and M. Hall-Arber, editors. Advancing an ecosystem approach in the Gulf of Maine. American Fisheries Society, Symposium 79, Bethesda, Maryland.

Runge, J. A. 2012. Observations and analysis of change in the Gulf of Maine: present status and future directions. Pages 103–114 *in* R. L. Stephenson, J. H. Annala, J. A. Runge, and M. Hall-Arber, editors. Advancing an ecosystem approach in the Gulf of Maine. American Fisheries Society, Symposium 79, Bethesda, Maryland.

Runge, J. A., and E. Braasch. 2005. Modeling needs related to the regional observing systems in the Gulf of Maine. Regional Association for Research on the Gulf of Maine, RARGOM Report 0501, Durham, New Hampshire.

Runge, J. A., and R. J. Jones. 2012. Results of a collaborative project to observe coastal zooplankton and ichthyoplankton abundance and diversity in the western Gulf of Maine: 2003–2008. Pages 345–359 *in* R. L. Stephenson, J. H. Annala, J. A. Runge, and M. Hall-Arber, editors. Advancing an ecosystem approach in the Gulf of Maine. American Fisheries Society, Symposium 79, Bethesda, Maryland.

Sinclair, M., and G. Valdimarsson, editors. 2003. Responsible fisheries in the marine ecosystem. Food and Agriculture Organization of the United Nations, Rome.

Smith, P. C., N. R. Pettigrew, P. Yeats, D. W. Townsend, and G. Han. 2012. Regime shift in the Gulf of

Maine. Pages 185–203 *in* R. L. Stephenson, J. H. Annala, J. A. Runge, and M. Hall-Arber, editors. Advancing an ecosystem approach in the Gulf of Maine. American Fisheries Society, Symposium 79, Bethesda, Maryland.

Stephenson, R. L. 2012. A perspective on advancing ecosystem research for the Gulf of Maine. Pages 409–415 *in* R. L. Stephenson, J. H. Annala, J. A. Runge, and M. Hall-Arber, editors. Advancing an ecosystem approach in the Gulf of Maine. American Fisheries Society, Symposium 79, Bethesda, Maryland.

Stephenson, R. L., and J. Annala. 2012. Progress on implementing ecosystem-based management in the Gulf of Maine. In G. H. Kruse, H. I. Browman, K. L. Cochrane, D. Evans, G. S. Jamieson, P. A. Livingston, D. Woodby, and C. I. Zhang editors. Global progress in ecosystem-based fisheries management. Alaska Sea Grant, University of Alaska, AK-SG-12-01PDF, Fairbanks.

Wallace, G. T., and E. F. Braasch, editors. 1997. Proceedings of the Gulf of Maine ecosystem dynamics scientific symposium and workshop. Regional Association for Research on the Gulf of Maine, RARGOM Report 97-1, Hanover, New Hampshire.

Walmsley, J. 2012. State-of-the-environment reporting for the Gulf of Maine. Pages 61–72 *in* R. L. Stephenson, J. H. Annala, J. A. Runge, and M. Hall-Arber, editors. Advancing an ecosystem approach in the Gulf of Maine. American Fisheries Society, Symposium 79, Bethesda, Maryland.

Westhead, M. 2012. Tools for integrated policy and management in the Gulf of Maine: a summary of major advancements since 1996. Pages 47–60 *in* R. L. Stephenson, J. H. Annala, J. A. Runge, and M. Hall-Arber, editors. Advancing an ecosystem approach in the Gulf of Maine. American Fisheries Society, Symposium 79, Bethesda, Maryland.

Symbols and Abbreviations

The following symbols and abbreviations may be found in this book without definition. Also undefined are standard mathematical and statistical symbols given in most dictionaries.

A	ampere
AC	alternating current
Bq	becquerel
C	coulomb
°C	degrees Celsius
cal	calorie
cd	candela
cm	centimeter
Co.	Company
Corp.	Corporation
cov	covariance
DC	direct current; District of Columbia
D	dextro (as a prefix)
d	day
d	dextrorotatory
df	degrees of freedom
dL	deciliter
E	east
E	expected value
e	base of natural logarithm (2.71828...)
e.g.	(exempli gratia) for example
eq	equivalent
et al.	(et alii) and others
etc.	et cetera
eV	electron volt
F	filial generation; Farad
°F	degrees Fahrenheit
fc	footcandle (0.0929 lx)
ft	foot (30.5 cm)
ft^3/s	cubic feet per second (0.0283 m^3/s)
g	gram
G	giga (10^9, as a prefix)
gal	gallon (3.79 L)
Gy	gray
h	hour
ha	hectare (2.47 acres)
hp	horsepower (746 W)
Hz	hertz
in	inch (2.54 cm)
Inc.	Incorporated
i.e.	(id est) that is
IU	international unit
J	joule
K	Kelvin (degrees above absolute zero)
k	kilo (10^3, as a prefix)
kg	kilogram
km	kilometer
l	levorotatory
L	levo (as a prefix)
L	liter (0.264 gal, 1.06 qt)
lb	pound (0.454 kg, 454g)
lm	lumen
log	logarithm
Ltd.	Limited
M	mega (10^6, as a prefix); molar (as a suffix or by itself)
m	meter (as a suffix or by itself); milli (10^{-3}, as a prefix)
mi	mile (1.61 km)
min	minute
mol	mole
N	normal (for chemistry); north (for geography); newton
N	sample size
NS	not significant
n	ploidy; nanno (10^{-9}, as a prefix)
o	ortho (as a chemical prefix)
oz	ounce (28.4 g)
P	probability
p	para (as a chemical prefix)
p	pico (10^{-12}, as a prefix)
Pa	pascal
pH	negative log of hydrogen ion activity
ppm	parts per million
qt	quart (0.946 L)
R	multiple correlation or regression coefficient
r	simple correlation or regression coefficient
rad	radian
S	siemens (for electrical conductance); south (for geography)

SD	standard deviation
SE	standard error
s	second
T	tesla
tris	tris(hydroxymethyl)-aminomethane (a buffer)
UK	United Kingdom
U.S.	United States (adjective)
USA	United States of America (noun)
V	volt
V, Var	variance (population)
var	variance (sample)
W	watt (for power); west (for geography)
Wb	weber
yd	yard (0.914 m, 91.4 cm)
α	probability of type I error (false rejection of null hypothesis)
β	probability of type II error (false acceptance of null hypothesis)
Ω	ohm
μ	micro (10^{-6}, as a prefix)
′	minute (angular)
″	second (angular)
°	degree (temperature as a prefix, angular as a suffix)
%	per cent (per hundred)
‰	per mille (per thousand)

American Fisheries Society Symposium 79:1–8, 2012

Ecosystem Services in the Gulf of Maine

Stephen S. Hale
Atlantic Ecology Division, National Health and Environmental Effects Research Laboratory, Office of Research and Development, U.S. Environmental Protection Agency Narragansett, Rhode Island 02882, USA

Maxine Westhead
Fisheries and Oceans Canada, Maritimes Region Bedford Institute of Oceanography 1 Challenger Drive, B500, Dartmouth, Nova Scotia B2Y 2V9, Canada

Abstract.—The primary goal of ecosystem-based management (EBM) is to sustain the long-term capacity of the natural world to provide ecosystem services. A technical workshop was held on October 5, 2009 at the 2009 Gulf of Maine Symposium, with the objective of moving toward identifying, mapping, quantifying, and valuing ecosystem services in the Gulf of Maine. Ecosystem services are the benefits humans derive from ecosystems—the things we need and care about that we get from nature. Making the benefits of biodiversity and ecosystem services more apparent to environmental managers and society at large is necessary to pave the way for more efficient policy and management. Ecosystem services can provide a framework for assessing and resolving trade-offs among potentially conflicting human activities. Many of the scientific and technical elements necessary to move forward with ecosystem services approaches and EBM in the Gulf of Maine are already in place and have been applied in other areas. Currently, what is lacking is a policy and regulatory framework. Outstanding research questions include a more complete understanding of *all* ecosystem services, how they can be valued, and how the links within and among social-ecological systems influence their delivery. To implement ecosystem services and EBM in the Gulf of Maine, we need a clear vision, institutions with clear mandates, EBM science infrastructure, and integrative and interdisciplinary partnerships. Infrastructure for United States–Canada science coordination is in place through the Gulf of Maine Council and RARGOM (the Regional Association for Research on the Gulf of Maine). However, management issues are more difficult because we have bilateral agreements only on fish stock management, and we need formal agreements to work together on broader ecosystem elements.

Introduction

The goal of the Ecosystem Services Workshop on October 5, 2009 at the 2009 Gulf of Maine Symposium was to move toward identifying, mapping, quantifying, and valuing ecosystem services in the Gulf of Maine. Sustaining the long-term capacity of systems to deliver ecosystem services is the core goal of ecosystem-based management (EBM) for the oceans (McLeod and Leslie 2009). Seven ecologists and economists gave presentations, and then the panel discussed the benefits that an ecosystem services approach would bring to management of the Gulf of Maine, how ecologists and economists can work together, and what implementation strategies are most likely to succeed.

Ecosystems and biodiversity are essential for human well-being. Ecosystem services (also referred to as ecosystem goods and services) are the

* Corresponding author: hale.stephen@epa.gov

benefits humans derive from ecosystems—the things we need and care about that we get from nature. This approach makes the importance of a healthy environment more obvious and relevant to politicians, economists, business people, and the public. It is hoped that this will motivate conservation and sustainability. Often, ecosystem services—other than goods provided that we can easily put a dollar value on—are not factored into important decisions that affect ecosystems and are not included in cost–benefit analyses. This damages the ability of nature to provide services, making human society and the environment poorer. Making the benefits of biodiversity and ecosystem services more visible to economies and society is necessary to pave the way for more efficient policy and management. Investing in natural capital can be more cost-effective than man-made solutions and can yield high returns (TEEB 2009).

The Gulf of Maine provides humans with many ecosystem services. While seafood, recreation, and esthetics are highly visible to the public, there are numerous other services that need recognition. The Millennium Ecosystem Assessment (MEA 2005) identified four categories of ecosystem services based on function (Figure 1). These services provide a link between ecosystems and human well-being, which in turn motivates human decisions regarding ecosystems (Figure 2).

The Ecosystem Services Workshop convened researchers from Canada and the United States, who are focused on evaluating ecosystem services, to share methods, techniques, and experiences. Workshop speakers and panel members identified further research needed in the Gulf of Maine to transform the way we account for the type, quality, and magnitude of nature's goods and services so they can be valuated and considered in management decisions. We discussed the need for data,

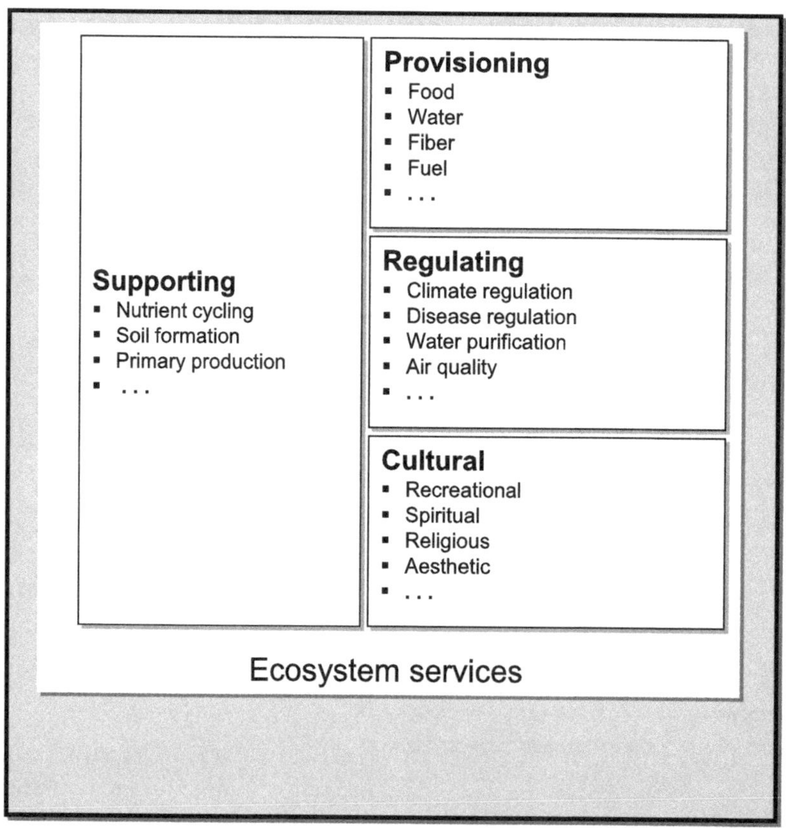

FIGURE 1. Categories of ecosystem services from the Millenium Ecosystem Assessment (MEA 2005).

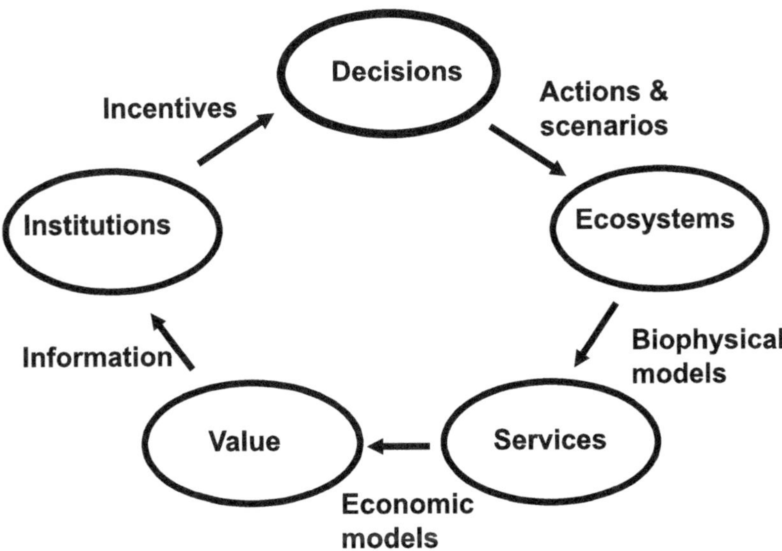

FIGURE 2. Links between ecosystems and human society. After Arkema and colleagues (Stanford University).

methods, and models to better understand and communicate the benefits of considering ecosystem services. We gauged interest in forming a partnership to identify, map, quantify, and valuate ecosystem services in the Gulf of Maine, including estuaries and coastal wetlands.

Summary of Presentations and Discussion

H. Leslie (Brown University) noted that as overfishing, climate change, pollution, habitat alteration, and other stressors continue to impact ocean ecosystems, national and international bodies have called for a shift toward more comprehensive and integrated management that recognizes the full array of benefits provided by coastal and marine ecosystems. Key elements of EBM include (1) place-based approach that considers the entire ecosystem, including people; (2) recognition of the interactive and cumulative effects of human activities; and (3) explicit analysis of trade-offs among different ecosystem services and human activities. A clear understanding of trade-offs among services is necessary for EBM because it is not possible to maximize all services in all places at all times. Tools are needed to assess these trade-offs and aid EBM. Ecosystem services can provide a framework for assessing and resolving trade-offs among different human activities in coastal and marine areas to maintain or restore healthy, resilient, and productive marine ecosystems. In order to implement EBM, a clear vision, institutions with clear mandates, EBM science infrastructure, and integrative and interdisciplinary partnerships among scientists and practitioners are needed.

A. DeMaio-Sukic and T. Souksanh (Fisheries and Oceans Canada) stated that accounting for ecosystem services is essential in our pursuit of economic prosperity if we want to accumulate sustainable wealth. Nonmarket valuation, such as stated preference methods, is a tool for quantifying, in monetary terms, the benefits of initiatives that improve our natural environment (e.g., species at risk). Governments play an important role in generating and disseminating knowledge that guides consumers' sustainable choices and in setting up policies that minimize the ecological footprint of economic activities and products. Interdisciplinary collaboration among scientists, resource managers, and economists is essential for developing analytical scenarios. Plausible science-based management scenarios are important for making the linkages between the provision of ecosystem services and their socioeconomic valuation and for ensuring policy-relevant results.

Fisheries and Oceans Canada undertakes a significant amount of economic analysis and research, with the goal of integrating ecosystem services considerations into policy decision making. But requirements for data and knowledge are large, and a conceptual framework is needed for systematically doing this. Examples include cost–benefit analysis and nonmarket valuation work for species at risk initiatives and pathways of effect models. For example, mean willingness to pay (WTP) for a marine protected area that would facilitate the recovery of several marine mammals in the St. Lawrence estuary for different scenarios ranged from Can$82 to $242 per household per year.

K. Arkema (Stanford University) and colleagues noted that the growing variety and intensity of human activities, coupled with impacts of climate change, threaten the sustained delivery of many important ecosystem services provided by coastal and ocean environments. Moreover, the processes and ecosystems that humans rely on for food, recreation, coastal protection, and other services are poorly understood, scarcely monitored, and often only appreciated after they are lost. The ultimate aim of EBM is to sustain the full suite of services; however, such a holistic approach requires knowledge about the true costs and benefits of policy decisions for humans and ecosystems. We need decision support tools and refined metrics for trade-off analysis, economic valuation, market development, and monitoring and evaluation. The Natural Capital Project, a partnership among Stanford University, The Nature Conservancy, and World Wildlife Fund, is developing a suite of spatially explicit ecosystem service models called Marine InVEST (Integrated Valuation of Ecosystem Services and Trade-Offs). Marine InVEST maps and values ecosystem services under current and future management and climate change scenarios and can be used with diverse habitats, policy issues, stakeholders, data limitations, and spatial and temporal scales. It includes modules for a variety of services such as food from fisheries and aquaculture, recreation, and coastal protection. Future models will quantify and value nursery habitat, transformation and sequestration of wastes, cultural services, and energy generation. It also includes process-based models that consist of a biophysical step where supply of the service is quantified, a use step where demand for the service is quantified, and an economic step for valuation in monetary terms. The models run in ArcGIS on input layers that describe ecosystem structure and human use (download InVEST at www.naturalcapitalproject.org). Scenarios are used to predict the effects of management decisions. The models are being tested in California, Puget Sound, and British Columbia. This work could inform integrated management in the Gulf of Maine by serving as an example of an ecosystem services scenario assessment.

T. Chopin (University of New Brunswick) and colleagues reported that to maintain the health of coastal waters, aquaculturists need more innovative, responsible, sustainable, and profitable practices that optimize efficiency, create diversification, and ensure the mitigation of the consequences of their activities. One of the innovative solutions of aquaculture for environmental sustainability (biomitigation), economic stability (product diversification and risk reduction), and societal acceptability (better management practices) is integrated multi-trophic aquaculture (IMTA). IMTA combines the cultivation of fed species (finfish, shrimps) with organic extractive species (suspension- and deposit-feeders) and inorganic extractive species (seaweeds) for a balanced ecosystem management approach. In Bocabec Bay, New Brunswick, the kelps kombu *Saccharina latissima* and awakame *Alaria esculenta* are grown together with mussels and salmon. Mussels were biofouling and a nuisance, and now they are a value-added co-cultured species. Research has shown that industry will make more money if IMTA is practiced. People would pay a 10% premium for mussels labeled as grown with sustainable practices. Experimental IMTA systems are being scaled up towards commercial levels. Extractive aquaculture not only produces biomass, it also renders provisioning, regulating, cultural, and supporting ecosystem services. The economic value of the environmental and societal services and benefits of extractive species needs to be recognized and accounted for to establish the true value of this component of IMTA. This would create even stronger incentives for aquaculturists to develop IMTA systems in which seaweeds and invertebrates could be traded for nutrient or carbon credits and coastal eutrophication reduction.

J. Grabowski (Gulf of Maine Research Institute) and colleagues noted that the decline of the oyster fishery in the eastern United States continues despite extensive restoration efforts. While other emergent biogenic estuarine habitats, such as sea grass beds and salt marshes, have received both legal and regulatory protection for performing valuable ecosystem functions (e.g., they act as nutrient sinks, stabilize sediments, and provide critical nursery grounds for fish and invertebrates), oyster reef habitat has been managed primarily as a resource to exploit. However, recently, coastal management plans recognized the need to further our understanding of the services provided by oyster reefs rather than permit continued overexploitation of oysters. In addition to producing oysters for commercial and recreational value, oyster reefs perform an array of ecosystem services. For instance, oyster reefs provide habitat for and augment the production of commercially and recreationally valuable fish, filter the water column, and consequently enhance sea grass habitat and stabilize shoreline erosion of salt marshes and other valuable habitats. An assessment of ecological and economic value of ecosystem services provided by oyster reefs suggests that the value of oyster reef services, provided by reef sanctuaries that persist for decades, likely far surpasses the value that could be derived from destructively harvesting reefs for oysters.

L. Leigh (Environment Canada) presented a proposed analytical framework for integrating the impact of policy decisions on ecological goods and services, as a part of cost–benefit analysis. Direct use values can presently be calculated from available physical and market data. Data and methods are becoming increasingly available for valuation of indirect use values, which represent the current and future benefits of natural capital use. Passive-use values, such as bequest and existence values, are the least tangible. A solid framework, methods for benefits valuation, data (i.e., social, economic value, and biophysical), and the capacity to quantify impacts of ecosystem services are needed. Additionally, since managers need to know the difference between options, good forecasting methods are needed. A framework was constructed that uses a total economic value approach to the valuation of ecosystem services (i.e., all direct and indirect uses). Valuation methods include revealed preference (e.g., tourist travel and time costs), stated preference (e.g., willingness to pay), and benefit transfer. The framework demonstrates the importance of a dynamic approach that measures the change in ecosystem services values due to policy decisions rather than the total stock of the services. The proposed framework will be applied to priority policy issues, including decisions about protected areas and sustainable management of water resources.

M. Fogarty (National Marine Fisheries Service) noted that consideration of trade-offs among potentially conflicting objectives is a central issue in EBM. In the Gulf of Maine, some key management and conservation concerns include the status of Atlantic cod *Gadus morhua* stocks, decline in bluefin tuna *Thunnus thynnus* condition, Atlantic herring *Clupea harengus* fishery development, and North Atlantic right whale *Eubalaena glacialis* recovery. Recently, important concerns have been raised about maintaining sufficiently high pelagic fish abundance to support predator populations. However, strategies to maintain high pelagic fish biomass may also have unintended consequences, including effects on cod recruitment, competition with right whales for their preferred prey—the copepod *Calanus finmarchicus*—and reduced condition factors for herring that lead to reduced condition of bluefin tuna. In the past three decades, herring biomass in the Gulf of Maine increased by an order of magnitude. Abundance and survival of late stage *C. finmarchicus* is inversely related to pelagic fish biomass. Intercalf interval of right whales is positively correlated with pelagic fish biomass. Calving success of right whales is positively correlated with late stage *C. finmarchicus* abundance. To confront these trade-offs, risks of alternative courses of action must be assessed and quantified. It is suggested that a formal decision analysis framework be used to provide a transparent approach to this problem. The approach specifies policy alternatives for a carefully bounded problem, defines a set of attributes against which management actions will be evaluated, assigns weights to the attributes reflecting both objectively defined characteristics and values and preferences, and assigns scores for each policy alternative against each attribute. This general decision framework can be set in the context of an explicitly adaptive management approach.

Panel Discussion

A panel consisting of H. Leslie (Brown University), T. Chopin (University of New Brunswick), A. DeMaio-Sukic (Fisheries and Oceans Canada), and M. Fogarty (National Marine Fisheries Service) discussed the benefits likely to result from an ecosystem services approach, the current state of this approach, with some examples and lessons learned, and guidance for next steps to ensure success.

Benefits of an Ecosystem Services Approach

Healthy ecosystems are the foundations of healthy economies. To develop policies and programs that ensure the viability of ecosystems while supporting sustainable development and the needs of human societies, we must document the economic value of ecosystem services. Accounting for ecosystem services is essential in our pursuit of economic prosperity if we want to accumulate sustainable wealth. Costs for replacing ecosystem services can be unacceptably high. Implementation of this approach requires clarification of our interests and goals. The approach accounts for cross-sectoral interests and cumulative impacts of activities that affect ecosystems. Benefits generated for each ecosystem service give decision makers a more complete picture of total costs and benefits so they can compare policy options. An ecosystem services approach would expand the variety of things we value and put all these on the playing field. From a governance standpoint, it gives certainty to industry by being explicit about what people value and why. It provides another tool for management decisions and a framework for assessing and resolving trade-offs.

Presently

We have many of the scientific and technical elements to move forward with ecosystem service approaches and EBM. Management for multiple services is underway at a range of scales globally. Models and other tools can facilitate the exploration of how ecosystem services are linked, and the trade-offs among them, and the identification of effective management strategies. Natural and social scientists can play vital roles in research, synthesis, and translation of knowledge to advance these efforts. We are currently engaged in recognition, awareness, and communication of the concept of ecosystem services and are conducting early valuations of ecosystem services and benefits. Any good or service will hold a variety of values for society; valuation helps frame trade-offs in monetary terms so they can be compared using the same unit of measurement. Some values show up as market values and some not. The Environmental Valuation Reference Inventory (www.evri.ca) is a valuation database that includes more than 2,200 studies. Individuals often have little information about the choices that are available to them. What is the ecological footprint of a certain good? How does it compare with that of a similar good? Governments and research institutions can reduce transaction costs by generating and disseminating the knowledge and information required for making sustainable choices and by setting up policies that minimize the ecological footprint of economic activities.

Examples

Management for multiple services is being applied in several places (e.g., California, Wadden Sea, and the Massachusetts Ocean Plan in the Gulf of Maine). In Puget Sound, efforts are underway to manage for multiple services using indicators and trade-off analysis tools. An ecosystem services approach was used to explore responses of coupled social-ecological systems (artisanal fishery and sport fishery tourism) to perturbations in Baja, California. Nonmarket solutions have been used in designation of marine protected areas in the St. Lawrence estuary and the listing and recovery of marine mammal species at risk (e.g., harbour seal *Phoca vitulina*, beluga whale *Delphinapterus leucas*, blue whale *Balaenoptera musculus*). The InVEST tool has been used for watershed land use planning in Hawaii; an upcoming application of Marine InVEST will be done on the west coast of Vancouver Island. Nutrients credits could be applied to IMTA in the Bay of Fundy. Social scientists on the IMTA team analyze attitudes and consumer opinion. Calculations have been made of monetary value from oyster reef restoration in Chesapeake Bay and North Carolina, to recover ecosystem services, including a tool to help with management decisions and measuring success of restoration. An ecosystem services framework has been applied to Canada's Northwest Territories' protected areas and Lake Winnipeg. Ecosystem services can provide a way to confront trade-offs between groundfish, pelagic fish, and right whales in the Gulf of Maine.

Lessons Learned

It is important to include ecosystem services in regulatory and management frameworks and policy. The benefits of ecosystem services need to be given full weight to balance economic considerations. Because regulatory and policy decisions can have massive implications for ecology and the economy, they need to be made in a more rigorous manner. An adaptive management approach is recommended. We need to keep it simple; however, we need to be specific in defining the exact ecosystem service. Large amounts of data are needed. Some situations may not require hard numbers, and a more theoretical or qualitative approach may be enough. We should be strategic and purposeful about our trade-offs. Markets could be used to reward sustainable behavior and good stewardship (e.g., pay more for ecolabeled seafood), but some methods of valuing ecosystem services remain unfamiliar or even controversial. We need a solid framework, methods for benefits valuation, data (social, economic value, and biophysical), and capacity to quantify impacts of ecosystem services. Interdisciplinary work is required between scientists and economists, but the disconnect between the disciplines remains a challenge. The objective is to achieve a dynamic balance between economy and ecosystem: a balance where natural productivity supports our needs and the economy does not pollute or degrade ecosystems to the extent that they cannot sustain our existence.

What Next?

There are many gaps in identifying, quantifying, and valuing ecosystem services that would benefit from more research. We need a more complete understanding of all ecosystem services and how they can be valued. How do the links within and among social-ecological systems influence the delivery of ecosystem services? How can emerging science be more effectively connected with management and policy processes, particularly in terms of trade-off analyses? How do we measure the success of this new approach? We could specify a manageable number of indicators that could fit into a driver-pressure-state-impact-response model of the Gulf of Maine. We could quantify what the scenario outcomes would be, then work together to develop the metrics that would inform decisions. Ecologists need to give economists physical measurements of the goods and services being provided by ecosystem functions. Dollar value is important, but there are other ways to get the message across, and we can be creative. It is important to understand training and perspectives between different camps of economists. Student ecologists and economists should take more courses from each other's fields. If aquacultured seaweeds and invertebrates are to be traded for nutrient or carbon credits and coastal eutrophication reduction, an agency (regional, national, or international) needs to be indentified to manage this. We should confront trade-offs by specifying policy alternatives for a carefully bounded problem. We need decision support tools and refined metrics for trade-off analysis, economic valuation, market development, monitoring, and evaluation. We need the perspectives of sociologists in addition to the views of ecologists and economists. Governance structures and a legislative basis for decisions are needed for the translation from science and policy into management.

To Ensure Success

We need clear vision, enabling institutions with clear mandate, EBM science infrastructure, and integrative and interdisciplinary partnerships. Pilot examples should be discrete projects with clear goals and vision. A pilot should start with local or regional groups with multi-disciplinary partnerships. While the Gulf of Maine could be manageable, it would be difficult to get a lot done at this scale, especially with bi-national issues to compound the work. Therefore, a United States–Canada comparative study may be better. It would be best to start with high-level processes because systems are complex. Practitioners should have realistic expectations and work in stepwise fashion. They should pick a project for a pilot where a decision is needed. We need governance structures and a legislative basis for decision making; the governance structures must be right to have a successful project. Governments need to implement policies for ecosystem services taxes and credits. Canada is better positioned with a national oceans policy but needs EBM policy and governance structures. The United States needs new structures for governance committees that can look at cross-sectoral issues (such as can be done with Canada's Oceans Act). For United States–Canada collaboration, effective infrastructure for science coordination is in place through the Gulf of Maine Council and RARGOM (the Regional Association for Research on the Gulf of Maine). However, the management issues are more difficult because we have bi-lateral agreements on fish stock management but on no other components of EBM. We need formal agreements to work together on broader ecosystem elements.

Acknowledgments

Special thanks to the seven workshop presenters and four panelists from whom we freely borrowed for this summary. Thanks to Rob Stephenson, Lara Cooper, and the others on the Gulf of Maine Science Symposium organizing committee for a stimulating conference and for ensuring that all went smoothly. Although the research described in this article has been funded in part by the U.S. Environmental Protection Agency, it has not been subjected to agency review. Therefore, it does not necessarily reflect the views of the agency. This is contribution number AED-10–023 of the U.S. Environmental Protection Agency, Office of Research and Development, National Health and Environmental Effects Research Laboratory, Atlantic Ecology Division, Narragansett, Rhode Island.

References

McLeod, K., and H. Leslie. 2009. Ecosystem-based management for the oceans. Island Press, Washington, D.C.

MEA (Millennium Ecosystem Assessment). 2005. Ecosystems and human well-being: synthesis. Island Press, Washington, D.C.

TEEB (The Economics of Ecosystems and Biodiversity). 2009. The economics of ecosystems and biodiversity for national and international policy makers—summary: responding to the value of nature. United Nations Environment Programme, Nairobi, Kenya.

American Fisheries Society Symposium 79:9–25, 2012

Linking Ecosystem Health and Services to Inform Marine Ecosystem-Based Management

Katie K. Arkema*
The Natural Capital Project, Stanford University
c/o NOAA Fisheries, Northwest Fisheries Science Center
2725 Montlake Boulevard East, Seattle, Washington 98112, USA

Jameal F. Samhouri
Pacific States Marine Fisheries Commission
205 SE Spokane Street, Suite 100, Portland, Oregon, USA
and
Conservation Biology Division, Northwest Fisheries Science Center
National Marine Fisheries Service, National Oceanic and Atmospheric Administration
2725 Montlake Boulevard East, Seattle, Washington 98112, USA

Abstract.—A growing variety and intensity of human activities threaten the health of marine ecosystems and the sustained delivery of services provided by oceans and coasts. The Gulf of Maine (GoM) is no exception to this trend, and as such, an ecosystem-based approach to managing the region has gained traction in recent years. The ultimate aim of marine ecosystem-based management (EBM) is to maintain ecosystem health (i.e., structure and function) and to sustain the full suite of ecosystem services on which people rely. Maintaining ecosystem health and sustaining services are related goals, both from a scientific and management perspective, yet in some cases, the interplay between the two is not well understood. Here, we examine relationships between attributes of ecosystem health and ecosystem services. In particular, we explore how outputs from ecosystem models, originally developed for ecosystem-based fisheries management (EBFM), can be used to quantify and value services of particular relevance to the GoM environments and human populations. We highlight services, such as the provisioning of food from fisheries, that ecosystem models are well equipped to inform and reveal where more work is needed to value other services, such as the protection from erosion and inundation afforded by coastal habitats. EBM also requires knowledge about the costs and benefits of management decisions for humans and ecosystems. We demonstrate how ecosystem models can be used to explicitly illustrate trade-offs between attributes of ecosystem health and ecosystem services that result from alternative management scenarios. By bridging the gap between models developed for EBFM and ecosystem service models, we identify existing science and future needs for informing an ecosystem approach to managing the GoM.

Introduction

Coastal and ocean environments provide humanity with many important benefits or ecosystem services (MEA 2005). However, a growing variety and intensity of human activities threaten the health of marine ecosystems and the sustained delivery of ecosystem services. A greater awareness of the need to manage for multiple uses and interconnectedness of marine ecosystems has led to a shift toward an ecosystem approach to manage-

* Corresponding author: karkema@stanford.edu

ment in the United States, Canada, and elsewhere (Christensen et al. 1996; Arkema et al. 2006; Smith et al. 2007; Levin et al. 2009; McLeod and Leslie 2009). Ecosystem-based management (EBM) is an integrated approach that focuses on the influence of multiple human activities on ecosystems. The ultimate aim of EBM is to maintain ecosystem health in order to sustain the services people desire from it (McLeod et al. 2005). These two components, maintaining ecosystem health and sustaining the delivery of ecosystem services, are not independent of one other. Indeed, the provisioning of ecosystem services depends on a well-functioning ecosystem, which in turn can be modified by the demand for various services (MEA 2005). In spite of this interconnectedness, the relationships between various attributes of ecosystem health and the delivery of specific services of interest are not well defined.

"Ecosystem health" is an oft-used term in regulatory and policy documents and in the public media (POC 2003; USCOP 2004; McLeod et al. 2005; MEA 2005). Here, we operationalize and define ecosystem health as attributes of ecosystem structure (e.g., species composition, physical habitat characteristics) and function (e.g., biological and physical processes) relevant to the management goal of maintaining a productive and resilient ecosystem. Our understanding is based on various avenues of ecological research and theory, including the relationship between diversity and ecosystem function (reviewed by Srivastava and Vellend 2005), food web theory (e.g., Pimm 2002; Vermaat et al. 2009), the ecosystem-based fisheries management literature (e.g., Link et al. 2002; Fulton et al. 2005), and ecosystem ecology (e.g., Odum 1969, 1985; Likens 1985; Carpenter et al. 1995). From this literature has emerged an enormous list of attributes that scientists use to characterize an ecosystem, including the relative abundance and diversity of species with ecological and/or conservation importance, rates of primary and secondary production, community energetics, nutrient and trophic cycling, and resilience (Samhouri et al. 2009). These attributes describe the status of the ecosystem irrespective of the human dimension.

In contrast with ecosystem health, the goal of maintaining ecosystem services focuses on endpoints that relate directly to human needs and desires (NRC 2005; MEA 2005). Scientific understanding of the relationship between management decisions, ecosystem processes, and the delivery of benefits for humans is improving. These benefits include not only provisioning services of fish and shellfish, but also regulating services, such as control of damage from storms or floods, supporting services of nursery habitats and forage fish, and cultural, aesthetic, and recreational values (Daily 1997; MEA 2005). The Millennium Ecosystem Assessment (MEA) of the status of ecosystem services around the globe (MEA 2005) spawned a flurry of academic research and action by nongovernmental organizations and some international governments to increase understanding and awareness of the linkages between our activities and what benefits we can expect to reap from ecosystems (e.g., Hayhoe et al. 2004; examples in Turner and Daily 2008). Although the MEA raised awareness of ecosystem services on a global scale, much scientific work remains to be done. Decision support tools, predictive models, and more detailed understanding of specific services and the relationships among them are needed to inform management decisions on local, regional, and national scales. Moreover, until recently, the focus of ecosystem services research has largely been on terrestrial ecosystem services, although advances are being made in the marine arena (Ruckelshaus and Guerry 2009; reviewed in Kareiva et al. 2011). Quantitative approaches developed for ecosystem-based fisheries management (EBFM) may be able to provide an initial step toward quantifying the delivery of at least a subset of services provided by coastal and ocean ecosystems under alternative management scenarios.

In this paper, we explore the relationship between ecosystem health and ecosystem services and seek to bridge the gap between ecosystem models developed for EBFM and modeling currently in development to map and value ecosystem services. First, we outline relationships between ecosystem attributes and ecosystem services, highlight which services existing ecosystem models are well equipped to inform, and reveal where more work is needed. Second, we use an ecosystem model to elucidate trade-offs that arise among ecosystem health attributes, among services, and among attributes and services under representative management strategies. Throughout the paper, we focus on ecosystem health attributes and services of importance in the Gulf of Maine (GoM). We discuss implications of the relationship between health and services and identify future needs for an ecosystem-based approach in the region.

Models for Understanding Ecosystem Health

The past few decades have seen the development of a variety of marine ecosystem models, especially in the domain of fisheries science and management (reviewed by Fulton et al. 2003; Plaganyi 2007; Fulton 2010; Rose et al. 2010). These models attempt to incorporate the dynamics of species from many trophic levels (e.g., bacteria, primary producers, primary consumers, intermediate consumers, and apex predators), in addition to capturing some of the physical processes, fluxes, and/or forcings (e.g., habitat heterogeneity, nitrogen cycling, and sea surface temperature) most relevant to the biota. Plaganyi (2007) identified at least five classes of marine ecosystem models that fit our definition and differ in their data requirements, complexity, biological detail, handling of uncertainty, integration of climatic and oceanographic effects, incorporation of biogeochemistry, and inclusion of a spatial dimension. Many, if not all, of these frameworks initially were designed to inform EBFM, though they are increasingly capable of modeling new fisheries management scenarios (e.g., catch shares), other types of anthropogenic pressures (e.g., pollution, habitat loss, and climate change), socioeconomic dynamics, and the assessment and management processes themselves. In order to explain how such models can be used to examine relationships between attributes of ecosystem health and ecosystem services, the authors describe in more detail one of the most commonly used modeling frameworks, Ecopath with Ecosim (Polovina 1984; Christensen and Walters 2004), and another, Atlantis, which is gaining traction in a variety of locations worldwide because of its comprehensiveness and versatility (Fulton et al. 2005; Kaplan and Levin 2009; Link et al. 2010). Indeed, an Ecopath model for the Gulf of Maine was constructed nearly a decade ago (Heymans 2002) and has since been elaborated (Link et al. 2008) concurrently with the development of an Atlantis model for the northeast U.S. Continental Shelf (Link et al. 2010).

Ecopath with Ecosim (EwE) is composed of two linked parts. Ecopath, introduced by Polovina (1984), is a static, mass-balanced food web model consisting of trophically connected functional groups represented in terms of biomass density and fishing fleets that impose an additional source of mortality on a user-defined subset of the functional groups. Ecosim, developed by Christensen and colleagues (Christensen and Walters 2004; Christensen et al. 2005), is a dynamic simulation model driven by coupled equations that are initialized from the Ecopath mass-balance solution and influenced by specified predator–prey relationships (in addition to dynamic harvest rates, physical forcings, and recruitment processes, if desired). Though not a prerequisite, Ecosim can also represent multiple age-/size-classes for each functional group using Deriso-Schnute equations (Walters et al. 2000) and nontrophic interactions (such as the reduction in vulnerability to predation of species that seek shelter in structure-forming organisms like kelp). EwE is used widely to predict the direct and indirect effects of fisheries practices on targeted and nontargeted species throughout an ecosystem, the relative importance of top-down and bottom-up factors in structuring food webs, qualitative outcomes of alternative management strategies, and the reliability of candidate indicators of ecosystem health (Plaganyi and Butterworth 2004; Fulton et al. 2005; Field et al. 2006; Heymans et al. 2007; Österblom et al. 2007; Plaganyi 2007; Link et al. 2008; Mackinson et al. 2009; Samhouri et al. 2009; Shannon et al. 2009; Worm et al. 2009).

Atlantis, originally introduced as a biogeochemical box model (Fulton 2001; Fulton et al. 2004), consists of several submodels. At its core is a spatially explicit, biophysical submodel that tracks the transport of nutrients through functional groups in a three-dimensional ecosystem. Like EwE, the Atlantis biophysical submodel is driven by consumer–resource interactions, recruitment processes, and physical forcings, but it also allows for migration and habitat preferences to be influenced by physical habitat features (e.g., canyons). Also, as in EwE, Atlantis functional groups can be represented as biomass pools or as age-structured populations. The biophysical submodel is coupled to an oceanographic transport model, a human-impacts model (primarily fishing fleets, but also coastal development, pollution, etc.), a sampling and assessment submodel that simulates real-world monitoring programs, a management submodel consisting of decision rules and management actions, and a socioeconomics submodel that handles a range of features from compliance decisions to taxes to social networks among fishing vessels. Thus, a variety of nontrophic effects can drive the dynamics of an Atlantis ecosystem model. Because it treats such a wide range of biological, physical, and socioeconomic components of the ecosystem, Atlantis is quite useful for management strategy evaluation (MSE; Smith et al. 2007). MSE is used to project alternative management scenarios into the future, via simulation, to reveal insights into the ecosystem's dynamics, trade-offs among management objectives inherent to specific policy actions, and unintended consequences of particular management choices (Sainsbury et al. 2000; Levin et al. 2009). In addition, Atlantis has been used to evaluate the influence of model complexity on ecosystem understanding, identify robust indicators for monitoring programs, consider cumulative impacts of multiple human pressures, and examine the consequences of spatial management (Fulton et al. 2003, 2005; Kaplan and Levin 2009).

A major benefit of ecosystem models such as EwE and Atlantis is that they capture the dynamics of multiple attributes of ecosystem health in a single model domain. In this way, it is possible to examine changes in different aspects of ecosystem health while accounting for interactions and linkages between them. For example, the biomass of harvested species and the areal extent of biogenic habitat each may be considered ecosystem attributes, and if biogenic habitat serves as a structural refuge from predation for some species, it will also influence their biomass. Though many ecosystem attributes are challenging to measure empirically, these models capably predict changes in both ecosystem structure and function. For instance, Fulton et al. (2005), Shannon et al. (2009), and Samhouri et al. (2009) used EwE and Atlantis models to measure the effects of different types and magnitudes of human perturbation on changes in ecosystem attributes ranging from structural characteristics like mean trophic level and diversity to functional aspects such as trophic cycling, the ratio of ecosystem respiration to ecosystem biomass, and resilience. Examples of other attributes of ecosystem health, well represented by EwE and Atlantis, are listed in the columns of Table 1, though this list is by no means exhaustive. While some attributes, like the biomass of harvested species, are directly related to ecosystem services, like the food obtained from fisheries, in other cases, the relationships between ecosystem health attributes derived from ecosystem models and ecosystem services are indirect or unknown (e.g., the rate of nutrient cycling and recreational use of the ecosystem) (see Linking Health Attributes to Services section below).

Framework for Modeling Ecosystem Services

An important component of EBM is to understand how alternative management strategies will influence the full suite of services provided by marine ecosystems. A promising approach, recently forwarded in terrestrial ecosystems, has been the development of ecosystem service models that are sufficiently general to be applied systematically in various sites at temporal and spatial scales relevant to management (Nelson et al. 2009; Tallis and Polasky 2009; Kareiva et al. 2011). Such models must combine the rigor of small-scale studies, which tend to focus on one habitat type or ecosystem service, with the breadth of broad-scale assessments to integrate across multiple services. Here, the key aspects of such an ecosystem service modeling framework are described.

A fundamental component is the application of sound ecological models and understanding to develop "ecological production functions." Essentially, a production function specifies the output of services that are provided by an ecosystem, given its condition and characteristic processes (NRC 2005). Such a modeling approach relies on descriptions of the relationships between ecosystem structure (e.g., distribution and abundance of biogenic habitats and key species, water temperature, and concentrations of particulate organic matter and sediments) and function (e.g., sediment erosion and accretion, production and growth rates of

TABLE 1. Example attributes of ecosystem health (columns) and marine ecosystem services (rows). Existing ecosystem models capture the dynamics of all of the ecosystem attributes. Some of the ecosystem services can be valued with existing models (e.g., food from fisheries), whereas models for valuing some of the other services (e.g., control of erosion and inundation) are currently under development. Grey shading indicates which attributes of ecosystem health have a primary influence on the delivery of a specific service. These attributes would likely be included in simple production function models that describe the aspects of ecosystem structure and function that are most fundamental to the delivery of provisioning, regulating, and cultural services. Some of these attributes have been described as supporting services (MEA 2005).

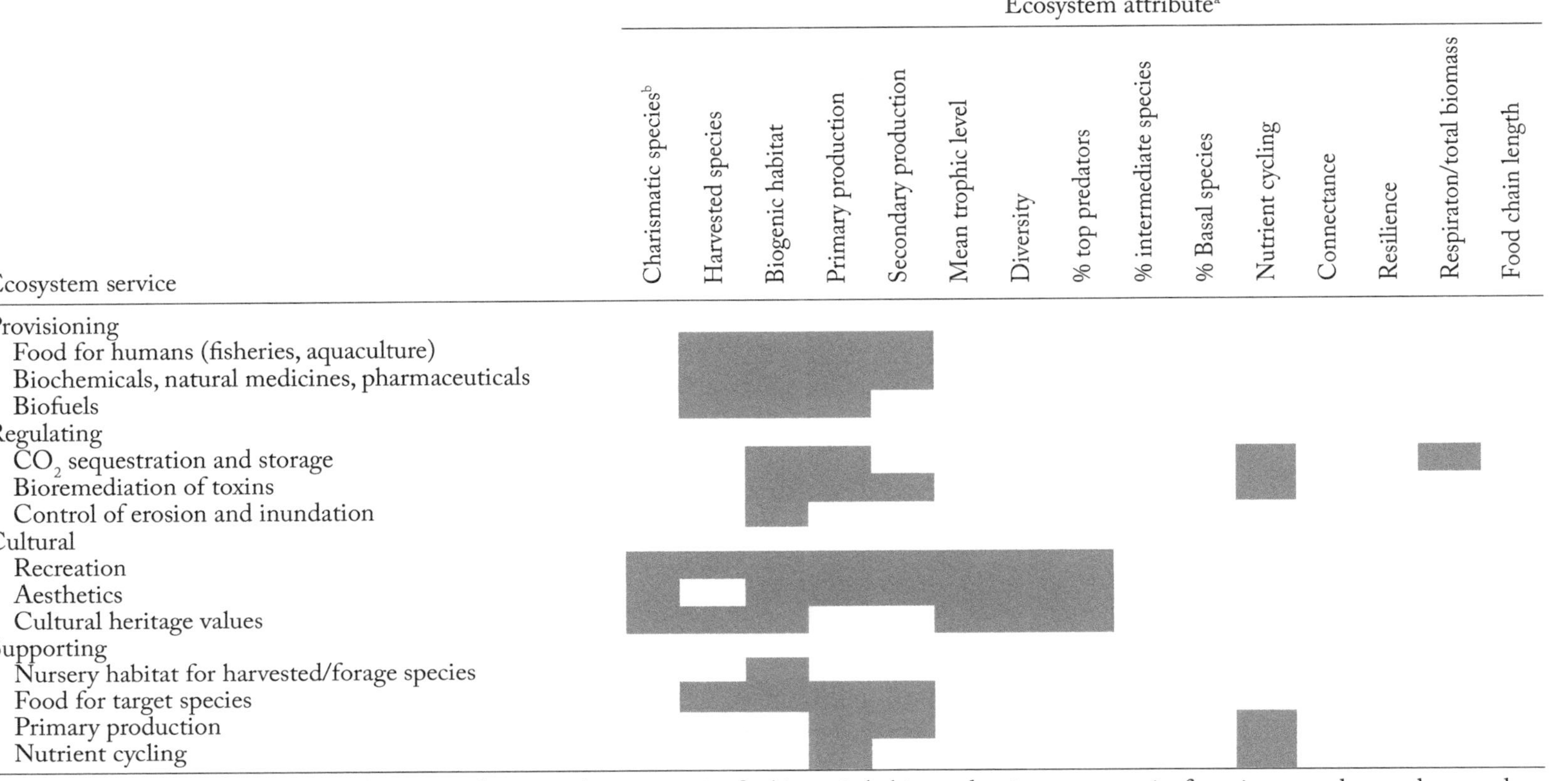

	Ecosystem attribute[a]														
Ecosystem service	Charismatic species[b]	Harvested species	Biogenic habitat	Primary production	Secondary production	Mean trophic level	Diversity	% top predators	% intermediate species	% Basal species	Nutrient cycling	Connectance	Resilience	Respiraton/total biomass	Food chain length
Provisioning															
Food for humans (fisheries, aquaculture)		■	■	■	■										
Biochemicals, natural medicines, pharmaceuticals		■	■	■	■										
Biofuels		■	■	■											
Regulating															
CO_2 sequestration and storage			■	■							■			■	
Bioremediation of toxins			■	■	■						■				
Control of erosion and inundation			■												
Cultural															
Recreation	■	■	■	■	■	■	■	■							
Aesthetics	■		■	■	■	■	■	■							
Cultural heritage values	■	■	■			■	■	■							
Supporting															
Nursery habitat for harvested/forage species			■												
Food for target species		■	■	■	■										
Primary production				■	■						■				
Nutrient cycling				■							■				

[a] Units for charismatic and harvested species = density or biomass; units for biogenic habitat = density or area; units for primary and secondary production = density, biomass, or rates; and units for trophic and nutrient cycling = rates. Percentages of different types of species refer to the percent of total ecosystem biomass. Resilience can be expressed as a return time following perturbation or as a proportional change in density/biomass of individual taxa relative to the entire ecosystem. Other attributes are labeled with their units or are unitless.

[b] Charismatic species may include marine mammals, sea turtles, and/or seabirds.

key species, wave attenuation, and consumption/filtration rates of organic matter) that are the basis for the delivery of a service (Tallis and Polasky 2009).

Ecosystem functions are fundamental to the delivery of benefits from natural environments, but ecosystem structure and function do not equate with services. Although critical for quantifying and valuing services, this distinction is often overlooked. Humans must benefit from an ecological process in order for it to be a service (Luck et al. 2009) and the value of a service depends both on *supply* and *demand*. Ecosystem functions influence the supply of ecosystem services, but they do not account for demand. Where are the people who enjoy services, and how much do they use? How will their use of a service change with different management decisions or policies? The combination of supply and demand generates use of ecosystem services. Defined quite broadly, the service is that portion of the ecosystem function that is used, not only via the consumption of physical goods, such as fish and timber, but also inclusive of the recreational and aesthetic appreciation of nature (i.e., nonconsumptive use values).

Management actions may influence service value through either their effects on supply or on demand (Vira and Adams 2009). Thus, models of ecosystem services must integrate analysis of the supply of the ecosystem function of interest with analysis of the location, type, and intensity of demand for services (Beier et al. 2008). Particularly useful outputs from ecosystem service models include maps of the availability and distribution of ecosystem services that indicate both areas where services are provided and where they are used (Tallis and Polasky 2009; Kareiva et al. 2011). For example, service maps that display the spatial distribution of metric tons of carbon sequestered have great utility in the context of climate change mitigation planning, while maps that display the spatial distribution of the volume of water yield available for hydropower production are critical to land use and development strategies. In some management contexts, it may be sufficient to quantify ecosystem services in biophysical units, as in the two preceding examples. For instance, some agencies make decisions about what activities will be allowed based on whether or not they meet a water quality standard or an allowable catch. Other decisions are tied to financial costs and benefits, and thus presenting policy alternatives in terms of the net benefits measured in a monetary currency is preferred. In these cases, it can be very useful to combine ecological production functions with economic valuation methods to estimate and report the monetary value of ecosystem services. The value of the changes in marketed services (e.g., fish, aquaculture, and avoided storm damage) can be quantified using market prices for marginal changes. Nonmarket valuation methods, including revealed preference and stated preference methods, can be used for ecosystem services that are not traded in markets (e.g., esthetic or cultural values; Freeman 2003; NRC 2005).

Integrating ecosystem and valuation modeling tools can help managers understand how trade-offs in ecosystem services may be modulated by trade-offs in the values of services. The framework discussed here could be described as a promising first step toward linking ecosystem health and services because it focuses on unidirectional effects of management decisions on the supply and demand of services. More work is needed to account for feedbacks between management decisions and ecosystem components by dynamically linking (1) models that deal with human behavioral responses to policies (e.g., dedicated behavior models) to (2) biophysical models that account for effects of management actions on the ecosystem functions that underlie the supply of services (Fulton 2010).

Linking Health Attributes to Ecosystem Services

The framework we use to link ecosystem health attributes to ecosystem services (Figure 1A) is based on connecting human actions to changes in ecosystem structure and function, ecosystem structure and function to the delivery of services, and the delivery of services to their values (NRC 2005; Daily et al. 2009). We focus on the extent to which outputs from ecosystem models can be used to generate information about the delivery of a suite of services under alternative management strategies. For the purposes of illustration, Figure 1 has been simplified so as not to include all the possible ecological and socioeconomic feedbacks and interac-

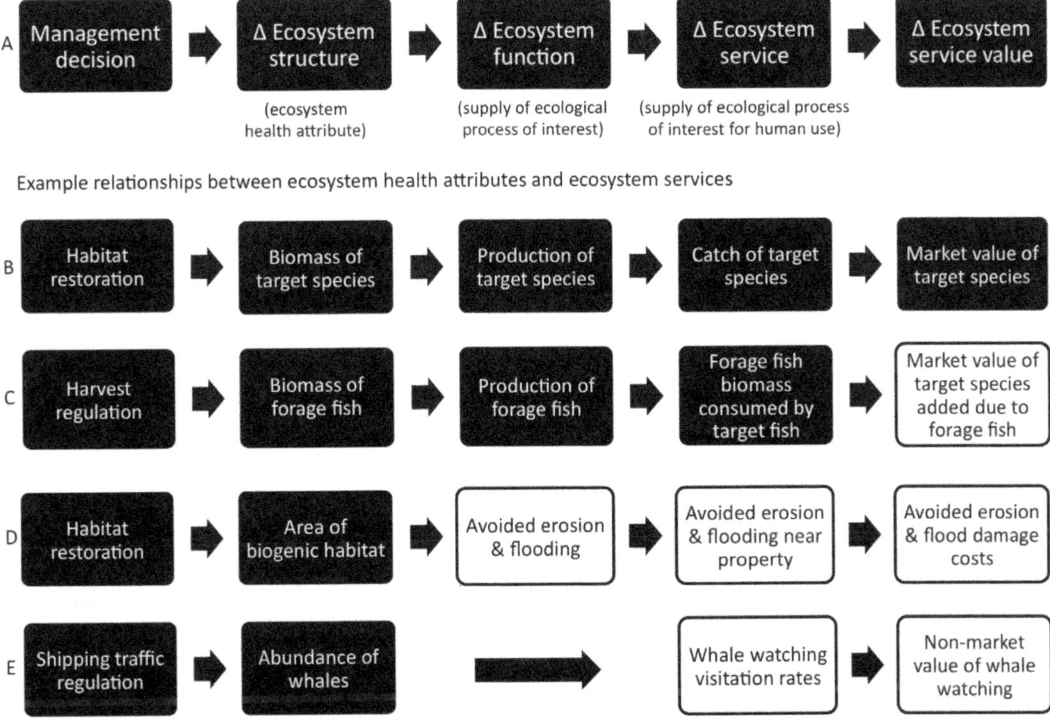

FIGURE 1. (A) A general framework for modeling ecosystem services (adapted from Daily et al. 2009). Note: in some cases, ecosystem health attributes are functional as well as structural (e.g., production of fish). (B–E) Example relationships between ecosystem health attributes and ecosystem services. Shaded boxes indicate that ecosystem models, developed largely for ecosystem-based fisheries management (EBFM), can be used to quantify a step in the logical chain. White boxes indicate a gap in current modeling capabilities. For the purposes of illustration, we have excluded the many possible ecological and socioeconomic feedbacks that may occur among steps.

tions that may occur among steps in the modeling framework (see last paragraph in previous section and Fulton [2010] for further discussion of feedbacks). We discuss three types of decisions (harvest regulation, habitat restoration, and shipping traffic) and four examples of services (catch of target species, production of forage species consumed by target species, the amount of erosion or inundation prevented via wave attenuation by biogenic habitat, and whale watching) that are significant in coastal and marine ecosystems. Our focus is on the relationship between management decisions and ecosystem attributes and services in light of examples relevant to the GoM ecosystem.

In the first example, we explore the influence of habitat restoration on the catch and market value of a species targeted for human food (Figure 1B). Our thought experiment is relevant in the GoM where Atlantic cod *Gadus morhua* young of the year recruit to structurally complex eelgrass and kelp habitat for protection from predation (Lazzari et al. 2003). An increase in the abundance of macrophytes and seaweeds, as a result of restoration activities, may lead to increases in the biomass and production of target species like Atlantic cod. Depending on current harvest regulations, the total catch and total market value of the target species should increase. Ecosystem models can predict outputs of the service in biophysical terms (i.e., kilograms of Atlantic cod) or in economic terms (i.e., the total market value of the catch). The health attribute in this example (biomass of target species) is closely related to the provisioning service (catch of target species), and ecosystem models are well equipped to assess changes in both factors. Worth noting, however, is that for examples like this one, where nontrophic effects are important, ecosystem models such as Atlantis and EwE may require enhancements to account for the influence of biogenic habitat on the production of target fish species.

Related to the catch of target fish and shellfish is the supporting service of forage species for higher trophic levels. Recent work has considered the trade-offs between harvest of a commercially valuable fish species (Pacific sardine *Sardinops sagax*) and its importance as prey for other species of commercial, recreational, and ecological significance (Hannesson et al. 2009). In the second example, a harvest regulation (e.g., catch quotas, area and seasonal closures) directly influences the biomass and production of a target species that is also a forage fish (Figure 1C). As pointed out by Hannesson et al. (2009), indirect effects may play a role in determining the biomass of forage species (see Trade-offs among Ecosystem-Based Management Goals section below). But for the sake of illustration, the simplest expectation is that a reduction in harvest of the forage fish species will positively influence its biomass and production, and this effect should increase prey availability for forage fish predators, which may also be target species. Because EBFM models were largely developed to increase the understanding of community-wide impacts of harvest practices and are based on trophic relationships, they are well equipped to assess changes in the catch of a higher trophic level species as a result of changes in fishing effort for lower trophic level species. One limitation, however, is that some ecosystem models do not include the final step in the framework for modeling ecosystem services: valuation of forage fish as a supporting service (indicated by white box in Figure 1C). This quantity can be calculated as the added market value of the target fish species due to an increase in its biomass and catch as a result of the addition of forage food (similar to the approach of Hannesson et al. (2009)).

In the third example, restoration activities increase the area of biogenic habitat, an ecosystem health attribute that also provides a variety of services (Figure 1D). In addition to the supporting service of acting as nursery habitats, eelgrass beds, estuaries, and saltmarshes (which are all foci of restoration efforts in the GoM (Gulf of Maine Council Habitat Restoration Committee 2004) have the ability to attenuate wave action and protect coastal areas from storm surge and erosion processes (reviewed in Irish et al. 2008; Koch et al. 2009). Unlike the previously discussed services, ecosystem models generally do not include nearshore hydrodynamic processes, such as wave attenuation, sediment transport, and water flow across inland areas, that are essential for quantifying avoided erosion or flooding in coastal areas (Komar 1998; reviewed in Irish et al. 2008). Moreover, these models generally lack the capacity to value coastal protection services via avoided sea defense costs (Moller et al. 2001), damages to property (FEMA 2009), affected people (Nicholls 2004), and/or lives lost (Das and Vincent 2009; indicated by white boxes in Figure 1D).

The effects of shipping traffic regulation on whale watching provides another example of a relationship between ecosystem attributes and services, for which there is a gap in our modeling capacity. Recent reports have implicated vessel traffic for sustained injuries and mortality caused by ship strikes to whales (Vanderlaan and Taggart 2006). Thus, we might expect traffic regulations, such as the designation of "areas to be avoided" or speed caps on specific routes, to lead to an increase in the abundance of whales. One service provided by whales is simply their existence value, which is essentially the benefit people receive from knowing that a particular environmental resource, such as marine mammals, exists (Krutilla 1967; NRC 2005). In this case, the ecosystem attribute, whale abundance, is the same as the service. However, whales also provide a recreational service, which can be determined using both market and nonmarket valuation approaches (Fisheries and Oceans Canada 2008; L. H. Pendleton, Ocean Foundation unpublished manuscript). The ecosystem models discussed in this paper currently lack the socioeconomic approaches necessary for predicting changes in recreational services and values of whale watching. Mortality from whale strikes due to shipping traffic is quite relevant in the GoM where the U.S. and Canadian governments have changed shipping routes in and out of Boston and the Bay of Fundy in an attempt to reduce whale strikes, in particular to the North Atlantic right whale (Right Whale Recovery Team 2000; Kraus et al. 2005).

Here, we have discussed just a few of the health attributes that reflect the ecosystem structure and function that are fundamental to delivering a full suite of services and predicting the influence of alternative management actions. Numerous other attributes of ecosystem health may have a primary influence on specific and broad categories of ecosystem services. Examples of these are indicated with gray shading in Table 1. Shaded attributes are likely to be included in simple production function models describing the aspects of ecosystem structure and function that are most fundamental to the delivery of provisioning, regulating, and cultural services. Some of these attributes have been described as supporting services (MEA 2005). The ecosystem health attributes that are not shaded may also be related to the delivery of services. In some cases, we may lack the scientific understanding to directly link changes in these attributes with services. In other cases, such as resilience, the ecosystem attribute may relate to temporal variability in the delivery of services. Increasingly complex ecosystem services models should be able to take into account such attributes in order to forecast fluctuations in services as a result of various perturbations.

Trade-offs among Ecosystem-Based Management Goals

The implementation of EBM raises policy questions about how one human use will affect others and how a policy focused on addressing one EBM goal (e.g., maintaining ecosystem health) will influence other goals (e.g., sustaining the delivery of ecosystem services) (Rosenberg and Sandifer 2009). Existing ecosystem models can provide insights into how alternative management actions affect such trade-offs among attributes of ecosystem health, among ecosystem services, and among ecosystem attributes and services (Walters and Martell 2004; Gerber et al. 2009; Kaplan and Levin 2009; Samhouri et al. 2010). Introduced below is a heuristic example of trade-offs among EBM goals using an EwE model for northern British Columbia (2000 AD; Ainsworth et al. 2008). The British Columbia model consists of 53 trophically linked functional groups (including 4 marine mammal, 32 fish, 12 invertebrate, 1 seabird, 2 primary producer, and 2 detritus groups). We chose to use this model because we are familiar with it (Samhouri et al. 2009, 2010); because it includes a biogenic habitat group in the food web (kelps and eelgrass), which allowed us to examine the effects of nearshore habitat management decisions; and because many of the functional groups have analogs in the GoM ecosystem. A similar approach could be employed using the recently developed Atlantis model for the northeast U.S. Continental Shelf (Link et al. 2010) within the GoM.

We measured the change in four attributes of ecosystem health and three ecosystem services under three different management scenarios. The ecosystem attributes we examined were drawn from the columns of Table 1 and include the abundance of charismatic species (specifically, the biomass of seals, sea lions, and seabirds); resilience, measured as the amount of change in the biomass of individual species (or functional groups) relative to the change in biomass of the entire ecosystem over the course of each simulation (Samhouri et al. 2009); one measure of the abundance of harvested species, groundfish biomass; and, the ratio of ecosystem respiration to ecosystem biomass, a measure of the maintenance costs in the ecosystem (Odum 1985). The ecosystem services we evaluated fall under the "food for humans" and "recreation" categories in Table 1 and include the recreational catch of lingcod *Ophiodon elongatus*, the recreational catch of Pacific halibut *Hippoglossus stenolepis*, and the catch of Pacific herring *Clupea pallasii* by the commercial gill-net and seine fisheries.

We compared the status quo management (i.e., harvest rates estimated for the year 2000; Ainsworth 2006) to three alternative management scenarios at the end of 25-year simulations. The management scenarios included (A) a 50% reduction in fishing mortality induced by the groundfish trawl fleet, (B) a 50% increase in kelp production (e.g., due to nearshore habitat restoration, Reed et al. 2006), and (C) the combined effects of reduced fishing as in (A) and increased kelp production as in (B). We highlight two points regarding these simulations. First, in this EwE model the groundfish trawl fleet targets lingcod and Pacific halibut, in addition to 14 other functional groups. Second, as in Samhouri et al. (2010), in scenarios B and C the authors incorporated a nontrophic, positive effect of macrophytes on several functional groups (juvenile herring and rockfishes, small crabs, shallow water benthic fish, and adult lingcod) that use the structural complexity of these habitat-forming organisms as a refuge from predation.

Reducing mortality induced by the groundfish trawl fishery by half led to a 36% decline in that fleet's catch relative to status quo management but only a 3% change in the groundfish stock size (i.e., biomass; Figure 2A). This somewhat counterintuitive result occurred largely because predator–prey interactions among the 16 functional groups constituting the groundfish guild reduced the expected beneficial effects of a decline in fishing-induced mortality. For instance, in this EwE model Pacific halibut make up a substantial portion of the diet of lingcod, but the reverse is not true (Ainsworth 2006). Thus, lingcod biomass increased in part due to increased availability of Pacific halibut prey when both species were released from trawl fishing pressure, whereas Pacific halibut biomass declined, though less so, for the same reason. Correspondingly, this management scenario caused a 7% decline in the recreational catch of Pacific halibut and a 34% increase in the recreational catch of lingcod (Figure 2A). Additionally, because several of the groundfish functional groups that benefited from a reduction in the trawl fishery (e.g., Pacific cod, lingcod, and turbot) prey on herring, the commercial catch of this forage species declined by13% as well. These responses illustrate how a single management decision focused on one sector, the groundfish trawl fishery, can, through direct and indirect effects, produce trade-offs among other ecosystem services, in this case the recreational catch of lingcod and Pacific halibut and the commercial catch of herring.

In terms of attributes of ecosystem health, the 50% reduction in fishing mortality induced by the groundfish trawl fleet caused an increase in resilience; a slight decline in the biomass of seals, sea lions, and seabirds; and no change in the ratio of ecosystem respiration to ecosystem biomass relative to status quo management (Figure 2A). The 33% increase in resilience implies that restricting the groundfish trawl fishery caused a smaller change in the abundances of individual functional groups relative to the overall change in ecosystem biomass than would have occurred under status quo management. The 4% decline in the abundance of charismatic species is a consequence of prey depletion (these groups forage on herring and a subset of the groundfish; Ainsworth 2006). The lack of response in the respiration to biomass ratio suggests that ecosystem maintenance costs were not altered by the change in the groundfish trawl fishery. As in the case of the ecosystem services, taken together these responses demonstrate that a management decision focused on one component of the ecosystem can lead to trade-offs among ecosystem health attributes (e.g., resilience versus charismatic species abundance). The responses also reveal trade-offs among ecosystem attributes and services (e.g., commercial groundfish and herring catches versus resilience).

The second management scenario, in which we simulated a 50% increase in kelp production, had marginal effects on the ecosystem attributes and services (Figure 2B). Indeed, only two of the ecosystem attributes and none of the ecosystem services differed by more than 2% from the status quo scenario. Resilience increased somewhat, by 13%, as did the respiration to biomass ratio (4%). Importantly, even changes of relatively small magnitude like these may be quite valuable to managers and stakeholders. Nonetheless, comparison of scenarios A and B suggests that either (1) changes in structure-forming nearshore habitat groups have small effects on the ecosystem relative to fishing, or (2) it is difficult to model accurately the nontrophic effects of macrophytes using EwE. Though we cannot distinguish between these two possibilities at this time, the Atlantis modeling framework treats both space and nontrophic effects of biogenic habitat explicitly and would be well suited to do so.

Finally, in the third management scenario, we explored the combined effects of reduced groundfish trawl fishing mortality and increased kelp production (Figure 2C). Because scenario B only had marginal effects on the ecosystem attributes and services, the responses of the ecosystem attributes and services appear qualitatively similar to those in scenario A (Figure 2A), which examined reduced fishing effects exclusively. Quantitatively, the combined effects of fishing and changes

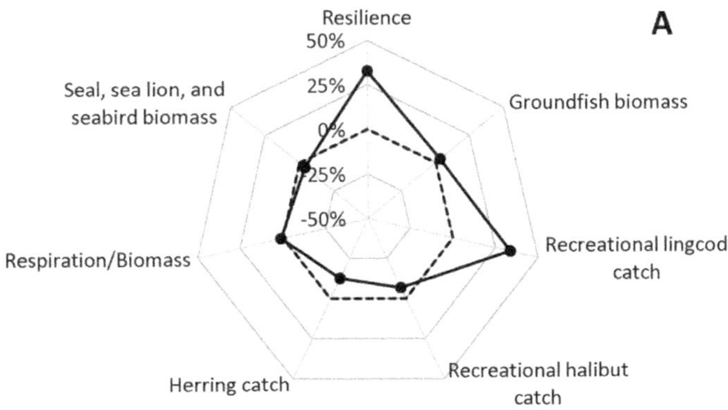

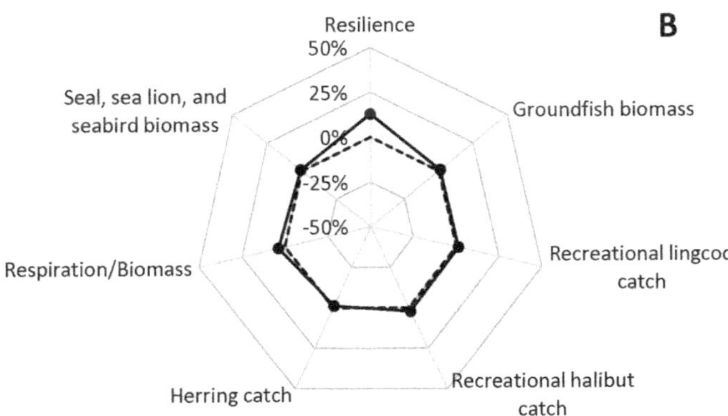

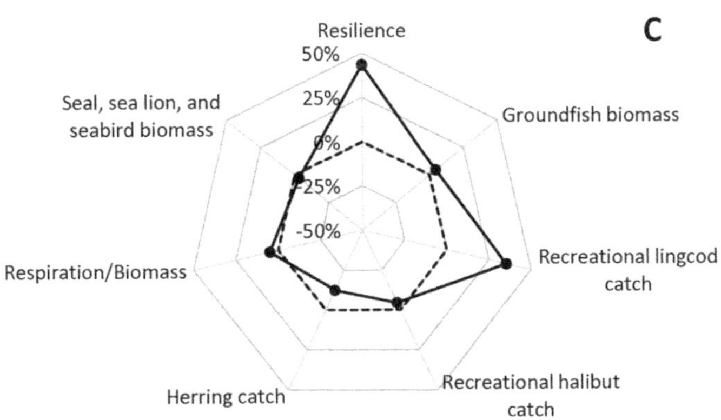

FIGURE 2. Spider plots depicting trade-offs among four attributes of ecosystem health (seal and bird biomass, resilience, groundfish biomass, and the ratio of ecosystem respiration to ecosystem biomass) and three ecosystem services (recreational lingcod catch, recreational halibut catch, and herring catch) predicted by an Ecopath with Ecosim model of northern British Columbia. The three scenarios represent (A) a 50% reduction in fishing mortality by the groundfish trawl fleet, (B) a 50% increase in kelp production (e.g., due to nearshore habitat restoration or reduced nutrient inputs), and (C) the combined effects of reduced fishing as in (A) and increased kelp production as in (B). Note that for each scenario, values of attributes and services are plotted as percentage differences from status quo management (indicated by the dashed line at zero) and rescaled so that positive values indicate improvements and negative values indicate declines.

in structure-forming nearshore habitat groups in scenario C were essentially additive. As the effects of multiple human impacts are frequently synergistic or antagonistic in nature (Crain et al. 2008), it is possible that the additivity seen here is a consequence of EwE model structure.

An Ecosystem Health and Services Approach for the Gulf of Maine

The GoM is well known as one of the world's richest marine ecosystems. Flanked by Massachusetts, New Hampshire, Maine, New Brunswick, and Nova Scotia to the north and west, and Georges Bank to the south and east, the GoM region has a long tradition of fishing, marine transportation, coastal development, and recreation. Indeed, seafood, recreational opportunities, and aesthetics are perhaps the most recognized services provided by coastal and marine ecosystems in the GoM. Additionally, a suite of important regulating and supporting services are provided by the diversity of environments (e.g., salt marshes, rocky intertidal, mudflat, eelgrass beds, sandy beaches, and nearshore and offshore subtidal habitats) characterizing the region (Gulf of Maine Council Habitat Restoration Committee 2004; Taylor 2008). As in other coastal areas around the world, a growing variety and intensity of human activities have increased the demands on the GoM marine ecosystem while threatening its historical functioning (Steneck et al. 2002) and overall health. Proposed and existing activities in the GoM include aquaculture, coastal development, discharge of sewage and other pollutants, energy production and distribution (wind farms, pipelines, and liquid natural gas terminals), fishing, tourism and recreation, seabed mining, telecommunications, and transportation (Taylor 2008). Balancing the numerous ocean uses, maintaining ecosystem health and services, and considering the current and future needs of stakeholder groups require comprehensive approaches.

Current EBM and marine spatial planning processes at the federal (e.g., Interagency Ocean Policy Task Force, www.whitehouse.gov/administration/eop/ceq/initiatives/oceans), regional (e.g., Massachusetts Ocean Management Plan, www.mass.gov), and local levels (e.g., Taunton Bay Management Plan, www.maine.gov/dmr/council/tauntonbay/index.htm) will benefit from incorporating an ecosystem services framework and an understanding of how attributes of ecosystem health interact with and sustain services (see also Hale and Westhead 2012, this volume). Spatial information about the provisioning, delivery, and use of ecosystem services can make explicit information regarding human activities and their effects on the coastal and ocean ecosystems of the GoM and identify target areas and beneficiaries for restoration. Ecosystem models can, and have, been used to assess trade-offs among a subset of services, especially related to fisheries, and among attributes of ecosystem health (Kaplan and Levin 2009; Samhouri et al. 2010). However, an assessment of trade-offs among a full set of ecosystem services will require tailoring models to predict changes in these services as a function of a variety of human uses of ocean ecosystems.

A variety of modeling approaches and efforts to synthesize information about marine services are underway at various institutions (e.g., Stanford University, World Wildlife Fund, The Nature Conservancy, University of Minnesota, University of Vermont, Conservation International, and United Nations Environmental Programme, to mention a few), and these may be of use to EBM efforts in the GoM. One suite of models, with which the authors are most familiar, is called Marine InVEST, and is being developed by the Natural Capital Project (Integrated Valuation of Ecosystem Services and Trade-Offs, Ruckelshaus and Guerry [2009]). Marine InVEST shares a number of key attributes with the original, terrestrial InVEST tool (Nelson et al. 2009; Tallis and Polasky 2009; Kareiva et al. 2011). The models allow users to map and value ecosystem services under current and future management and climate change scenarios. Highly flexible, for use with diverse habitats, policy issues, stakeholders, data limitations, and spatial and temporal scales, Marine InVEST comprises models for a variety of services (food from fisheries and aquaculture, recreation, coastal protection, and wave energy generation). The InVEST modeling framework includes key aspects of the approach discussed in this paper and, with site-specific calibration and parameterization, will be transferable to the GoM. A particularly powerful approach would be to links outputs from the recently developed

Atlantis model for the GoM (Link et al. 2010) with outputs for a full suite of services provided by Marine InVEST.

Implementing EBM in the GoM presents unique governance challenges, as doing so will require coordination among multiple states and across an international border. It is suggested that the ecosystem health and services models discussed in this paper can provide common ground and common currencies for tackling questions about the consequences of alternative management actions in the GoM. Key lessons from other systems in which these approaches have been used are as follows:

- Though they may not have a primary influence on endpoints of direct use to humans, attributes of ecosystem health are important for sustaining the delivery of ecosystem services. Existing models can be used to ask how different components of these two EBM goals change in relation to one another. A greater mechanistic understanding of the interplay between ecosystem health and services is a fertile area for future research.
- Existing data sources and models can provide guidance toward the mapping and valuation of ecosystem services but must be integrated across disciplines and synthesized within a cohesive framework to be useful for resource managers and policy makers.
- Ecosystem models like EwE and Atlantis, and ecosystem service models currently under development, can assist policy makers by exposing the trade-offs and counterintuitive results of alternative management decisions. Any one management action is likely to produce benefits for some attributes of ecosystem health or some ecosystem services and costs to others. The models discussed in this paper can alert managers to potential trade-offs and help them to reduce the likelihood of unintended consequences with detrimental effects on ecosystem structure and function. It is important to recognize and understand how single-sector management decisions alter species interactions and influence ecosystem health and services in direct and indirect ways (Mangel and Levin 2005; Tallis et al. 2008).

It is suggested that obtaining and integrating mechanistic information about the relationships between ecosystem health and ecosystem services into EBM science will produce compelling, useful, and transparent advice for resource managers and policy makers in the GoM region and beyond.

Acknowledgments

We thank C. Ainsworth, the author of the British Columbia EwE model, for allowing us to use his model and for helpful comments and suggestions. We also thank the organizers of the GoM symposium for inviting us to participate in the Technical Workshop on Ecosystem Services in the Gulf of Maine. We appreciate numerous discussions about modeling marine ecosystem services with members of the Marine Initiative of the Natural Capital Project. P. Levin, M. Ruckelshaus, and M. Plummer provided valuable feedback and inspiration. We thank Steve Hubbell for facilitating this collaboration and Risso for maintaining it.

References

Ainsworth, C. H. 2006. Strategic marine ecosystem restoration in northern British Columbia. Doctoral dissertation. University of British Columbia, Vancouver.

Ainsworth, C. H., T. J. Pitcher, J. J. Heymans, and M. Vasconcellos. 2008. Reconstructing historical marine ecosystems using food web models: northern British Columbia from pre-European contact to present. Ecological Modelling 216:354–368.

Arkema, K. K., S. C. Abramson, and B. M. Dewsbury. 2006. Marine ecosystem-based management: from characterization to implementation. Frontiers in Ecology and the Environment 4:525–532.

Beier, C., T. Patterson, and F. Chapin. 2008. Ecosystem services and emergent vulnerability in managed ecosystems: a geospatial decision-support tool. Ecosystems 11:923–938.

Carpenter, S. R., S. W. Chisholm, C. J. Krebs, D. W. Schindler, and R. F. Wright. 1995. Ecosystem experiments. Science 269:324–327.

Christensen, N. L., A. M. Bartuska, J. H. Brown, S. Carpenter, C. D'Antonio, R. Francis, J. F. Franklin, J. A. MacMahon, R. F. Noss, and D. J. Parsons. 1996. The report of the Ecological Society of America Committee on the scientific basis for ecosystem management. Ecological Applications 6:665–691.

Christensen, V., and C. J. Walters. 2004. Ecopath with Ecosim: methods, capabilities and limitations. Ecological Modelling 172:109–139.

Christensen, V., C. Walters, and D. Pauly. 2005. Ecopath with Ecosim: a user's guide. University of British Columbia, Fisheries Centre, Vancouver. Available: www.ecopath.org. (February 2012)

Crain, C. M., K. Kroeker, and B. S. Halpern. 2008. Interactive and cumulative effects of multiple human stressors in marine systems. Ecology Letters 11:1304–1315.

Daily, G. C., editor. 1997. Nature's services: societal dependence on natural ecosystems. Island Press, Washington, D.C.

Daily, G. C., S. Polasky, J. Goldstein, P. M. Kareiva, H. A. Mooney, L. Pejchar, T. H. Ricketts, J. Salzman, and R. Shallenberger. 2009. Ecosystem services in decision making: time to deliver. Frontiers in Ecology and the Environment 7:21–28.

Das, S., and J. Vincent. 2009. Mangroves protected villages and reduced death toll during Indian super cyclone. Proceedings of the National Academy of Sciences of the United States of America 106:7357.

FEMA (Federal Emergency Management Agency). 2009. Multi-hazard loss estimation methodology. Hurricane Model HAZUS®MH MR3 technical manual. Department of Homeland Security, Federal Emergency Management Agency, Mitigation Division, Washington, D.C.

Field, J. C., R. C. Francis, and K. Aydin. 2006. Top-down modeling and bottom-up dynamics: Linking a fisheries-based ecosystem model with climate hypotheses in the northern California Current. Progress in Oceanography 68:238–270.

Fisheries and Oceans Canada. 2008. Estimation of the economic benefits of marine mammal recovery in the St. Lawrence Estuary. Fisheries and Oceans Canada, Policy and Economics Regional Branch, Quebec.

Freeman, A. M. I. 2003. The measurement of environmental and resource values: theory and methods. Resources for the Future, Washington, D.C.

Fulton, E. A. 2001. The effects of model structure and complexity on the behavior and performance of marine ecosystem models. Doctoral dissertation. University of Tasmania, Hobart, Tasmania, Australia.

Fulton, E. A. 2010. Approaches to end-to-end ecosystem models. Journal of Marine Systems 81:171–183.

Fulton, E. A., A. D. M. Smith, and C. R. Johnson. 2003. Effect of complexity on marine ecosystem models. Marine Ecology Progress Series 253:1–16.

Fulton, E. A., A. D. M. Smith, and C. R. Johnson. 2004. Biogeochemical marine ecosystem models I: IGBEM—a model of marine bay ecosystems. Ecological Modelling 174:267–307.

Fulton, E. A., A. D. M. Smith, and A. E. Punt. 2005. Which ecological indicators can robustly detect effects of fishing? ICES Journal of Marine Science 62:540–551.

Gerber, L., L. Morissette, K. Kaschner, and D. Pauly. 2009. Should whales be culled to increase fishery yield? Science 323:880–881.

Gulf of Maine Council Habitat Restoration Subcommittee. 2004. The Gulf of Maine habitat restoration strategy. Gulf of Maine Council on the Marine Environment.

Hale, S. S., and M. Westhead. 2012. Ecosystem services in the Gulf of Maine. Pages 1–8 *in* R. L. Stephenson, J. H. Annala, J. A. Runge, and M. Hall-Arber, editors. Advancing an ecosystem approach in the Gulf of Maine. American Fisheries Society, Symposium 79, Bethesda, Maryland.

Hannesson, R. G., S. Herrick, Jr., and J. Field. 2009. Ecological and economic considerations in the conservation and management of the Pacific sardine (*Sardinops sagax*). Canadian Journal of Fisheries and Aquatic Sciences 66:859–868.

Hayhoe, K., D. Cayan, C. B. Field, P. C. Frumhoff, E. P. Maurer, N. L. Miller, S. C. Moser, S. H. Schneider, K. N. Cahill, E. E. Cleland, L. Dale, R. Drapek, R. M. Hanemann, L. S. Kalkstein, J. Lenihan, C. K. Lunch, R. P. Neilson, S. C. Sheridan, and J. H. Verville. 2004. Emissions pathways, climate change, and impacts on California. Proceedings of the National Academy of Sciences of the United States of America 101:12422–12427.

Heymans, J. 2002. The Gulf of Maine, 1977–1986. fisheries impacts on North Atlantic ecosystems: models and analyses. Fisheries Centre Research Reports 9:128–150.

Heymans, J. J., S. Guenette, and V. Christensen. 2007. Evaluating network analysis indicators of ecosystem status in the Gulf of Alaska. Ecosystems 10:488–502.

Irish, J. L., L. N. Augustin, G. E. Balsmeirer, and J. M. Kaihatu. 2008. Wave dynamics in coastal wetlands: a state-of-knowledge review with emphasis on wetland functionality for storm damage reduction. Shore and Beach 76:52–56.

Kaplan, I. C., and P. S. Levin 2009. Ecosystem-based management of what? An emerging approach for balancing conflicting objectives in marine resource management. Pages 77–96 *in* R. Beamish and B. Rothschild, editors. The future of fisheries science in North America. Springer Press, Secaucus, New Jersey.

Kareiva, P., H. Tallis, T. Ricketts, G. Daily, and S. Polasky. 2011. Natural capital: theory and practice of mapping ecosystem services. Oxford University Press, Oxford, UK.

Koch, E. W., E. B. Barbier, B. R. Silliman, D. J. Reed, G. M. E. Perillo, S. D. Hacker, E. F. Granek, J. H. Primavera, N. Muthiga, S. Polasky, B. S. Halpern, C. J. Kennedy, C. V. Kappel, and E. Wolanski. 2009. Non-linearity in ecosystem services: temporal and spatial variability in coastal protection. Frontiers in Ecology and the Environment 7:29–37.

Komar, P. D. 1998. Beach processes and sedimentation, 2nd edition. Prentice Hall, Englewood Cliffs New Jersey.

Kraus, S. D., M. W. Brown, H. Caswell, C. W. Clark, M. Fujiwara, P. K. Hamilton, R. D. Kenney, A. R. Knowlton, S. Landry, C. A. Mayo, W. A. McLellan, M. J. Moore, D. P. Nowacek, D. A. Pabst, A. J. Read, and R. M. Rolland. 2005. North Atlantic right whales in crisis. Science 309:561–562.

Krutilla, J. 1967. Conservation reconsidered. American Economic Review 57:777–786.

Lazzari, M., S. Sherman, and J. Kanwit. 2003. Nursery use of shallow habitats by epibenthic fishes in Maine nearshore waters. Estuarine Coastal and Shelf Science 56:73–84.

Levin, P. S., M. J. Fogarty, S. A. Murawski, and D. Fluharty. 2009. Integrated ecosystem assessments: developing the scientific basis for ecosystem-based management of the ocean. PLoS (Public Library of Science) Biology [online serial] 7:e14. DOI: 10.1371/journal.pbio.1000014.

Likens, G. E. 1985. An experimental approach for the study of ecosystems: the fifth Tansley lecture. The Journal of Ecology 73:381–396.

Link, J. S., J. K. T. Brodziak, S. F. Edwards, W. J. Overholtz, D. Mountain, J. W. Jossi, T. D. Smith, and M. J. Fogarty. 2002. Marine ecosystem assessment in a fisheries management context. Canadian Journal of Fisheries and Aquatic Sciences 59:1429–1440.

Link, J. S., E. A. Fulton, and R. J. Gamble. 2010. The northeast US application of ATLANTIS: a full system model exploring marine ecosystem dynamics in a living marine resource management context. Progress in Oceanography 87:214–234.

Link, J., W. Overholtz, J. O'Reilly, J. Green, D. Dow, D. Palka, C. Legault, J. Vitaliano, V. Guida, M. Fogarty, J. Brodziak, L. Methratta, W. Stockhausen, L. Col, and C. Griswold. 2008. The northeast U.S. Continental Shelf energy modeling and analysis exercise (EMAX): ecological network model development and basic ecosystem metrics. Journal of Marine Systems 74:453–474.

Luck, G. W., R. Harrington, P. A. Harrison, C. Kremen, P. M. Berry, R. Bugter, T. P. Dawson, F. de Bello, S. Diaz, C. K. Feld, J. R. Haslett, D. Hering, A. Kontogianni, S. Lavorel, M. Rounsevell, M. J. Samways, L. Sandin, J. Settele, M. T. Sykes, S. van den Hove, M. Vandewalle, and M. Zobel. 2009. Quantifying the contribution of organisms to the provision of ecosystem services. BioScience 59:223–35.

Mackinson, S., B. Deas, D. Beveridge, and J. Casey. 2009. Mixed-fishery or ecosystem conundrum? Multispecies considerations inform thinking on long-term management of North Sea demersal stocks. Canadian Journal of Fisheries and Aquatic Sciences 66:1107–1129.

Mangel, M., and P. S. Levin. 2005. Regime, phase and paradigm shifts: making community ecology the basic science for fisheries. Philosophical Transactions of the Royal Society B 360:95–105.

McLeod, K. L., and H. Leslie. 2009. Ecosystem-based management for the oceans. Island Press, Washington, D.C.

McLeod, K. L., J. Lubchenco, S. R. Palumbi, and A. A. Rosenberg. 2005. Scientific consensus statement on marine ecosystem-based management. Signed by 221 academic scientists and policy experts with relevant expertise and published by the Communication Partnership for Science and the Sea. Available: www.compassonline.org/sites/all/files/document_files/EBM_Consensus_Statement_v12.pdf. (February 2012)

MEA (Millennium Ecosystem Assessment). 2005. Ecosystems and human well-being: current state and trends. Island Press, Washington D.C.

Moller, I., T. Spencer, J. R. French, D. J. Leggett, and M. Dixon. 2001. The sea-defence value of salt marshes: field evidence from north Norfolk.

Journal of the Chartered Institution of Water and Environmental Management 15:109–116.

Nelson, E., G. Mendoza, J. Regetz, S. Polasky, H. Tallis, D. Cameron, K. Chan, G. Daily, J. Goldstein, P. Kareiva, E. Lonsdorf, R. Naidoo, T. Ricketts, and R. Shaw. 2009. Modeling multiple ecosystem services, biodiversity conservation, commodity production and tradeoffs at landscape scales. Frontiers in Ecology and the Environment 7:4–11.

Nicholls, R. J. 2004. Coastal flooding and wetland loss in the 21st century: changes under the SRES climate and socio-economic scenarios. Global Environmental Change Human and Policy Dimensions 14:69–86.

NRC (National Research Council). 2005. Valuing ecosystem services: toward better environmental decision-making. National Academies Press, Washington D.C.

Odum, E. P. 1969. Strategy of ecosystem development. Science 164:262–270.

Odum, E. P. 1985. Trends expected in stressed ecosystems. BioScience 35:419–422.

Österblom, H., S. Hansson, U. Larsson, O. Hjerne, F. Wulff, R. Elmgren, and C. Folke. 2007. Human-induced trophic cascades and ecological regime shifts in the Baltic Sea. Ecosystems 10:877–889.

Pimm, S. L. 2002. Food webs. University of Chicago Press, Chicago.

Plaganyi, E. E. 2007. Models for an ecosystem approach to fisheries. Food and Agriculture Organization of the United Nations, FAO Fisheries Technical Paper 477, Rome.

Plaganyi, E. E., and D. S. Butterworth. 2004. A critical look at the potential of ECOPATH with ECOSIM to assist in practical fisheries management. African Journal of Marine Science 26:261–287.

POC. 2003. America's living ocean: charting a course for sea change. A report to the nation. Pew Trusts, Washington, D.C.

Polovina, J. J. 1984. Model of a coral reef ecosystem. 1. The ecopath model and its application to French Frigate Shoals. Coral Reefs 3:1–11.

Reed, D. C., S. C. Schroeter, and D. Huang. 2006. An experimental investigation of the use of artificial reefs to mitigate the loss of giant kelp forest habitat. A case study of the San Onofre Nuclear Generating Station's artificial reef project. California Sea Grant College Program. University of California, San Diego, California.

Right Whale Recovery Team. 2000. Canadian North Atlantic Right Whale Recovery Plan. Prepared for World Wildlife Fund Canada, Toronto and Fisheries and Oceans Canada, Ottawa.

Rose, K. A., J. I. Allen, Y. Artioli, M. Barange, J. Blackford, F. Carlotti, R. Cropp, U. Daewel, K. Edwards, and K. Flynn. 2010. End-to-end models for the analysis of marine ecosystems: challenges, issues, and next steps. Marine and Coastal Fisheries: Dynamics, Management, and Ecosystem Science 2:115–130.

Rosenberg, A. A., and P. A. Sandifer. 2009. What do managers need? Pages 13–30 *in* K. McLeod and H. Leslie, editors. Ecosystem-based management for the oceans. Island Press, Washington, D.C.

Ruckelshaus, M. H., and A. Guerry. 2009. Valuing marine ecosystems? Marine Scientist 26:26–29.

Sainsbury, K. J., A. E. Punt, and A. D. M. Smith. 2000. Design of operational management strategies for achieving fishery ecosystem objectives. ICES Journal of Marine Science 57:731–741.

Samhouri, J. F., P. S. Levin, and C. J. Harvey. 2009. Quantitative evaluation of marine ecosystem indicator performance using food web models. Ecosystems 12:1283–1298.

Samhouri, J. F., P. S. Levin, and C. H. Ainsworth. 2010. Identifying thresholds for ecosystem-based management. PLoS (Public Library of Science) ONE [online serial] 5(1): e8907. DOI:10.1371/journal.pone.0008907.

Shannon, L. J., M. Coll, and S. Neira. 2009. Exploring the dynamics of ecological indicators using food web models fitted to time series of abundance and catch data. Ecological Indicators 9:1078–1095.

Smith, A. D. M., E. J. Fulton, A. J. Hobday, D. C. Smith, and P. Shoulder. 2007. Scientific tools to support the practical implementation of ecosystem-based fisheries management. ICES Journal of Marine Science 64:633–639.

Srivastava, D. S., and M. Vellend. 2005. Biodiversity-ecosystem function research: is it relevant to conservation? Annual Review of Ecology Evolution and Systematics 36:267–294.

Steneck, R. S., M. H. Graham, B. J. Bourque, D. Corbett, J. M. Erlandson, J. A. Estes, M. J. Tegner. 2002. Kelp forest ecosystems: biodiversity, stability, resilience and future. Environmental Conservation 29:436–59.

Tallis, H., Z. Ferdaña, and E. Gray. 2008. Linking terrestrial and marine conservation planning and threats analysis. Conservation Biology 22:120–130.

Tallis, H., and S. Polasky. 2009. Mapping and valuing ecosystem services as an approach for conservation and natural-resource management. Annals of the New York Academy of Sciences 1162:265–283.

Taylor, P. H. 2008. Gulf of Maine ecosystem-based management toolkit survey report. Gulf of Maine Council on the Marine Environment. Available: www.gulfofmaine.org/ebm. (February 2012).

Turner, R. K., and G. C. Daily. 2008. The ecosystem services framework and natural capital conservation. Environmental and Resource Economics 39:25–35.

USCOP (U.S. Commission on Ocean Policy). 2004. An ocean blueprint for the twenty-first century. USCOP, Washington, D.C.

Vanderlaan, A. S. M., and C. T. Taggart. 2006. Vessel collisions with whales: the probability of lethal injury based on vessel speed. Marine Mammal Science 23:144–156.

Vermaat, J. E., J. A. Dunne, and A. J. Gilbert. 2009. Major dimensions in food-web structure properties. Ecology 90:278–282.

Vira, B., and W. M. Adams. 2009. Ecosystem services and conservation strategy: beware the silver bullet. Conservation Letters 2:158–162.

Walters, C., D. Pauly, V. Christensen, and J. F. Kitchell. 2000. Representing density dependent consequences of life history strategies in aquatic ecosystems: EcoSim II. Ecosystems 3:70–83.

Walters, C. J., and S. J. D. Martell. 2004. Fisheries ecology and management. Princeton University Press, Princeton, New Jersey.

Worm, B., R. Hilborn, J. K. Baum, T. A. Branch, J. S. Collie, C. Costello, M. J. Fogarty, E. A. Fulton, J. A. Hutchings, S. Jennings, O. P. Jensen, H. K. Lotze, P. M. Mace, T. R. McClanahan, C. Minto, S. R. Palumbi, A. M. Parma, D. Ricard, A. A. Rosenberg, R. Watson, and D. Zeller. 2009. Rebuilding global fisheries. Science 325:578–585.

American Fisheries Society Symposium 79:27–33, 2012

Peopling the Marine Ecosystem

BONNIE J. MCCAY*
Department of Human Ecology, Rutgers University
New Brunswick, New Jersey 08901, USA

Abstract.—The study of ecosystems, such as the Gulf of Maine, and efforts to realize the objectives of ecosystem-based management are proceeding apace, but the so-called "human dimensions" need greater attention, rhetoric notwithstanding. They are mainly limited to representations of the anthropogenic effects of people and their artifacts and activities, such as overfishing, pollution, or drilling for oil. The relevant ecosystem has inputs from people but does not include them. As in standard fisheries management, the people are relegated to a single indicator, "F," fishing mortality, and perhaps, if we push it a bit, in "E" for effort, as in CPUE (catch per unit effort). Making it ecosystem-based adds small "e" to equations, or elaborates on matters such as predation, competition, sea surface temperatures, but does no more for the people involved. We tend to keep people out of the "ecosystem," despite much rhetoric to the contrary. If we want to take this seriously, we need to address not only the above notion of anthropogenic influences on a nonhuman ecosystem, but also "social and economic impact" and related analyses, and recent efforts at understanding "coupled natural and human systems." Throughout, we should not lose sight of the critical roles of people as actors—in tragedies, comedies, and other dramas of the commons—and as chroniclers and witnesses.

Endogenous or Exogenous to Marine Ecosystems?

"Ecosystems include people" statements are numerous. Two examples should suffice to remind the reader. In 1999, a National Oceanic and Atmospheric Administration (NOAA) panel, commissioned to develop principles for ecosystem-based management, wrote, "Today, humans are a major component in most ecosystems" and went on to usefully explain the "human component of the ecosystem" as including "the humans themselves, their artifacts and manufactured goods... and their institutions and cultures" (Ecosystem Principles Advisory Panel 1999:13). In 2009, NOAA's National Centers for Coastal Ocean Science program published an even stronger statement on their Web site from the report of a Council on Environmental Quality subcommittee: "Humans are integral to ecosystems, and the human dimensions of ecosystems are an integral focus of the science needed"; (Joint Subcommittee on Ocean Science and Technology 2007). The report went on to lay out the challenge: "Understanding the impact of humans on the ocean, the impact of the ocean on humans, and the human aspects of ocean governance provides the scientific basis for ensuring ocean health and quality of life for this and future generations" (Joint Subcommittee on Ocean Science and Technology 2007).

However, the scientific contribution of studies of human dimensions of the oceans is underdeveloped. The major focus to date has been on the first of the challenges, "understanding the impact of humans on the ocean," the anthropogenesis mode of thinking about people and the oceans. This is clear in a further statement of the Ecosystem Principles Advisory Panel (1999:13), where the focus is given to "activities that impact marine ecosystems":

* Corresponding author: mccay@aesop.rutgers.edu

> There are two requirements for managing human interactions with marine ecosystems. One is to develop an understanding of the basic characteristics and principles of these ecosystems—what patterns they exhibit and how they function in space and time. The second is to develop an ability to manage activities that impact marine ecosystems, consistent with both their basic principles and with societal goals concerning the kinds of behavior we would like ecosystems to exhibit (i.e., health and sustainability). (Ecosystem Principles Advisory Panel 1999:13)

The focus on humans and their activities as exogenous drivers of change in marine ecosystems is a noteworthy endeavor, and in recent times, it has been exemplified by efforts to document and model cumulative effects of multiple anthropogenic drivers (Halpern et al. 2008), as well as to examine the effects of specific human activities such as fishing, dumping of waste, collecting or destroying corals, and the like.

In this dominant approach to the question of where the people are in ecosystem analyses, the human dimension is usually treated as outside rather than inside the ecosystem of interest. People are no longer integral to ecosystems except as sources of disruption and governance. And when they and their activities are analyzed within the ecosystem, it is in very limited ways. Moreover, until recently, in ecosystem-based fisheries management, the human dimension is often focused on fishing, to the exclusion of other activities that could affect the ecosystem and its services, partly because of the oft-bemoaned fragmentation of ocean governance (Ekstrom et al. 2009). But even within the fisheries domain, the focus is constrained to the "F" of stock assessments, that is, an estimate of how many deaths of marine organisms are caused by fishing.

Thus, the human dimension appears as little more than some estimation of fishing mortality ("F"). To be sure, "F" may rely on data reflecting human activity, such as measures of catch and effort, and indicators such as CPUE (catch per unit effort) can be extremely important where other data are limited. Moreover, the documents used in fisheries policy deliberation may include social and economic data, but in the final analysis, people are represented by little more than the "F" used in stock assessments and the more diffuse politics of management decision making. Making fisheries management ecosystem-based adds small "e" to equations (Marasco et al. 2007), or elaborates on matters such as predation, competition, sea surface temperatures, but it typically does no more for the people involved, who continue to be treated as outside the system (exogenous) rather than a part of it (endogenous).

Impact of the Oceans on People

Two kinds of policy-related discourse and research concern the matter of marine ecosystems affecting people, shifting the arrow in the other direction (but still as exogenous to the ecosystems per se). One discourse comes mainly from the environmentalist and biological communities, although it requires the expertise of economists and other social scientists: the estimation of ecosystem services (Arkema and Sambouri 2012; Hale and Westhead 2012; both this volume). This requires assigning monetary and nonmonetary value to what nature offers, linking ecological outcomes to social values and economic demand, and enabling some empirical basis for the trade-offs that are inherent in ecosystem-based management (EBM; Barbier 2009; Wainger and Boyd 2009). The other discourse is squarely in the domain of social science, the long-standing enterprise of social and economic "impact analysis" (Griffin et al. 1993; National Marine Fisheries Service 1994). Impact analysis is legally mandated in federal fisheries management via the Magnuson-Stevens Act and the National Environmental Policy Act, among other laws. It underscores the value of social and economic realities alongside biological ones. Heretofore, it has focused mostly on assessing the impacts of management alternatives for fishery-dependent businesses and communities, the latter enhanced by an amendment to the Magnuson-Stevens Act that added a national standard for the protection of fishing communities (Olson 2005). Federal management measures that derive from EBM efforts will be no less subject to the impact requirements than are current single-species-based management efforts.

A focus on social and economic impacts of changes in access to, and the availability and quality of, marine ecosystem services such as fish is extremely important, particularly in enabling the articulation of trade-offs. However, note that it retains the unidirectionality of the anthropogenetic approach. Instead of focusing on how human activities affect biophysical dimensions of the ecosystems, this approach looks at how changes in regulations, designed to mitigate the problems, affect individuals, businesses, families, communities, industries. Again, the critical "human dimensions" are viewed as outside the ecosystem of interest, and it is hard to see how the humans are integral to the ecosystem.

People and Participation

The human dimension of governance appears in prescriptions for improved EBM that emphasize who participates and how, as well as in discussions of what goes wrong with fisheries management and, by implication, ecosystem-based fisheries management. If people are not really endogenous to the marine ecosystem, they are at least implied as included in the governance of that system.

Ecosystem-based management is sometimes held to require more and broader citizen participation. In the United States, EBM arose initially in the domain of forest management, where it responded both to shifting societal values and major conflicts and thereby included a call for greater public participation in management, including a broader set of actors and sources of knowledge, as well as more extensive consultation (Yaffee 1994). When applied to the marine domain, EBM also includes participatory recommendations. For instance, Marasco et al. (2007) include, within their recommendations for practical steps toward ecosystem-based fisheries management, a broader stakeholder base and adjustments to advisory systems to broaden representation of civic and environmental groups, as well as attention to finding mechanisms for joint decision making (see also Pikitch et al. 2004). Although not specific to EBM, this is reflected in the proliferation of research programs involving collaboration between fishers and scientists (Johnson 2007), as well as the increased presence of members of environmental groups on management councils and their committees.

On the other hand, the recent histories of resistance to the adoption of more precautionary, science-based management measures in New England and elsewhere give support to those who observe that the engagement of fisheries people in fisheries management is risky. People—especially commercial fishers—are deemed more likely to take risk-prone rather than risk-averse approaches to uncertainty. Indeed, it has been said that "the root cause of the lack of sustainability is the socio-politically biased response of management to intrinsic uncertainty" (Botsford et al. 1997). Accordingly, fisheries management has been subject to the ratchet effect. Managers, under constant political pressure for greater harvests because of their short-term benefits to society (jobs and profits), allow harvests to increase when fishery scientists cannot specify with certainty that the next increase will lead to overfishing and collapse (Ludwig et al. 1993). The situation reflects the "paradoxes of transparency" (Wilson 2009) in fisheries science, whereby calls for greater transparency, through improved participation of a broader group of people, results in challenges for science-based policy. Recent changes to the Magnuson-Stevens Act are designed to check this tendency by requiring that decisions on the upper bounds of catch limits be made by science bodies (committees of the regional management councils) and that they are adjusted for scientific, as well as implementation, uncertainty, thus, in effect, reducing the scope of decisions that can be directly affected by industry and public participation.

Expanding the Scope of the System: Coupled Systems

A way to make good on the promise of viewing humans as parts of marine ecosystems is to expand the scope of the systems being analyzed. This appears in the notions of "socio-environmental systems" (SES) and "coupled human and natural systems" (CHAN). Some of the key concepts in the study of human dimensions are captured in a recent paper on SES (Ostrom 2009). Figure 1 shows the close overlap of the human and nonhuman dimensions of an SES. The nonhuman side represents the nonhuman biological and physical aspects of the system, which provide resources for humans but are also affected by human activities. The human side would encompass both the material aspects of humans—their biological and physical roles as consumers, producers, and transformers—and the social, cultural, economic, political, and even scientific and technological aspects of humans, all of which influence their interactions with the nonhuman components as well as a critical set of subsystems, marked on Figure 1 with rectangular boxes. Following Ostrom (2009), those subsystems are the resource units (or ecosystem services) generated by the nonhuman components, the human resource users, subsets of the larger human populations, and the human institutions that comprise governance: government, markets, and civil society, insofar as they bear upon the particular "resource" interactions. All of these interact; there are inputs and outcomes. The features of governance systems (e.g., property rights systems, operational rules), user groups (e.g., social heterogeneity, leadership), and resource units (e.g., economic value, behavior of fish) that give rise to different ecological and social outcomes can be quantitatively or qualitatively assessed using well-developed methodologies (Ostrom 2009), just as biological and physical features of the ecosystem can be assessed with other tools. Such a framework can provide an analytic tool for addressing questions such as when the users of a resource will invest time and energy to avert a "tragedy of the commons" and thus produce a sustainable fishery and fishing community.

The closely related conceptual framework of CHAN touches on similar issues but with a stronger focus on emergent properties, systemic mechanisms, nonlinear processes, and feedback or reciprocal effects (Liu et al. 2007). A critical question is whether the feedback loops are positive or negative, "whether they lead to acceleration or deceleration in rates of change of both human and natural components as well as their interactions" (Liu et al. 2007:640). There are questions about temporal and spatial, as well as organizational couplings, thresholds, and resilience in coupled systems. For example, whether couplings between human and natural components of a marine ecosystem are direct and local, or indirect and global, can be a key question influencing how humans respond to signs of change in the ecosystem, and an increasingly important one with the globalization of both human and natural processes (Berkes et al. 2006).

The "coupling" and "feedback" metaphors contrast with the unidirectional notions of anthropogenic drivers of environmental change or regulatory or environmental "drivers" of socioeconomic "impacts." The coupled systems approaches

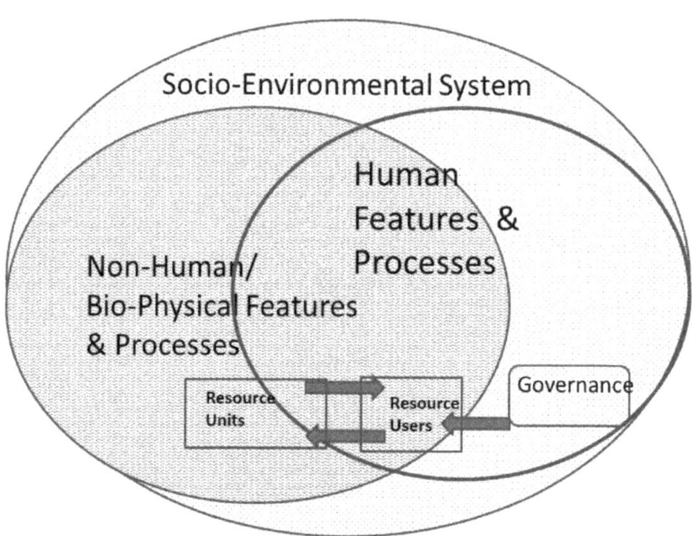

Figure 1. Core subsystems in a framework for analyzing social-ecological systems (Ostrom 2009).

direct us to efforts to examine and study points of articulation between humans, their institutions, and natural phenomena, with a view toward understanding interactive, not just one-way, processes. The systems approach can be helpful in directing attention, for example, to conditions under which people and their institutions respond to changes in the condition of the natural components of the system by changing behavior (through individual or policy decisions) in ways that make things worse, so-called "positive feedback." This is one way to explain the collapse of Atlantic cod *Gadus morhua* and other groundfish stocks of Newfoundland in the 1980s and the subsequent reorganization of both the resource and the socioeconomic components of the system (Finlayson and McCay 1998; McCay 2000; McCay et al. 2011). In contrast would be processes of negative feedback or deviation compensation, where signs of disruption to the flow of ecosystem services (i.e., fish landings) are responded to in ways that allow the fish populations to rebound, as we have identified among a set of fishery cooperatives in Mexico (Ponce-Diáz et al. 2009; McCay et al. 2011).

Conclusion

Humans are not exogenous to marine ecosystems but are part of them and must be studied as such. It is equally important to keep in sight the critical roles of people as chroniclers and witnesses—through their work as historians, scientists, artists, and poets—and, of course, as actors—in tragedies (Hardin 1968), comedies (Rose 1994), and other "dramas of the commons" (Ostrom et al. 2002). The oceans are "peopled seascapes" where people are not just the culprits, but also "co-operative, recuperative, restorative agents of ocean change" (Shackcroff et al. 2009). This perspective opens the scientific lens to the notion that human action is not just a source of singular or cumulative negative effects on marine environments, but is potentially corrective, as well, as sources of comedies and not just tragedies of the marine commons (McCay 2000).

Building upon such a conclusion does not come easily. The effort to incorporate human dimensions into ecological analyses and EBM in the United States is rooted in the work of the U.S. Man and the Biosphere Program and its focus on the Florida Everglades (Ogden 2008) and efforts to implement adaptive management for the Columbia River basin (Lee 1993), as well as conflicts over old-growth forests in the Pacific Northwest (Yaffee 1994, 1996). As Laura Ogden points out, the reality of socioecological approaches to EBM is shaped by legislative mandates, the composition and politics of interagency networks and both nongovernmental and quasigovernmental organizations that emerge with foci on complex, large-scale ecosystems and other institutional contexts that are "sites of political contest, having their own cultures, histories, and terrains of expertise to defend and stake out" (Ogden 2008:30). In the Everglades case, and probably for many others, the social component of socioecological systems analysis has been given short shrift in that process, and only time will tell how it will fare in efforts to realize coupled systems approaches in marine domains.

Moreover, the coupled systems approach calls for truly integrated multi- and interdisciplinary work. Concepts such as resilience (Davidson-Hunt and Berkes 2003), scale (Young 2002; Wilson et al. 2005), complexity (Wilson 2002) and learning (Armitage et al. 2007) are central and apply to both sides of coupled systems. A classic misapprehension is that social scientists are there to provide environmental education or appease the discontented. For this and other reasons, social scientists, and the science they bring to multidisciplinary endeavors, are often marginalized, as recently observed for marine protected areas projects (Christie et al. 2003; Blount and Pitchon 2007). Effective scientific contributions to EBM require instead that they work closely with other scientists to generate science-based understanding of the workings of the socioeconomic, political, and cultural dimensions of the system. Both social and natural scientists must give up their naiveté, becoming more knowledgeable about both the socioeconomic and the biophysical dimensions of marine systems and the methodologies required to delineate the ways that people really play roles as full members of marine ecosystems.

Acknowledgments

I appreciate the invitation to participate in the Perspective session of the Gulf of Maine Symposium '09 on Advancing Ecosystem Research for the Future of the Gulf and thank the many participants in that meeting in St. Andrews, New Brunswick, as well as three anonymous reviewers, for questions and ideas that informed this revision of the remarks that I gave there. I also thank the New Jersey Agricultural Experiment Station and the National Science Foundation for support of research that contributed to these remarks.

References

Arkema, K. K., and J. F. Samhouri. 2012. Linking ecosystem health and services to inform marine ecosystem-based management. Pages 9–25 *in* R. L. Stephenson, J. H. Annala, J. A. Runge, and M. Hall-Arber, editors. Advancing an ecosystem approach in the Gulf of Maine. American Fisheries Society, Symposium 79, Bethesda, Maryland.

Armitage, D., F. Berkes, and N. Doubleday. 2007. Adaptive co-management: collaboration, learning and multi-level governance. University of British Columbia Press, Vancouver.

Barbier, E. B. 2009. Ecosystem service trade-offs. Pages 129–144 *in* K. L. McLeod and H. M. Leslie, editors. Ecosystem-based management for the oceans. Island Press, Washington, D.C.

Berkes, F., T. P. Hughes, R. S. Steneck, J. A. Wilson, D. R. Bellwood, B. Crona, C. Folke, L. H. Gunderson, H. M. Leslie, J. Norberg, M. Nyström, P. Olsson, H. Österblom, M. Scheffer, and B.Worm, B. 2006. Globalization, roving bandits, and marine resources. Science 311:1557–1558.

Blount, B. J., and A. Pitchon. 2007. An anthropological research protocol for marine protected areas: creating a niche in a multidisciplinary cultural hierarchy. Human Organization 66:103–111.

Botsford, L. W., J. C. Castilla, and C. H. Peterson. 1997. The management of fisheries and marine ecosystems. Science 277:509–515.

Christie, P., B. J. McCay, M. L. Miller, C. Lowe, A. T. White, R. Stoffle, D. L. Fluharty, L. T. McManus, R. Chuenpagdee, C. Pomeroy, D. O. Suman, B. G. Blount, D. Huppert, R.-L. Villahermosa Eisma, E. Oracion, K. Lowry, and R. B. Pollnac. 2003. Toward developing a complete understanding: a social science research agenda for marine protected areas. Fisheries 28(12):22–26.

Davidson-Hunt, I. J., and F. Berkes. 2003. Nature and society through the lens of resilience: toward a human-in-ecosystem perspective. Pages 53–82 *in* F. Berkes, J. Colding and C. Folke, editors. Navigating social-ecological systems: building resilience for complexity and change. Cambridge University Press, Cambridge, UK.

Ecosystem Principles Advisory Panel. 1999. Ecosystem-based fishery management: a report to Congress by the Ecosystem Principles Advisory Panel, National Marine Fisheries Service. Available: www.nmfs.noaa.gov/sfa/EPAPrpt.pdf. (February 2012)

Ekstrom, J. A., O. R. Young, S. D. Gaines, M. Gordon, M., and B. J. McCay. 2009. A tool to navigate overlaps in fragmented ocean governance. Marine Policy 33:532–535.

Finlayson, A. C., and B. J. McCay. 1998. Crossing the threshold of ecosystem resilience: the commercial extinction of northern cod. Pages 311–337 in C. Folke and F. Berkes, editors. Linking social and ecological systems: institutional learning for resilience. Cambridge University Press, Cambridge, UK.

Griffin, W. L., D. Tolman, and C. Oliver. 1993. Economic impacts of TEDs on the shrimp production sector. Society and Natural Resources 6:291–308.

Hale, S. S., and M. Westhead. 2012. Ecosystem services in the Gulf of Maine. Pages 1–8 *in* R. L. Stephenson, J. H. Annala, J. A. Runge, and M. Hall-Arber, editors. Advancing an ecosystem approach in the Gulf of Maine. American Fisheries Society, Symposium 79, Bethesda, Maryland.

Halpern, B. S., S. Walbridge, K. A. Selkoe, C. V. Kappel, F. Micheli, C. D'Agrosa, J. F. Bruno, K. S. Casey, C. Ebert, H. E. Fox, R. Fujita, D. Heinemann, H. S. Lenihan, E. M. P. Madin, M. T. Perry, E. R. Selig, M. Spalding, R. Steneck, and R. Watson. 2008. A global map of human impact on marine ecosystems. Science 319:948–952.

Hardin, G. 1968. The tragedy of the commons. Science 162:1243–1248.

Johnson, T. 2007. Integrating fishermen and their knowledge in the science policy process: case studies of cooperative research in the northeastern U.S. Doctoral dissertation. Rutgers the State University, New Brunswick, New Jersey.

Joint Subcommittee on Ocean Science and Technology. 2007. Charting the course for ocean science in the United States for the next decade: an ocean research priorities plan and implementation strategy. Available: www.whitehouse.gov/sites/default/files/microsites/ostp/orppfinal.pdf. (February 2012)

Lee, K. 1993. Compass and gyroscope: integrating science and politics for the environment. Island Press, Washington, D.C.

Liu, J., T. Dietz, S. R. Carpenter, M. Alberti, C. Folke, E. Moran, A. N. Pell, P. Deadman, T. Kratz, J. Lubchenco, E. Ostrom, Z. Ouyang, W. Provencher, C. Redman, S. H. Schneider, and W. W. Taylor. 2007. Complexity of coupled human and natural systems. Science 317:1513–1516.

Ludwig, D., R. Hilborn, and C. Walters. 1993. Uncertainty, resource exploitation, and conservation: lessons from history. Science 260:17–18.

Marasco, R. J., D. Goodman, C. B. Grimes, P. W. Lawson, A. E. Punt, and T. J. Quinn, II. 2007. Ecosystem-based fisheries management: some practical suggestions. Canadian Journal of Fisheries and Aquatic Sciences 64:928–939.

McCay, B. J. 2000. Property rights, the commons, and natural resource management. Pages 67–82 *in* M. D. Kaplowitz, editor. Property rights, economics, and the environment. JAI Press, Stamford, Connecticut.

McCay, B. J., W. Weisman, and C. F. Creed. 2011. Coping with environmental change: systemic responses and the roles of property and community in three fisheries. Pages 381–400 *in* R. Ommer, I. Perry, P. Cury, and K. Cochrane, editors. World fisheries: a social-ecological analysis. Wiley-Blackwell Scientific Publications, Oxford, UK.

National Marine Fisheries Service. 1994. The guidelines and principles for social impact assessment. National Oceanic and Atmospheric Administration, National Marine Fisheries Service, NMFS-TM-SPO-16, Silver Spring, Maryland.

Ogden, L. 2008. The Everglades ecosystem and the politics of nature. American Anthropologist 110:21–32.

Olson, J. 2005. Re-placing the space of community: a story of cultural politics, policies, and fisheries management. Anthropological Quarterly 78:247–267.

Ostrom, E. 2009. A general framework for analyzing sustainability of social-ecological systems. Science 325:419–422.

Ostrom, E., T. Dietz, N. Dolšak, P. C. Stern, S. Stonich, and E. U. Weber, editors. 2002. The drama of the commons. National Academy Press, Washington, D.C.

Pikitch, E., C. Santora, E. A. Babcock, A. Bakun, R. Bonfil, D. O. Conover, P. Dayton, P. Doukakis, D. Fluharty, B. Heneman, E. D. Houde, J. Link, P. A. Livingston, M. Mangel, M. K. McAllister, J. Pope, and K. J. Sainsbury. 2004. Ecosystem-based fishery management. Science 306:1891–1892.

Ponce-Díaz, G., W. Weisman, and B. J. McCay. 2009. Co-responsabilidad y participación en el manejo de pesquerías en México: lecciones de Baja California Sur. [Co-management and participation in fisheries management in Mexico: lessons from Baja California Sur] Pesca y Conservación 1(1):1–9.

Rose, C. M. 1994. The comedy of the commons: custom, commerce, and inherently public property. Pages 105–162 *in* C. M. Rose, editor. Property and persuasion; essays on the history, theory, and rhetoric of ownership. Westview Press, Boulder, Colorado.

Shackeroff, J. M., E. L. Hazen, and L. B. Crowder. 2009. The oceans as peopled seascapes. Pages 33–54 *in* K. L. McLeod and H. M. Leslie, editors. Ecosystem-based management for the oceans. Island Press, Washington, D.C.

Wainger, L.A., and J. W. Boyd. 2009. Valuing ecosystem services. Pages 92–111 *in* K. L. McLeod and H. M. Leslie, editors. Ecosystem-based management for the oceans. Island Press, Washington, D.C.

Wilson, D. C. 2009. The paradoxes of transparency: science and the ecosystem approach to fisheries management in Europe. Amsterdam University Press, Amsterdam.

Wilson, D. C., M. Ahmed, S. V. Siar, and U. Kanagaratnam. 2005. Cross-scale linkages and adaptive management: fisheries co-management in Asia. Marine Policy 30:523–533.

Wilson, J. 2002. Scientific uncertainty, complex systems, and the design of common-pool institutions. Pages 327–359 *in* E. Ostrom, T. Dietz, N. Dolšak, P. C. Stern, S. Stonich and E. U. Weber, editors. The drama of the commons. National Academy Press, Washington, D.C.

Yaffee, S. L. 1994. The wisdom of the spotted owl: policy sessions for a new century. Island Press, Washington, D.C.

Yaffee, S. L. 1996. Ecosystem management in practice: the importance of human institutions. Ecological Applications 6:724–727.

Young, O. 2002. Institutional interplay: the environmental consequences of cross-scale interactions. Pages 263–291 *in* E. Ostrom, T. Dietz, N. Dolšak, P. C. Stern, S. Stonich and E. U. Weber, editors. The drama of the commons. National Academy Press, Washington, D.C.

American Fisheries Society Symposium 79:35–40, 2012

We Really Can Get There from Here: Important Steps Toward Ecosystem-Based Fishery Management

Sally McGee*
The Nature Conservancy
55 Church Street, New Haven, Connecticut 06510, USA

Abstract.—There is a strong desire to move toward ecosystem-based fisheries management (EBFM) in New England. However, there are many other problems in U.S. fisheries that have prevented timely adoption of EBFM. Changes in the Magnuson Stevens Fishery Conservation and Management Act, requiring catch limits for federally managed fisheries, may help speed the movement toward ecosystem-based approaches to fishery management. In the meantime, the term "integrated fishery management" (IFM) is being used by the New England Fishery Management Council to mean incorporating flexibility between fishery management plans to allow, and account for, mixed catch. If IFM is going to provide a step toward EBFM, there are several actions that must be taken: (1) adequate monitoring across all fisheries must be provided, (2) flexibility across existing management plans must encourage reporting and landing rather than discarding, (3) interjurisdictional coordination within and among states and federal management bodies must improve, and (4) application of the U.S./Canada Resource Sharing Understanding must expand.

Introduction

Ecosystem-based fisheries management (EBFM) has been on the to-do list of fishery managers in New England for a long time. The problem in implementation has been the need to address urgent problems, especially in the context of U.S. federal fisheries, before the big picture, the ecosystem that fisheries depend upon, can be considered adequately.

The good news is that with changes made to the Magnuson-Stevens Fishery Conservation and Management Act (MSA) in 2007, requiring scientifically determined catch limits for federally managed fisheries, many of the most urgent problems should be addressed in the coming years (Magnuson-Stevens Fishery Conservation and Management Act as amended 2007; Westhead 2012, this volume) As a result of the new requirement for accurate accounting of catch, there is renewed integrated fishery management (IFM). Currently, the term "integrated fishery management" is being used by the New England Fishery Management Council (NEFMC) to mean incorporating flexibility between fishery management plans to allow, and account for, mixed catch (e.g., yellowtail flounder in the sea scallop fishery; river herring and groundfish species in the sea herring fishery). For the purposes of this discussion, I will use the term "integrated fishery management" (IFM) to mean a step toward EBFM. IFM is an evolving process in the context of the NEFMC that has the potential to move the region toward an ecosystem-based approach to fishery management.

If this new approach (IFM) is going to provide a step toward EBFM, there are several key actions that must be taken: (1) adequate monitoring across all fisheries must be provided, (2) flexibility across existing management plans must encourage reporting and landing rather than discarding, (3) interjurisdictional coordination within and among states and federal management bodies must improve, and (4) application of the U.S./Canada Resource Sharing Understanding must expand. Each of these elements is a step toward improving and

* Corresponding author: smcgee@tnc.org

integrating NEFMC-managed fisheries, and each represents an important step toward EBFM.

History

In recent years, New England has taken a tentative approach to exploring and implementing EBFM. In 2005, the NEFMC received a small amount of federal funds for an ecosystem pilot project in the region. Oversight of this project was given by the council to its habitat committee (a subset of the full council, which I chaired at the time). The committee, at the direction of the full council, decided to spend a significant portion of these pilot funds to learn about stakeholder views of EBFM. The council also actively solicited input from the National Marine Fisheries Service (NMFS) Northeast Fisheries Science Center (NEFSC) on incorporating ecosystem knowledge in fisheries management (www.nefmc.org/ecosystems/index/html).

In 2005, technical work from the NEFSC was well received by the council's habitat committee. The committee held public meetings on EBFM and had advanced significantly in understanding the value of incorporating ecosystem-based approaches in fishery management. Understanding the complexity of making a fundamental change in management strategy, the committee recommended that the council take a first step toward EBFM by delineating ecosystems in New England waters. This would lay the foundation for EBFM, and perhaps a fisheries ecosystem plan, with no associated regulatory action. The committee agreed that this step should be something the council would be comfortable with in terms of exploring potential change in approach to management. The recommended action would simply describe the ecosystems in which New England fisheries operate.

Although, the habitat committee recommended action, the full council did not take its advice. Unfortunately, many council members and fishery stakeholders view the work of the habitat committee as a low priority for New England fishery management. As a result, no action was taken by the council to advance the committee recommendation to delineate New England ecosystems and move toward EBFM (New England Fishery Management Council 2005). Also, influencing the willingness and ability of the NEFMC to adopt ecosystem approaches has been the continued urgency of legally mandated requirements of the MSA. Simply, the council has been in the mode of addressing the most urgent problems (overfished stocks and continued overfishing) rather than other important issues on its agenda, such as implementing ecosystem approaches to management. By virtue of urgency, legal requirements, and lack of adequate understanding of the importance of taking a holistic approach to management, the council has taken a short-term view rather than an ecosystem view of New England fishery resources.

Urgent Problems

The NEFMC has often operated in crisis mode at the expense of incorporating consideration of ecosystem function. In part, the delay in implementation of ecosystem approaches has been the result of the continued pressing need to stop overfishing and rebuild depleted stocks. In the groundfish fishery, there has been chronic overfishing for decades. The response to a constant chain of emergencies has been near constant micromanagement (USOFR 2011). In the sea scallop fishery, the problem was chronic overfishing as well. The council and NMFS closed certain areas to protect groundfish, and protected sea scallop populations as an unintended benefit. This allowed the depleted scallop resource to come back to abundance (Hart and Rago 2006). However, other problems in the scallop fishery remain (namely, bycatch). In the sea herring fishery, the urgent problem in recent years has been bycatch of groundfish (haddock Melanogrammus aeglefinus). But comprehensive monitoring is still largely lacking in this fishery. Is there a problem in terms of bycatch of haddock or other species such as blueback herring *Alosa aestivalis* and alewives *A. pseudoharengus*? We do not really know without a comprehensive sea herring monitoring program (2009 correspondence between Atlantic States Marine Fisheries Commission and U.S. Secretary of Commerce).

Lack of ecosystem approaches has not been seen as urgent in comparison to these fishery specific problems. Because of economic importance, cultural dependence, overcapitalization, and the hard political choices that these urgent problems force, many fishery-specific problems remain. This has prevented the region from shifting to a more holistic approach, despite the fact that all of these ongoing problems point to the need for an ecosystem-based approach to management.

The Good News

The good news is that several changes were made to the MSA in 1996 and in 2007 that directly encourage (but do not require) action that move fishery management toward ecosystem-based approaches. While there no mandates to implement an ecosystem approach, in 1996 the definition of optimum (yield) was amended as follows (new text in bold):

> The term "optimum," with respect to the yield from a fishery, means the amount of fish which— will provide the greatest overall benefit to the Nation, particularly with respect to food production and recreational opportunities, and taking into account **the protection of marine ecosystems** (Section 1802 in Magnuson-Stevens Fishery Conservation and Management Act as amended 2007).

And, in 2007, the findings of the act were amended to include the following:

> A number of the Fishery Management Councils have demonstrated significant progress in integrating ecosystem considerations in fisheries management using the existing authorities provided under this Act (Section 1801).

Also in 2007, the act expanded the scope of influence of the previously established Ecosystem Panel (new text in bold),

> Not later than 180 d after the date of enactment of the Sustainable Fisheries Act, the Secretary shall establish an advisory panel under this Act to develop recommendations to **expand the application of ecosystem principles** in fishery conservation and management activities (Section 1882).

These are all very positive statutory changes. But rather than relying on discretionary ecosystem-focused provisions, other changes in the most recent MSA reauthorization may move federal fisheries toward EBFM more effectively.

The MSA now requires that all fisheries establish by regulation annual catch limits (ACLs) and accountability measures (AMs). These regulations must be in place by 2010 for fisheries subject to overfishing and by 2011 for other species (Section 1853). Catch must be quantifiably limited, and all fishery management plans and associated regulations must establish an equally quantifiable means to account for those fish, including provisions to make up for fish caught above and beyond catch limits. Once ACLs and AMs are in place—and the NEFMC is well on its way to doing this—the most urgent problems in New England's fisheries should be addressed. Because these new provisions are resulting in significant advances in all of the New England management plans, these catch limit and accountability measure requirements will move New England toward EBFM.

Integrated Fishery Management

Partly as a result of these new MSA requirements for ACLs and AMs, there is renewed interest in interspecies or IFM. Currently, the term "integrated fisheries management" is being used by the NEFMC to mean incorporating flexibility across fishery management plans to allow and, most importantly, account for, mixed catch (e.g., yellowtail flounder Limanda ferruginea in the sea scallop fishery). IFM is a step toward EBFM. And accurate accounting of catch and bycatch is fundamental to understanding the impact of fisheries on ecosystems in which they operate.

The IFM is distinct from the approach the NEFMC pursued but abandoned in 2005. At that time, the council was explicitly considering delineating ecosystems as a precursor to development of a fisheries ecosystem plan. Now, the council is considering integrating fishery management plans. While this may seem to be a barely discernible step in the direction of EBFM, this integrated approach across existing plans includes ecosystem principles. And in a political context, IFM is supported by a majority of current council members, unlike ecosystem delineation in 2005.

With the new statutory requirements for ACLs and AMs, thorough accounting of all catch is necessary. Inconsistencies and incomplete accounting across fishery management plans that were previously ignored are now highlighted. The bar is now higher, requiring an accounting of catch across fisheries. Integrating fishery management plans will provide for an accounting of all catch including discards. Integration will allow for better understanding of direct and indirect impacts of all fisheries.

These new legal requirements force issues like bycatch to be addressed directly. Bycatch problems are widely acknowledged but poorly understood. And they are highlighted by the requirement to account fully for catch. There are continued bycatch problems in New England—groundfish in the scallop fishery; goosefish (also known as monkfish) *Lophius americanus* in the scallop fishery; groundfish in the groundfish fishery (weak stocks and strong stocks mixed); and groundfish, river herring, and other species in the sea herring fishery. In reality, they are all intertwined. But New England fisheries are currently managed in isolation.

Clearly, each fishery is not isolated ecologically. Therefore, the chair of NEFMC has convened an interspecies committee to address the need to integrate these separate management plans. The interspecies committee is comprised not only of NEFMC members, but also representatives from the Atlantic States Marine Fisheries Commission (ASMFC), the Mid Atlantic Fishery Management Council, and NMFS, including staff from the NEFSC (New England Fishery Management Council Interspecies Committee, www.nefmc.org/interspecies/interspecies.html).

As of April 2010, this committee has met twice with the intent to address the requirement established by MSA in 2007, to account for catch and discards, and to set catch limits across multiple plans and multiple sectors and jurisdictional boundaries. During its two meetings, the committee has discussed these issues as they relate to New England and the mid-Atlantic regions, as well as the potential to develop an overall vision statement to guide a more holistic management system. This is a slowly evolving process that has the potential to move the region toward an ecosystem-based approach to management.

Key Ingredients

The following observations are offered from my perspective as a member of the NEFMC interspecies committee. Specifically, if the council is to succeed in IFM, and if IFM is to serve as a step toward EBFM, several key aspects of fishery management at the council level must be addressed, including

1. improved monitoring and improved bureaucratic flexibility,
2. interjurisdictional coordination within U.S. waters, and
3. interjurisdictional coordination with Canada.

The first step toward implementing an ecosystem approach is to establish comprehensive fishery monitoring systems. Comprehensive monitoring means accurate accounting of all catch. To date, the cost of monitoring in New England has largely been borne by the government. In New England, there have been significant new federal investments in groundfish and sea herring monitoring in 2010 and 2011. However, the region cannot expect this funding to be available in perpetuity. Quota set-asides to support industry funded programs have promise in many fisheries. Cost sharing with dealers and other shore support businesses has also been proposed. Affordable monitoring systems must be developed, including provisions for video instead of human observers and partial monitoring rather than 100% coverage (at sea or dockside) when practical and effective (Turris and McElderry 2008).

The second key to implementing an ecosystem approach by integrating fishery management plans is to improve flexibility across plans. Fisheries can no longer be bureaucratically isolated. Managers should not be forced to allocate based strictly on economic factors. For example, they should not be compelled to allocate yellowtail flounder to the sea scallop fishery because it is worth more in dollars grossed as bycatch (essentially providing access to more high value scallops) than it is worth as directed catch in the groundfish fishery. This approach does not address the need for catch accountability. In fact, rigid allocation schemes only encourage discarding (National Marine Fisheries Service 2011). What is needed is flexibility across management plans. Let fishermen largely determine who catches fish, either as bycatch or directed catch. Allow for integrated, cross-fishery quota management. At the same time, require all fishermen to account fully for all catch through ample monitoring systems.

Increased flexibility and increased monitoring will allow values other than dollars to be incorporated into fisheries (e.g., a diverse groundfish fleet and cooperation between fisheries). Increased flexibility will also discourage discarding.

The third key to IFM as a step toward EBFM is improving coordination across jurisdictions within U.S. waters. Battles across state lines and state/federal lines need to be addressed more directly than they have under current statutes. In the United States, this means addressing inconsistencies between MSA and the Atlantic Coastal Fisheries Cooperative Management Act (the statute that governs ASMFC). While accommodating states' rights, these two statutes need to be amended to be consistent in terms of catch limits and accountability measures, bycatch, and protections for essential fish habitat (Atlantic Coastal Fisheries Cooperative Management Act 1993). Consistent, adequate standards to determine impacts of fisheries across political boundaries will allow integration of management plans and improve ecosystem considerations that cross those boundaries.

The fourth key to making progress toward an ecosystem approach is to strengthen the U.S./Canada Resource Sharing Understanding. This bilateral "understanding" was established through an agreement between the two countries in 1995. Each year, the U.S.–Canada Transboundary Steering Committee exchanges scientific information and determines quota-based allocations of three transboundary fish stocks (Atlantic cod *Gadus morhua*, haddock, and yellowtail flounder) on Georges Bank in the Gulf of Maine (Transboundary Management Guidance Committee 2002). An even more holistic approach to shared resources must be brought to the U.S./Canadian negotiating table, including all transboundary stocks. Sea herring, and perhaps other species, should be added to those currently managed through the understanding. This does not mean exemption from statutory requirements for U.S. fisheries. As with state/federal boundaries, this does mean explicit acknowledgment of the geographic scope of fisheries that span international boundaries.

Last, some obsolete assumptions about ecosystem-based approaches must be addressed. The assumption that EBFM means that fundamentals like catch limits and accountability measures will be unnecessary or impractical is false. The assumption that the biomass targets for rebuilding depleted stocks will need to be lowered for all managed species under an ecosystem-based approach is also false. To the contrary, an ecosystem-based approach means improved productivity, flexibility, and economic return. An ecosystem-based approach means building more robust fisheries that are necessarily based on firm catch limits and accountability measures to ensure accounting and understanding fishery impacts on ecosystems and to ensure optimum productivity of those ecosystems.

Why Does This Matter?

With the higher bar set in MSA in terms of catch limits and accountability measures, New England is making strides to integrate its fisheries and, as a result, taking important steps in the direction of EBFM. An integrated approach to fishery management, spurred on by these new standards for catch limits and accountability measures, represents a step toward EBFM. If successful, this will provide flexibility for fishermen, who should achieve more profit with less uncertainty in management. Scientists should have more and better data and less variance in scientific output. And managers will no longer be limited by artificial, bureaucratic structures. Instead, fishermen, scientists and managers will be able to incorporate the realities of natural systems into management systems. We really can get there from here, and we are on the way.

References

Atlantic Coastal Fisheries Cooperative Management Act. 1993. U.S. Code, volume 16, section 71.

Hart, D. R., and P. J. Rago. 2006. Long-term dynamics of U.S. Atlantic sea scallop *Placopecten magellanicus* populations. North American Journal of Fisheries Management 26:490–501.

Magnuson-Stevens Fishery Conservation and Management Act as amended. 2007. U.S. Code, volume 16, sections 1801–1884.

New England Fishery Management Council. 2005. New England Fishery Management Council Habitat/MPA/Ecosystems Committee meeting summary. March 28, 2005, Newport, RI. Available: www.nefmc.org/habitat/meetsum/habitat_mar05.pdf (November 2011).

National Marine Fisheries Service. 2009. Weekly yellowtail flounder bycatch from sea scallop access areas, FY2009, closed area II. Available: http://www.nero.noaa.gov/ro/fso/Reports/ScallopProgram/CA2_YT.pdf (November 2011).

Transboundary Management Guidance Committee. 2002. Development of a sharing allocation proposal for transboundary resources of cod, haddock and yellowtail flounder on Georges Bank. Fisheries and Oceans Canada, Fisheries Management Regional Report 2002/01. Available: www.mar.dfo-mpo.gc.ca/science/tmgc/background/FMR%202002_01.pdf (November 2011).

Turris, B., and H. McElderry. 2008. Evaluation of monitoring and reporting needs for groundfish sectors in New England. Available: www.gmri.org/upload/files/groundfishmonitoringneedsfinalreportfinal.pdf (June 2012).

USOFR (U.S. Office of the Federal Register). 2011. Management measures for the NE multispecies and monkfish fisheries. Code of Federal Regulations, Title 50, Part 648, Subpart F. U.S. Government Printing Office, Washington, D.C.

Westhead, M. 2012. Tools for integrated policy and management in the Gulf of Maine: a summary of major advancements since 1996. Pages 47–60 *in* R. L. Stephenson, J. H. Annala, J. A. Runge, and M. Hall-Arber, editors. Advancing an ecosystem approach in the Gulf of Maine. American Fisheries Society, Symposium 79, Bethesda, Maryland.

American Fisheries Society Symposium 79:41–45, 2012

Fishermen without Borders

JEAN GUY D'ENTREMONT*
Scotia Harvest Seafoods Inc.
#2066 Route 335, Lower West Pubnico, Yarmouth County, Nova Scotia, B0W 2C0, Canada

Abstract.—The Gulf of Maine (GoM) may have defined borders to some, but to the Canadian fishing industry, it carries a flow of larvae, nutrients, and other resources that help sustain the fishery from Georges Bank to the West Scotian Slope to the Bay of Fundy and all points in between. The GoM provides a source of wealth to people and communities, as well as supplying what may be one of the last natural foods on the planet. The fishing industry has been using the GoM for centuries, yet it is only recently that monitoring and data gathering has been taking place. In my opinion, we can extract much more value from the fisheries than we presently do. If the fisheries resource of the GoM is not delivering its full potential, who is ultimately responsible and accountable? In the past decade, transboundary groundfish resources from Georges Bank have been successfully managed through the Transboundary Management Guidance Committee. We can improve decision making even further in a greater ecosystem context, recognizing that decisions have to be made with the information available. An ecosystem approach to fisheries proposes a pragmatic view based on assessing the risk of not meeting agreed objectives.

Introduction

As a commercial fishing vessel skipper who has fished the Gulf of Maine (GoM) at different periods since 1985, and who has had a family fishing business with vessels and processing for 25 years, I have seen changes take place in the industry. Also, with my involvement as co-chair of two international North Atlantic responsible fishing conferences (North Atlantic Responsible Fishing Conference, St John's, Newfoundland and Labrador, Canada 2000 and North Atlantic Responsible Fishing Conference, Yarmouth, Nova Scotia, Canada 2003), I have learned much from the experiences of other fishermen in different areas of the world, including Scotland, Iceland, and England. Since 1998, I have been a member of the Fisheries Resource Conservation Council and, more recently, chairman of the council. Through all my experiences, and when my three sons grew up and began choosing their paths in life, I decided that my wish and objective was to provide them with an opportunity for a career in the seafood business of the future: one that is sustainable and landing high-quality seafood products for a world market at competitive prices and quality. I hope for a fishery that can sustain year-round employment and ensure adequate controls, monitoring, and surveillance in order to achieve third party sustainability certification such as that provided by the Marine Stewardship Council (MSC). Frankly, I have been very disappointed in the progress of the groundfish fisheries in North America in the past quarter century, especially when I see the possibilities and opportunities in other jurisdictions around the world. Iceland, for example has nearly all its fisheries on an individual transferable quota (ITQ)-type management that allows the fishermen to not only balance the fishing capacity with the available resource, but to innovate in areas of both fish quality and efficiency. Consequently, the Icelandic brand has developed a quality standard that has gained international recognition. For the most part, Iceland receives a premium price on world markets for its fish and fish products. Also,

* Corresponding author: jeanguy@scotiaharvest.com

Iceland has developed a funding system where fishermen pay into a national fisheries fund from a small portion of their fish sales. These funds are used by the Icelandic minister of fisheries to innovate in areas of quality, efficiency, and marketing, in both the processing and harvesting sectors. So, instead of simply talking about opportunities, I made a commitment and chose to take a path, a risky path at times, to develop a business strategy and plan that hinges on a secure, stable access to the fisheries resource and the building of modern technology in vessels to bring the fish ashore at the highest quality practical for the fresh fish market and follow that by upgraded processing and marketing to ensure an "ocean to plate" (or how I refer to it as a "boat to throat") approach to fish harvesting. In Canada, with the current management regime and the Department of Fisheries and Oceans' (DFO) move towards policies that promote sustainability, prosperity, and application of the precautionary approach, I have been successful. It has not come without challenges though. In my mind, the next steps will be critical for the fisheries of the future in the GoM to be sustainable, and I will discuss in this paper where I see areas needing improvement.

Although I call this paper "Fishermen without Borders," I think we all know that fishermen/women in the GoM have plenty of rules, regulations, and borders, though the fish that occupy it do not. The GoM may have defined borders to some, but to the fishing industry, it carries a flow of larvae, nutrients, and other resources that help sustain the fishery, from Georges Bank to the West Scotian Slope to the Bay of Fundy and all points in between. The fish that occupy her waters provide (1) an economic driver to harvesters and communities; and (2) a large source of natural food, what might be the last natural food on the planet, that is, without additives and preservatives.

Why are fish important to me? I have been involved in harvesting groundfish all my life. I have always been amazed at the sea and its beauty and bounty. So much goes on under the waves in the form of predator–prey relationships and life cycles of organisms, both small and large, that one has to respect the complexities of the ocean system. My entire family has been blessed to earn a respectable living from the sea since 1945, being involved both in the harvesting and processing of fish resources.

Evolution of Fisheries of the Gulf of Maine

When I started my career fishing in the mid-1980s, the Canadian government was only beginning to look at quotas and limiting catches. I remember a meeting in Yarmouth, when DFO came to say that they needed to implement a quota on the North American Fisheries Organization division 4X cod of 30,000 metric tons. We have since lived through periods of overfishing that reduced fish populations, in addition to all kinds of management strategies to balance the fishing capacity with the available resource, but it was not until 1991, when ITQs were established, that this strategy was given a chance to succeed. There has been much pain endured to reduce vessels, fish plants, and numbers of people and their communities, but we are finally to a point where we see groundfish stocks starting to recover, and while the fishing has changed in where the fish were being landed, the methods of handling the catch, and thus the catch quality, for the most part, have not improved substantially. At least, quality has not matched improvements around the rest of the North Atlantic. Because the Canadian and U.S. fisheries and markets are so closely linked, it is apparent that the low quality of the catch and fish products was similar in both countries.

The ITQs enabled the industry to balance the fishing capacity with the available resource so that fishers could start planning a more sensible harvesting, processing, and marketing of our fish. Since 2006, I have embarked on a personal goal of quality in all aspects of harvesting, processing, and marketing. I started by ensuring top-quality fish on board the vessel, by designing and building a state-of-the-art vessel with "super-chilling" technology. The result has been the development of the Léry Charles Brand wild-caught premium haddock Melanogrammus aeglefinus sold mainly in Ontario and Quebec. Demand exceeds supply of this fish mainly because of its high quality (www.scotiaharvest.com). The bottom line is that a value-added approach to fish harvesting requires a secure source of raw material; that is, "if we take care of the fish, the fish will take care of us."

Current Issues Facing Fisheries in the Gulf of Maine

In the past decade, I feel that the fishing industry has improved by looking at a more ecosystem approach; however, it has been on an ad hoc basis. Fishers in the GoM have only begun dealing with the issue of bycatch and other physical impacts of fishing gear. In my case with groundfish, in 1993, after considerable testing with cameras of the separator trawl in the GoM, we established that with the proper gear modifications, selectivity can be achieved and bycatch reduced. Bycatch is a concern that needs to be recognized before efforts can be put towards reducing it, and some fisheries are only just realizing the extent of the importance of reducing/eliminating bycatch. To illustrate, one example is the lobster fishery with its cusk bycatch. With COSEWIC (Committee on the Status of Endangered Wildlife in Canada) listing more species for consideration under Canada's Species at Risk Act, this legislation will certainly have a profound effect on current fisheries practices. Cusk is under consideration for listing and is taken incidentally as a bycatch in the lobster fishery and is thought not to survive when released from traps. Another issue facing fisheries in the GoM is that on the Canadian side, our government has increased funding to science but reduced funding towards stock assessments. I understand that in order to better understand the ecosystem and its functions, more effort needs to go towards understanding predator–prey relationships, but we cannot forget that understanding stock status is also very important. Ecosystem-based approach is needed, but I feel that stock assessments and other measures of basic biology must not be neglected. I feel that we are dropping the stock assessment work too soon. Markets for fish are also being threatened by retailers demanding third party eco-labels. Fisheries that are found not to be third-party certifiable may be very limited in where they can sell their product. In my case with Nova Scotia haddock, we expect MSC certification in the summer of 2010, so we will be able to demonstrate that we are sustainable; however, many Canadian fisheries lack monitoring, control, and surveillance and will have great difficulty being certified. This eco-certification is a huge change from a decade ago.

As a member of the Fisheries Resource Conservation Council (FRCC), I have seen that when fishermen are faced with determining quotas or trip limits for the coming year, short-term solutions or consultations are seldom ones that provide long-term benefits. From my experience, long-term planning is always difficult in fisheries where fishermen do not know for whom they are conserving the fish. If the decision is made to provide access to a different group in the coming years, then in all fairness, individual fishermen who lose access or quotas cannot plan long term in terms of investments or markets. In Canada, a large portion of the groundfish is allocated by ITQs or enterprise allocation, which is similar to the catch share program now being proposed in the United States. For those fisheries where access to the resource is more secure, generally the industry has longer-term goals and sustainability is achievable. Also, usually the organizational structure of the fisheries participants improves with more security of access because there is much less infighting for access in those groups. But the fishing industry still has shortcomings with value added by quality and, therefore, I ask you, who is responsible for loss of potential in the fishery resource in the GoM? When one lands poor quality fish and sells to a market demanding lower quality, it means that you will experience lower value than if you had high quality product. The landed value in Nova Scotia of marine wild-caught fish (including lobsters, which make up nearly half of this value) averages around Can$1 billion annually. I believe the value should be, and could be, significantly higher. So who is responsible for the shortfall? Is it the fishing industry? Governments? Institutions? TRAC (Transboundary Resources Assessment Committee), GOMAC (Gulf of Maine Advisory Committee), National Marine Fisheries Service, a combination, all, or others? In most instances, my opinion is that fishers act the way that the management system around them permits, so I put the bulk of the responsibility on the regulator/government to ensure that the proper fisheries objectives are defined and adequate monitoring, control, and surveillance is in place. The fishing industry has codes of conduct for responsible fisheries operations, so I feel they too need to be accountable; however, to improve the resource value, it is for the government to decide what it is willing to accept.

A Vision for an Ecosystem Approach to Management

First we need to secure the resource, and we do this by developing long-term, sustainable fishery plans involving all stakeholders, and this means all stakeholders, involving different fishery groups, nongovernmental organizations, and politicians. Then, we build these long-term plans on the four pillars of sustainability: ecological, economic, social, and institutional. Furthermore, we balance these pillars with lots of discussions. Striking the delicate balance is critical: if you place too much emphasis on one pillar, others will suffer. The GoM borders two countries that may have different ideas of what sustainability means. In the case of Georges Bank transboundary groundfish stocks, countries jointly determined consistent harvest strategies, so we have set a good precedent. Also, we must include an ecosystem consideration matrix that highlights risks that we know today and that will assist decision making. We must start now and build from there.

Here, I have outlined a five-step program to achieve an ecosystem approach to management (EAM) as used in Australia and adopted by the Food and Agriculture Organization of the United Nations. The first step is to determine the scope of what it is that is being managed. I feel that the GoM is too large and complex for this approach to be effective, so it is likely better to start by determining the scope of each fishery, such as GoM herring or groundfish, and including people from both fleet sectors to assist in developing plans for the other. The second step is to identify the issues related to retained species and nonretained species or, in other words, directed and bycatch species, ecosystem impacts, community wellbeing, and others. It means considering the four pillars of sustainability for each issue around the species being studied. Once you have determined all the issues, you need to work through the third step, which is determining the risks associated with harvesting activities and the risks to the four pillars. For example, what are the risks of not reaching economic or social or ecological or institutional sustainability? Risk is determined using a risk matrix table, which helps determine the relationship between the probability of an event happening and its impact. Using the risk matrix, we can study the probability of an event happening (i.e., is there a remote, unlikely, possible, or likely chance that it will happen?) and what is the impact of that event? (minor, moderate, major, or extreme?) If the probability of an event happening is possible and the impact is major, then the matrix determines the risk is high, and it means you need to reduce the risk by taking the necessary action. The fourth step would be to develop a management response for issues that are presenting medium and high risks (e.g., limit reference points). The fifth and final step is to develop and implement operational plans for management from step four (e.g., monitoring, reporting plans). The risk matrix approach encompasses all that is directly and indirectly linked to the target stock and quickly shows where effort should be applied to reduce risks. One should not spend too much effort working on issues that are in the low or negligible risk category. The FRCC asked Dr. Rick Fletcher from Australia to assist the council in going through the process and exercise using Gulf of St. Lawrence Atlantic herring *Clupea harengus* (FRCC 2009). What we found was that there exists a lack of clear, precise fishery objectives, which is the case in most fisheries worldwide. When you can determine clear objectives, you can manage your fishery much better. If you do not know where you are going, you will never know if you get there or not. The "Fletcher" EAM (R. Fletcher, presentation at the Fisheries Resource Conservation Council meeting, 2007) is both a process and a living document. This approach can be used right away with existing knowledge, and as you determine the medium and high risk areas and find solutions, you simply update your plans.

Are we ready to take this approach? Yes we are, and we have a good example of a start in the Georges Bank groundfish as a transboundary resource. After competing for access to the resource in the late 1980s and the early 1990s, Canada and the United States were overfishing the groundfish on Georges Bank until the Transboundary Resources Assessment Committee (TRAC) was set up in 1998. The TRAC invited scientists and fishermen from both Canada and the United States to jointly determine stock status. Management advice was developed by the same principles of having scientists, fishermen, and managers from both countries. Stocks have been comanaged by the Transboundary Management Guidance Committee (TMGC) for the past decade, which provides the advice to GOMAC for approval. Both countries usually agree and accept the recommendations of TMGC. It is not a perfect system, but I have no doubt that we are much better at managing the Georges Bank groundfish resource now than we were a decade ago. For more information on how this approach works, look at the Web site at www.mar.dfo-mpo.gc.ca/science/TRAC/TRAC.html.

In conclusion, I suggest that we must have two major priorities: get more from less—we need to draw more value from the resource and we may find we need less resource to satisfy the fishery's needs, and secure the resource—by setting clear objectives and assessing their risk to sustainability. More details on this approach are provided on the FAO Web site (FAO 2006).

References

FAO (Food and Agriculture of the United Nations). 2006. Fisheries topics: ecosystems. Available: www.fao.org/fishery/topic/2880. (December 2009)

FRCC (Fisheries Resource Conservation Council). 2009. Fishing into the future: the herring fishery in eastern Canada. FRCC, Ottawa.

American Fisheries Society Symposium 79:47–60, 2012

Tools for Integrated Policy and Management in the Gulf of Maine: A Summary of Major Advancements Since 1996

Maxine Westhead*
Fisheries and Oceans Canada, Maritimes Region
Bedford Institute of Oceanography
1 Challenger Drive, B500, Dartmouth, Nova Scotia B2Y 4A2, Canada

Abstract.—This paper briefly summarizes some of the regional developments and advances in the theme "Tools for Integrated Policy and Management" that were presented at the October 2009 Gulf of Maine Science Symposium held in St. Andrews, New Brunswick. Tools for integrated policy and management meet a wide range of functions and originate from a variety of fields, from the physical and biological sciences to social sciences and organizational studies. Some of the tools presented at the 2009 Gulf of Maine symposium are discussed in this paper. Knowledge gaps still exist regarding impacts of development and industrial activities, which becomes apparent when trying to evaluate and quantify cumulative impacts. Geographic information system-based tools and elements of marine spatial planning such as human use mapping can help establish better cumulative impact assessments. While it is recognized that advances in integrated policy and management have occurred since the previous (1996) Gulf of Maine science symposium, evaluation of that progress is still in development. Tools such as state of the environment reporting, integrated assessments/ecosystem overviews, frameworks for implementation of ecosystem-based management (EBM), or approaches developed by the EBM Tools Network all assist in measuring and evaluating our progress towards full implementation of an ecosystem approach within the context of management activities and actions.

Context

The October 2009 Gulf of Maine Science Symposium, held in St. Andrews, New Brunswick, explored a variety of topics, ranging from evaluation of ecosystem services to benthic habitat mapping techniques to anthropogenic and external influences on the Gulf of Maine. One of the major themes, "Tools for Integrated Policy and Management," explored the management and policy tools and approaches that are required to implement integrated management in the Gulf of Maine. In particular, poster authors and oral presenters addressed one or more of the following topics:

- Development of new scientific approaches, tools, and analysis methods for an ecosystem approach to management, including integrated ecosystem models; management strategy evaluation approaches, including marine protected areas as a spatial conservation tool;
- Integration of the above approaches, tools, and methods into policy and management;
- Coordination of science and management scales (the scale for observation and the scale for feedback) for multiple ecosystem services and sectors—identifying temporal and spatial mismatches between biological systems and human institutions; and
- Assessment and measurement of management success from a coupled social-ecological perspective.

This paper briefly summarizes some of the regional developments and advances in tools for

* Corresponding author: maxine.westhead@dfo-mpo.gc.ca

integrated policy and management that were presented at the symposium. In reviewing advances since the last Regional Association for Research in the Gulf of Maine Symposium in 1996, it is important to note that the developments discussed in this paper are not an exhaustive list. As well, this paper only briefly touches on some of the major advancements since 1996 and does not go into great detail, but references for more comprehensive reading are provided where available. The tools presented at the 2009 Gulf of Maine Symposium included those used for organizing and evaluating information, such as tools to better classify ecosystem features; tools for assessing the implementation of management objectives, such as ecosystem approach to management frameworks and state of the environment reporting; and tools for communication, such as adult learning models and new media technology.

Setting the Stage

We have known for decades that the Gulf of Maine ecosystem is complex. It is a tidal ecosystem with linkages beyond provincial, state, and national borders. The tidally driven and sediment-laden waters of the Bay of Fundy continually mix with the waters of the Gulf of Maine proper. As a semi-enclosed sea, renowned as one of the richest marine ecosystems, it is a challenging and unique environment within which to work. These ecosystem complexities challenge our ability to monitor and assess the effectiveness of management decisions and actions.

The geopolitical structure surrounding the Gulf of Maine is also very complex, comprised of two countries, three states, two provinces, federal regulators, and several First Nations/tribal governments. The development of institutional arrangements, collaborative relationships, and regulatory regimes within and among the various structures is very complicated. On top of this are strong communities of place where local citizens have passionate ties and deep connections to their local coastal and marine environments. Many of these communities also have rich Aboriginal heritage, in addition to diverse industrial sectors, all operating within the Gulf of Maine.

At the same time, the Gulf of Maine is a well-researched area with a high level of interest in management and resource use issues. For instance, a Google Scholar search for "Gulf of Maine" yields more than 14,000 published articles since the last Gulf of Maine science symposium in 1996. The same tool searching for "Bay of Fundy" finds more than 10,000 items since 1996. The Gulf of Maine Council on the Marine Environment alone has produced more than 300 products since the late 1990s. Fuelling that research are numerous universities along the Gulf of Maine coast, formal scientific research stations such as the Gulf of Maine Research Institute, the laboratories of the National Oceanic and Atmospheric Administration's Northeast Fisheries Research Institute (Woods Hole Laboratory and Maine Field Station), and the St. Andrews Biological Station and Bedford Institute of Oceanography (both Fisheries and Oceans Canada facilities). Indeed, there is a wealth of scientific interest in the area.

The Concept of "Tools" for Integrated Policy and Management

A tool is something that is necessary to complete a task or something that makes a task easier or faster. For example, medicine is a tool to combat illness or disease. Geographic information system (GIS)-based information can be used as a tool to inform the location of a project or the extent of effects. Monitoring data can be a tool to inform managers if policies or regulations are being effective at a local level. Integrated policy and management in this paper's context implies close linkages and relationships between science and policy, science and management, and management and policy, in addition to the more intricate layers of sociological, political, and legal connections and influences that are involved in decision making.

Tools for integrated policy and management range in scope and discipline (e.g., scientific or sociological research, classification systems, governance models, planning, GIS-based data collection, identification of cultural heritage, environmental impact assessment, regulations, and enforcement). Ultimately, tools for integration cross disciplines or broaden the scope of more traditional approaches (e.g., single species fisheries management plans moving to integrated fisheries management plans or assessment of single activities moving to cumulative impact assessment).

This paper explores some tools that are broadly defined to capture a range of different approaches to advance integrated policy and management in the Gulf of Maine context.

Tools for integrated policy and management meet a wide range of functions and originate from a variety of fields, from the physical and biological sciences to social sciences and organizational studies. Some of the tools presented at the 2009 Gulf of Maine Symposium, and discussed in this paper, include advancements in (1) legislation and policy; (2) government collaboration; (3) marine spatial planning; (4) resource management; and (5) reporting, communication, education, and challenges and future directions.

Advancements in Legislation and Policy

Domestic Fisheries Management:

In the United States, the 1996 Sustainable Fisheries Act (SFA) amended the Magnuson-Stevens Act with a requirement to identify Essential Fish Habitat (EFH), and to make progress towards reducing, avoiding, or mitigating both fishing and nonfishing impacts to these habitats (NOAA 2010). The 1996 SFA also introduced three new national standards to address fishing vessel safety, fishing communities, and bycatch (while also revising some of the existing seven), by which all fishery management plans must comply (NMFS 2003).

Following adoption of the SFA, Amendment 7 to the Magnuson-Stevens Act reduced days at sea and closed some areas to fishing. The Magnuson-Stevens Act was also amended in 1996 to strengthen the role of science for the identification of "essential fish habitat" (EFH). Defined the same year as the previous Gulf of Maine symposium in 1996, EFH is now required to be identified for each federally managed fish species in the United States (Kurland 1998). With respect to defining these habitats, New England is far ahead of other U.S. regions due to long-term and intensive study specific to EFH identification. As a case in point, D. Preble (New England Fishery Management Council) noted during his keynote speech that the New England Fisheries Management Council has a database of publications with more than 400 citations specific to EFH. In 1998, EFH descriptions were basic text with coarse maps. Today, they include scientifically based descriptions with GIS-based maps that also consider abiotic elements. As L. A. McGee (Batelle Memorial Institute) and C. Demarest (National Marine Fisheries Service) suggested in their presentation, this opens the door to using EFH descriptions in broader GIS-based marine spatial planning exercises, more effectively integrating fisheries management with broader ocean-use planning.

In 1997, amendments to the Magnuson-Stevens Act in the United States brought in new guidelines for the national standards. How these legislative changes have been implemented, and what they have achieved across the country, is outlined in a report by the National Marine Fisheries Service (NMFS 2003). By 1998, all regional fishery management councils (including the New England Fishery Management Council in the Gulf of Maine region) were required to adopt stricter measures to control overfishing of groundfish and to reduce bycatch. Shortly thereafter in 1999, Amendment 9 defined "optimum yield," "overfishing," "overfished," and "biological reference points." All of these legislative changes described above have imposed stricter measures to prevent overfishing and rebuild overfished fisheries. Each fishery management plan now requires specific objectives and criteria for determining when a stock is overfished, and measures to rebuild the stock must be established.

Most recently, the amendments in 2006 have resulted in the use of different controls, or tools, to improve the management of fish catches. President Bush signed the amendments of the Magnuson-Stevens Fishery Conservation and Management Reauthorization Act of 2006 on January 12, 2007. This was significant new legislation that builds upon the U.S. Ocean Action Plan in five key areas: (1) mandating the use of annual catch limits and accountability measures to end overfishing in America by 2011, (2) providing for widespread market-based fishery management through limited access privilege programs, (3) strengthening enforcement and fishing laws, (4) improving information and decisions about the state of ocean ecosystems, and (5) calling for increased international cooperation (NOAA 2010).

Fisheries management within Canada has also taken a noteworthy shift since 1996, when a more single-species focus was the standard. New integrated fisheries management plans (IFMPs) go beyond the focus of quotas and also identify goals related to conservation, management, enforcement, and science. The plans also require a periodic review of the current state of the fishery against the plan's objectives. As a document, IFMPs are intended to provide a clear summary of the characteristics of the fishery, scientific elements, management objectives, and measures and criteria for obtaining those objectives. Although IFMPs will play a strong role in the implementation of the sustainable fisheries framework (described immediately below), they will also incorporate a rules-based approach to decision making that is more transparent, rigorous, and systematic (DFO 2010). Existing fisheries management plans will be converted to the new IFMP format over time.

Ongoing efforts to update the Canadian Fisheries Act legislation (enacted in 1868) has resulted in the creation of new policies and strategies to manage Canadian waters, in particular the sustainable fisheries framework (SFF), released in 2009. The primary goal of the SFF is to ensure that Canada's fisheries are environmentally sustainable while supporting economic prosperity. The framework has three associated policies: a fishery decision-making framework incorporating the precautionary approach, a policy for managing impacts of fishing on sensitive benthic areas, and a policy on new fisheries for forage species (DFO 2009). The SFF pulls together both new and existing fisheries management policies and program tools to assist Fisheries and Oceans Canada (DFO) in the identification of gaps in fisheries management program delivery and to guide the continual improvement of more sustainable fisheries (DFO 2009). The framework is designed to facilitate a more rigorous, consistent, and transparent approach to decision making across all key fisheries in Canada. It is anticipated that the framework will be implemented using adaptive management principles and continue to evolve as lessons are learned and new policies and tools are created (DFO 2009).

Joint Canada–U.S. Fisheries Management

In 1984, the International Court of Justice delineated the maritime boundary that separates Canadian and U.S. waters in the Gulf of Maine. During the 1980s, increased cross-boundary fishing effort on Georges Bank prompted both countries to consider complimentary approaches to resource management and enforcement. Canada and United States of America: Agreement on Fisheries Enforcement (1990) was signed as a result, supporting efforts to prosecute unauthorized fishing on both sides of the boundary. A series of regional bilateral discussions between Canada and the United States resulted in the creation of the Canada–U.S. Transboundary Steering Committee in 1995.

The steering committee was established to explore joint Canada–U.S. management of groundfish stocks and quota allocation (eastern Georges Bank Atlantic cod *Gadus morhua*, eastern Georges Bank haddock *Melanogrammus aeglefinus*, and yellowtail flounder *Limanda ferruginea*) and clearly recognized the importance of bilateral collaboration. This joint fisheries management work is now carried out by several committees, all of which report to the steering committee that is cochaired by the regional director general of DFO Maritimes Region in Canada and the regional administrator of the National Oceanic and Atmospheric Administration's (NOAA) Northeast Regional Office. The Transboundary Resources Assessment Committee (www2.mar.dfo-mpo.gc.ca/science/trac/TRAC.html), who first met in 1998, conducts the fisheries resource assessments. The Transboundary Management Guidance Committee (www2.mar.dfo-mpo.gc.ca/science/tmgc/introduction.html), who first met in 2000, developed a resource-sharing agreement for quota on Georges Bank, discusses quota allocation between the two countries, and provides nonbinding guidance to the steering committee.

In recognition of the limitations of single-species stock assessments, each country's efforts to move towards an ecosystem approach to management, and the opportunities to build on this bilateral structure, the steering committee intentionally broadened its scope in the mid-2000s. Three additional working groups were created in an effort to move towards a more comprehensive and integrated approach to oceans management. These were the Species at Risk Working Group established in 2003, the Habitat Working Group established in 2004, and the ad-hoc Oceans

Working Group established in 2005. This demonstrates an evolution towards a more ecosystem-based approach to management on both sides of the border.

Oceans Management

Developments in resource management outlined above have gone hand in hand with changes in the overall legislative framework for managing Canadian waters. In Canada, the Oceans Act was passed in 1996, the same year as the previous Gulf of Maine symposium, followed by Canada's Oceans Strategy in 2002 (DFO 2002), which outlined Canada's approach to implementation of the Oceans Act. The prominent thrust of the Oceans Act mandates the Department of Fisheries and Oceans, on behalf of the Government of Canada, to lead and facilitate the development of integrated management (IM) plans for coastal and marine waters, and create a national system of marine protected areas (MPAs). Integrated management is the coordinated management between stakeholders and regulators of all human activities in a given management area, so that negative human-ecosystem effects and human-human conflicts can be minimized. The principles of IM in Canada are an ecosystem approach to management, sustainable development, and the precautionary approach (Oceans Act 1996).

To move forward on Oceans Act implementation, northeast of the Gulf of Maine, the Government of Canada has developed, in collaboration with participants, the Eastern Scotian Shelf Integrated Management (ESSIM) Plan (DFO 2007). The ESSIM initiative is comprised of a planning office (DFO-led), a stakeholder advisory council, and multiple working groups that provide the support and means for collective action. In the early 2000s, ESSIM participants (federal and provincial governments, fishing industry, academia, conservation organizations, etc.) worked diligently to come to agreement on and create Canada's first integrated oceans management plan. Although not legally binding, the ESSIM Plan outlines objectives and strategies agreed to by all participating regulators and stakeholders (for more detail, see Rutherford et al. 2005; DFO 2007). Through the Oceans Act, Canada has also established seven MPAs across the country, one of which is in the Bay of Fundy/Gulf of Maine ecosystem (Musquash Estuary MPA, legally designated December 14, 2006).

The Government of Canada passed the Species at Risk Act in June of 2003 (prohibitions in effect June 2004), which is a federal commitment to prevent wildlife species from becoming extinct, and defines necessary actions for their recovery. It also provides for the legal protection of wildlife species, similar to the U.S. Endangered Species Act of 1973. Transboundary species, such as the North Atlantic right whale *Eubalaena glacialis*, have become a collaborative effort between Canada and the United States, and in the case of species with even greater ranges, such as leatherback turtles *Dermochelys coriacea*, Mexico also plays a role in recovery efforts.

In the United States, President Obama established an Interagency Ocean Policy Task Force, led by the White House Council on Environmental Quality on June 12, 2009. On December 14, 2009 the task force released an interim framework for effective coastal and marine spatial planning (MSP) for a 60-day public review and comment period (Interagency Ocean Policy Task Force 2009). On July 19, 2010, the final recommendations from the Interagency Ocean Policy Task Force were released and include four key elements: (1) a national policy for the stewardship of the ocean, coasts, and Great Lakes; (2) a policy coordination framework; (3) an implementation strategy; and (4) a framework for effective coastal and marine spatial planning. President Obama's more recent executive order of July 19, 2010 adopted the final recommendations and established a National Policy for the Stewardship of the Ocean, Coasts, and Great Lakes, as well as a National Ocean Council. The order also directs federal agencies to take appropriate steps towards implementation of the policy and the establishment of regional planning bodies, with appropriate advisory mechanisms, to develop a regional coastal and marine spatial plan in 3 to 5 years (Council on Environmental Quality, www.whitehouse.gov/administration/eop/ceq/initiatives/oceans).

Created in 2005, the Northeast Regional Ocean Council (NROC) is a state and federal partnership whose terms of reference provide the following mission: "To assist the region's Governors to identify coastal and ocean management priorities that require a coordinated regional response and to foster collaboration that effectively addresses these issues" (NROC 2010a). The council's coordinated approach reaches across state boundaries to find and implement solutions to the region's most pressing ocean and coastal issues. The three priority issues identified in 2007 include ocean ecosystem health and marine spatial planning, coastal resiliency and climate change adaptation, and ocean energy (K. Leyden, Maine Coastal Program and Northeast Regional Ocean Council, presentation at the California and the World Ocean Conference). The scope of NROC includes the Gulf of Maine and extends beyond, from Maine to Connecticut, and NROC's aim is to use the impetus of the 2010 Executive Order to position the region in realizing tangible results for coastal and marine spatial planning (NROC 2010b). How advances in MSP unfold and develop, and the relationship between the regional planning body and NROC, is currently being discussed (NROC 2010b).

Ecosystem-Based Management

The ecosystem approach to management (EAM) or ecosystem-based Management (EBM), used interchangeably in this paper, involves considerations of how human activities impact the ecosystem and an understanding of how the ecosystem in turn affects human activities (Arkema et al. 2006; Arkema and Samhouri 2012; Fogarty et al. 2012; McGee 2012; all this volume). The emphasis is on management of the human activities to control ecosystem impacts and, at the same time, recognizes how the ecosystem influences the way the activity should be managed. Most importantly, EBM recognizes humans as an integral part of the ecosystem (McCay 2012, this volume), as we take ecosystem services and our activities influence ecosystem processes (Levin et al. 2009). Application of the concept requires consideration of how an activity will impact all ecosystem components, such as ecosystem goods and services, and not simply the resource itself (McLeod et al. 2005). Thus, a key feature of EBM is recognition of and conscious decisions regarding trade-offs among and between ecosystem services and management goals.

With little practical scientific advice to inform management authorities on how to select specific management measures to achieve EBM goals, effective implementation still faces significant difficulties (Taylor 2008; Levin et al. 2009). To tackle this problem, the best available scientific knowledge and understanding must be compiled and organized in a way that informs decisions at multiple scales and across sectors. The United States has proposed integrated ecosystem assessments (Fluharty et al. 2006; Levin et al. 2009) and Canada has developed ecosystem overview and assessment reports. From these assessments, or baseline documents, the key to moving EBM from concept to practice lies in defining ecosystem-level decision criteria, or reference points, that trigger management actions (Samhouri et al. 2010).

While we have a wealth of data available to us as a result of historical emphasis on reductionist investigations (Wilson 1998), an understanding of the whole ecosystem and how they respond to human activities is still the missing link (Levin et al. 2009). Although these new EBM frameworks and ecosystem assessment documents promote sustainability and an ecosystem approach to management, it is too soon to comment on whether or not they have been effective in the Gulf of Maine. Another contemporary challenge is how to measure success of management efforts. The challenge is to determine what the measurable ecological, social, and institutional outcomes could be to achieve the goals of EBM, and which management measures will contribute to those outcomes (McLeod and Leslie 2009).

Advancements in Government Collaboration

Although there is now a better understanding of the overall requirements for broader oceans management and legislation, and policy has been developed to facilitate the implementation of EBM, the current operating environment in the Gulf of Maine does not lack challenges. Oceans management on both sides of the Canada–U.S. border is complex, with many different jurisdictions responsible for managing fisheries and other aspects of the marine environment in the Gulf of Maine. Preble commented during his keynote speech that this complexity has been compounded by the use of the court system to address fisheries management issues, particularly in the United States.

On the Canadian side of the border, the federal and provincial governments with an interest and/or mandate in managing ocean activities have established a Regional Committee on Coastal and Ocean Management (RCCOM). Rotating co-chairs from federal and provincial departments facilitate discussions and actions required for collaborative and informed management of activities in marine waters. The RCCOM is a senior executive forum for federal and provincial governments, with some mandate or jurisdiction in the marine environment to provide oversight, coordination, and direction to the planning and management processes related to integrated oceans and coastal management in the Maritime Provinces of Canada (New Brunswick, Nova Scotia, and Prince Edward Island) (T. Hall, Fisheries and Oceans Canada, personal communication). The RCCOM could be considered the Canadian counterpart to NROC in the United States.

The Gulf of Maine Council on the Marine Environment (GoMC) spans political boundaries and is a U.S.–Canada partnership comprised of government and nongovernment organizations working to maintain and enhance the environmental quality of the Gulf of Maine. They organize conferences and workshops, offer grants and recognition awards, conduct environmental monitoring, provide science translation to management and the public, raise public awareness about the Gulf, and make connections between people, organizations, and information. The GoMC celebrated its 20th anniversary in 2009, and member agencies continue to show dedication and commitment to the council's achievements.

Advancements in Marine Spatial Planning

As both P. H. Taylor (Waterview Consulting) and V. DeLauer (COMPASS) stated in their presentations, marine spatial planning is a process that relies on GIS-based tools and mapping to assist with the visualization of possible development and resource-use scenarios. Marine special planning is a comprehensive, adaptive, integrated, ecosystem-based, and transparent spatial planning process based on sound science (EOP 2009) that assists with on-the-water implementation of integrated management. MSP processes should be designed to bring together multiple users of ocean space (e.g., energy, fishing, regulators, conservation, and recreation) to help make informed and coordinated decisions regarding the use of marine resources and space. Using maps, MSP can create a more comprehensive picture of marine space by identifying where and how an area is used and can visually display the location of renewable and nonrenewable resources and habitats. These maps are usually GIS-based for ease of manipulation, layering, and analysis and can incorporate a range of data from geo-referenced species- or habitat-specific data to more personal local and traditional knowledge. By building comprehensive, regional maps, one can better plan for a more integrated approach to ocean use and space allocation. Approaches to MSP vary widely and can be tailored to suit regional needs (Gilliland and Laffoley 2008; Pomeroy and Douvere 2008). Although MSP does not necessarily impart any change in existing authorities or jurisdictions to participating parties, it has been suggested that legal authority and political support are important factors for success (Ehler 2008).

MSP is a tool that can provide the appropriate anthropogenic and ecological context to enable integrated, forward-looking, consistent decision making regarding the use of ocean space. Many countries around the world have been working towards MSP plans for years and in particular places, such as Germany, Belgium, and The Netherlands in the North Sea and the UK's Irish Sea, are quite advanced (for synopsis, see Douvere and Ehler 2008). It is important to recognize that these European seas face much greater competition for space than currently exists in the Gulf of Maine, which could be the impetus for MSP advancement in the overseas areas mentioned above.

As outlined in the Oceans Management Section, the interim framework for MSP in the United States from the Interagency Ocean Policy Task Force (EOP 2009) provides recommendations for ocean policy, which include MSP. Some individual states are already undertaking MSP-based initiatives, which could be used as case studies to learn from, as MSP is advanced in the Gulf of Maine. Closest to home is the Massachusetts Ocean Act of 2008 that requires a comprehensive ocean management plan, which supports ecosystem health and economic vitality, balances current ocean uses, and considers future needs. California and Oregon are also examples of MSP in action. More detailed U.S. descriptions are outlined on NOAA's Coastal and Marine Spatial Planning Web site (www.msp.noaa.gov). The application and future of MSP in Canadian waters is currently being discussed, but many elements of the MSP process are already in practice, such as human use mapping (Breeze and Horsman 2005), the identification of ecologically and biologically significant areas (Buzeta et al. 2003; Buzeta and Singh 2008), fundamental scientific descriptions of geographic planning units (Kostylev et al. 2001), and, in some cases, multi-stakeholder advisory bodies, to come to consensus on management goals and objectives for a given area (DFO 2007).

Significant advancement of MSP in the Gulf of Maine in the near future would be a strong proactive step towards preventing future conflicts as the number of ocean uses and users expands in the region. There are new challenges facing the Gulf of Maine, such as wind energy, tidal energy, drilling for oil reserves should current moratoriums be lifted, and new protected areas—not to mention existing space conflicts such as traditional fisheries activities and aquaculture lease siting. Pursuing MSP now, before serious conflicts arise and escalate, could help poise decision makers and users favorably for planning the future of the region.

Advancements in Reporting, Communication, and Education

Since 1996, there has been significant technological and organizational evolution towards the use and development of more effective and efficient tools for reporting, communicating, and educating. In particular, advances in social media and Web-based technology have provided delivery mechanisms that allow for greater access to information, knowledge, and data.

With respect to the Gulf of Maine, the Gulf of Maine Summit was held in St. Andrews in October 2004, for which the centerpiece report was entitled "Tides of Change Across the Gulf: An Environmental Report on the Gulf of Maine and Bay of Fundy" (Pesch and Wells 2004). That report set the stage for the summit meeting, which sought to produce a shared vision for the gulf's future. The report, focused on three themes, summarized results from meetings and provided in-depth information on the themes of land use (coastal development), contaminants and pathogens, and fisheries and aquaculture. As a result of the summit meeting, the Ecosystem Indicators Partnership (ESIP), a committee of the Gulf of Maine Council on the Marine Environment, has formed to develop indicators and integrate regional data for a new Web-based reporting system for marine ecosystem monitoring. ESIP (www.gulfofmaine.org/esip) has been working towards placing data and information directly into the hands of decision makers regarding the three key priority issues above, in addition to three others, which are climate change, eutrophication, and aquatic habitats.

P. Boudreau of the Coastal and Ocean Information Network Atlantic initiative (COINAtlantic) summarized this initiative, which aims to provide one-stop access to online data for those responsible for implementing integrated management. This is a Canadian initiative of the Atlantic Coastal Zone Information Steering Committee, which has a mandate to develop, implement, and sustain a network of data providers and users that will support secure access to data, information, and applications for decision making by coastal and ocean managers and users. The initiative does not host data but aims to serve as a primary access point for those seeking GIS-based data and provide information on how to acquire it.

Tools and approaches now exist that are used to better organize, evaluate, and share existing information, with the goal of implementing EBM. As summarized in presentations by both Taylor and DeLauer, the EBM Tools Network (www.ebmtools.org) is an alliance of EBM tool users, providers, and researchers that looks to promote the use and development of EBM in coastal and marine environments. In the spring of 2009, the network launched a road map for EBM practitioners that illustrates the eight core elements of EBM, focus questions, case studies, and tools for implementation in a real-world context. Key to this effort is the embracement and successful use of a host of new media technologies to bring training and learning opportunities to practitioners' desktops, from webinars to e-mail listservs to Web-based information about current research and its application.

Science translation in the Gulf of Maine is an ongoing effort and a valuable tool to get technical data from the hands of few into user friendly information and into the hands of many. A suggestion voiced during plenary discussion at the 2009 Gulf of Maine Science Symposium was that greater value should be placed on social science expertise to make stronger linkages between public communications material and desired change. The possible overestimation of the motivational power of science products for managers and the public was also suggested. Along those lines, suggested by both Fuert (2012, this volume) and DeLauer (2012, this volume) during their presentations at the symposium, was a shift in perception of decision makers from receptacles for information to valuable partners that could and should be engaged.

There are many networks of people and institutions, both formal and informal, either solely focused on the Gulf of Maine/Bay of Fundy or with overlaps in the region. A positive, yet challenging development that has occurred since the 1996 symposium is the proliferation of organizations and networks focused on the Gulf of Maine marine environment and/or watershed (discussed during plenary at the 2009 symposium). The number of organizations with an interest in the gulf demonstrates the importance of the region to a wide range of stakeholders, including politicians and regulatory authorities, students and researchers, resource users, and those living in coastal communities. At the same time, this poses communication and governance challenges. The multitude of diverse interests brings the necessary diversity of perspective to discussions but also makes timely integration of policy, management, and science a challenge. The means by which all can work together effectively without some duplication of effort remains an ongoing challenge.

The application of developmental psychology and adult learning models to oceans management and the emphasis on exploring novel means of communicating are relatively new considerations for an integrated approach to management. As groups move to develop EBM/EAM in the Gulf of Maine, they are challenged to consider the varying characteristics of those involved with integrated management, in addition to the diverse ways that individuals or organizations can contribute (DeLauer 2012). Time spent building capacity and understanding among those involved with EBM may be, at the very least, as important as developing the technical tools to implement it. The potential for new technologies to work together, enabling greater collaboration, communication, and knowledge sharing and thereby permitting management to be more open and transparent, is great.

The Gulf of Maine Council on the Marine Environment has recognized the importance of reporting on the state of the environment (Pesch and Wells 2004; Thompson 2010; Walmsley 2012, this volume). These reports provide important context for developing policy and decision making and foster the use of science in these activities. A state of the environment report should provide an analysis of environmental trends and conditions, measure progress towards sustainability, contribute to informed and open decision making, contribute to public awareness about environmental health issues, and serve the public's right to know by providing scientific information about the environment in a way that is accessible to the general public (Walmsley 2012). A state of the Gulf of Maine report has just recently been released by the Gulf of Maine Council (Thompson 2010). The process of undertaking this type of reporting offers great opportunity, as it requires parties to bring together knowledge and information from a wide variety of sources, including traditional and local knowledge, into one place for synthesis and interlacing. These reports, along with the process of creating them, can broaden our scope, horizon, and understanding of what we know and what remains to be understood.

Operating Environment and Ongoing Challenges

Highlighted in this paper thus far are just a few of the recent advancements in legislation and policy, government collaboration, marine spatial planning, and reporting, communication, and education. There is great scientific interest in the region, and many scientists are both passionate about their work and having that work inform better management of the marine environment. There is also great personal interest in the region, with many citizens who feel a deep sense of pride and connection with the marine and coastal environment of the Bay of Fundy/Gulf of Maine.

Significant advances have been made with respect to fisheries science—stock assessments, gear impacts, bycatch reduction, and so forth—which helps to better manage our fishing efforts (Kostylev et al. 2002; Link 2002; Grizzle et al. 2009; Hourigan 2009). Knowledge gaps still exist regarding impacts of development and other industrial activities. This becomes apparent when trying to evaluate and quantify cumulative impacts. Effective measurement of ecosystem impacts of existing activities is still challenging, and new uses are knocking on the door of the Gulf of Maine, such as liquefied natural gas terminals, tidal power, offshore wind energy, and MPA planning. The GIS-based tools and elements of MSP, such as human use mapping, can help inform better cumulative impact assessments.

Although not discussed at length at the Gulf of Maine science symposium, MPA planning deserves mention as both Canada and the United States strive to meet international targets over the next decade. MPAs are geographically discrete areas that can range in scope and regulatory power from voluntary seasonal avoidance of fishing gear types to fully closed marine reserves. The science and management of MPAs share some important overlaps and common goals with EBM, but to date, the efforts have remained mostly separate (Halpern et al. 2010). Most nations around the world have committed to set aside 10–30% of their waters as MPAs by 2012 as a result of the 2002 World Summit on Sustainable Development and the Conference of the Parties in 2006 (WSSD 2006). While only a few countries are close to this goal (Wood et al. 2008), governments around the world are in the process of exploring ways to meet international commitments. The historical division between resource management and biodiversity conservation has resulted in little connection and little guidance on how to use MPAs within an EAM (Halpern et al. 2010). As more MPAs are established to reach international conservation targets, it is important to start making stronger and complimentary connections with the broader context of ocean management and spatial planning for all activities.

Political will and interest is also a factor in federal and state/provincial funding decisions, which ties into a topic raised at the 2009 Gulf of Maine Science Symposium. Within the realm of the general public, despite past efforts for education as described above, general consensus at the symposium was that there is still a low level of general knowledge and understanding of our oceans (or ocean literacy). The voting public can be a strong voice in political arenas and are a largely untapped resource for encouraging change from the ground up, encouraging politicians and managers who make decisions regarding budgets and policy directions. If the public is not aware of problems or issues in the marine environment, issues will not be raised with politicians and political staff at town hall meetings or other venues, and finding solutions could remain a low priority. Taylor and DeLauer stressed during their presentations that one way to mitigate this is by placing more emphasis on science translation (discussed above) and the human elements of marine resource management—engaging stakeholders in decision making, communicating management processes, and using stakeholders and the public as resources for ideas and creativity.

While much has been published about the shift of resource management from single-species to an ecosystem context over the past decades (e.g., Jennings et al. 2001; Link 2002; Fogarty et al. 2012; Lawton et al. 2012; both this volume), a paramount challenge for implementing a full ecosystem approach to management is the lack of established methods for implementing EBM. This lack of methodology was noted by both Taylor and DeLauer during their presentations. There is a plethora of information on the theoretical nature of, and considerations for, applying EBM but very little published on actual implementation. Within the Canadian federal government, DFO is currently drafting a framework for implementation that includes setting objectives, defining ecosystem attributes, and developing strategies and tactics for managed activities. Based on underpinnings from a national workshop in 2001 (Jamieson et al. 2001), this framework is still under discussion and development but, in concert with ecosystem overview and reports, provides a solid baseline for regional consideration. On the U.S. side, NOAA's integrated ecosystem assessment process provides the counterpart to the Canadian approach. Through the ad-hoc Oceans Working Group of the Canada–USA steering committee, an ecosystem overview for the Gulf of Maine has been drafted and is planned for release in 2011 (Hall, personal communication) With this report, recent U.S. commitment for MSP planning and implementation, ongoing research, and efforts of the Gulf of Maine Council, the region is poised to advance the implementation of EBM. This will require ongoing commitment and collaborative efforts of both government and stakeholders to steer future actions and usher in a new era.

Future Directions and Priorities

Both Canada and the United States have made progress in understanding and implementing EBM, but better tools for evaluating and managing cumulative effects are still warranted (Halpern et al. 2008). New activities in the Gulf of Maine need to be considered against the backdrop of the current uses of space, time, and resource availability. As competition for space changes and grows, and in order to fully implement and realize EBM, collaborative efforts to develop tools and approaches for understanding cumulative effects will be required.

While it is recognized that advances in integrated policy and management have occurred since the 1996 Gulf of Maine Science Symposium, an overall evaluation of those advances is developing. The application of state of the environment reporting, integrated assessments/ecosystem overviews, frameworks for implementation of EBM, or approaches developed by the EBM Tools Network all assist in measuring and evaluating our progress towards full implementation of an ecosystem approach within the context of management activities and actions.

Improved communication among those involved in research, management, education, and policy in the Gulf of Maine is occurring. Raising the levels of ocean literacy within the general public is also desirable through ongoing work, such as the many Gulf of Maine Council's science translation projects. A suggestion raised during the 2009 symposium plenary session was to include the communication of research results as a significant component of the research planning cycle and financial and human resources (i.e., dedicated staff time) and for this activity to be built into all research programs. Furthermore, it was recommended at the symposium that researchers developing tools or approaches for integrated policy and management aim to engage those who will use or benefit from research at the earliest possible stage (DeLauer et al. 2012; Fuert 2012). The collaboration among decision makers, scientists, and others working towards the integration of policy and management in developing tools and approaches will aid in making tools most practical, pragmatic, and, above all, applicable to the implementation of EBM in the Gulf of Maine.

There are a multitude of challenges with respect to integrating policy and management, and ultimately EBM, in the Gulf of Maine and around the world. Some challenges are ecological and many are institutional, but all are interconnected. To adequately address the challenges, practical lessons learned and the ever-increasing body of knowledge available to managers and resource users must be integrated. There is no single individual, organization, or jurisdiction that can do this—it must be a shared endeavor and one that carries challenges that all must all embrace. Individuals, organizations, and management authorities may need to step beyond traditional approaches and explore new ways of thinking, planning, and implementing for the future of the region.

References

Arkema, K. K., S. C. Abramson, and B. M. Dewsbury. 2006. Marine ecosystem-based management: from characterization to implementation. Frontiers in Ecology and the Environment 4:525–532.

Arkema, K. K., and J. F. Samhouri. 2012. Linking ecosystem health and services to inform marine ecosystem-based management. Pages 9–25 *in* R. L. Stephenson, J. H. Annala, J. A. Runge, and M. Hall-Arber, editors. Advancing an ecosystem approach in the Gulf of Maine. American Fisheries Society, Symposium 79, Bethesda, Maryland.

Breeze, H. and T. Horsman. 2005. The Scotian Shelf: an atlas of human activities. Fisheries and Oceans Canada (Maritimes Region), Oceans and Coastal Management Division, DFO/2005-816; Dartmouth, Nova Scotia.

Buzeta, M-I., R. Singh, and S. Young-Lai. 2003. Identification of significant marine and coastal areas in the Bay of Fundy. Canadian Manuscript of Fisheries and Aquatic Sciences 2635.

Buzeta, M-I., and R. Singh. 2008. Identification of ecologically and biologically significant areas in the Bay of Fundy, Gulf of Maine. Volume 1: Areas identified for review, and assessment of the Quoddy region. Canadian Technical Report of Fisheries and Aquatic Sciences 2788.

Canada and United States of America: Agreement on Fisheries Enforcement. 1990. United Nations Treaty Series, no. 31531.

DeLauer, V., S. Ryan, I. Babb, P. Taylor, and P. DiBona. 2012. Linking science to management and policy through strategic communication. Pages 89–101 *in* R. L. Stephenson, J. H. Annala, J. A. Runge, and M. Hall-Arber, editors. Advancing an ecosystem approach in the Gulf of Maine. American Fisheries Society, Symposium 79, Bethesda, Maryland.

DFO (Fisheries and Oceans Canada). 2002. Canada's oceans strategy. DFO, Oceans Directorate, Ottawa.

DFO (Fisheries and Oceans Canada). 2007. Eastern Scotian Shelf integrated ocean management plan. DFO, Oceans and Habitat Branch, DFO/2007-1229, Dartmouth, Nova Scotia.

DFO (Fisheries and Oceans Canada). 2009. Sustainable fisheries framework. Available: www.dfo-mpo.gc.ca/fm-gp/peches-fisheries/fish-ren-peche/sff-cpd/overview-cadre-eng.htm. (December 2010).

DFO (Fisheries and Oceans Canada). 2010. Preparing an integrated fisheries management plan (IFMP). Guidance document. February 15, 2010. Available: www.dfo-mpo.gc.ca/fm-gp/peches-fisheries/ifmp-gmp/guidance-guide/preparing-ifmp-pgip-elaboration-eng.htm. (December 2010).

Douvere, F., and C. Ehler. 2008. New perspectives on sea use management: initial findings from European experience with marine spatial planning. Journal of Environmental Management 90:77–88.

Ehler, C. 2008. Conclusions: benefits, lessons learned, and future challenges of marine spatial planning. Marine Policy 32:840–843.

Endangered Species Act of 1973. U.S. Code, volume 16, sections 1531 et seq.

Feurt, C. 2012. Collaborative learning strategies to overcome barriers to ecosystem management in coastal watersheds of the Gulf of Maine. Pages 73–88 *in* R. L. Stephenson, J. H. Annala, J. A. Runge, and M. Hall-Arber, editors. Advancing an ecosystem approach in the Gulf of Maine. American Fisheries Society, Symposium 79, Bethesda, Maryland.

Fluharty, D., M. Abbott, R. Davis, M. Donahue, S. Madsen, T. Quinn, J. Rice, and J. Sutinen. 2006. Evolving an ecosystem approach to science and management throughout NOAA and its partners. Final Report of the external review of NOAA's ecosystem research and science enterprise – a report to the NOAA Science Advisory Board. Available: www.sab.noaa.gov/Reports/eETT_Final_1006.pdf. (December 2010).

Gilliland, P. M., and D. Laffoley. 2008. Key elements and steps in the process of developing ecosystem-based marine spatial planning. Marine Policy 32:787–796.

Grizzle, R. E., L. G. Ward, L. A. Mayer, M. A. Malik, A. B. Cooper, H. A. Abeels, J. K. Greene, M. A. Brodeur, and A. A. Rosenberg. 2009. Effects of a large fishing closure on benthic communities in the western Gulf of Maine: recovery from the effects of gillnets and otter trawls. Fishery Bulletin 107:308–317.

Halpern, B. S., S. E. Lester, and K. L. McLeod. 2009. Placing marine protected areas onto the ecosystem-based management seascape. Proceedings of the National Academy of Science of the United States of America 107:18312–18317.

Halpern, B. S., K. L. McLeod, A. A. Rosenberg, and L. B. Crowder. 2008. Managing for cumulative impacts in ecosystem-based management through ocean zoning. Ocean and Coastal Management 51:203–211.

Hourigan, T. F. 2009. Managing fishery impacts on deep-water coral ecosystems of the USA: emerging best practices. Marine Ecology Progress Series 397:333–340.

Interagency Ocean Policy Task Force. 2009. Interim framework for effective coastal and marine spatial planning. The White House Council on Environmental Quality, Washington, D.C.

Jamieson, G., R. O'Boyle, J. Arbour, D. Cobb, S. Courtenay, R. Gregory, C. Levings, J. Munro, I. Perry, and H. Vandermeulen. 2001. Proceedings of the national workshop on objectives and indicators for ecosystem-based management. Fisheries and Oceans Canada, Canadian Scientific Advisory Secretariat, Proceedings Series 2001/009, Ottawa.

Jennings, S., M. J. Kaiser, and J. D. Reynolds. 2001. Marine fisheries ecology. Blackwell Scientific Publications, Oxford, UK.

Kostylev, V. E., R. C. Courtney, G. Robert, and B. J.

Todd. 2002. Stock evaluation of giant scallop (*Placopecten magellanicus*) using high-resolution acoustics for seabed mapping. Fisheries Research 60:479–492.

Kostylev, V. E., B. J. Todd, G. B. J. Fader, R. C. Courtney, G. D. M. Cameron, and R. A. Pickrill. 2001. Benthic habitat mapping on the Scotian Shelf based on multibeam bathymetry, surficial geology and sea floor photographs. Marine Ecology Progress Series 219:121–137.

Kurland, J. M. 1998. Implications of the essential fish habitat provisions of the Magnuson-Stevens Act. Pages 104–106 *in* E. M. Dorsey and J. Pederson, editors. Effects of fishing gear on the sea floor of New England. Conservation Law Foundation, Boston.

Link, J. 2002. What does ecosystem-based fisheries management mean? Fisheries 27(4):18–21.

Levin, P. S., M. J. Fogarty, S. A. Murawski, and D. Fluharty. 2009. Integrated ecosystem assessments: developing the scientific basis for ecosystem-based management of the ocean. PLoS (Public Library of Science) Biology [online serial] 7(1):e1000014. DOI: 10.1371/journal.pbio.1000014

Magnuson-Stevens Fishery Conservation and Management Reauthorization Act of 2006. Public Law 479, 109th Congress (12 January 2007).

McCay, B. J. 2012. Peopling the marine ecosystem. Pages 27–33 *in* R. L. Stephenson, J. H. Annala, J. A. Runge, and M. Hall-Arber, editors. Advancing an ecosystem approach in the Gulf of Maine. American Fisheries Society, Symposium 79, Bethesda, Maryland.

McLeod, K. L., J. Lubchenco, S. R. Palumbi, and A. A. Rosenberg. 2005. Communication Partnership for Science and the Sea scientific consensus statement on marine ecosystem-based management. Available: www.compassonline.org/sites/all/files/document_files/EBM_Consensus_Statement_v12.pdf. (February 2012).

McLeod, K. L., and H. M. Leslie, editors. 2009. Chapter 17, State of practice. Pages 314–321 *in* K. L. McLeod and H. M. Leslie, editors. Ecosystem-based management for the oceans. Island Press, Washington, D.C.

NMFS (National Marine Fisheries Service). 2003. Implementing the Sustainable Fisheries Act: achievements from 1996 to the present. National Oceanic and Atmospheric Administration, Silver Spring, Maryland.

NOAA (National Oceanic and Atmospheric Administration). 2010. Magnuson-Stevens Fishery Conservation and Management Act reauthorized. Available: www.nmfs.noaa.gov/msa2005. (December 2010).

NROC (Northeast Regional Ocean Council). 2010a. Terms of reference. Available: http://collaborate.csc.noaa.gov/nroc/about/nroc-terms/default.aspx. (December 2010).

NROC (Northeast Regional Ocean Council). 2010b. Advancing regional coastal and marine spatial planning. November 9, 2010. Workshop proceedings. Roger Williams University, Portsmouth, Rhode Island.

Oceans Act. 1996. Statutes of Canada 1996, chapter 31.

Pesch, G., and P. Wells. 2004. Tides of change across the gulf: an environmental report on the Gulf of Maine and Bay of Fundy. Gulf of Maine Council on the Marine Environment and Global Programme of Action Coalition for the Gulf of Maine. Available: www.gulfofmaine.org/council/publications/tidesofchangeacrossthegulf.pdf. (February 2012).

Pesch, G. G., and P. G. Wells, editors. 2004. Tides of change across the gulf. An environmental report on the Gulf of Maine and Bay of Fundy. Prepared for the Gulf of Maine Summit: Committing to Change, Fairmont Algonquin Hotel, St. Andrews, NewBrunswick, Canada, October 26-29th, 2004. Gulf of Maine Council on the Marine Environment and the Global Programme of Action Coalition for the Gulf of Maine. Available: http://www.gulfofmaine.org/council/publications/tidesofchangeacrossthegulf.pdf. (July 2012).

Pomeroy, R., and F. Douvere. 2008. The engagement of stakeholders in the marine spatial planning process. Marine Policy 32:816–822.

Rutherford, R. J., G. H. Herbert, and S. S. Coffen-Smout. 2005. Integrated ocean management and the collaborative planning process: the Eastern Scotian Shelf Integrated Management (ESSIM) Initiative. Marine Policy 29:75–83.

Samhouri, J. F., P. S. Levin, and C. H. Ainsworth. 2010. Identifying thresholds for ecosystem-based management. PLoS (Public Library of Science) [online serial] 5(1):e8907. DOI: 10.1371/journal.pone.0008907.

Taylor, P. H. 2008. Gulf of Maine ecosystem-based management toolkit survey report. Gulf of Maine Council on the Marine Environment. Available: www.gulfofmaine.org/ebm (November 2011).

Thompson, C. 2010. The Gulf of Maine in context: state of the Gulf of Maine report. Report for the Gulf of Maine Council on the Marine Environment. Available: www.gulfofmaine.org/state-of-the-gulf/docs/the-gulf-of-maine-in-context.pdf. (February 2012).

Walmsley, J. 2012. State-of-the-environment reporting for the Gulf of Maine. Pages 61–72 *in* R. L. Stephenson, J. H. Annala, J. A. Runge, and M. Hall-Arber, editors. Advancing an ecosystem approach in the Gulf of Maine. American Fisheries Society, Symposium 79, Bethesda, Maryland.

Wilson, E. O. 1998. Integrated science and the coming century of the environment. Science 279:2048–2049.

WSSD (World Summit on Sustainable Development). 2006. Plan of implementation, paragraph 31(c). Available: www.un.org/jsummit/html/documents/summit_docs/2309_planfinal.htm. (December 2010).

Wood, L. J., L. Fish, J. Laughren, and D. Pauly. 2008. Assessing progress towards global marine protected targets: shortfalls in information and action. Oryx 42:340–351.

American Fisheries Society Symposium 79:61–72, 2012

State-of-the-Environment Reporting for the Gulf of Maine

Jay Walmsley*
Fisheries and Oceans Canada, Oceans and Coastal Management Division
Bedford Institute of Oceanography
Post Office Box 1006, Dartmouth, Nova Scotia B2Y 4A2, Canada

Abstract.—The Gulf of Maine Council on the Marine Environment, established in 1989 as a regional entity to manage environmental quality in the Gulf of Maine, recently recognized the importance of state-of-the-environment reporting as a management tool. Although participating members are individually taking steps to catalogue the collective understanding of the Gulf of Maine, until recently there has been no gulf-wide synthesis of pressures on the environment, biophysical and socioeconomic status and trends, and responses to identified issues. After a 9-month process to develop a scope for the project, in December 2009, the council approved the compilation of the "State of the Gulf of Maine Report," to be launched in June 2010. The main objective of the "State of the Gulf of Maine Report" will be to provide information on the issues affecting the gulf in a form that is easily accessible and readable without compromising scientific validity. The reporting framework to be used will be the driving forces-pressure-state-impacts-response framework. This framework lends itself most easily to reporting on an issue-by-issue basis, so that the pressures, state, impacts, and responses are described for each issue in turn. The "State of the Gulf of Maine Report" will be a modular document that comprises an upfront section or "context document" that provides the background and context to the Gulf of Maine and a series of issue or theme papers that focus on priority areas of the council. It is also envisaged that a wiki site would be useful for informal reporting by interested parties. The main challenges to developing a state-of-the-environment reporting system for the Gulf of Maine are meeting target audience's expectations, providing the correct level of detail, providing adequate facts in an understandable manner, processing and management of information, development and use of indicators, and funding and human resource capacity.

Introduction

In the past three decades, the concept of *sustainable development* has been promoted as the most appropriate approach to achieving long-term security for humans. Sustainable development has been defined as "development that meets the needs of the present without compromising the ability of future generations to meet their own needs" (WCED 1987) or "improving the quality of life while living within the carrying capacity of supporting ecosystems" (Monro and Holdgate 1991). It refers to development that aims for equity within and between generations and adopts an approach where the economic, social, and environmental aspects of development are considered in a holistic fashion.

In 1992, the Agenda 21 program of the United Nations, which resulted from the World Summit on Sustainable Development in Rio de Janeiro, identified the need for environmental reporting to support sustainable development. Agenda 21 stated that

> [s]pecial emphasis should be placed on the transformation of existing information into forms more useful for decision

* Corresponding author: jaywalmsley@ns.sympatico.ca

> making and on targeting information at different user groups. Mechanisms should be strengthened or established for transforming scientific and socioeconomic assessments into information suitable for both planning and public information.

This was supported through The Convention on Access to Information, Public Participation in Decision-Making, and Access to Justice in Environmental Matters (the Aarhus Convention), drawn up by the Economic Commission for Europe's Committee on Environmental Policy, in Aarhus, Denmark, in June 1998. This convention recognizes that, to be able to assert this right, citizens must have access to information, be entitled to participate in decision making and have access to justice in environmental matters. Furthermore, the convention recognizes that, in the field of the environment, improved access to information and public participation in decision making enhance the quality and the implementation of decisions, contribute to public awareness of environmental issues, give the public the opportunity to express its concerns, and enable public authorities to take due account of such concerns. Since then, numerous regional, national, and local efforts have been made to compile state-of-the-environment reports (SOERs; see www.grida.no).

The Gulf of Maine Council on the Marine Environment (GOMC), established in 1989 as a regional entity to manage environmental quality in the Gulf of Maine (GOMC 2007), has recognized the importance of state-of-the-environment (SOE) reporting as a management tool. The Gulf of Maine Council on the Marine Environment Action Plan 2007–2012 (GOMC 2007) states that the council needs to "respond[e] to managers' needs for state-of-the-environment reporting and ecosystem indicators" (GOMC 2007). The council has also adopted the strategy for indicator development outlined in the Mills (2006) document. The strategy is implemented through the Gulf of Maine Ecosystem Indicator Partnership (ESIP; www.gulfofmaine.org/esip).

Participating council members have individually taking steps to catalog the collective understanding of the Gulf of Maine (e.g., ecosystem overview report, Northwest Atlantic bioregional assessment, ESIP, etc.), and there are also many fine examples of reports that address aspects of an SOER (e.g., Pesch and Wells 2004; ACZISC Secretariat and Dalhousie University Marine and Environmental Law Institute 2006; New Hampshire Estuaries Project 2006; Wake et al. 2006; Taylor 2008). However, until recently, there was no initiative that provided a gulf-wide synthesis of pressures on the environment, biophysical and socioeconomic status and trends, and responses to identified issues. After a 9-month process to develop a scope for the project, which included a review of SOE initiatives from around the world, in December 2009, the council approved the compilation of the "State of the Gulf of Maine Report," to be launched in June 2010.

This paper presents an overview of SOE reporting; summarizes some of the reporting requirements for the Gulf of Maine, provides a reporting framework for the "State of the Gulf of Maine Report," and identifies some of the challenges.

State of the Environment Reporting

The two main purposes of SOERs are to foster the use of science in policy and decision making and to report to the public on the condition of the environment (Environment Canada 2009). A SOER report should provide a comprehensive analysis of environmental conditions and trends, measure progress towards sustainability, contribute to informed and open decision making, contribute to public awareness about environmental health and what can be done about it, and serve the public's right to know by providing scientific information about the environment in an easily understandable form (BC Government 2009).

In general, SOERs attempt to answer five key questions (Environment Canada 2009):

- What is happening in the environment (i.e., what are the environmental conditions and trends)?
- Why is it happening (i.e., how are human activities and other stresses linked to the issue in question)?
- Why is it significant (i.e., what are the ecological and socioeconomic effects)?
- What is being done about it (i.e., how is society responding to the issues through govern ment and industry action and voluntary initiatives)?

- Is this sustainable (i.e., are human actions depleting environmental capital and causing deterioration of ecosystem health)?

Successful preparation of SOERs is reliant on existing knowledge and information. The SOER process often brings together data and information that were previously available but fragmented. The process also needs to take advantage of ongoing monitoring and research and should seek to establish ongoing partnerships in data and information collection and sharing. An important aspect of SOE reporting is also the identification of data gaps and possible monitoring requirements.

Reporting Requirements for the Gulf of Maine

Identification of reporting requirements is an important step in the development of a SOER. For the Gulf of Maine, previous literature that outlined some of the user requirements was used (e.g., Della Valle 2006) In addition, several members of the council, the working group, and ESIP were interviewed with respect to reporting requirements for the Gulf of Maine and possible ideas for the future SOER.

Target Audience

In compiling a SOER, it is critical that the target audience be identified. This allows the document or products to be customized to accommodate their needs and requirements. The GOMC (2007) report identified eight primary target audiences for the work of the council: premiers and governors, coastal law makers, coastal decision makers, coastal managers, academics, gulf residents and visitors, marine-dependent industries, and the science community. Listening sessions undertaken in 2005 and 2006, under the auspices of ESIP (Della Valle 2006), indicated that the most important audiences for reporting purposes generally fell into the first four categories: premiers and governors, coastal law makers, coastal decision makers, and coastal managers. However, it has also been recognized by the council that the general public should be considered as a target audience, particularly due to their ability to influence coastal policy makers and decision makers.

The document requirements of these audiences differ considerably. In the case of high-level government officials, the documentation may take the form of a briefing note, while information to the general public is often provided in the more generally appealing form of fact sheets or brochures. In both these cases, the professional staff of government and nongovernment organizations are required to take the information provided and modify it to serve the needs of the audience. Thus, the professional staff in organizations and agencies around the Gulf of Maine, both in the United States and Canada, have been identified as the primary audience for the "State of the Gulf of Maine Report," while coastal decision makers and the general public are considered to be secondary audiences.

User Needs

Della Valle (2006) identified user needs for a future SOER for the Gulf of Maine as part of the 2005–2006 listening sessions. These sessions were complemented by the interviews of council, working group, and ESIP members in 2009 (Walmsley 2009a). The following were identified as important aspects of a SOER:

- Information should be provided in a form that is easily accessible and readable, without compromising scientific validity. As well as providing information on "state" of the environment, the report should describe pressures on the ecosystem and societal responses to those pressures. The report should discuss efforts to address ecosystem health concerns and concrete actions to maintain or improve conditions.
- The main objective of the Gulf of Maine report will be to inform stakeholders on the issues affecting the gulf. The focus should be on those issues that had already been identified as priorities by the council. The six focal areas that had previously been identified include climate change, fisheries and aquaculture, coastal development, contaminants, eutrophication, and aquatic habitats. This does not preclude identifying and understanding some of the emerging issues that may impact the future of the Gulf of Maine, nor does it limit the SOER to those issues.

- To provide commonality in reporting, the SOER should discuss regional indicators, including measures of societal response, where available. There has already been considerable effort by ESIP to develop indicators for the six focal areas (www.gulfofmaine.org/esip).
- Where possible, evaluation of the health or state of the gulf should be tied to specific, measurable goals and threshold responses should be identified. The GOMC report (GOMC 2007) provides three overarching goals that the SOER should address as appropriate:
  - coastal and marine habitats are in a healthy, productive and resilient condition;
  - environmental conditions in the Gulf of Maine support ecosystem and human health; and
  - Gulf of Maine coastal communities are vibrant and have marine-dependent industries that are healthy and globally competitive.
- The reporting system should be flexible. In the mid- to late 1990s, when many countries were developing their first SOERs, a comprehensive approach was used to develop them. This took vast resources (time, human, and financial), and few of these have been updated in the same format since. Since the aim of SOERs is to monitor trends over time, a more appropriate approach, which allows updating and does not place too much pressure on limited resources, was required.
- The historical context should be provided in a SOER if possible. This may be in the form of anecdotal information, which may also be considered to be as important as quantitative information. Case studies can play an important role in SOE reporting, particularly to tie the issues into the lives of the people living in the Gulf of Maine region.
- The boundaries of the area or issue being reported on should be clearly delineated in the SOER. Although the physical boundaries of the Gulf of Maine and its watershed are fairly well defined, the areas of influence or impact surrounding a particular issue may change according to the issue.
- Consideration should be given to new methods of accessing and using information. For instance, if appropriately formatted, younger generations of people are far more at ease reading information from a screen rather than hard copy, and accessing data over the Internet.

Reporting Framework and Structure

Driving Forces-Pressures-State Impacts-Response Framework

An important principle in SOE reporting is that it be guided by a conceptual framework that facilitates development of information and makes the linkages between the environment and socioeconomic factors. Such a framework should bring order and convergence to the structure of the presentation and analysis of SOE information (NETCAB 1999). The framework to be used for the "State of the Gulf of Maine Report" is the driving forces-pressures-state-impacts-response (DPSIR) framework (Figure 1). The DPSIR framework is viewed as providing a systems-analysis view of the relation between the environmental system and the human system (Smeets and Weterings 1999). According to this framework, social and economic developments and natural conditions (driving forces) exert pressure on the environment, and as a consequence, the state of the environment changes (Fogarty et al. 2012, this volume). This leads to impacts on human health, ecosystems, and materials, which may elicit a societal response that feeds back on all the other elements (see Figure 1). Although the DPSIR framework was developed as an extended cause-effect-response model, the framework is most useful in describing the origins and consequences of environmental problems.

This framework lends itself most easily to reporting on an issue-by-issue basis, so that the pressures, state, impacts, and responses are described for each issue in turn. When this approach is used, it does not really matter where you start reporting. The starting point should generally reflect the point you would like to emphasize most. In this way, the environmental issues are given prominence as the issues of concern, and the human influences are highlighted as causes. By using the framework, the relationship between humans and the natural environ-

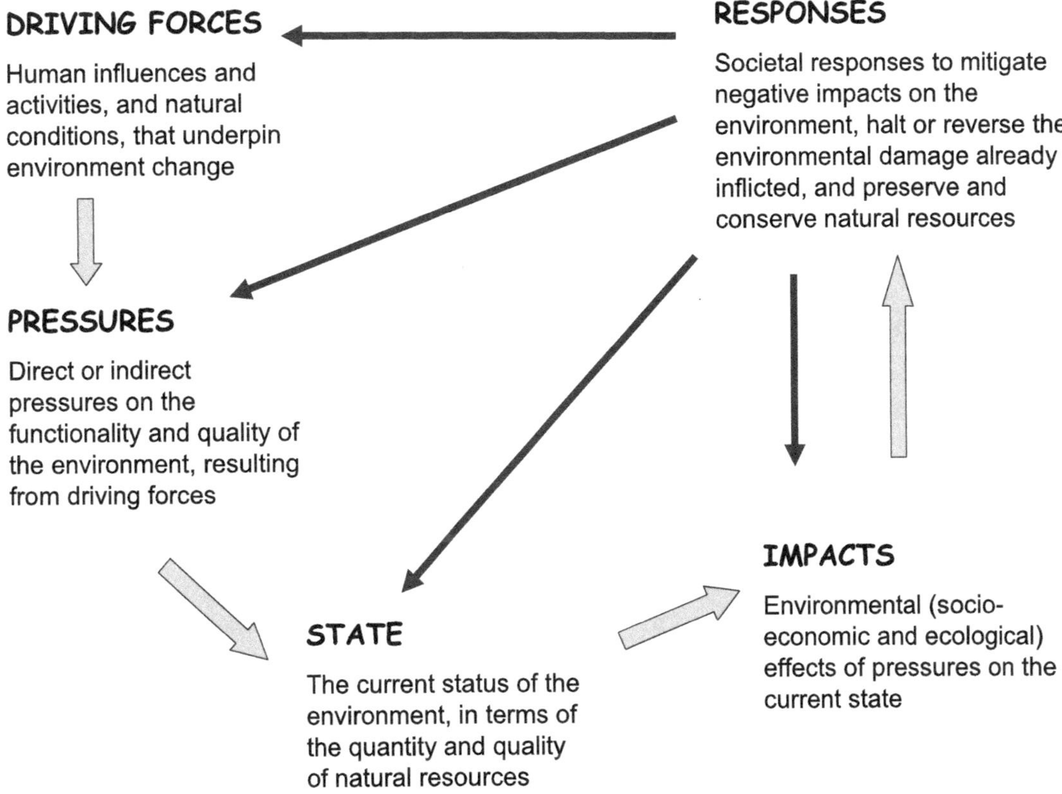

FIGURE 1. Interlinkage among the three recognized aspects of sustainability.

ment is seen to be more integrated, and sustainable development needs and actions are clearer to the reader.

Figure 2 provides a DPSIR analysis for the Gulf of Maine, providing an indication of some of the aspects that could be included in the "State of the Gulf of Maine Report."

Report Structure

State-of-the-environment reporting does not necessarily result in a single product. In fact, the current trend is for a SOE reporting system or process to be established that allows flexibility of use and links to a variety of initiatives that are SOE-related. A similar approach will be taken in the Gulf of Maine, with the following as products of the SOE reporting system: a modular report that can be readily updated as required (formal SOE document) and a Web portal to be used as a communication tool and that provides linkages to other products and initiatives. It is also envisaged that a wiki site would be useful for informal reporting by interested parties.

Modular state-of-the-environment report.—The "State of the Gulf of Maine Report" will be a modular document whose parts can be readily updated at different time intervals. It will include an up-front section or "context document" that provides the background and context to the Gulf of Maine and a series of issue or theme papers that focus on priority areas of the council or its members.

The context document will be a relatively static document that provides an overview of the Gulf of Maine. The information provided in this document will be the type of information that rarely changes, such as the boundaries of the area under discussion, the size of the watershed, and history of the watershed. It will also provide an overview of the driving forces (i.e., natural conditions and macroeconomic and social conditions)

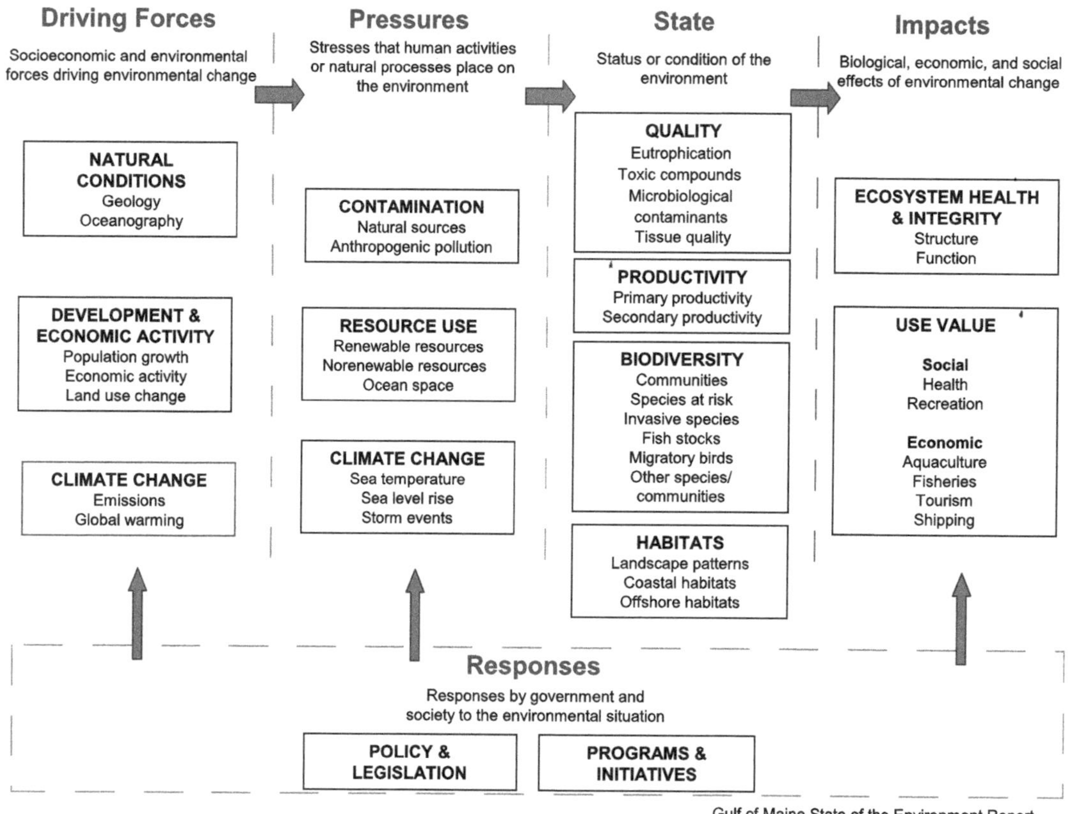

FIGURE 2. Driving forces-pressures-state-impacts-response analysis for the Gulf of Maine.

and address the current environmental governance in the Gulf of Maine.

The theme papers will provide an evaluation of priority issues that are of interest in the Gulf of Maine. An incremental approach will be taken to develop the theme papers, with priority theme papers being developed first. It is envisaged that the theme papers will be developed over an 18-month to 2-year period. After this, the theme papers will be regularly updated at a time interval appropriate for each issue. For instance, for theme papers that make use of census data, this may be every 5 years, while theme papers that rely on water quality data may be updated biennially or annually. This complements the incremental approach for indicator development that is currently being taken by ESIP (Mills 2006). Because the theme papers will be part of the formal reporting mechanism for the gulf, their structure will be based on the reporting framework discussed previously. In addition, it will be critical that they are compiled in such a way that they take into consideration the user's needs.

A list of theme papers, based on the six ESIP priority areas, was identified by the council working group in 2009 (Table 1). At the time of writing this paper, five of the theme papers were being compiled, including "Climate Change and Coastal Hazards"; "Climate Change and Its Effect on Ecosystems, Habitats, and Biota"; "Coastal Ecosystems and Habitats"; "Marine Invasive Species"; and "Emerging Issues." The latest version of the "State of the Gulf of Maine Report" is available at www.gulfofmaine.org/state-of-the-gulf/.

SOE Web portal.—The SOE Web portal is a communication mechanism to deliver the "State of the Gulf of Maine Report." The council already hosts a Web site that provides ease of access to documents for members and the public alike. A SOE Web portal should provide access to all SOE-re-

TABLE 1. List of theme papers for "State of the Gulf of Maine Report."

Priority area	Theme paper
Climate change	"Climate Change and Its Effects on Humans" "Climate Change and Its Effects on Ecosystems, Habitats, and Biota"
Fisheries and aquaculture	"Aquaculture in the Gulf of Maine" "Commercial Fisheries and Fish Stock Status"
Coastal development	"Land Use and Coastal Development"
Contaminants	"Toxic Contaminants" "Microbial Pathogens and Toxins"
Eutrophication	"Eutrophication"
Aquatic habitats	"Coastal Ecosystems and Habitats" "Offshore Ecosystems and Habitats" "Watershed Status"
Other	"Marine Invasive Species" "Species at Risk" "Emerging Issues"

lated documents, including the formal "State of the Gulf of Maine Report" and related initiatives such as the ESIP site and the *Gulf of Maine Times*. Examples of SOE reports that use this mechanism to report on the environment are the Norwegian State of the Environment Report (see Norway 2012) and the South African state of the environment report (see South Africa 2012).

Gulf of Maine wiki.—A wiki is a Web page that anyone can edit. It exists as a tool for collaborative content creation and is designed to allow multiple authors to add, remove, and edit content (text and graphics) (Government of Canada 2009). Wikis have been a dominant Web-based collaboration tool for several years, have seen service in many environments, and have been customized for many tasks (Government of Canada 2009). The best example of a wiki is Wikipedia, the online encyclopedia (see www.wikipedia.org).

Wiki is a Hawaiian word meaning "quick." Wikis, and the software supporting them, are designed so that content can be made available quickly and easily. By its very nature, a wiki sees the author and the audience participant as the same thing, and it is the exception rather than the rule that an audience member will only have read access to the wiki. The multiple author capability of a wiki makes it an effective tool for mass collaborative authoring. Wikis allow anyone to edit pages, but keep a complete track of version history. As such, it is easy to compare differences across versions and to roll back changes. While it seems counterintuitive, this opening up of editing capability to everyone typically allows the best possible quality content to emerge. Where people spot something they believe to be incorrect, they can fix it immediately, rather than going through a lengthy editorial process. Advantages and disadvantages are listed in Table 2 below.

The Gulf of Maine has numerous interested parties undertaking various initiatives at different levels. Often, the information generated by these initiatives is at a local level and is generally not reported in formal SOERs, except in the form of case studies. In the case of the Gulf of Maine, there may not be resources to capture all this work in the form of theme papers, although some is captured by the Gulf of Maine Times. Many of these initiatives are aimed at meeting the goals outlined in the GOMC (2007) report, and it would be valuable to have these documented in some form that does not place an unnecessary financial burden on the council and its members.

Table 2. Advantages and disadvantages of using wikis (adapted from Government of Canada 2009).

Advantages	Disadvantages
• Encourages collaborative document creation and sharing, as articles are created by the community through collaboration. • Facilitates the creation of a shared knowledge base. • Broadens participation and reach and builds horizontal community, particularly where people are separated geographically and where relationship building is important. • By linking together stakeholders that would not otherwise connect and communicate, it becomes possible to harness learning and knowledge, leading to a greater capacity for innovation and organizational agility. • Allows the creation of unique content. The content is typically not the type to be found on Intra- or Internet sites. • Fosters innovation and the development of new ideas. • Makes use of expertise that may otherwise not be available (e.g., may harness traditional knowledge). • Wiki pages are dynamic (new articles added continuously) and content appears instantaneously. • Changes to the text are attributed to the author(s). • Reduces document version conflicts. Previous versions are archived and it is easy to revert to a previous version if required. • Wiki formatting syntax is easy to learn. • Content of pages can be semi-protected to protect pages from vandalism or violations of content policies by anonymous users.	• All members can add or edit the work of others and disputes may arise. • False information may be included. • Difficult to cite as an "authoritative source." • It is a challenge to keep content and links current (similar to Web sites and blogs). • Scientist or others who wish to publish their data in more formal reports may not wish to share data on a wiki. • Concerns about copyright for some people.

Thus, the establishment or use of a wiki site that anybody with an interest (and a story) can contribute to has been recommended to the council. Because the membership of the council is predominantly government authorities, the appetite for developing a wiki as part of the "State of the Gulf of Maine Report" is small, and it remains as only a possibility as part of the future reporting mechanism for the Gulf of Maine.

Environmental Indicators

One method of providing information in a format that is usable by policy makers and decision makers is through the use of sustainability indicators. An indicator is a parameter that provides information about an environmental issue with a significance that extends beyond the parameter itself (OECD 1993). Mills (2006) provides a more ecosystem-focused definition of indicators that is used in the context of the Gulf of Maine: "Indicators are quantitative or qualitative measures that provide information about the status of or changes in natural, cultural, and economic aspects of an ecosystem." Agenda 21 (Chapter 40) states that "indicators of sustainable development need to be developed to provide solid bases for decision-making at all levels and to contribute to the self-regulating sustainability of integrated environmental and development systems." This has led to the acceptance of sustainability indicators as basic tools for facilitating public choices and supporting policy implementation (Von Meyer 2000). They provide information on relevant issues, identify development-potential problems and perspectives, analyze and interpret potential conflicts and synergies, and assist in assessing policy implementation and impacts (Von Meyer 2000). In essence, they allow us to better organize, synthesize, and use information.

The main goal of establishing indicators is to measure, monitor, and report on progress towards sustainability. Within this, indicators have numerous uses and potential for improving environmental management, including (Hammond et al. 1995; Walmsley and Pretorius 1996)

- monitoring and assessing conditions and trends on a national, regional, and global scale;
- comparing situations;
- assessing the effectiveness of policy making;
- marking progress against a stated benchmark;
- monitoring changes in public attitude and behavior;
- ensuring understanding, participation, and transparency in information transfer between interested and affected parties;
- forecasting and projecting trends; and
- providing early warning information.

Indicators should have three essential qualities; they should be "simple, quantifiable and communicable" (Walmsley and Pretorius 1996). The most effective environmental indicators are

- scientifically valid (i.e., credible and accepted by experts in the field),
- representative of key issues and broader impacts or effects,
- appropriate to the reporting time scale,
- useful for prediction,
- relevant to policy makers and enable individuals to make meaningful decisions, and
- compatible with other indicators to present an overall picture.

The development of meaningful environmental indicators is complex. A set of indicators should be broad enough to represent the overall environment, yet be few enough to present an understandable picture of environmental quality.

Developing good indicators is a continuous process of improvement. There are several indicator development and reporting initiatives within the Gulf of Maine (see Mills 2006), the only gulf-wide one being ESIP. Since 2006, ESIP has been active developing a set of core indicators for the six priority areas for the Gulf of Maine. The intention is that these indicators will feed into the SOE reporting process but not necessarily drive it. Indicators that have been identified as core indicators by ESIP are summarized in Table 3.

The development of SOERs generally assists in identification of indicators that can be used on an ongoing basis for an area. The compilation of the theme papers may require the use of indicators in addition to those identified as core indicators by ESIP. These may not have compatible data gulf-wide but are still valuable for reporting on status and trends. These indicators will complement the core ESIP indicators in providing an understanding of the state of the environment. In addition, consideration will be given to complex indicators that link environmental impacts with human activities.

Overcoming Challenges in Developing the "State of the Gulf of Maine Report"

There are several challenges in producing a SOER (Walmsley 2009b), including

Meeting target audience's expectations—Due to several challenges, the expectations of all the target audiences for a SOER are rarely met. This occurs mainly because target groups are not adequately profiled or interacted with before production. In many cases, the target audience identified is too broad to be able to focus the product. Even for the "State of the Gulf of Maine Report," it is highly unlikely that the needs of all

TABLE 3. Ecosystem Indicator Partnership indicators.

Priority area	Indicator
Aquatic habitats	Extent of eelgrass Extent of salt marsh Locations of tidal restrictions
Climate change	Sea level change Precipitation trends and anomalies Air temperature trends and anomalies
Coastal development	Point sources Population density Employment density Impervious surface coverage
Contaminants	Sediment triad data Shellfish sanitation data Gulf watch/mussel watch data
Eutrophication	Nitrogen loading Secchi depth Dissolved oxygen Chlorophyll *a*
Fisheries and aquaculture	Production/area for aquaculture Economic value of aquaculture Mean length of all sampled fish Economic value of fisheries Proportions of stock at or above targeted biomass

the target audiences will be met simultaneously. However, it is hoped that the primary target audience, professional staff in or ganizations around the gulf, will use the material provided in the document to develop spin-off products that meet the needs of a wide audience, whether they be the general public or decision and policy makers.

Level of detail and comprehensiveness—Many SOER processes have been aimed at producing comprehensive, detailed reports that require numerous resources (human and financial) to complete (see Government of Canada 1996; State of the Environment Advisory Council 1996). It is now recognized that a more balanced approach, which focuses on the most important themes or issues, is appropriate. In the case of the Gulf of Maine, the context document will provide a brief overview, an introduction to the area, without providing extensive detail. However, the theme papers will provide more detail on specific topics and should point interested readers to more comprehensive sources of information.

Authorship—Authorship and review of the theme papers is considered critical to the success of the "State of the Gulf of Maine Report." Authors for theme papers should be technically competent (having some recognized expertise in the subject matter) and have the ability to write well, but they need not be technical experts in their field. Institutional authorship will not be encouraged, with the exception of council committees. A review process will be put in place to ensure that the theme papers are of the highest technical quality, that they provide facts rather than opinion, that language used in the papers is appropriate for the target audience, and that the publication requirements of the council are met. The review process will include peer review by technical experts in the field. In this way, a balance between scientific validity and readability should be obtained.

Processing and management of information—The most frequent limiting factor for SOER production is the availability of relevant and suitable information. This is a particular problem when dealing with multiple jurisdictions, each of which has different data availability, format, scale, and so forth. It is often more appropriate to pick the "low-hanging fruit" when it comes to selecting information for SOERs by using information that is readily available. Because of the multijurisdictional nature of the Gulf of Maine, this may be a particular problem that is not readily overcome. Where gulf-wide data are not available, province- or state-specific information may need to be used until such time as compatible data sets are developed. The "State of the Gulf of Maine Report" should also be seen as an opportunity to identify gaps in in formation.

Indicators—All SOERs are dependent on the use of indicators (qualitative and quantitative). In many cases, SOERs can get sidetracked because of the time and focus devoted towards the selection of ideal indicators rather than the presentation of readily available information. Many SOERs are a compilation of readily available information, and the development of indicators is an iterative process. The fact that ESIP has undertaken considerable development of core indicators for the Gulf of Maine will be of great benefit to SOE reporting for the area; however, for many of the theme papers, available proxy indicators may have to be used rather than ideal indicators.

Funding and human resource capacity—Funding and human resource capacity are always a concern for any large undertaking such as SOE reporting. Both will continue to be issues of concern for the council. The project will require some dedicated funding to initiate and maintain the SOE reporting system. Also, due to the fact that the "State of the Gulf of Maine Report" is not a once-off product, but rather a long-term, ongoing process aimed at progressively understanding the environment and its management, long-term commitments by the council will be required, including maintenance of the Web portal and updating of theme papers, as required. However, because of the modular nature of the report, it is expected that this commitment will not be too onerous for the council and its members.

References

ACZISC Secretariat and Dalhousie University Marine and Environmental Law Institute. 2006. Overview of the current governance in the Bay of Fundy/Gulf of Maine: transboundary collaborative arrangements and initiatives. Oceans and Coastal Management Division. Fisheries and Oceans Canada, Dartmouth, Nova Scotia.

BC Government. 2009. State of environment reporting. Available: www.env.gov.bc.ca/soe/about.html. (January 2012).

Della Valle, E. A. 2006. Gulf of Maine indicators. Final Report of the Listening Sessions and Evaluation of Tides of Change Report. Gulf of Maine Ecosystem Indicators Partnership. Gulf of Maine Council on the Marine Environment, Falmouth, Maine.

Environment Canada. 2009. Federal state of the environment reports, a vision for federal state of the environment (SOE) reporting in Canada. Environment Canada, Ottawa.

Fogarty, M. J., K. D. Friedland, L. Col, R. Gamble, J. Hare, K. Hyde, J. S. Link, S. Lucey, H. Liu, J. Nye, W. J. Overholtz, D. Richardson, B. Rountree, and M. Taylor. 2012. Status of the northwest U.S. Continental Shelf large marine ecosystem: an indicator-based approach. Pages 139–165 *in* R. L. Stephenson, J. H. Annala, J. A. Runge, and M. Hall-Arber, editors. Advancing an ecosystem approach in the Gulf of Maine. American Fisheries Society, Symposium 79, Bethesda, Maryland.

GOMC (Gulf of Maine Council on the Marine Environment). 2007. Gulf of Maine Council on the Marine Environment Action Plan 2007–2012. Available: www.gulfofmaine.org (January 2009).

Government of Canada. 1996. The state of Canada's environment. Environment Canada, Ottawa.

Government of Canada. 2009. GCpedia. Available: www.gcpedia.gc.ca (February 2009).

Hammond, A., A. Adriaanse, E. Rodenburg, D. Bryant, and R. Woodward. 1995. Environmental indicators: a systematic approach to measuring and reporting on environmental policy performance in the context of sustainable development. World Resources Institute, Washington, D.C.

Mills, K. E. 2006. A strategy for Gulf of Maine ecosystem indicators and state of the environment reporting. Gulf of Maine Council on the Marine Environment. Available: www.gulfofmaine.org/esip/docs/esipstrategy.pdf (February 2009).

Monro, D., and M. Holdgate, editors. 1991. Caring for the earth: a strategy for sustainable living. IUCN (World Conservation Union), UNEP (United Nations Environment Programme), and WWF (World Wide Fund for Nature), Gland, Switzerland.

NETCAB (Networking and Capacity Building Programme). 1999. SOE Info, SOE reporting conceptual frameworks. NETCAB, World Conservation Union Regional Office for Southern Africa (IUCN-ROSA), Harare, Zimbawbwe.

New Hampshire Estuaries Project. 2006. State of the estuaries 2006. Available: http://prep.unh.edu/resources/pdf/ 2006_state_of_the-nhep-06.pdf. (February 2009)

Norway. 2012. State of the environment Norway. Available: www.environment.no/topmenu/about-soe-norway. (January 2012).

OECD (Organisation for Economic Co-Operation and Development). 1993. Environmental indicators: basic concepts and terminology. OECD, Background Paper No. 1, Paris.

Pesch, G., and P. Wells. 2004. Tides of change across the gulf: an environmental report on the Gulf of Maine and Bay of Fundy. Gulf of Maine Council on the Marine Environment and Global Programme of Action Coalition for the Gulf of Maine, Falmouth, Maine.

Smeets, E., and R. Weterings. 1999. Environmental indicators: typology and overview. European Environment Agency, Technical Report No. 25, Copenhagen.

South Africa. 2012. State of the environment South Africa. Available: http://soer.deat.gov.za/frontpage.aspx?m. (January 2012)

State of the Environment Advisory Council. 1996. Australia: state of the environment 1996. An independent report presented to the Commonwealth Minister for the Environment. Commonwealth of Australia, Canberra.

Taylor, P. H. 2008. Salt marshes in the Gulf of Maine: human impacts, habitat restoration, and long-term change analysis. Gulf of Maine Council on the Marine Environment, Falmouth, Maine.

Von Meyer, H. 2000. Territorial indicators for sustainable development. Why? And how? Pages 150–157 *in* Frameworks to measure sustainable development. Organisation for Economic Co-operation and Development, Paris.

Wake, C, E. Burakowski, G. Lines, K. McKenzie, T. Huntington, and W. Burtis. 2006. Cross border indicators of climate change over the past century: northeastern United States and Canadian Maritime Region. The Climate Change Task Force of the Gulf of Maine Council on the Marine Environment in cooperation with Environment Canada and Clean Air-Cool Planet, Falmouth, Maine.

Walmsley, J. 2009a. Gulf of Maine state of the environment reporting. Scoping document. Final Report. Fisheries and Oceans Canada, Dartmouth, Nova Scotia.

Walmsley, R. D. 2009b. Methodology. Pages 10–14 *in* Our coast, live, work, play, protect: the 2009 state of Nova Scotia's coast technical report. Provincial Oceans Network, Nova Scotia.

Walmsley, R. D., and J. P. R. Pretorius. 1996. Environmental indicators. Department of Environmental Affairs and Tourism, State of the Environment Series Report No. 1, Pretoria.

WCED (World Commission on Environment and Development). 1987. Our common future. Oxford University Press, Oxford, UK.

American Fisheries Society Symposium 79:73–88, 2012

Collaborative Learning Strategies to Overcome Barriers to Ecosystem Management in Coastal Watersheds of the Gulf of Maine

Christine Baumann Feurt*
University of New England, Center for Sustainable Communities
11 Hills Beach Road, Biddeford, Maine 04005, USA
and
Wells National Estuarine Research Reserve, Coastal Training Program
342 Laudholm Farm Road, Wells, Maine 04090, USA

Abstract.—Marine resource managers and environmental policy makers trained in disciplines grounded in the natural sciences learn quickly that some of the biggest challenges to the practice of ecosystem management are social ones. While ecosystem theory provides a conceptual framework for integrating the ecological, socioeconomic, cultural, and institutional elements of environmental problems, the *practice* of ecosystem management remains elusive. A potential model for addressing these challenges of ecosystem management is a U.S. "science to management" initiative of the National Oceanic and Atmospheric Administration implemented within the National Estuarine Research Reserve System. The Coastal Training Program initiative provided the context for this study's development of innovative interdisciplinary approaches supporting ecosystem management to mitigate land-based pollution impacts in coastal watersheds in the Gulf of Maine.

Introduction

Using an educational approach to design coastal decision-maker workshops and trainings, the coastal training program (CTP) was envisioned to support the transfer and delivery of science-based information to managers working on issues critical to each region of the host National Estuarine Research Reserve (NERR; Kennish 2004). Each of the National Estuarine Research Reserve System (NERRS) coastal training programs uses interdisciplinary place-based approaches to determine managers' needs for scientific information and preferred methods for communicating the information. This chapter presents an action research case study of the development of the CTP at the Wells, Maine NERR. The research evaluated the application of social science methodologies to create a program based upon collaborative learning principles applied within the framework of community based ecosystem management. The author conducted stakeholder and institutional analysis of the people and organizations responsible for water management in southern Maine to reveal the elements of the complex system and facilitate understanding of the context of municipal decision making about water. Results of these methodologies, combined with instructional design and collaborative learning methodologies, contributed to the formation of collaborative partnerships in coastal watersheds in southern Maine that transcended the original model of delivery of expert knowledge to a passive audience. The partnership model facilitated cross-disciplinary dialogue about the practice of community-based ecosystem management, deliberation about the appropriate application of science to management, and implementation of best practices.

* Corresponding author:cfeurt@une.edu

Ethnographic analysis of the cultural roots of conflict, motivational forces guiding ecosystem management, and perceived barriers to collaboration guided the place-based design of this national science to management initiative as it was adapted for the Wells NERR. The social science methodologies applied in this case study yielded surprising and valuable perspectives about the social system influencing community-based ecosystem management. Research revealed that fundamental assumptions about managers and municipal officials, as potential receptacles of science-based knowledge, failed to appreciate the rich kaleidoscope of expertise actively engaged in managing water. The knowledge and professional practices applied within this kaleidoscope of expertise was better conceived of as an important resource to be accessed using collaborative learning methodologies. The conceptual framework for integrating social science into ecosystem management developed through this research is proposed as one model for a fundamental perspective shift in the meaning of interdisciplinarity at the science-management–policy interface.

The Search for New Tools to Facilitate Learning in Ecosystem Management

Global environmental change presents unprecedented challenges for 21st century scientists, policy makers, and environmental managers (NRC 2009). The complexity and interconnectedness of the social and ecological systems that underlie environmental change are forcing the redefinition of issues, fostering new liaisons that transcend traditional boundaries, and transforming environmental management (Gunderson and Holling 2002). Nowhere is this change more evident than in coastal and estuarine systems. Here, population pressure and the environmental waste outputs of human economic and social systems deposited into land, water, and atmosphere are concentrated and delivered by the hydrologic cycle. Ecological systems responses include harmful algal blooms, eutrophication, hypoxia and accumulation of toxins, all of which reflect back to the human system through health effects, economic loss, and consequences for future generations (USCOP 2004; Fluharty et al. 2006).

Integrative theories in ecology and ecosystem management propose frameworks that encompass understanding of ecological, economic, and institutional systems and the dynamic, cross-scale interactions that contribute to unpredictability and complexity (Allen and Hoekstra 1992; Gunderson and Holling 2002; Meffe et al. 2002). Recognition of the importance of resilience in ecosystems, fluid and responsive institutions, and management linked to learning evolve from practices aligned with these new theoretical frameworks (Lee 1993; Machlis et al. 1997; Wondolleck and Yaffee 2000; Beatley 2009). Trends in ecological research, ecosystem management and environmental policy increasingly incorporate systems approaches, adaptive management, and innovative policy strategies developed through collaborative processes (Walters 1986; Allen and Hoekstra 1992; Gunderson et al. 1995; Berkes and Folke 2000; Fiorino 2001; NRC 2001a, 2001b, 2002; McLeod and Leslie 2009). These transdisciplinary approaches engage the people involved in environmental problem solving in deliberative processes to foster social learning and civic science (Lee 1993; NRC 1996; Costanza 1998; Endter-Wada et al. 1998; Lubchenco 1998; Boesch 1999, 2001; Visser 2004). The action research paradigm is particularly suited to transdisciplinary situations (Greenwood and Levin 1998). Action research is embedded in the system where the research questions arise. The people with a stake in problem identification and solution are engaged in research that aims to better understand the root causes of a situation in order to develop effective solutions. The wisdom of the people closest to the situation is treated as a knowledge and problem-solving resource. Ecosystem management benefits from this orientation to research, and this chapter offers a case study that demonstrates how understanding and engaging diverse stakeholders is accomplished through collaborative learning.

Learning through adaptive management is the cornerstone for theories and practices that embrace uncertainty by framing policy and management decisions as experiments (Holling 1978; Walters 1986; Walters and Holling 1990; Lee 1993; Gunderson et al. 1995; Gunderson and Holling 2002). In his essay on learning in the edited volume *Barriers and Bridges to the Renewal of Ecosystems and Institutions* (Gunderson et al. 1995), social psychologist Donald Michael calls for profound learning that includes an examination of the role that beliefs, unconscious needs, and motives play in personal, organizational, and social change directed toward the goal of environmental sustainability (Michael 1995). Learning through adaptive management is difficult. Research examining the application of adaptive management in watershed management and in business practice identifies both individual and institutional resistance to underlying premises and theory. Adaptive management seems to be easier said than done (Argyris and Schon 1996; Allan 2004; Allan and Curtis 2005). Genuine learning associated with adaptive management is constrained by strongly entrenched habits of practice or what Allan (2004) calls "imperatives." Imperatives include an orientation to action and progress over reflection, the need to control and simplify complex human and social systems, and self-deception to maintain the status quo rather than challenge established practices. An example of an imperative guiding water researchers is the strongly held idea that biophysical scientific documentation of water-quality degradation communicated in reports delivered to municipal officials will result in changes in policy and behavior. A very busy town manager who has been the recipient of a number of these reports pleaded, "Just tell me what you want me to do!" His more immediate need was for prescriptive knowledge about actions to be taken and, just as importantly, the ability to find the resources to build municipal capacity and pay for those actions (Feurt 2007).

This action research case study focuses on learning in the decision-making arena of coastal watershed management in the Gulf of Maine. The Gulf of Maine shares management challenges common in watersheds across the United States. Municipal officials, environmental management agencies, and the public make land-use decisions that affect coastal waters. Local land-use practices and development contribute to coastal ecosystem degradation from nonpoint source pollution caused by sediment, nutrients, toxins, and microbial contaminants (USCOP 2004; Fluharty et al. 2006). This coupling of land use and coastal water quality provides a litmus test for land management and locally instituted environmental practices. The communication of scientific findings to decision makers is considered vital to the practice of ecosystem management (Lubchenco 1998; Meffe et al. 2002; Fluharty et al. 2006). Institutions, like the NERRS, generating science-based information, focus attention on municipalities and local governments in an effort to foster the incorporation of ecosystem management principles into decision making and policy. Scientists, technology developers, regulators, and environmental nongovernmental organizations have information and prescriptions for effective local action. Municipal officials can feel bombarded by these prescriptions when they are added to the already overwhelming task of "running their towns" (Feurt 2007).

The National Oceanic and Atmospheric Administration's Coastal Training Program Puts a Face and a Place on Ecosystem Management

The CTP of the NERRS represents a manifestation of *the new social contract for science* (Lubchenco 1998; NOAA 2003). The CTPs at 28 reserves engage in science translation, professional training, and implementation of unique collaborative processes supporting ecosystem management (Cook et al. 2002; NOAA 2008a, 2008b, 2009). Most CTPs focus on municipal and regional applications of science to management and policy. Habitat conservation and restoration, land use, stormwater management, and climate change adaptation are addressed. At the scale where the CTP operates, abstract theories and principles of ecosystem management assume a face and a place in the form of the people who implement the practices of ecosystem management in their work in tangible landscapes (Meffe et al. 2002). The CTP works directly with municipal planners, citizen land-use boards, land trusts, and public works directors who strive to take actions to improve water quality, conserve locally valued habitat, and maintain ecosystem services in the communities where they live work and play (Feurt 2007). The CTPs have become trusted sources of scientific information, valued partners in environmental problem solving, and architects of community-based ecosystem management (GLEFC 2004; Feurt 2007; GEARS 2009).

The Wells NERR CTP developed and piloted cultural models-based collaborative learning to frame and implement community-based ecosystem management as a way to address the need to understand the social system within which biophysical science knowledge would be applied. Both the need for this kind of systems understanding and the potential for cultural models research to contribute had been identified but not tested (Kempton et al. 1995; Lubchenco 1998; Krum and Feurt 2002).

The study of cultural models has evolved during the past two decades of research in cultural and cognitive anthropology. Cultural models are shared perceptions and attitudes about how the world works. They are implicit, taken for granted, and operate below the level of consciousness (Holland and Quinn 1987; Strauss and Quinn 1997). The cultural models studied in this research play a role in the production, transmission, and application of knowledge related to water management. Results of research and the development of the Wells NERR CTP provided an understanding of the knowledge and cultural models of water used by water managers in southern Maine and the ways that knowledge and cultural models influenced the use of science in decision making (Feurt 2003, 2006, 2007). Collaborative learning is a stakeholder engagement process designed to make progress to improve environmental problems (Daniels and Walker 2001). The practice of collaborative learning employs a toolkit of techniques to stimulate creative discussion, foster dialogue despite conflict and controversy, and develop group-generated implementation strategies. Collaborative learning is especially amenable to issues involving conflict and scientific uncertainty (Daniels and Walker 2001).

Ethnographic research to improve science translation for nonpoint source pollution control documented shared values among diverse stakeholders, a rich resource for collaborative problem solving. Knowledge about the cultural models used by southern Maine water managers in environmental decision making, combined with the process and strategies of collaborative learning, facilitated watershed stewardship, including creation of watershed councils, to support collaboration across municipal boundaries, implementation of watershed management plans, development of a collaborative conservation plan, identification of action strategies to reduce nonpoint source pollution, and adoption of innovative stormwater management technologies (Feurt 2003, 2006, 2007). The results of 7 years of research and evaluation of cultural models based on collaborative learning is synthesized and presented in a practitioner's guide and PowerPoint presentation that has been used in trainings and conference presentations (Feurt 2008, 2009).

Understanding the Role of Culture in Science Translation for Ecosystem Management

Cultural models concerning water's importance, causes of pollution, and appropriate ways to protect water were identified through ethnographic research and used in the development of the Wells NERR CTP (Feurt 2007, 2008). Examples of these cultural models appear below in Figure 1. Each of the propositions represents a cognitive key that unlocks doors leading to complex mental libraries where ideas, attitudes, values, and perceptions are organized. Psychologists and educational theorists call these units in our mental libraries mental models (Gentner and Stevens 1983; Collins and Gentner 1987). Mental models function like maps, templates, and field guides as we move through the world, allowing us to unconsciously recognize the familiar, categorize without thinking, and link novel experiences to what we already know. Our mental models allow us to recognize a borzoi as a dog the first time we see one. When we order lunch, eat, and pay the check in a restaurant, we draw from script-like mental models that guide and constrain our behavior (Gentner and Stevens 1983; Holland and Quinn 1987).

Cultural anthropologists are interested in the ways mental models are learned and transformed within a social group to become shared *cultural models*. Cultural models are taken for granted and are implicit within the social groups where they are shared (Holland and Quinn 1987; Kempton et al. 1995, 2001; Kempton 2001; Quinn 2005). Although they are used unconsciously, they cause us to pay attention to select aspects of our surroundings, recognize objects and patterns, and assign meaning to our experiences. Cultural models have motivational force and guide our behavior (D'Andrade and Strauss 1992; D'Andrade 1995).

Cultural models research examines the complex interaction of attitudes, values, and modes of

- Water is the basis for life on earth.
- Nature makes water.
- Water is a resource to be used.
- Water is a commodity.
- Water is landscape.
- Water is waste.

FIGURE 1. Cultural models of water.

understanding surrounding an array of environmental issues, including global climate change (Kempton 1991a, 1991b, 1993, 1997; Kempton et al. 1992; Kempton and Craig 1993), protected areas management (Pfeffer et al. 2001), landscape conservation (Dailey 1999), blue crab *Callinectes sapidus* management (Paolisso 2002), and non-point source pollution (Bunting-Howarth 2001; Feurt 2007). Cultural models of environmental issues have been the focus of increased research attention for more than a decade (Kempton et al. 1995; Feurt 2003; Power and M. Paolisso 2005).

Environmental conflicts can arise from cultural differences associated with values, beliefs, and knowledge (Corbett 2006). An understanding of conflicts arising from different cultural models can be used to improve dialogue. Science represents only one way of knowing about environmental issues. Research has shown that the cultural models of nature held by farmers and watermen demonstrate an understanding of the resilient and chaotic attributes of nature that is in line with modern complexity theory. Perspectives of these people who are in daily contact with nature are unique and valuable for collaborative learning applied in the context of comanagement of natural resources (Paolisso and Maloney 2000; Paolisso and Chambers 2001; Paolisso 2002; Feurt 2007).

Coastal and estuarine-related cultural models research has been used to determine perceptions of effective coastal planning (Christel et al. 2001), stakeholder and public perceptions of toxic dinoflagellate blooms (Falk et al. 2000; Kempton and Falk 2000; Paolisso and Maloney 2000; Paolisso and Chambers 2001), farmers' understanding of nutrient enrichment in the Chesapeake Bay (Paolisso and Maloney 2000), and perceptions of watermen about the role of science and regulation in management of the Chesapeake Bay blue crab fishery (Paolisso 2002). Understanding the cultural models used by the lay public has helped scientists and resource managers communicate with important stakeholder groups and has facilitated collaborative learning and public participation in decision making related to nutrient management plans for Delaware coastal bays (Bunting-Howarth 2001), implementation of watershed management plans in the Gulf of Maine (Feurt 2007), and comanagement of the blue crab fishery in the Chesapeake Bay (Paolisso 2002). This research has the broad goal of understanding how humans make sense of, and understand, environmental issues and how this understanding is translated into decision making and action. Applying an understanding of conflicting cultural models to participatory and collaborative processes can improve dialogue among stakeholders and create policies and environmental solutions that benefit from a combination of different kinds of knowledge (Bunting-Howarth 2001; Paolisso 2002; Feurt 2007). Ethnographic research techniques, including interviews, transcription and coding of discourse, and participant observation, are used in this study to make explicit the divergent cultural models that contribute to conflict among stakeholder groups (Senge 1990; Strauss and Corbin 1990; Weiss 1994; Hammersley and Atkinson 1995; Bernard 1998; Bernard and Ryan 2010).

Results: A Conceptual Framework for *Science to Management* that Incorporates Knowledge of Cultural Models and Decision-Support Processes Like Collaborative Learning

Cultural models-based collaborative learning is an expert practice for bridging the systems of science and management. The graphic designs in Figure 2 are used to illustrate the conceptual framework presented in the *Collaborative Learning Guide for Ecosystem Management* and Power Point presentations for the project. The role of collaborative learning as a bridge connecting the system of science and science products with the system of management and policy depends upon analysis of the perceptual, institutional, and disciplinary barriers separating the two systems. Social science methodologies and theories, like the cultural models research associated with this study, facilitate effective bridging by collaborative learning.

Because community-based ecosystem management is based upon a place-based, shared vision of desired future outcomes, its principles and practices connect social capital and natural capital (Meffe et al. 2002). The cultural models research for this study provided qualitative evidence for

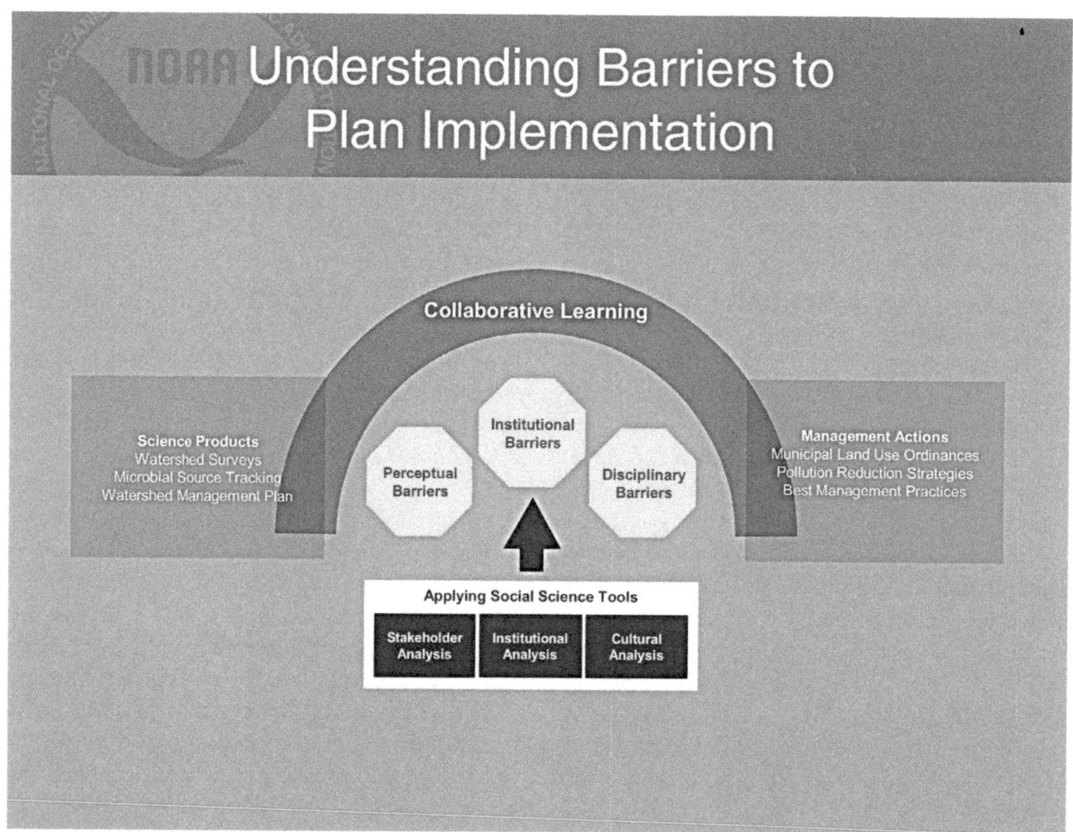

FIGURE 2. Collaborative learning is a science to management bridge (Feurt 2007).

the structure of the municipal system responsible for land use and water-quality protection. Shared values about the importance of clean water motivate a social system of practitioners and advocates in southern Maine. Like the tributaries of a watershed contributing flows that determine the quality of water downstream, each of the elements of this social system affects different aspects of land use and management to influence the quality and quantity of water that collects and flows to estuaries at the land–sea interface (Figure 3). The eight elements of this social system, shown below (Figure 3), interact to protect the values associated with clean water. Work within each element of the system is guided by professionally established best practices and a set of values and ethics that define the culture of the group. Science plays a role in this social system by shaping best practices and providing feedback about the ability of actions taken by each group to achieve goals in alignment with group values and professional ethics. The Center for Watershed Protection has analyzed and synthesized scientific papers about best practices for water quality protection to produce the "eight tools for watershed protection," providing guidance for stormwater engineers and land-use planners (CWP 2000). The professional expertise of people working within this social system is augmented by commitment to the communities they serve and attachment to the places where they work, play, and raise their families (Feurt 2007).

"Discovering" the Kaleidoscope of Expertise

Knowledge about how people in southern Maine value water came from a series of interviews conducted in 2004 in coastal watersheds with people involved in municipal water management. The initial intent guiding the interviews was to identify gaps in the knowledge of municipal officials that could be addressed by providing science-based information as portrayed in the traditional science translation model in Figure 4. However, after only a few interviews, the limitations of the original model became clear. The delivery of information model fails to incorporate the knowledge and expertise of stakeholders in the system.

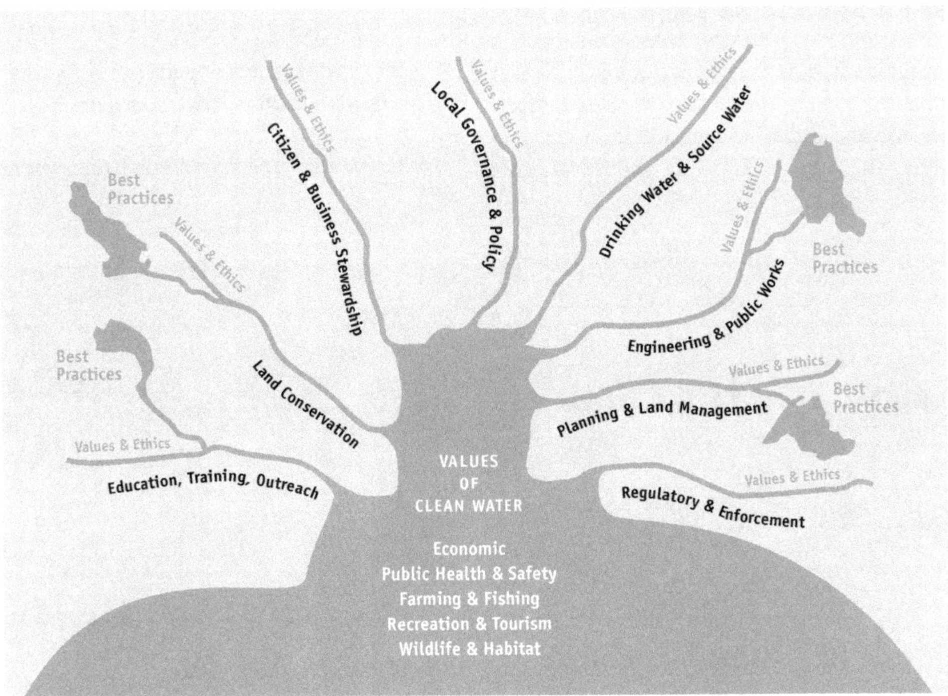

FIGURE 3. The social landscape of municipal water management in southern Maine (Feurt 2008).

The collaborative learning model represents the transformation of the linear concept of information delivery (Figure 4) to a systems understanding of the knowledge and expertise residing in the municipal system for protecting and managing water (Figure 5). The metaphor of the *kaleidoscope of expertise* came from the realization that each person views water and their role in managing water through an individual lens affected by their education, training, work experience, and the requirements of their job. Taken together, the combined knowledge and expertise of the people responsible for water is a resource for learning and problem solving. Collaborative learning is a way to tap this resource and fills the limitations in the original model by engaging all participants in co-creation of knowledge (Daniels and Walker 2001; Feurt 2007).

Knowledge in the Kaleidoscope

Seven types of knowledge or *ways of knowing* about water emerged from analysis of the interviews of southern Maine water managers. The interviewees included people involved in scientific research and implementation of state regulatory programs, as well as municipal officials. During each open-ended interview, informants talked about their work, how they valued water, and their ideas about threats to water, the causes and effects of those threats, and what could be done to protect water. The ways of knowing about water are described below and coded into the *kaleidoscope of expertise* model. Informants draw from multiple ways of knowing in their work.

1. Ecological Knowledge **ECO**: Understanding of the structure and functions of a watershed, the hydrologic cycle, connections between groundwater and surface water, and the value of ecosystem services provided by a watershed.
   People who use this knowledge: ecologists, farmers, and hydrologists
2. Governance Knowledge **GOV**: Understanding the interrelationships among policy, regulations, government hierarchy, planning documents, and ordinances, and the structures and processes in place to execute them.
   People who use this knowledge: town planners, code enforcement officers, elected officials, and regulators
3. Land Use Knowledge **LAN**: Understanding the ways land management and conservation and the design of infrastructure and development can influence water quality and quantity, and the ways that the economic value and ecological value of land can be balanced.
   People who use this knowledge: Town planners, farmers, developers, public works directors, and water district managers
4. Educational Practices Knowledge **EDU**: Understanding the ways knowledge is generated

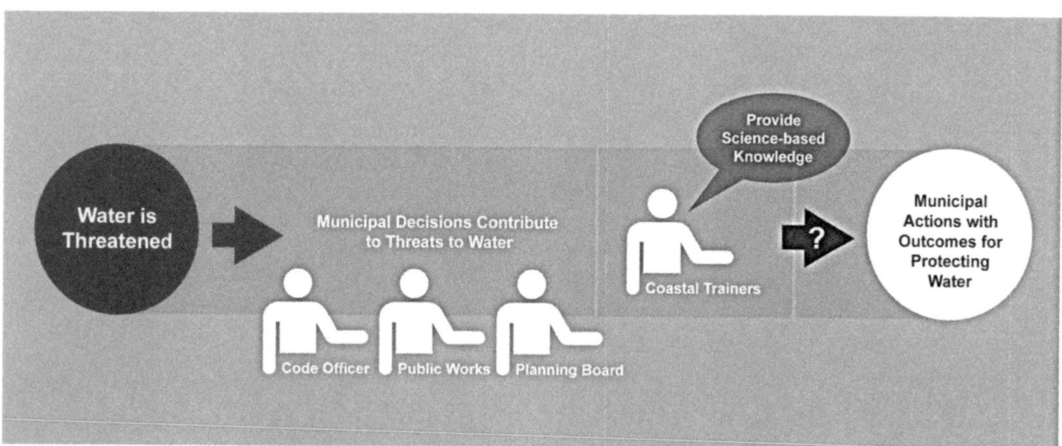

FIGURE 4. A model of traditional science translation (Feurt 2008).

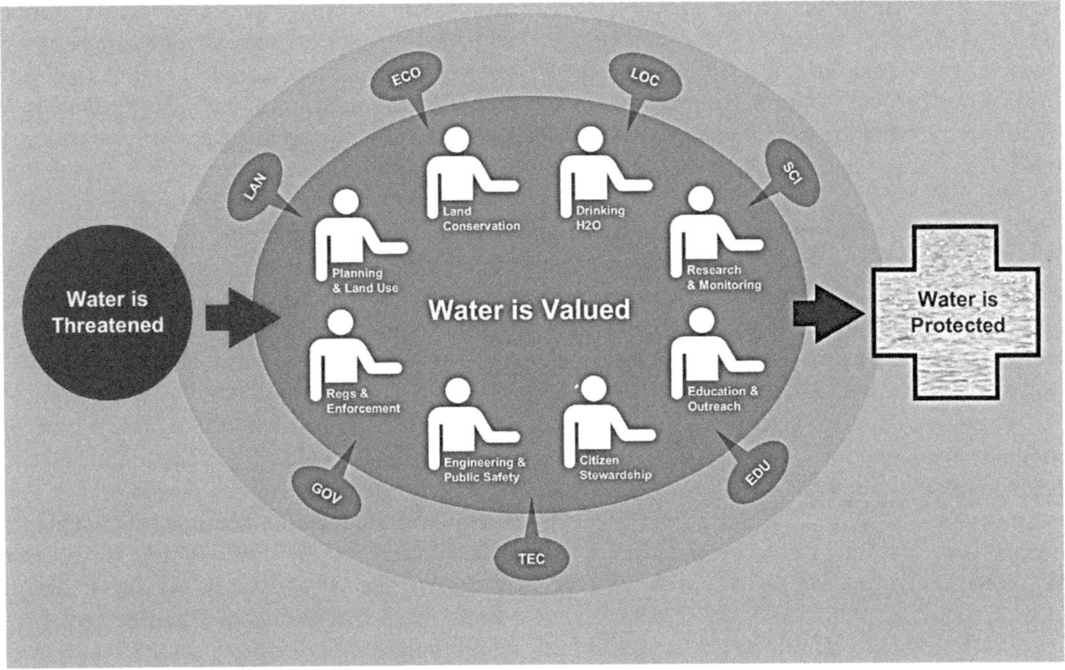

FIGURE 5. The collaborative learning model with the kaleidoscope of expertise–the municipal system is a resource not a receptacle (Feurt 2008).

and transferred among the knowledge domains and designing and evaluating the effectiveness of education and outreach strategies.
People who use this knowledge: Education and outreach specialists, trainers, science translators, and town planners.

5. Science Knowledge **SCI**: Understanding the factors influencing water quality and quantity for the purpose of documenting conditions, monitoring change, understanding cause and effect relationships, and evaluating the effectiveness of management practices and policies.
People who use this knowledge: Biophysical and social scientists, water quality monitors, and regulators
6. Technology Knowledge **TEC**: Understanding the use and application of engineering and computer technologies to the protection of water, mitigation of impacts, implementation of best management practices, and restoration of lost structure and function in the watershed.
People who use this knowledge: Engineers, public works directors, and geographic information system specialists
7. Local Knowledge **LOC**: Understanding the connections between the people and places in the community, including familiarity with town history, values, and conflicts.
People who use this knowledge: Town planner, public works director, elected officials, farmers, and land developers.

Discussion

This case study introduces collaborative learning as one tool for structuring and assisting the complex dialogues that drive ecosystem management. Dialogue and deliberation inherent in collaborative learning allow the diverse stakeholders engaged in assessing the ecological, socio-economic, and cultural consequences of ecosystem management to develop a shared vision of future outcomes and engage in adaptive management to determine if actions undertaken as experiments are progressing toward the desired vision.

One example of a difficult challenge facing managers and decision makers is how to execute the practice of ecosystem management in the face of climate change. The National Research Council (NRC) recognizes the unprecedented decision-making challenges associated with global climate change (NRC 2009). In its publication *Informing Decisions in a Changing Climate*, the National Academy of Sciences Panel on Strategies and Methods for Climate-Related Decision Support cautions that environmental conditions, stressors, and responses will be shaped by events that have not previously existed, and society will need new strategies for adapting to these changes. These thoughts echo the needs identified a decade earlier by Lubchenco's (1998) call for a new social contract for science. The NRC report recommends that government agencies and the scientific community organize their decision support efforts around six principles of effective decision support. Each of the six principles are addressed by the cultural models-based collaborative learning methodology presented in this chapter: begin with users' needs, give priority to process over products, link information producers and users, build connections across disciplines and organizations, seek institutional stability, and design processes for learning.

Collaborative learning has the potential to be used by practitioners working on the front lines of climate change adaptation, a venue where the application of ecosystem management theory, principles, and framework are critically important. The front lines of climate change adaptation are the same places described in this chapter. Communities where planners, public works directors, stormwater engineers, and researchers negotiate meanings, pose questions, and work. These are the places where ecosystem management has a face and a place. People live, work, and play in the natural systems affected by daily decisions. When scientific findings give shape and meaning to the changes people see or cannot see in the places they care about, collaborative learning is one way to foster the analysis and deliberation required to incorporate that knowledge into policies and management actions.

Collaborative learning is part of the rapidly growing interdisciplinary field of environmental communication. This field encompasses scholarly research and practical applications dealing with the ways society understands and responds to environmental messages and events. Environmental communication marries participatory and collaborative approaches with traditional environmental education and interpretation, taking the most effective practices and principles from the craft and framing them within interdisciplinary theories of learning and behavior. This rich interdisciplinary field addresses the communication of science and environmental risk, multi-stakeholder collaboration, public participation, conflict resolution, social marketing, environmental journalism, the representation of nature in popular culture, and environmental advocacy campaigns (Cox 2006). Decision support, as a discreet practice, is emerging in a critical role for environmental communication in the context of climate change (NRC 2009). These diverse approaches provide tools to educate, alert, persuade, mobilize, and engage people in ecosystem management.

Three other approaches to environmental communication are highlighted here for comparison with collaborative learning: public participation/civic engagement, social marketing, and technology transfer/science translation. Because each approach is based upon different theoretical and applied traditions, they vary in their orientation to social learning, behavior change, knowledge flow, and goals.[1] These approaches can be applied within a matrix of scale and temporal conditions. The scale of this research was a small group of two dozen municipal officials working in their hometown watersheds. The temporal aspect of this project is engagement over an extended period. The Protecting Our Children's Water project is framed as 2005–2025 to foster intergenerational thinking. During this project, collaborative learning was also applied by the Wells NERR CTP, with groups gathering for a one-time strategy session to develop consensus and overcome conflict in order to make progress on shared goals. The three other forms of environmental communication that contribute to ecosystem management are described below for comparison with collaborative learning.

[1] *The Collaborative Learning Guide for Ecosystem Management* (Feurt 2008) contains a conceptual model of environmental communication for ecosystem management.

Public Participation and Civic Engagement

The history of public participation has strong roots in the New England town meeting. Public participation in federal environmental policy increased with the passage of the National Environmental Policy Act in 1969. The relationship among federal environmental agencies and public participation has evolved through a series of laws, policies, and on-the-ground experience during the past four decades. Political theorist Kai Lee (1993) captured the link between public participation (civic engagement) and ecosystem management using the metaphor of the compass and gyroscope. Science is the compass that guides ecosystem management toward goals of sustaining ecosystem structure and function. Civic engagement is the gyroscope providing course corrections related to societal goals and priorities. Public participation consists of a rich collection of approaches aimed at improving the quality and legitimacy of decisions and increasing the capacity of federal agencies and their constituencies to engage in long-term policy dialogues. The collaborative learning process is one technique that can be adapted to facilitate public participation, which has been shown by to be especially important for ecosystem management (Meffe et al. 2002; McLeod and Leslie 2009)

Community-Based Social Marketing

The second approach to environmental communication is community-based social marketing (CBSM; McKenzie-Mohr and Smith 1999). Drawing from theory and practices associated with social psychology and marketing, CBSM can be applied to foster practices that support ecosystem management objectives. Conserving and restoring riparian buffers is an example of a project identified as an action item by the watershed councils in the Protecting Our Children's Water project. CBSM uses a rigorous four-step method that can be used to design and implement behavior change projects. The key to success with this method is preliminary research to identify the barriers and benefits of the desired action. Combining this knowledge with behavior change tools documented as effective by social psychology research results in a strategy designed to reduce barriers and increase benefits. The strategy for behavior change is tested and improved through a pilot project, then implemented in a broader context. Evaluation of behavior change success depends upon knowledge of baseline conditions before the project.

While CBSM is most frequently applied to implement local projects that support ecosystem management, the approach can be adapted to a broader scale as a means of identifying and overcoming barriers to building capacity for ecosystem management. The Protecting Our Children's Water project applied elements of CBSM to initiate and support the development of watershed councils in southern Maine (Feurt 2007, 2008).

Technology Transfer and Science Translation

The third, and perhaps most familiar, type of environmental communication that supports ecosystem management is technology transfer or science translation. Knowledge flows from a source of expertise toward a user in the communication model for both technology transfer and science translation. Extension and outreach also follow this basic model of knowledge dissemination. The eloquent and well-documented theory of the diffusion of innovations provides rich empirical evidence for the elements and variables that interact to support effective diffusion and adoption of knowledge, practices and policies (Rogers 1995, 2002). Facilitating technology transfer and science translation in support of ecosystem management requires that users perceive an advantage to the new knowledge or practice and that it is consistent with their values and experiences. Potential users need to recognize the relevance of new knowledge to their work. The complexity of new knowledge needs to be reduced to a point where the user can try the new idea (The Collaborative Learning Guide aims to do that for this research). Even more effective are opportunities to observe or participate with others using the new knowledge or practice. Collaborative learning can be used as a means to introduce scientific findings or best practices with relevance to ecosystem management to groups of stakeholders. Collaborative learning supports decision making by providing a structured opportunity for groups to analyze and deliberate about scientific findings and discuss the application of that information to management or policy (Feurt et al. 2009; NRC 2009). It can also facilitate ongoing relationships among policy makers, regulators, managers, and scientists to address shared goals for ecosystem management.

Conclusion—A New Paradigm Evolving

Conceiving of ecosystem management as a social science adaptable to addressing global environmental change requires a paradigm shift in thinking by resources users, managers, policy makers, and scientists. Paradigm shifts are not easy or quick, as documented in some of the keystone literature of science and ecosystem management (Kuhn 1962; Lee 1993; Gunderson and Holling 2002). The need for a paradigm shift in the way science is made available and used by society was clearly articulated by Dr. Jane Lubchenco, current director of the National Oceanic and Atmospheric Administration, in an article published in Science in 1998.

> The whole system of science, society and nature is evolving in fundamental ways that cause us to rethink the way science is deployed to help people cope with a changing world. Scientists should be leading the dialogue on scientific priorities, new institutional arrangements, and improved methodologies to disseminate and utilize knowledge more quickly (Lubchenco 1998:496).

The cultural models-based collaborative learning approach described in this chapter is one such methodology (Feurt 2007). This research builds upon the work of Daniels and Walker (2001) by using ethnographic methods to develop knowledge of the cultural models used in decision making about water. This knowledge enhanced the systems understanding required for robust collaborative problem-solving processes. This research demonstrated the potential for integrating natural and social science to achieve the goals of ecosystem management. Coupling the science generation system and the science utilization system was a guiding principle for this research. Working within both systems benefited the development of a product that was tested weekly in the environment where it would be applied. Coupling the generation of knowledge and the use of that knowledge to solve problems is the foundation of the Action Research paradigm (Greenwood and Levin 1998).

This project developed, tested and disseminated a two-component conceptual model of the *kaleidoscope of expertise* and the *collaborative learning bridge* as a way of using social science to implement communication objectives of ecosystem management. The kaleidoscope of expertise is a metaphor for the way members of a system view an environmental issue from many perspectives. The system within which an environmental issue is embedded includes the people experiencing the issue in the places where they live, work and play. These are places they care about, places they are responsible for, and places where they connect regularly with other people who share a stake in the condition of that place. Knowing the components and boundaries of the kaleidoscope of expertise for a specific issue puts a place and a face on ecosystem management. This knowledge is critical for collaborative problem solving.

Acknowledgments

Funding and support for this project was provided by CICEET, the Cooperative Institute for Coastal and Estuarine Environmental Technology. A partnership of the National Oceanic and Atmospheric Administration and the University of New Hampshire, CICEET develops tools for clean water and healthy coasts nationwide.

References

Allan, C. 2004. Improving the outcomes of adaptive management at the regional scale. Doctoral dissertation. Charles Sturt University, Albury, New South Wales, Australia.

Allan, C., and A. Curtis. 2005. Nipped in the bud: why regional scale adaptive management is not blooming. Environmental Management 36:414–425.

Allen, T., and T. Hoekstra. 1992. Toward a unified ecology. Columbia University Press, New York.

Argyris, C., and D. Schon. 1996. Organizational

learning II: theory, method and practice. Addison-Wesley, Reading, Massachusetts.

Beatley, T. 2009. Planning for coastal resilience, best practices for calamitous times. Island Press, Washington, D.C.

Berkes, F., and C. Folke, editors. 2000. Linking social and ecological systems: management practices and social mechanisms for building resilience. Cambridge University Press, New York.

Bernard, H., editor. 1998. Handbook of methods in cultural anthropology. Altamira Press, New York.

Bernard, H. R., and G. W. Ryan. 2010. Analyzing qualitative data: systematic approaches. Sage Publications, Thousand Oaks, California.

Boesch, D. 1999. The role of science in ocean governance. Journal of Ecological Economics 31:189–198.

Boesch, D. 2001. Science and integrated drainage basin coastal management: Chesapeake Bay and Mississippi Delta. Pages 37–50 *in* B. Von Bundungen and R. Turner, editors. Science and integrated coastal management. Dahlem University Press, Berlin.

Bunting-Howarth, K. 2001. Cultural models, public participation and policy: *Pfiesteria piscicida* and non-point source pollution in Delaware's inland bays. Doctoral dissertation. University of Delaware, Newark.

Christel, D., W. Kempton, and J. Harris. 2001. The effects of values and cultural models on policy: an anthropological approach to environmental policy in Tampa Bay. Proposal prepared for EPA Workshop: Understanding Public Values and Attitudes to Ecological Risk Management. University of Delaware, College of Marine Studies, Newark.

Collins, A., and D. Gentner. 1987. How people construct mental models. Pages 243–269 *in* D. Holland and N. Quinn, editors. Cultural models in language and thought. Cambridge University Press, Cambridge, UK.

Cook, D., N. Diamond, and P. Ticco. 2002. National Estuarine Research Reserve System Coastal Decision Maker workshops and the Coastal Training Program: an assessment with recommendations for future activities. National Oceanic and Atmospheric Administration, Washington, D.C.

Corbett, J. 2006. Communicating nature: how we create and understand environmental messages. Island Press, Washington, D.C.

Costanza, R. 1998. Principles of sustainable governance of the ocean. Science 281:198–199.

Cox, R. 2006. Environmental communication and the public sphere. Sage Publications, Thousand Oaks, California.

CWP (Center for Watershed Protection). 2000. The practice of watershed protection: techniques for protecting and restoring urban watersheds. Center for Watershed Protection, Baltimore, Maryland.

D'Andrade, R. 1995. The development of cognitive anthropology. Cambridge University Press, Cambridge, UK.

D'Andrade, R., and C. Strauss. 1992. Human motives and cultural models. Cambridge University Press, Cambridge, UK.

Dailey, M. 1999. Cultural models of forests and ecological change on the Appalachian Plateau, 1750–1840. Doctoral dissertation. University of Georgia, Athens.

Daniels, S., and G. Walker. 2001. Working through environmental conflict: the collaborative learning approach. Praeger. Westport, Connecticut.

Endter-Wada, J., D. Blahna, R. Krannich, and M. Brunson. 1998. A framework for understanding social science contributions to ecosystem management. Ecological Applications 8:891-904.

Falk, J., F. Darby, and W. Kempton. 2000. Understanding mid-Atlantic residents' concerns, attitudes and perceptions about harmful algal blooms: Pfiesteria piscicida. University of Delaware, Sea Grant DEL-SG-05-00, Newark.

Feurt, C. 2003. Cultural models: a tool for enhancing communication and collaboration in coastal resources management. Report submitted to the NOAA/UNH Cooperative Institute for Coastal and Estuarine Environmental Technology, Durham, New Hampshire.

Feurt, C. 2006. Science translation for non-point source pollution control: a cultural models approach with municipal officials. Final Report submitted to the NOAA/UNH Cooperative Institute for Coastal and Estuarine Environmental Technology, Durham, New Hampshire.

Feurt, C. 2007. Protecting our children's water, using cultural models and collaborative learning to frame and implement ecosystem management. Doctoral dissertation. Antioch University New England, Keene, New Hampshire.

Feurt, C. 2008. Collaborative learning guide for ecosystem management. The Wells National Estuarine Research Reserve. Available: http://swim.

wellsreserve.org/ctp/Collaborative%20Learning%20Guide.pdf. (November 2011).

Feurt, C. 2009. Collaborative learning strategies to overcome barriers to science translation in coastal watershed management. Report submitted to NOAA/UNH Cooperative Institute for Coastal and Estuarine Environmental Technology, Durham, New Hampshire.

Feurt, C., T. Smith, and Z. Steele. 2009. Headwaters, a collaborative conservation plan for Sanford, Maine. Prepared for the Sanford Planning Board by the Wells National Estuarine Research Reserve in collaboration with Southern Maine Regional Planning Commission. Available: http://swim.wellsreserve.org/ctp/Sanford%20Conservation%20Plan%2009.pdf (November 2011).

Fiorino, D. 2001. Environmental policy as learning: a new view of an old landscape. PublicAdministration Review 61:322–334.

Fluharty, D., M. Abbott, R. Davis, M. Donahue, S. Madsen, T. Quinn, J. Rapporteur, and J. Sutinen. 2006. Preliminary report: External Ecosystem Task Team report to NOAA Science Advisory Board: evolving an ecosystem approach to science and management thoughout NOAA and its partners. National Oceanic and Atmospheric Administration, Silver Spring, Maryland.

GEARS (Global Evaluation and Applied Research Solutions). 2009. NOAA Coastal Services Center Training Program and NOAA National Estuarine Research Reserve System Coastal Training Program joint external review of training activities: findings and improvement options. Final Report - March 2009. GEARS, Decatur, Georgia.

Gentner, D. and A. Stevens. 1983. Mental models. Lawrence Erlbaum Association, Hillsdale, New Jersey.

GLEFC (Great Lakes Environmental Finance Center). 2004. Trends analysis of coastal training programs in the National Estuarine Research Reserve System. Prepared for the Estuarine Reserves Division of the National Oceanic and Atmospheric Administration by GLEFC, Cleveland State University, Cleveland, Ohio.

Greenwood, D., and M. Levin. 1998. Introduction to action research, social research for social change. Sage Publications, Thousand Oaks, California.

Gunderson, L., and C. Holling. 2002 Panarchy understanding transformations in human and natural systems. Island Press, D.C.

Gunderson, L., C. Holling, and S. Light, editors. 1995. Barriers & bridges to the renewal of ecosystems and institutions. Columbia University Press, New York.

Hammersley, M., and P. Atkinson. 1995. Ethnography: principles in practice. Routledge, New York.

Holland, D., and N. Quinn. 1987. Cultural models in language and thought. Cambridge University Press, Cambridge, UK.

Holling, C. S., editor. 1978. Adaptive environmental assessment and management. John Wiley & Sons, New York.

Kempton, W. 1991a. Lay perspectives on global climate change. Global Environmental Change 1:183–208.

Kempton, W. 1991b. Public understanding of global warming. Society and Natural Resources 4:331–345.

Kempton, W. 1993. Will public environmental concern lead to action on global warming? Annual Review of Energy and the Environment 18:117–145.

Kempton, W. 1997. How the public views climate change. Environment 39(9):12–21.

Kempton, W. 2001.Cognitive anthropology and the environment. Pages 49–71 *in* C. L. Crumley, editor. New directions in anthropology and environment. Alta Mira Press, Walnut Creek, California.

Kempton, W., J. Boster, and J. Hartley. 1995. Environmental values in American culture. MIT Press, Cambridge, Massachusetts.

Kempton, W., and P. Craig. 1993. European thinking on global climate change. Environment 35(3):16–20.

Kempton, W., J. Darley, and P. Stern. 1992. Psychology and energy conservation. American Psychologist 47:1213–1223.

Kempton, W., and J. Falk. 2000. Cultural models of Pfiesteria: toward cultivating more appropriate risk perceptions. Coastal Management 28:273–285.

Kempton, W., D. C. Holland, K. Bunting-Howarth, E. Hannan, and C. Payne. 2001. Local environmental groups: a systematic enumeration in two geographical areas. Rural Sociology 66:557–578.

Kennish, M., editor. 2004. Estuarine research, monitoring and resource protection. CRC Press. Washington, D.C.

Krum, C., and C. Feurt. 2002. Wells National Estuarine Research Reserve Coastal Training Program: market analysis and needs assessment. Executive summary. Available: www.wellsreserve.org/sup/downloads/ctp_exec_sum.pdf. (November 2011).

Kuhn, T. S. 1962. The structure of scientific revolutions, 1st edition. University of Chicago Press, Chicago.

Lee, K. 1993. Compass and gyroscope: integrating science and politics for the environment. Island Press, Washington, D.C.

Lubchenco, J. 1998. Entering the century of the environment: a new social contract for science. Science 279:491–497.

Machlis, G. E., J. E. Force, and W. R. Burch, Jr. 1997. The human ecosystem part I: the human ecosystem as an organizing concept in ecosystem management. Society and Natural Resources 10:347–367.

McLeod, K. L., and H. Leslie. 2009. Ecosystem-based management for the oceans. Island Press, Washington, D.C.

Meffe, G., L. Nielsen, R. Knight, D. Schenborn. 2002. Ecosystem management, adaptive, community-based conservation. Island Press, Washington, D.C.

McKenzie-Mohr, D., and W. Smith. 1999. Fostering sustainable behavior, an introduction to community-based social marketing. New Society Press, Gabriola Island, British Columbia.

Michael, D. 1995. Barriers and bridges to learning in a turbulent human ecology. Pages 461–489 *in* L. Gunderson, C. Holling, and S. Light, editors. 1995. Barriers and bridges to the renewal of ecosystems and institutions. Columbia University Press, New York.

NOAA (National Oceanic and Atmospheric Administration). 2003. National Estuarine Research Reserve System strategic plan, 2003–2008. NOAA, Silver Spring, Maryland.

NOAA (National Oceanic and Atmospheric Administration). 2008a. At-a-glance: the National Estuarine Research Reserve Coastal Training Program. Available: www.nerrs.noaa.gov/Doc/PDF/Training/CTPAGlance.pdf. (November 2011).

NOAA (National Oceanic and Atmospheric Administration). 2008b. Improving decisions affecting our coasts through training and assistance. NERRS Coastal Training Program. Available: www.nerrs.noaa.gov/Doc/PDF/Training/NERRSCTPPublication.pdf. (November 2011).

NOAA (National Oceanic and Atmospheric Administration). 2009. What is the Coastal Training Program? Available: www.nerrs.noaa.gov/CTPDefault.aspx?ID=381. (November 2011).

NRC (National Research Council). 1996. Understanding risk, informing decisions in a democratic society. National Academy Press, Washington, D.C.

NRC (National Research Council). 2001a. Grand challenges in environmental sciences. National Academy Press, Washington, D.C.

NRC (National Research Council). 2001b. Marine protected areas: tools for sustaining ocean ecosystems. Report from the Committee on the Evaluation, Design, and Monitoring of Marine Reserves and Protected Areas in the United States, Ocean Studies Board. National Academy Press, Washington, D.C.

NRC (National Research Council). 2002. New tools for environmental protection, education, information, and voluntary measures. National Academy Press, Washington, D.C.

NRC (National Research Council) 2009. Informing decisions in a changing climate. The National Academies Press. Washington, D.C.

Paolisso, M. 2002. Blue crabs and controversy on the Chesapeake Bay: a cultural model for understanding watermen's reasoning about blue crab management. Human Organization 61:226–239.

Paolisso, M., and E. Chambers. 2001. Culture, politics, and toxic dinoflagellate blooms: the anthropology of *Pfiesteria*. Human Organization 60:1–12.

Paolisso, M., and S. Maloney. 2000. Recognizing farmer environmentalism: Nutrient runoff and toxic dinoflagellate blooms in the Chesapeake Bay region. Human Organization 59:209–221.

Pfeffer,M, J. Schelhas, and L. Day. 2001. Forest conservation, value conflict and interest formation in a Honduran national park. Rural Sociology 66:382–402.

Power, L., and M. Paolisso, 2005. Linking estuarine ecology and community heritage: a socio-cultural needs assessment of Monie Bay Reserve. Report prepared for The National Estuarine Research Reserve, Maryland Department of Natural Resources, Annapolis, Maryland.

Quinn, N., editor. 2005. Finding culture in talk, a collection of methods. Palgrave Macmillan, New York

Rogers, E. 1995. Diffusion of innovations, 4th edition. The Free Press, New York.

Rogers, E. 2002. The nature of technology transfer. Science Communication 23:323–341.

Senge, P. 1990. The fifth discipline. Currency Doubleday, New York.

Strauss, A., and J. Corbin. 1990. Basics of qualitative research: grounded theory procedures and techniques. Sage, Newbury Park, California.

Strauss, C., and N. Quinn. 1997. A cognitive theory of cultural meaning. Cambridge University Press, Cambridge, UK.

Visser, L., editor. 2004. Challenging coasts: transdisciplinary excursions into integrated coastal zone development. Amsterdam University Press, Amsterdam.

Walters, C. 1986. Adaptive management of renewable resources. Macmillan, New York.

Walters, C. J., and C. S. Holling. 1990. Large-scale management experiments and learning by doing. Ecology 71:2060–2068.

Weiss. R. 1994. Learning from strangers: the art and method of qualitative interview studies. The Free Press, New York.

Wondolleck, J., and S. Yaffee. 2000. Making collaboration work: lessons from innovation in natural resource management. Island Press, Washington, D.C.

American Fisheries Society Symposium 79:89–101, 2012

Linking Science to Management and Policy through Strategic Communication

Verna DeLauer*
Clark University, George Perkins Marsh Institute
950 Main Street, Worcester, Massachusetts 01610, USA

Susan Ryan
Gulf of Maine Census of Marine Life, University of Southern Maine
350 Commercial Street, Portland, Maine 04101, USA

Ivar Babb
National Undersea Research Center, University of Connecticut at Avery Point
1080 Shennecossett Road, Groton, Connecticut 06340, USA

Peter Taylor
Waterview Consulting, Yarmouth, Maine

Pam DiBona
The New England Aquarium, Central Wharf, Boston Massachusetts 02110, USA

Abstract.—In this paper, we synthesize the current thinking of scientists involved in the 2009 Gulf of Maine Symposium about successes and challenges in communicating the science underlying ecosystem-based management (EBM) and summarize the evolution of this issue as reflected in past Gulf of Maine symposia. We present survey results from scientists, representing a range of disciplines, that indicate a critical need for strategic communication as part of EBM. We focus primarily on the transfer of science to policy makers and managers, both of whom play an influential role in shaping current and future public opinion and practice. We illustrate best practices to enhance communications about EBM and offer recommendations for using communication techniques to speak to people from their cultural lens and knowledge base. We assert that "strategic communication" is necessary to enable EBM, and scientists, managers, and communication professionals must work together to employ its principles within a coherent framework.

Introduction

Striving to sustain the integrity of a large marine ecosystem, such as the Gulf of Maine, presents the challenge of creating a shared research and management vision among multiple jurisdictions and stakeholders. The past decade's growth in information technology has allowed for an unprecedented scope of information exchange that is changing the way research is conducted, synthesized, and translated. It is largely recognized that this rapid information exchange is occurring on multiple scales, within and among disciplines and across spatial boundaries (Green and Gilbert 1995; Treise and Weigold 2002). As the ability to transfer scientific information increases and connections across jurisdictions strengthen, the need to interpret this information grows as well. Particularly important is science translation and synthesis to enable and encourage shared knowledge and insight among scientists, policy makers, and managers, referred to

* Corresponding author: vernadelauer@yahoo.com

throughout as *decision makers*. The transfer of scientific research and tools was a main topic at the 2009 Gulf of Maine Symposium, which brought together more than 200 individuals to "make recommendations on the knowledge required to move forward with an integrated management approach/ ecosystem approach to management" (www.rargom.org/Symposium2009). The general sentiment was that strategic use of diverse and effective communication styles and approaches can enable more thoughtful and comprehensive exchanges of information for an integrated approach.

While the themes throughout this paper are applicable elsewhere, here we reflect on the past decade of work throughout the Gulf of Maine that strived to link science to policy and management processes. We synthesize the results of the 2009 Gulf of Maine Symposium, conducted by the authors to understand how the thinking among scientists has evolved to link science more effectively with management and policy. We create a context for discussing strategic communication based on feedback we received from scientists from a range of disciplines and by reflecting on past recommendations from this same scientific community. Like other contributors to this volume (Feurt 2012; Hall-Arber et al. 2012; Lawton et al. 2012; Runge 2012; all this volume), we assert that strategic communication is an essential element of EBM, not an add-on activity, and needs to be applied in the Gulf of Maine to enable EBM. We offer recommendations for using communication techniques from other disciplines, such as psychology, to speak to people "where they're at" rather than where we think they should be. We stress the need for, and offer insights toward, a comprehensive and coordinated regional communication plan. The communication among scientists and nonscientists deserves its own set of management criteria. We recognize that communication among scientists and decision makers is only one connection within a grid of many needed network connections. We make it a focus of this paper, in direct response to the concerns and needs identified by these audiences in our survey.

Progress in Linking the Islands of Science and Policy

Contributors throughout this volume assert that the Gulf of Maine is arguably one of the most studied ecosystems in the world (Grabowski and Hart 2012; Hale 2012; Hare and Kane 2012; Lawton et al.; Smith et al. 2012; all this volume). Research activities have been presented periodically at several events, including the 1991 Gulf of Maine Scientific Workshop (Woods Hole, Massachusetts), the 1996 Gulf of Maine Ecosystem Dynamics Scientific Symposium (St. Andrews, New Brunswick), and, most recently, the 2009 Gulf of Maine Symposium (St. Andrews, New Brunswick). Each of these research-focused events highlighted the need for increased dialogue between the research and decision-maker communities. The 1996 symposium was specifically designed to provide opportunities for managers and scientists to interact and identify research topics of shared importance. Participants recognized that one of the key ingredients to ensure that science met the needs of management, and vice versa, was communication (Wallace and Braasch 1997). We strive to better clarify what more efficient, relevant, and successful communication might entail.

There have also been efforts specifically devoted to improve the interaction between researchers and decision makers in the Gulf of Maine region. For a comprehensive summary of the evolution of science-based management in the region, we looked to Ernst (2004), the National Research Council's Ocean Studies Board, and their workshop "Improving Interactions between Coastal Science and Policy" (NRC 1995), as well as the Regional Association for Research on the Gulf of Maine's "Mechanisms for Improving the Integration of Science and Management in Decisions Affecting the Environmental Quality of the Gulf of Maine" (Ballard et al. 1997), among other relevant sources. Suggestions emanating from these national and regional efforts include (1) a recognition of the value of a long-term, overarching vision for the Gulf of Maine that includes ongoing input from all stakeholders to guide policy, management, and research; (2) a better understanding of how and when scientific information is used in decision making; (3) the need to establish reward mechanisms that value scientific contributions to societal needs; and (4) the development of programs that emphasize new skills and relationships required for integration of science and management. A tangible offshoot of these efforts was the Gulf of Maine Science Translation Project that sought to accelerate the transfer of scientific findings and techniques to resource managers, planners, policy makers, and other coastal decision makers in the region (Gulf of Maine Council on the Marine Environment, www.gulfofmaine.org/science_translation).

In addition to seeking to improve communications among the research, management, and policy-making communities, underlying themes of the 1996 symposium were ecosystem modeling and transboundary linkages between terrestrial and marine ecosystems. Similarly, the 1997 workshop emphasized the need to

> [d]evelop and apply an integrated ecosystem perspective in addressing environmental issues...that bear on the structure, functioning of the marine system and the societal stresses on it.

This recognition of the need for an ecosystem perspective was the precursor to the current intellectual shift toward an ecosystem-based approach to management that is widely accepted within the research and management communities today. It was also indicative of an international move toward the approach.

Integration across management sectors is one of many basic tenets of EBM. However, before the EBM concept was broadly understood, the proliferation of data and information about the Gulf of Maine resulted in efforts to develop networks to transfer data and research syntheses among scientists and to policy makers and managers (New England Aquarium 1999). Participants at a workshop, convened by the New England Aquarium (1999), recognized that a regional electronic data and information exchange system that could provide a mechanism for communication and information exchange would help improve resource management decisions. Recently, the establishment of the Northeast Coastal and Ocean Data Partnership (www.necodp.org) has meant that physical and biological data sets dating back to the 1960s can be integrated and made accessible to users. Through metadata standards and shared data portals, coupled with modeling and visualization tools, managers and policy makers can integrate information into their decision making that was unimaginable two decades ago.

We reflect on the past to highlight that the decades of investment in research, management, and policy initiatives in the Gulf of Maine region have culminated in numerous reports and proceedings to inform a holistic, ecosystem-based approach to sustain this vital ecosystem. One could argue, however, that despite these efforts, some of which are highlighted here, there remains a need for a broad visioning process for the Gulf of Maine that includes input from all stakeholders to guide policy, management, and research. Lacking a regional vision of EBM, it is essential to look at the notion of strategic communication as a possible path toward regional coordination, recognizing that improved communication alone is one of many critical variables. Looking to the recent past, one can see great strides toward EBM have been made at the state, local, and regional scales within the Gulf of Maine.

Coastal New England states such as Rhode Island, Massachusetts, New Hampshire, and Maine share a strong maritime history, particularly around recreational and commercial fishing, and the act of balancing competing ocean uses. As early as 1983, Rhode Island had legislatively authorized special area management plans to better address competing priorities among decision makers. More recently, Massachusetts became the first state in the nation to mandate integrated management in state waters. Across the international border, in 2006 the Eastern Scotian Shelf Integrated Ocean Management Plan, under the Canada Oceans Act, was approved, which, among other strategies, included an ecosystem-based management framework. Most recently, the Northeast Regional Ocean Council has conducted several meetings of regional decision makers and scientists to discuss a regional framework for coastal and marine spatial planning. The region is poised to become a leader in the management of marine systems. To make this happen—and especially to reach successful management outcomes—will require a strong feedback loop between science and management. Communication is the linkage in that feedback loop. Without effective two-way communication linkages between science and management, the Gulf of Maine region will have limited capacity for adaptive management, for example, which is a core element of ecosystem-based management (Taylor and De-

Lauer 2009) Science, in relation to management, should be looked upon as a social contract from which to gauge the implications of environmental decisions (Wilsdon et al. 2005).

Using Strategic Communication to Create Shared Meaning

In 2005, more than 200 recognized marine researchers throughout the United States signed on to a consensus statement for an ecosystem-approach to management.

> Ecosystem-based management is an integrated approach to management that considers the entire ecosystem, including humans. The goal of ecosystem-based management is to maintain an ecosystem in a healthy, productive and resilient condition so that it can provide the services humans want and need. Ecosystem-based management differs from current approaches that usually focus on a single species, sector, activity or concern; it considers the cumulative impacts of different sectors. (McLeod et al. 2005)

By breaking down these sentences, one discovers additional meaning inherent in the definition. "Ecosystem-based management is an integrated approach to management that considers the entire ecosystem." Underlying this sentence is the sentiment that the human perspective of coastal oceans must be broadened (McCay 2012, this volume). To fully appreciate this perspective, one might consider that the human species has been part of the marine ecosystem in the Gulf of Maine region for 14 centuries and integral to changes in the ecosystem for nearly four centuries.

"The goal of ecosystem-based management is to maintain an ecosystem in a healthy, productive and resilient condition so that it can provide the services humans want and need." Underlying this sentence is the perspective that it is a responsibility to conserve and sustain ecological value in order for humans to continue to benefit from ecosystems. Likewise, we need to acknowledge the conflict between human wants and needs and ecosystem integrity (Arkema and Samhouri 2012; Hale and Westhead 2012; both this volume).

"Ecosystem-based management differs from current approaches that usually focus on a single species, sector, activity or concern; it considers the cumulative impacts of different sectors." Implicit in this sentence is the need for dialogue and deliberation among sectors, and reflection within sectors, about impacts to ecosystem goods and services. Both as a scientific community and larger society, we need to see human activity as part of the natural system while we seek to modify our behavior to minimize impacts. As the jargon and terminology used to describe EBM is dissected, more explicit meanings emerge. In short, the definition of EBM is calling for individuals to take and hold three perspectives at the same time: a holistic perspective on the marine environment, an introspective perspective on one's responsibility to it, and a reflective perspective on one's actions and interactions in relation to others.

With the plurality of meaning inherent in the EBM definition comes an even greater diversity of meaning when other audiences and EBM stakeholders are considered, given their current knowledge, values, and opinions. The underlying components of EBM must be articulated in ways that speak to this plurality. Scientists and communication professionals must become partners in relaying messages about ecosystem interconnections to decision makers and others, since what is coherent to one individual may not be identifiable to another (Bohm 1996). Scientists and communication professionals must find ways to create shared meaning among themselves and their audiences, thus creating a strategic alliance (Yankelovich 1999). Pielke's honest broker is a good example of this, in which scientists articulate and clarify the complexity of meaning for decision makers, enabling them to envision the choices/alternatives and determine which most suits their interests and values or those of their organization or constituents (Pielke 2007). The same kinds of issues partnerships and strategic alliances are needed between managers and communications professionals so that messages about management issues and challenges are conveyed successfully to scientists, creating shared meaning and closing the feedback loop for adaptive management.

The EBM entails engaging stakeholders with disparate interests in very different ways than sector-based management systems. In a sector-based approach, there is a tendency for people to share an implicit background of assumptions, values, expectations, and routines for decision making; with an EBM approach, none of the above is necessarily true. In fact, the lack of shared or taken-for-granted background is essential to robust EBM, which makes the process less stable and more dynamic. We can no longer look at one aspect of human impact on the marine environment without looking at it in relation to others. This requires engagement and coordination across sectors, which adds an additional layer of complexity to an already complex decision-making process and adds new communication demands for those trying to implement it. It may be necessary to provide more explicit communication to diverse audiences about trade-offs among a set of management options, for example.

Although more demanding, the process of communication across sectors (and interests) provides for more inclusivity and transparency. The rationale for one approach over another must be communicated with greater clarity and rise above the routine jargon and inherent bias in a single-sector process. These trade-offs, in essence, are nothing more than the complex and messy world of human compromise (i.e., politics). Scientists often find themselves in the unique position of needing to explicitly communicate the bargaining chips in management and policy processes.

In considering how scientific findings can help to articulate the EBM perspective, it must be considered how the information is being communicated, to whom, and for what purpose. As trusted sources of information, scientists have the unique responsibility of communicating the ways in which humans impact the ecosystems in which they live and how the structure and functioning of the ecosystem influences human behavior and perception. Communication that is two-way, inclusive, articulate, and consistent among messengers (i.e., scientists) is a critical but often forgotten objective for enabling a more holistic approach to management. Science, when presented in audience-appropriate ways, has the capacity to bridge disparate viewpoints and enable change. These points illustrate how strategic communication is defined in relation to EBM. Similar to the thought, effort, and resources devoted to selecting ecological indicators to measure the integrity of the system, we must identify ways of measuring the clarity and appropriateness of the scientific messages received by policy makers and managers and the public conversation that ultimately ensues (Jasanoff 2010).

Strategic communication conveys deliberate message(s) through the most suitable media to the designated audience(s) at the appropriate times to contribute to, and achieve, the desired long-term effect(s). In other words, one should create a specific plan to bring about a desired outcome and the identification of an influential message and/or methodology (Cox 2010). Strategic communication also involves commitment to a long-term plan of delivering consistent messages that resonate across diverse sectors, cultures, and audiences that is integrated vertically and horizontally across all those involved in the goals of a core mission (DoD SC 2008). The advantages of this approach are many when applied to science communication, both within the Gulf of Maine research community and between scientists and decision makers. However, it is also stressed that strategic communication need not only be about achieving desired outcomes, but, in addition, creating communication processes which feel legitimate to all interests involved. Next, we discuss how the most recent 2009 Gulf of Maine Symposium points to a fundamental need for strategic communication for EBM in the Gulf of Maine to strengthen both the way science is conveyed and how scientists and communication professionals build stronger alliances with their audiences.

Focusing a Strategic Communication Lens on the Gulf of Maine

In 2007, 75 scientists, policy makers, and managers from throughout the Gulf of Maine watershed gathered to set a regional agenda for applying science to policy and management processes (www.gulfofmaine.org/ebm/meeting2007). One call to action was the need to better connect scientists and communication professionals to actively catalyze and support the development of science to address marine conservation policy needs. Scientists specifically asked for communication tools, networks, ideas, and support to ensure that their research results be made accessible and relevant to a wide range of audiences. They wanted venues to be able to discuss their work with other scientists and to connect with those who could provide outreach and education on the relevance of their work to EBM. To respond to that need, a regional EBM communication work group was formed—a network of professionals dedicated to interpreting, synthesizing, and translating scientific findings with credibility and respect—including the authors of this paper.

In preparation for the 2009 Gulf of Maine Symposium, we surveyed symposium presenters to capture participants' perspectives on interactions between science and policy/management in the Gulf of Maine region (see Appendix A). Our goal was to synthesize the results into a snapshot of EBM communication successes and challenges. We were particularly interested in the perspectives of symposium participants because many find themselves in positions in which they are asked to provide a unique, credible, and trustworthy source of scientific information, and/or they are users of such information. The survey asked respondents to describe the relevance of their work to EBM; provide examples of how they have attempted to link science, policy, and/or management; and share their ideas for improving that linkage.

Results of the 2009 Gulf of Maine Symposium Presenter Survey

An online, open-ended survey was sent to 108 individuals presenting their information via presentation, poster, or technical session. Fifty-eight responded to basic questions that asked about their professional background, the relevance of their work to EBM, examples of how they bridged science with education and/or science with policy, and challenges and recommendations in communicating science to nonscientists. Last, participants were asked to indicate how best to achieve a regional, coordinated strategy for communicating ecosystem science. The following describes six key observations that we believe speak to the infrastructure already in place to truly develop a regional strategic communication plan.

Observation #1: Gulf of Maine Regional Networks Are in Place

All respondents indicated an affiliation with a regional network of researchers, managers, or policy makers, such as the Gulf of Maine Census of Marine Life project (42%), Regional Association for Research in the Gulf of Maine (42%), and the Gulf of Maine Council on the Marine Environment (26%). Forty-four percent indicated affiliation with large data networks such as the Northeastern Regional Association of Coastal Ocean Observing Systems and its partners or established subregional networks such as the Bay of Fundy Ecosystem Partnership. A smaller percentage of respondents were affiliated with networks dedicated to communication, such as Communication Partnership for Science and the Sea (16%), Canadian Healthy Oceans Network (13%), and the New England Ocean Science Education Collaborative (11%). While these affiliations might prove useful and strategic for the individual respondents, the affiliations alone do not lead to more strategic communication toward EBM. In addition, the initiatives and projects listed above continually struggle to better integrate their membership, goals, objectives, and resources.

Observation #2: Promise of Research to Advance Ecosystem Approaches Is in Place

Fifty-one respondents illustrated the promise of their work to advance understanding of the Gulf of Maine ecosystem. Although the responses are too diverse to summarize here, nearly all indicated a dedicated commitment toward ecosystem relationships, data sharing, and integrated approaches to research and management. It is important to differentiate, however, between a scientific understanding of ecosystems and the conceptual awareness one must have to integrate that knowledge into an approach to management action.

Observation #3: Activities for Bringing Science to Future Decision Makers Are in Place at the Individual and Local Scales

Thirty-two respondents offered examples of how they have bridged science and education through their work in higher education, K–12 classrooms, informal education projects, and science centers. Although not a focus of this paper, the responses indicate that most education efforts are done at the local level and developed through one-on-one relationships and short-term interactions between scientists and their audience.

Observation #4: The Building Blocks for Interpreting and Translating Science to Policy Makers and Managers Are in Place

Thirty-seven respondents replied to a question about bridging science and policy. The answers varied, with many offering specific or local examples, several commenting that they could not say with certainty but believe their work has contributed to policy changes, a few "awaiting policy changes," and five respondents responding "none to date." Those who offered successful examples demonstrated elements of strategic communication (i.e., direct two-way dialogue between scientists and decision makers, synthesis of research results within and across disciplines, and the meaningful visualization of data over time and space).

Observation #5: A Coordinated and Consistent Messaging Approach Is Needed

Thirty respondents each identified one challenge and one recommendation to bridge gaps between scientists and nonscientists. Time and funding to work toward these efforts was mentioned, but the majority indicated the need for more clarity and consistency of messages, better coordination and delivery of information, and deliberate opportunities to create knowledge and build consensus among stakeholders, an important attribute of an ecosystem approach. The significance of these responses was not that decision makers need more information, but rather, a more coordinated effort among scientists is needed to communicate the information in more consistent and user-relevant ways.

Observation #6: A Regional Strategic Communication Plan for Ecosystem-Based Management Could Be Well Received

In the last part of the survey, respondents were asked to indicate the value of nine proposed communication strategies toward developing a regional coordinated communication plan, and all 58 responded. These results indicate that for those responding to our survey, science communication to all sectors is valued, but pathways bridging science to policy are more valued over pathways between science and education, both formal and informal. The top five preferred pathways point to a regional need for strategic—not more—communication and include (1) fostering scientist/decision-maker relationships; (2) promoting and strengthening networks between scientists, educators, and decision makers; (3) collaborating to share resources; (4) improving communication between disciplines; and (5) synthesizing research for multiple audiences.

Being strategic about communication means intentionally bringing scientific research into policy making, management, and the public discourse in ways that are timely and relevant. Since the two go hand in hand, the lack of an overarching research and management vision for the Gulf of Maine region impedes the articulation of a coordinated, consistent approach to communications to policy makers and managers. Disjointed, inconsistent messaging among a range of sources could greatly hinder this audience's understanding and ability to act on issues needing an ecosystem approach. Next, we use the findings of the survey to suggest ways to use strategic communication to enhance regional coordination for EBM in the Gulf of Maine.

Strategic Communication Recommendations

Based on the evolution of coastal ocean management in the region, and the recent innovative management strategies at the state and regional levels, it is believed that the Gulf of Maine is poised to promote and institutionalize interactive, transparent, and inclusive communication among scientists and decision makers.

Strategic science communication about EBM is not about needing more—more information, more tools, more collaboration, or more stakeholder engagement. Instead, EBM communication must be strategically and intentionally designed with scientists, managers, and communication professionals in partnership. It is only through strategy, intentionality, and consistency that people will be able to mentally take stock of the social, economic, and ecological drivers affecting our coastal ocean (DeLauer 2009; McGuigan et al. 2009). Based on the top five strategies listed above, we make the following recommendations toward a more comprehensive and coordinated strategic communications plan for policy-maker and manager audiences in the region and support our recommendations with quotes and examples from our survey and literature within the social sciences. We are not implying that the following recommendations are not already being employed in some capacity. Nor do we suggest that strategic communication is easy or straightforward and that engaging in any of the following practices will ensure knowledge will be co-constructed and/or meaningful to the audience. Rather, we highlight these elements as essential starting points of a regional strategic (interactive) communication plan.

Recommendation #1: Nurture Scientist/Decision Maker One-on-One Relationships

Identify the problem at hand together.—Scientists and decision makers should engage in a two-way dialogue to try to understand the issue from each other's perspective. What kinds of decisions are policymakers and managers trying to make and why? To whom do they feel most responsible? How do they articulate their values, concerns, and struggles in choosing one course of action over another? What do scientists see as the most critical environmental problems and the drivers, and how does that relate to what managers see? Two-way (at least) communication allows for frequent questions and clarifications to wade through scientific jargon and complexity. It also offers a view of someone else's reality, a key aspect of strategic communications. In general, scientists seek answers to better understand "how the world works," whereas decision makers engage the issue from the perspective of taking action—"what should we do" (Baron 2010). We believe that communication that is interactive, inclusive, and considers the context in which exchange of ideas takes place offers a promising way toward the acquisition and use of scientific information and knowledge (Lewenstein 1995).

However, one must also be considerate of the different ways in which individuals value, acquire, reason about, and use scientific knowledge. One might look to the literature on learning theory to better understand this phenomenon. For example, King and Kitchener's (1994) reflective judgment model illustrates the diverse ways individuals digest data and theory within a particular context and form a judgment. Their model groups problems into well-structured (e.g., single correct answers such as 2 + 2 = 4) and ill-structured problems that pose a range of potential alternative solutions with a high degree of uncertainty (e.g., mitigating climate change impact on coasts). King and Kitchener study the development of "epistemic cognition" to illuminate the ways in which individuals evaluate knowledge claims, differentiate between ill-structured and well-structured problems, and ultimately define their stance on a particular issue (in our case, the science underlying ecosystem-based management). King and Kitchener are not alone. There is a growing body of researchers from different branches of psychology and sociology who are interested in understanding environmental decision making, especially as the world's problems become more complex and urgent. Despite that, there are few, if any, sociologists and cognitive scientists engaged in regional EBM discussions.

Employ sequencing.— It is easy for any expert to forget that the rest of the world may not be up to speed on the latest findings of their field. Scientists, particularly those developing decision-support tools, should ensure that they communicate the basics behind a complex topic such as EBM before explaining models, maps, and other spatial analysis resources, as needed. It is critical that decision makers first understand how an ecosystem approach to management is relevant to their agenda. Scientists should frame EBM in a way that helps managers see their dilemma as part of a bigger picture that includes a range of possible solutions and concrete examples.

One respondent reflects on the inadequacies of current communication channels to reach the broadest range of stakeholders.

One recommendation is to expand the forms of Web-based outreach and communication to encompass a broader array of outlets, including somewhat unconventional (to us) approaches such as Twitter and Facebook.

This poses two challenges. As new media and social media gain popularity, regional scientists should be equipped with a basic working knowledge of these technologies. In addition, these technological interfaces must be designed in ways that employ sequencing (i.e., communication that progressively introduces new and more complex ideas and that provides a context from which to understand the information being presented). This points to another potential area for collaboration among scientists and communication professionals.

Recommendation #2: Promote and Strengthen Networks between Scientists, Decision Makers, and Educators

Consider the sphere of influence.— The survey results call for expanded venues, not only for information exchange, but for real dialogue among prime circles of influence—science, policy, and education. Although science communication to decision makers is the focus of this paper, coordinated approaches to educators about EBM concepts are critical to achieving and maintaining public support for funding and policy. We consider educators in the broadest sense, including formal (e.g., higher education and schools), informal (e.g., aquaria, science centers), and public education through major media sources (e.g., journalism, public broadcasting). Scientist respondents cited several ways in which they engage with formal and informal education as part of National Science Foundation broader impacts component. Scientists also work with local teachers to enlist students in monitoring fieldwork, and some make guest appearances in middle school, junior high and high school science classes. None of the respondents pointed to a good working relationship with media professionals, and they expressed the need to enhance these connections.

Scientist respondents shared many examples of their work being applied to policy and management decisions. Mechanisms include participating on science advisory councils, attending public meetings as an invited expert, producing white papers for managers, and synthesizing basic science into management recommendations and protocols. A respondent from Canada shared that his "research has led to the closure of softshell clam harvesting during spawning season in the Kouchibouguac National Park, New Brunswick, in order to improve the recruitment success of the species." Similarly, a scientist from the United States notes that "early work helped the state of New Hampshire to change the bacterial indicators they use to classify shellfish and swimming beach waters. Recent work is helping to establish new guidelines for toxic contaminants and naturally occurring human pathogens."

Based on the responses, there is still a need for more substantive interactions among scientists and media professionals. There are several opportunities for scientists to participate in communication training aimed at making science matter to journalists (e.g., COMPASS [Communication Partnership for Science and the Sea]), yet scientists are not always aware of the opportunities, and sustained funding to maintain these types of capacity-building initiatives remains a challenge.

Establish more formal mechanisms for collaboration.— There are several research organizations in the Gulf of Maine (e.g., Regional Association for Research in the Gulf of Maine, Northeastern Regional Association for Coastal Ocean Observing Systems), as well as communications networks (Communication Partnership for Science and the Sea, New England Ocean Science Education Collaborative) that have already established mechanisms for data and information sharing, science synthesis, and translation. Partnerships among scientists, decision makers, and these organizations can strive to become more formalized and more inclusive with ocean users.

Recommendation #3: Collaborate to Share and/or Obtain Resources

Fifty-fifty split.—Scientists and communication professionals can use regional networks to build partnerships that enhance outcomes, minimize redundancy, and further regional goals. These partnerships can support one another in writing collaborative grant proposals that seek to give as much effort to the communication of scientific results as the scientific process and discovery. At the same time, the process of developing such proposals will create an opportunity for knowledge to be created, not just at the individual level, but also at the group and regional level.

Recommendation #4: Improve Communication between Disciplines to Achieve More Consistent Messaging for Multiple Audiences

Truly engage the human dimension.—In all levels of interaction, natural scientists and communication professionals can benefit by engaging those who study human populations. In particular, psychologists and social psychologists offer a window on the challenges underlying individual and group processing of complex scientific information. We need to look inward, at ourselves, as the key for understanding current environmental problems and solutions such as climate change (CRED 2009). To be effective at strategic communication, we must engage scientists who study human behavior who can offer best practices for connecting with diverse audiences. For example, researchers affiliated with the American Psychological Association recently issued a report entitled "Psychology and Global Climate Change: Addressing a Multi-Faceted Phenomenon and Set of Challenges" (APA 2010). This report provides a set of recommendations and a call to action for psychologists of different areas of expertise to engage in the issue of climate change.

Explore other disciplines.—This past year, many psychological reports have been released that inform our approach (Thaler and Sunstein 2008; APA 2009; CRED 2009). Natural and social scientists need to find ways to connect more consistently if we are to fully understand the human dimensions of EBM, though these venues are not always available or convenient. At the very least, a good understanding of current literature can help scientists better understand the mindsets of their audiences so that their messages are framed appropriately. For example, one can look to current research on mental models—ways in which our brains organize new and existing information (Moxnes and Saysel 2009). Current research looking at decision makers' mental models of EBM suggests that managers can recite the tenets of the concept quite well, yet embodying those tenets in practice is quite difficult; there is a disconnect between descriptively understanding EBM characteristics and conceptually understanding its complexity (DeLauer 2009). This information can equip scientists with a variety of ways to frame their messages once they gauge the extent to which their audience can integrate the scientific complexity. It is crucial that those communicating these messages help audiences better perceive which facts are most relevant and how social and natural variables interact to influence the larger system. The goal is to help audiences better gauge and interpret facts that make the greatest difference in maintaining coastal and ocean ecosystems through a better understanding of what is already known, what is valued, and what substantive information is missing or being misinterpreted (Morgan et al. 2002; Paolisso 2007).

Synthesize research for multiple audiences.—Scientists and communication professionals can work together to develop core messages about EBM for use with policymakers and managers. These audiences are bombarded with messages about EBM-related constructs such as marine spatial planning, ocean zoning, and ecosystem services. A coordinated strategic communication plan can link research, policy, and management goals to these constructs in consistent ways to alleviate confusion caused by mixed messages. Consistency also has a greater potential to engender trust and credibility from those outside these sectors. In developing a regional plan, we recommend a visioning process that identifies the core social, economic, and ecological messages of EBM and offers messages and delivery techniques targeted to the diverse range of decision makers. For example, a visioning process might help identify how a changing natural environment affects policy makers' constituents and what they value most. It might help scientists clarify what aspects of ecosystem services science are most relevant to managers. It might help translate the concept of resilience in terms of both social and ecological change. It might enable the establishment of a baseline of best communication practices between scientists and decision makers and offer new insights about evolving that connection.

Discussion

The Gulf of Maine faces an onslaught of short- and long-term impacts from climate change, pollution, overexploitation, and invasive species while, at the same time, funding for research and management is constrained by competing national priorities. Yet, we stand at a fortuitous junction, having both the historical insight from past Gulf of Maine symposia combined with the impetus and infrastructure needed to create a shift toward ecosystem-based management. We believe that the region would greatly benefit through a joint effort among scientists, communication professionals, and decision makers to develop a comprehensive strategic communications plan that mirrors the findings of the research community, balances the constraints of the management community, and addresses the values and interests of the public. This paper is a call to action for regional science groups (e.g., RARGOM and Northeastern Regional Association of Coastal and Ocean Observing Systems) and communication experts (e.g., COMPASS and New England Ocean Science Education Collaborative) to continue this conversation. These recommendations are only a start toward lessening the complexity inherent in an EBM approach. They point to the need to alleviate the uncertainty involved in moving away from sector-based management by making explicit the importance of meaningful human interaction and respectful exchange of information among scientists, decision makers, and communications professionals.

References

APA (American Psychological Association). 2009. Psychology and global climate change: Addressing a multi-faceted phenomenon and set of challenges. A report by the APA Task Force on the Interface Between Psychology and Global Climate Change. APA, Washington, D.C.

Arkema, K. K., and J. F. Samhouri. 2012. Linking ecosystem health and services to inform marine ecosystem-based management. Pages 9–25 *in* R. L. Stephenson, J. H. Annala, J. A. Runge, and M. Hall-Arber, editors. Advancing an ecosystem approach in the Gulf of Maine. American Fisheries Society, Symposium 79, Bethesda, Maryland.

Ballard, S., M. Connor, and D. Gordon, editors. 1997. Mechanisms for improving the integration of science and management in decisions affecting the environmental quality of the Gulf of Maine. Regional Association for Research on the Gulf of Maine 97:2.

Baron, N. 2010. Escape from the ivory tower: a guide to making your science matter. Island Press, Washington, D.C.

Bohm, D. 1996. On dialogue. Routledge, London.

Cox, R. 2010. Environmental communication and the public sphere, 2nd edition. Sage, Thousand Oaks, California.

CRED (Center for Research on Environmental Decisions). 2009. The psychology of climate change communication: a guide for scientists, journalists, educators, political aides, and the interested public. CRED, New York.

DeLauer, V. 2009 The mental demands of marine ecosystem-based management: a constructive developmental lens. Doctoral dissertation, University of New Hampshire, Durham.

Ernst, M. 2004. A survey of coastal managers' science and technology needs prompts a retrospective look at science-based management in the Gulf of Maine. Gulf of Maine Council on the Marine Environment. Available: www.gulfofmaine.org/council/publications/coastalmanagerssurveyreport.pdf. (January 2012)

Feurt, C. B. 2012. Collaborative learning strategies to overcome barriers to ecosystem management in coastal watersheds of the Gulf of Maine. Pages 73–88 *in* R. L. Stephenson, J. H. Annala, J. A. Runge, and M. Hall-Arber, editors. Advancing an ecosystem approach in the Gulf of Maine. American Fisheries Society, Symposium 79, Bethesda, Maryland.

Grabowski, J. H., and T. Hart. 2012. Seafloor mapping for ecosystem management in the Gulf of Maine. Pages 273–280 *in* R. L. Stephenson, J. H. Annala, J. A. Runge, and M. Hall-Arber, editors. Advancing an ecosystem approach in the Gulf of Maine. American Fisheries Society, Symposium 79, Bethesda, Maryland.

Green, K. C., and S. W. Gilbert. 1995. Great expectations: content, communications, productivity, and the role of information technology in higher education. Change 27(2):8–18.

Hale, S. S. 2012. Spatial patterns of subtidal benthic invertebrates and environmental factors in the nearshore Gulf of Maine. Pages 167–183 *in* R. L. Stephenson, J. H. Annala, J. A. Runge, and M. Hall-Arber, editors. Advancing an ecosystem approach in the Gulf of Maine. American Fisheries Society, Symposium 79, Bethesda, Maryland.

Hale, S. S., and M. Westhead. 2012. Ecosystem services in the Gulf of Maine. Pages 1–8 *in* R. L. Stephenson, J. H. Annala, J. A. Runge, and M. Hall-Arber, editors. Advancing an ecosystem approach in the Gulf of Maine. American Fisheries Society, Symposium 79, Bethesda, Maryland.

Hall-Arber, M., J. Pederson, and P. Wells. 2012. Anthropogenic and external influences on the Gulf of Maine: workshop summary. Pages 235–241 *in* R. L. Stephenson, J. H. Annala, J. A. Runge, and M. Hall-Arber, editors. Advancing an ecosystem approach in the Gulf of Maine. American Fisheries Society, Symposium 79, Bethesda, Maryland.

Hare, J. A., and J. Kane. 2012. Zooplankton of the Gulf of Maine—a changing perspective. Pages 115–137 *in* R. L. Stephenson, J. H. Annala, J. A. Runge, and M. Hall-Arber, editors. Advancing an ecosystem approach in the Gulf of Maine. American Fisheries Society, Symposium 79, Bethesda, Maryland.

Jasanoff, S. 2010. Testing time for climate science. Science (May 7):328.

King, P. M., and K. S. Kitchener. 1994. Developing reflective judgment: understanding and promoting intellectual growth and critical thinking in adolescents and adults. Jossey-Bass Publishers, San Francisco.

Lawton, P., L. S. Incze, and S. L. Ellis. 2012. Representation of biodiversity knowledge within ecosystem-based management approaches for the Gulf of Maine. Pages 375–391 *in* R. L. Stephenson, J. H. Annala, J. A. Runge, and M. Hall-Arber, editors. Advancing an ecosystem approach in the Gulf of Maine. American Fisheries Society, Symposium 79, Bethesda, Maryland.

Lewenstein, B. V. 1995. Science and the media. Pages 343–359 *in* S. Jasanoff, G. E. Markle, and J. C. Peterson, editors. Handbook of science and technology studies. Sage Publications, Thousand Oaks, California.

McCay, B. J. 2012. Peopling the marine ecosystem. Pages 27–33 *in* R. L. Stephenson, J. H. Annala, J. A. Runge, and M. Hall-Arber, editors. Advancing an ecosystem approach in the Gulf of Maine. American Fisheries Society, Symposium 79, Bethesda, Maryland.

McLeod, K. L., J. Lubchenco, S. R. Palumbi, and A. A. Rosenberg. 2005. Scientific consensus statement on marine ecosystem-based management. Available: www.compassonline.org/sites/all/files/document_files/EBM_Consensus_Statement_v12.pdf. (January 2012)

McGuigan, R., P. Popp, and V. DeLauer. 2009. The complexity of climate change and conflict resolution: are we up to the challenge? Conflict Resolution (Fall):16–19.

Morgan, M. G., B. Fischhoff, A. Bostrom, and C. J. Atman. 2002. Risk communication: a mental models approach. Cambridge University Press, Cambridge, UK.

Moxnes, E., and A. K. Saysel. 2009. Misperceptions of global climate change: information policies. Climatic Change 93:15–37.

New England Aquarium. 1999. Out of the fog: furthering the establishment of an electronic environmental information exchange for the Gulf of Maine. New England Aquarium, Aquatic Forum Series Report 99:1, Boston.

NRC (National Research Council). 1995. Improving interactions between coastal science and policy. National Academy Press, Washington, D.C.

Paolisso, M. 2007. Cultural models and cultural consensus of Chesapeake Bay blue crab and oyster fisheries. National Association of Practicing Anthropology Bulletin 28:123–133.

Pielke, R. A. 2007. The honest broker: making sense of science in policy and politics. Cambridge University Press, Cambridge, UK.

Runge, J. A. 2012. Observations and analysis of change in the Gulf of Maine: present status and future directions. Pages 103–114 *in* R. L. Stephenson, J. H. Annala, J. A. Runge, and M. Hall-Arber, editors. Advancing an ecosystem approach in the Gulf of Maine. American Fisheries Society, Symposium 79, Bethesda, Maryland.

Smith, P. C., N. R. Pettigrew, P. Yeats, D. W. Townsend, and G. Han. 2012. Regime shift in the Gulf of Maine. Pages 185–203 *in* R. L. Stephenson, J. H. Annala, J. A. Runge, and M. Hall-Arber, editors. Advancing an ecosystem approach in the Gulf of Maine. American Fisheries Society, Symposium 79, Bethesda, Maryland.

Taylor, P. H. and V. DeLauer. 2009. EBM roadmap. Waterview Consulting, COMPASS and EBM Tools Network. Available: www.ebmtools.org/?q=roadmap/overview.html. (September 2010).

Thaler, R. H., and C. R. Sunstein. 2008 Nudge: improving decisions about health, wealth, and happiness. Yale University Press, New Haven, Connecticut.

Treise, D., and M. R. Weigold. 2002. Advancing science communication: a survey of science communicators. Science Communication 23:310–322.

Wilsdon, J., B. Wynne, and J. Stilgoe. 2005. The public value of science or how to ensure that science really matters. Demos, London.

Yankelovich, D. 1999. The magic of dialogue: transforming conflict into cooperation. Simon and Schuster, New York.

Appendix I: Survey Questions

Name

Title affiliation

Network affiliations

1. Please provide a brief summary of your research and its contribution to our understanding the Gulf of Maine (i.e., two or three sentences explaining why your research is important to our region).
2. Do you have an example of how you have bridged science and education? (e.g., a successful project or partnership with a science education center or a school district).
3. To help move our communication strategy forward, please identify one challenge and one recommendation in bridging these gaps.
4. Please indicate the value of the following ideas toward developing a regional strategy for communication of ecosystem science:
    - Promote and strengthen the networks between researchers, educators, and policymakers in the region
    - Summarize research for multiple audiences through new media
    - Develop a shared portal to multiple Web sites with a general name (e.g., "Discover the Gulf of Maine")
    - Collaborate to share resources or obtain funding
    - Improve communication between disciplines (within marine sciences and with related fields, such as archaeology, historical ecology, and local knowledge)
    - Foster scientist/policy-maker relationships (e.g. briefings or policy forums)
    - Support scientist/educator relationships and (e.g. research opportunities for educators or on-line collaboration)
    - Develop resources for K–12 (formal education)
    - Develop resources for science centers, aquaria, and Web sites (informal education)
    - Other

American Fisheries Society Symposium 79:103–114, 2012

Observations and Analysis of Change in the Gulf of Maine: Present Status and Future Directions

JEFFREY A. RUNGE*
School of Marine Sciences, University of Maine, Gulf of Maine Research Institute 350 Commercial Street, Portland, Maine 04101, USA

Abstract.—This chapter summarizes contributions to a theme session of the 2009 Gulf of Maine Science Symposium held in St. Andrew's, New Brunswick in October, 2009. The session highlighted the present status of science required to observe, interpret, and predict changes in the Gulf of Maine ecosystem in the context of strategies for regional implementation of an ecosystem approach to management (EAM). Perspectives on present ecosystem approaches to Gulf of Maine fisheries management contrast the integrated ecosystem assessment approach by the U.S. National Oceanic and Atmospheric Administration, with the more incremental advancement to EAM based on traditional fisheries management practices undertaken by Fisheries and Oceans Canada. A section on contributions from the broader research community provides perspectives on observations and different approaches to analysis, including coupled physical biological modeling as a tool for the integration, interpretation, and prediction of multidisciplinary environmental data. The Atlantic Zonal Monitoring Program has established an observing system for physical and biological characteristics of Canadian coastal waters, and NERACOOS (the Northeast Regional Association of Coastal Ocean Observing Systems) is developing infrastructure for coordination of U.S. regional observing activities. A common theme is the need for more sustained time series of critical physical and biological variables that document change, especially in nearshore, coastal, and benthic habitats. Additionally, there is a need to development and maintain bridges to transfer new research knowledge, understanding, and analysis tools to the state, provincial, and federal agencies and fisheries management councils where EAM will be implemented.

Introduction

The concept of an ecosystem approach to management (EAM) of marine ecosystems has evolved over at least the past 50 years (Sissenwine and Murawski 2004) and has been widely adopted internationally. Important international milestones, with respect to the management of fisheries in particular, include the Reykjavik Declaration on Responsible Fisheries in the Marine Ecosystem (2001) and an FAO technical document (FAO 2003), which provides guidelines for making an ecosystem approach to fisheries management (or ecosystem-based fisheries management, EBFM; these terms will be used interchangeably here) operational. Implementation of EAMs proceeds slowly, owing to the complexity of a process that involves government agencies, native tribes, industry and resource users, community and nongovernmental organization interest groups, academic researchers, and the general public (Holliday and Gautam 2005). Each nation must meet this challenge in its own way and also work with neighboring countries to establish effective management strategies for ecosystems that transcend international boundaries. In the Gulf of Maine, the governance authority for EAM is different between the United States and Canada (Westhead 2012, this volume); while Fisheries

* Corresponding author: jeffrey.runge@maine.edu

and Oceans Canada (DFO) has the lead role in managing the effects of human activities in marine waters, the authority in the United States is distributed among a variety of agencies and levels of government, requiring cross-sector collaboration (Holliday and Gautam 2005).

The theme session entitled "Observations, Data Collection, Analyses and Tools for an Ecosystem Approach to Management in the Gulf of Maine: Present Status and Future Directions," represents a contribution to the incremental development of EAM in the Gulf of Maine. The session highlighted the present status of science required to observe, interpret, and predict changes in the Gulf of Maine ecosystem in the context of strategies for regional EAM implementation. The range of subjects applicable to this theme is evident in the diversity of presentations. The theme's plenary session provided an unusual opportunity for Gulf of Maine experts from across disciplines to discuss issues related to the observations and analyses needed for implementation of EAM. While no single session and gathering of experts can, by itself, provide comprehensive answers to the questions posed in this symposium, this chapter, nevertheless, offers a review of the theme presentations and plenary discussion that addresses, at least in part, the overarching symposium goal to synthesize and advance the ecosystem research that supports the future management of the Gulf of Maine.

The chapter is organized into three sections. First, a section on ecosystem approaches to fisheries management focuses on perspectives and science conducted at the respective federal fisheries science centers. A section on contributions from the broader research community provides perspectives on observations and different approaches to the analysis and interpretation of observing data, with a focus on nearshore, coastal, and benthic habitats. A common theme in both sections is the need for long time series of critical physical and biological variables that document change due to both climate and anthropogenic forcing. In the final section, the symposium goal is addressed in the form of a synthesis of responses to three questions posed in the plenary discussion. How far have we come? How well situated is the science to inform management? What must we do next?

Ecosystem Approaches to Fisheries Management: Perspectives from the Federal Fisheries Science Centers

In light of the need to implement ecosystem-based management for U.S. marine ecosystems, the National Oceanic and Atmospheric Administration (NOAA) has developed the concept of integrated ecosystem assessments (IEAs) as a formal structure for integrating multidisciplinary scientific analysis into the management process (Levin et al. 2009). The IEAs inform resource managers about the cumulative impacts of human activities and environmental forcings on marine ecosystems and thereby contribute to management decisions with regard to defined ecosystem and fisheries objectives. The effectiveness of management actions are then monitored in the ecosystem to provide feedback for further management actions. The effectiveness of the IEA depends, in part, on the development of a set of critical indicators that capture the ecosystem state and its responses to change. M. Fogarty et al. (2012, this volume) presented the results of Northeast Fisheries Science Center (NEFSC) efforts to provide an indicator-based assessment of the northeast U.S. Continental Shelf large marine ecosystem. A driver-pressure-state-impact-response framework (OECD 2003) is employed to assess patterns of change in the ecosystem, as well as management responses to impacts of this change. From a suite of more than 80 variables, the NEFSC determined a set of indicators representing climate and human drivers, pressures on the physical system (e.g., temperature change) resulting from the climate drivers, pressures resulting from human activities (e.g., fishing effort and catch), ecosystem state (e.g., phytoplankton and zooplankton biomass indices, and a suite of indicators of living marine resources), and impacts (e.g., amount and value of catch). Major fisheries management interventions over the past several decades are reviewed to illustrate management responses and subsequent ecosystem reactions. Methods of analysis included a minimum/maximum autocorrelation factor analysis for detection of common trends, chronological cluster analysis for identification of possible ecosystem shifts, and multivariate multiple regression analysis for examination of relationships among drivers and pressures and among pressures and ecosystem states. The results of the analysis show large-scale perturbations since the 1960s, particularly in the 1980s, due to both human activities and climate. A major conclusion from this analysis is that top-down effects due to fishing appear to be the dominant influence on living marine resources.

In addition to regulatory authority over fisheries, DFO was given, by the Oceans Act in 1997, the lead role in managing the effects of all human activities in marine waters. S. Gavaris (Fisheries and Oceans Canada, St. Andrews Biological Station) discussed the steps that DFO is taking to make an ecosystem approach operational in eastern Canadian waters. The DFO framework represents an incremental evolution to EAM from traditional fisheries management, with its focus on controlling exploitation of harvested resources. It is a pragmatic approach, building on what is familiar and drawing in the wealth of expertise developed for traditional management. Three broad conservation objectives have been identified: maintain productivity, preserve biodiversity, and protect habitat. In each case, the emphasis is on controlling human activity so that it does not cause unacceptable reduction or modification in any of these attributes, recognizing that there are other influencing forces over which there is no control. These three broad objectives are translated into a parsimonious suite of strategies to control associated pressures, for example, keep fishing mortality (a pressure) below a specified reference point, evaluated by the measurement of impacted ecosystem attributes (such as yield, biomass, and recruitment). The joint effects of multiple pressures are considered, and management measures (tactics) are undertaken to control the pressures. Strategies may be adjusted to take into account for changes (e.g., climate-driven) unrelated to fishing, even though it may not be possible to do anything about those changes. A pilot test of this framework, at present limited to fisheries (the dominant activity), is being undertaken on Georges Bank. After establishment of a governance structure that lends itself to effective governance in the context of EAM, a matrix of strategies, fishing activities, and key identified pressures within identified spatial management units, reference points, and impacted attributes to be assessed is assembled. This framework is then used to set the tactics to execute EAM in this region. A critical factor for successful implementation is the science for understanding relationships between a pressure and tactics and between a pressure and its attributes, and the external forcing that may cause these relationships to change over time.

In a complementary presentation on the use of resource monitoring programs to inform EBFM, J. Fisher (Fisheries and Oceans Canada, Bedford Institute of Oceanography) and R. Brown (National Marine Fisheries Service, Northeast Fisheries Science Center) highlighted the value of federal, state, and provincial trawl and acoustic surveys to the observation of trends in abundance of living resources. As these time series extend over years and decades, they become more and more valuable as sources of information to relate to environmental and anthropogenic drivers on the ecosystem. They also argued that functionally significant traits, such as body size, species distribution and trophic structure, may be as important as assessment of species diversity for detection and understanding of change.

As two examples of Canadian and U.S. federal collaboration in the use of trawl surveys to study forcing of the Gulf of Maine ecosystem, N. Shackell (Fisheries and Oceans Canada, Bedford Institute of Oceanography) and colleagues and J. Nye (U.S. Environmental Protection Agency) and colleagues reported on analysis of long-term trends in fish species and climate and ecosystem indicators. Using three different techniques for statistical analysis of multiple time series, Nye et al. (2010) looked for synchrony among four regions of the Gulf of Maine area in relative biomass trends in 19 fish species assessed by Canadian and U.S. fisheries surveys between 1960 and 2007. They also investigated correlation of fish biomass trends with total landings as an index of fishing pressure and four environmental indicators: mean Gulf of Maine area sea surface temperature (SST), the number of days with winds greater than 12.5 m/s (an index of storm events), the NAO index in winter, and the Atlantic multidecadal oscillation (AMO). They found coherence in biomass trends at some level in all fish species. There was coherent decline in relative biomass of five fish species and an increase in two species across all surveys during the time period. Correlated with these trends was an overall decline in the index of fishing pressure and an increase in the AMO index. The results indicate that both environment and fishing pressures influence populations and that responses are species-dependent. Shackell et al. (2012) analyzed correlation and trends among a broader range of large-scale and regional-scale climate drivers and biological indicator responses, including lower trophic level, macroinvertebrate biomass, fish biomass, and fish size/community structure. The results indicate a similar multivariate temporal trend occurring between 1980 and 1990, after which there were differences between northern (eastern and western Scotian Shelf) and southern (Georges Bank, Gulf of Maine) regions. The AMO was the only climate index correlated with the primary temporal trend across all ecosystems. There is no mechanistic understanding to support this correlation. Fishing pressure, combined with latitudinal gradients in environmental variables, is hypothesized to be the primary determinants of the observed ecosystem trends, especially in the higher trophic levels. Consistent results across regions and national borders strengthen the case that large-scale trends are real and should be factored in to EBFM.

Perspectives on Ecosystem Approaches to Management from the Broader Research Community

As would be expected, the broader research community, comprising experts from universities, state and provincial agencies, and the private sector, as well as federal scientists, contributed a diversity of science relevant to the concept of EAM.

A number of presentations reported on observations and analysis that are directly relevant to fisheries research and management. K. Stokesbury (University of Massachusetts Dartmouth) and colleagues reported on a video survey method for estimating the absolute number of sea scallops in closed and open areas on Georges Bank and in the Middle Atlantic Bight. The survey, conducted on fishing vessels annually since 2003, yields estimates of abundance and population characteristics of sea scallops, covering the sea scallop resource in northeastern U.S. waters. The video images also capture substrate characteristics and observations of other benthic taxa. A time series compilation of these data allows tracking of changes in population abundance and habitat structure for inclusion in analyses of impacts on the benthic ecosystem. C. Bergeron (University of Maine) and colleagues reported on development of a growth model of American lobster *Homarus americanus* that incorporates bottom temperature time series to account for spatial differences in growth trajectories, providing more realistic analysis of size at age in stock assessment models. N. Makris (Massachusetts Institute of Technology) and colleagues described new technological advances for detecting large herring schools, and their behavior, using a new low-frequency acoustic sensing system (ocean acoustic waveguide remote sensing). Research conducted in concert with the National Marine Fisheries Service's (NMFS) annual herring survey indicate that the development of vast herring shoals in the Gulf of Maine occur just prior to synchronized spawning. J. Becker and J. Kanwit (Maine Department of Marine Resources) presented a survey of bycatch of haddock *Melanogrammus aeglefinus* in the northeast U.S. herring fishery. These activities represent examples of state-supported or university-based research and university–industry collaborations that contribute directly to EAM for major regional fisheries.

J. R. Morrison (Northeastern Regional Association of Coastal Ocean Observing Systems [NERACOOS]) gave an overview of NERACOOS, one of the 11 regional associations in the United States whose function is to develop and serve as the coastal component of the Integrated Ocean Observing System. Established as an independent, nonprofit organization in 2008, NERACOOS currently sponsors observation, modeling, and outreach activities at five New England state universities, the Gulf of Maine Research Institute, the Woods Hole Oceanographic Institution, the Bedford Institute of Oceanography, NMFS, the Gulf of Maine Council, and two industry partners. It operates seven instrumented buoys in the Gulf of Maine, including two deep buoys located in the Northeast Channel, four coastal buoys that are widely spaced down the Maine and Massachusetts coasts, and a buoy that monitors water quality in waters off Boston. NERACOOS operates an HF RADAR network that provides data for estimation of surface currents that are used in the surface current forecasting system employed by the U.S. Coast Guard Search and Rescue program and form part of the National Surface Current Measurement program. NERACOOS supports development of the Inundation Forecast System of the Northeast Coastal Ocean Forecast System to provide warning of coastal flooding, to facilitate evacuation and other emergency measures, and to enable rationale planning for sustainable land use and potential impacts of climate-related sea level rise. Through a strategic planning process, NERACOOS has identified four focus areas that require the maintenance and future development of regional observation capacities: ocean and coastal ecosystem health, coastal hazards resilience, ocean energy planning and management, and maritime security/safety. These priority themes are aligned with regional coastal and ocean management priorities identified by the Northeast Regional Ocean Council, an advisory body consisting of representatives from the five New England coastal states and six federal government agencies, with which NERACOOS has a strategic partnership. Understanding and mitigating impacts of climate change is a central issue across all of the priority themes.

P. Smith et al. (2012, this volume) provided an example of NERACOOS-supported observations used to detect change in physical drivers of the Gulf of Maine ecosystem. At some point between 2000, when a previous mooring located in the Northeast Channel of the Gulf of Maine was removed, and 2004, when a new mooring was installed in the same area as part of what is now NERACOOS, a shift occurred in the flow of water in and out of the Gulf of Maine. Observations spanning the period from 1975 to 2000 had indicated mean inflow of shelf and slope water inshore on the Scotian Shelf off Cape Sable and on the eastern side of the Northeast Channel (at all depths), with mean outflow in the surface layer and at mid-depth on the western side of the Northeast Channel. Between 2004 and 2008, however, a much different current pattern was observed, in which there were sustained deep-layer outflows on the eastern side, especially in the winters of 2004–2005 and 2006–2007, when outflow currents also extended to the surface layer. Estimation of water transport into and out of the Gulf of Maine indicate reversal in net deep inflow transport and an increase in fresh Scotian Shelf inflow at Cape Sable by about 40%, representing a dramatic shift in sources of water for the Gulf of Maine ecosystem. This shift has potentially important implications for nutrient balance and primary and secondary production in the Gulf of Maine ecosystem.

The potential for acquisition of new physical and biological oceanographic data over a range of space and time scales warrants new approaches for assessing and forecasting climate change impacts on the Gulf of Maine marine ecosystem. C. Davis (Woods Hole Oceanographic Institution) and colleagues reported on the development of three-dimensional biological-physical models of the Gulf of Maine, initially for application to interpretation of physical and plankton data collected in association with the U.S. GLOBEC (Global Ocean Ecosystem Dynamics) Georges Bank/Northwest Atlantic Program. The models couple high resolution physical circulation models, providing hydrographic and circulation fields, to the modeled dynamics of nutrients and plankton, including the early life stages of fish and commercially important invertebrates (e.g., Ji et al. 2008a, 2008b, 2009; Kristiansen et al. 2009; Churchill et al. 2011). The results of these models yield insight and information on environmental conditions for recruitment and spatial connectivity among populations with planktonic stages (e.g., Runge et al. 2010). In general, these coupled multidisciplinary models can serve to integrate multiple data sets for interpretation and prediction of climate impacts on the Gulf of Maine, and the approach is applicable to a variety of management issues, including recruitment variability, connectivity among populations, timing of harmful algal blooms, biogeographic shifts of key native and invasive species, and nutrient and primary production dynamics in estuaries, coastal regions, and the deeper shelf sea.

Observation and analysis of change in the Gulf of Maine pelagic ecosystem, including implementation of a coupled physical-biological modeling approach to interpretation and forecast of climate change impacts, require collection of temporal and spatial data on abundance and diversity of plankton. Planktonic populations respond relatively quickly to environmental variability and therefore are sensitive indicators of change, for example, ecosystem shifts that have been detected in the Gulf of Maine in the late 1980s (e.g., Nye et al. 2010; Hare and Kane 2012, this volume). C. Johnson (Fisheries and Oceans Canada, Bedford Institute of Oceanography) and J. Hare (National Marine Fisheries Service, Northeast Fisheries Science Center) reviewed the status of plankton monitoring in the Gulf of Maine (Johnson and Hare 2012, this volume). They highlighted the need for consistent, repeated sampling for identification of changes in zooplankton abundance, community composition, and seasonal timing of production cycles (Ji et al. 2010). Of 18 zooplankton observing programs started since 1910, four are still ongoing at some level: the Continuous Plankton Recorder (since 1961), the National Marine Fisheries Service MARMAP and Econom programs (since 1977), the Massachusetts Water Resources Authority time series (since 1992), and the Canadian Atlantic Zonal Monitoring Program (AZMP; since 1998). To monitor change in seasonal production cycles and cover gaps in coastal time series, they recommend establishing monitoring using the same AZMP protocol at sentinel coastal sites in the Gulf of Maine.

Several contributions provided time series observations and approaches for monitoring and assessment of ecosystem components in the coastal Gulf of Maine, including nearshore bays and estuaries. C. E. Tilburg (University of New England) and colleagues presented results from the Saco River Coastal Observing System, designed to characterize the spatial and temporal variation in the Saco River Plume. N. Poulton (Bigelow Laboratory for Ocean Sciences) and colleagues presented results of weekly observations over 9 years (2001–2009) of phytoplankton, bacterioplankton, and eukaryotic heterotrophs in Boothbay Harbor. The results illustrate the potential of technological advances (flow cytometry and FlowCAM, an in-flow imaging system) to assess change in coastal microbial communities. B. Thompson (University of Maine) and colleagues reported on results of a weekly phytoplankton time series program, established in 2002, in the Damariscotta Estuary. Measured variables include size fractionated chlorophyll a, temperature, salinity, dissolved inorganic nutrients, and major taxonomic groups. A. Diamond (University of New Brunswick) presented results from 15 years of monitoring of seabirds on Machias Seal Island, on the border between the Gulf of Maine and the Bay of Fundy. Dramatic shifts in seabird diet were observed. During this period, the continent's largest colony of Arctic terns was lost, apparently through a combination of oceanographic changes affecting food supply, predation by gulls, and unintended consequences of changes to management policies. C. M. Tilburg (Gulf of Maine Council on the Marine Environment) presented the Ecosystem Indicator Partnership (ESIP), the GOMC initiative to identify and follow indicators to monitor Gulf of Maine ecosystem health. In contrast to the federal focus on fisheries management, ESIP also emphasizes monitoring for coastal issues, such as coastal development, contaminants, coastal eutrophication, and aquaculture. The difficulty in finding appropriate data to assess these issues was highlighted.

Three contributions focused on monitoring and analysis of the coastal benthos in the Gulf of Maine ecosystem. G. Pohle and L. Van Guelpen (Huntsman Marine Science Centre) reported on the Bay of Fundy component of the National Geography In Shore Areas project (NaGISA), a worldwide collaboration to inventory and monitor coastal benthic biodiversity. The common protocol involves sampling along 30-m transects from the high intertidal zone to a depth of 10 m and focuses on rocky shore and soft bottom sea grass communities. The data are compiled in the Census of Marine Life Ocean Biogeography Information System. As an alternative or complement to quantitative determination of benthic biodiversity at fixed locations, J. Grant (Dalhousie University) and colleagues proposed a new multivariate statistical approach for assessing the health of coastal benthic environments. They described a latent health factor index that can be developed from analysis of process-oriented variables, such as tidal exchange and residence times, sedimentation rate, and nutrient upwelling, characterizing the specific local environment. Another approach, involving estimation of the depth of the redox potential discontinuity by sediment profile imaging, can serve to indicate eutrophication or other forms of organic loading, such as aquaculture waste, and can be rigorously employed as a tool for community participation in local monitoring (e.g., Grant 2010). M.-J. Abgrall (University of

New Brunswick) and colleagues reported on their study of the role of sediment geochemistry, in particular the formation of hydrogen sulfide, to the survival of settling stages of softshell clams, suggesting that indices derived from sediment geochemistry could be an important tool for assessing the health of substrates for this ecologically important species.

The recognition that study of regional marine biodiversity can inform decision making about marine and coastal issues was highlighted in a contribution by S. Ellis (University of Southern Maine) and colleagues, who compared results of four ecosystem level studies of marine biodiversity, including the Gulf of Maine Area Census of Marine Life. Assessment of variation in diversity of harvested fish and invertebrates across the Gulf of Maine was explored by A. Cook and A. Bundy (Fisheries and Oceans Canada, Bedford Institute of Oceanography), who combined demersal trawl survey data and food habits data to investigate variation in species distribution and abundance.

B. Branton (Dalhousie University) and E. Bridger (Gulf of Maine Research Institute) provided an overview of the Ocean Tracking Network (OTN), a large international effort based in Canada to track movements of large marine animals, such as salmon, tuna, sharks, seals, crabs, and whales, using acoustic telemetry. The purpose of OTN is to gain a greater understanding of large animal movement and migration patterns in relation to ocean conditions and properties. This knowledge will inform conservation management decisions particularly needed for species such as bluefin tuna *Thunnus thynnus*, whose abundances have dropped over the past several decades. A particular challenge addressed in this presentation is the data policy and management of large amounts of data that will be generated from this program.

Present Status and Future Directions

How Far Have We Come?

Since the last Gulf of Maine science symposium, which took place at the same venue in St. Andrews, New Brunswick, 14 years earlier, there is evidence from the range and content of contributions to this session of significant progress in development of capacity to observe and interpret change in the Gulf of Maine, and in advancement toward regional ecosystem management. For example, progress has been made in the establishment of regional observing systems to record changes in physical and ecosystem properties of the Gulf of Maine. The Canadian AZMP began in 1998, supporting the Halifax line and a fixed station on the Scotian Shelf off Halifax and in the Bay of Fundy, providing sustained funding for observation of physical and biological variables at the upstream boundary of the Gulf. The Gulf of Maine Ocean Observing System (GoMOOS) was established in the late 1990s, resulting in the installation of a series of observational moorings along the Maine and Massachusetts coasts for weather and hydrographic measurements and limited phytoplankton sampling. In 2009, the transition from GoMOOS to NERACOOS was made after several years of planning by a multidisciplinary advisory committee of regional experts. The geographic range of NERACOOS includes not only the Gulf of Maine, but also the southern New England Bight and Long Island Sound. This regional infrastructure has advanced the capacity for coordinated observation of change, although the extent of sustained funding and range of variables to be observed is not yet established.

A number of time series that began earlier than 1996 have continued. Some new time series have started while others have had to be discontinued (e.g., Johnson and Hare 2012). In addition to annual and biannual fish trawl surveys conducted by both NMFS and DFO, NMFS has sustained the Continuous Plankton Recorder line between Cape Sable and Boston, as well as the EcoMon surveys that are conducted six times each year. Satellite-derived sea surface chlorophyll time series began in 1997, and presently two satellites (SeaWifs and MODIS) cover the Gulf of Maine each day.

The duration of these time series is beginning to yield information on interannual and interdecadal variability and trends. The long time series of observations of hydrography, zooplankton, and fish species biomass, and subsequent analyses, have revealed a significant shift in the Gulf of Maine ecosystem starting in the late 1980s and continuing through the 1990s. Changes include fresher surface salinity (e.g., Greene and Pershing 2007; Mountain and Kane 2010), shifts in the relative composition of the zooplankton community to smaller copepods (Pershing et al. 2005; Hare and Kane 2012), increases in abundance of Atlantic herring *Clupea harengus* (e.g., Overholtz et al. 2004), and coherent trends in a number of fish species (Nye et al. 2010). Causal mechanisms put forward for these involve increased input of fresher water, perhaps originating from the Arctic via the Labrador Sea (Greene and Pershing 2007), increased stratification and phytoplankton in fall enhancing fall production of small zooplankton (Greene and Pershing 2007), and fishing pressure as a dominant influence on fish biomass (e.g., Nye et al. 2010; Fogarty et al. 2012). Evidence was presented at this session of the possibility of a new regime shift forced by dramatic changes in input Scotian Shelf water, starting sometime between 2004 and 2008. These observations and analyses set up challenges for the research and management community in the face of both continued fishing pressure and predictions of unidirectional forcing due to global atmospheric warming. As stated by Greene and Pershing (2007), "The resilience of northwest Atlantic shelf ecosystems is being tested by climate forcing from the bottom up and predator overexploitation from the top down. Predicting the fate of these ecosystems will be one of oceanography's grand challenges for the 21st century."

To meet these challenges, new tools for integrating and analyzing observational data have been developed, with specific applications for the Gulf of Maine. Notable are great advances in coupled physical-biological modeling, supported in part by the U.S. GLOBEC Georges Bank/ Northwest Atlantic Program. These biophysical models have the capacity to integrate across multidisciplinary data sets, from hydrography to the planktonic early life stages of fish, to provide mechanistic interpretation and predictions based on the multiple data sets collected by observing systems. These biophysical models can be linked to spatially explicit life history models of fish populations (e.g., Runge et al. 2010) to provide insights into both bottom-up and top-down forcing impacts on stock abundance and distribution. Trophic models of energy and material flux across the ecosystem food web and time series and multivariate statistical approaches web have also advanced significantly (e.g., Link and Bundy 2012, this volume). These approaches have been facilitated by the tremendous increase in computer capacity over the past decade.

In terms of EAM, both Canadian and U.S. federal agencies responsible for management of harvested resources have been working toward the definition of management goals and operational objectives and strategies to achieve these objectives within the framework of ecosystem management, including integration of observations into information or assessment tools for management decisions. The U.S. strategy involves development of IEAs, as presented by Fogarty et al. (2012) within a framework of evaluating drivers (e.g., large-scale climate forcing), pressures (e.g., temperature increase), ecosystem states (from observations and indicators), assessment of impacts (using ecosystem models), and determination of responses. The IEAs will be used to guide implementation of EBFM in spatial management units, for example the Gulf of Maine and Georges Bank as separate units, under the premise that lower trophic level production capacity in each region will determine a finite amount of overall fishery production potential for that region. The developing Canadian strategy is more incremental in approach, building on existing fisheries management practices within a framework of defining objectives (e.g., maintain productivity, preserve biodiversity, and protect habitat), determining strategies and specifying tactics to achieve objectives.

How Well Situated is the Science to Inform Management?

The establishment of the AZMP in Canada and NERACOOS in the United States, and the Continuous Plankton Recorder and Ecomon surveys, combined with ongoing living marine resource surveys by federal and state agencies and research-industry partnerships, provide a foundation for providing information to management about change in the Gulf of Maine. As we have seen above, knowledge of past and present change is substantially improved, and new developments in ecosystem modeling and multivariate statistical and time series analysis offer new tools to integrate and interpret multidisciplinary data. Different scientific approaches using different types of ecosystem modeling to assess change and upper trophic level interactions are being developed, involving considerable thought and effort.

Nevertheless, the science is not realizing its potential to contribute to management. The range of ecosystem variables that are currently measured in time series collections, and the spatial scale of time series observing in the Gulf of Maine, is inadequate. The NERACOOS is woefully underfunded relative to its potential to serve as an entity to organize observing data collection in the region. High-frequency sentinel coastal time series stations in the Gulf of Maine, to complement the existing AZMP time series, are lacking. An integrated scientific strategy engaging integrative biological physical models to inform federal-level management about ecosystem change has not yet been achieved.

An example where science could do a better job to inform management could be taken from the current U.S. planning to implement EBFM. Sustainable harvesting of a portion of fishery production capacity, based on regional area estimates of primary production, may be an achievable objective if the monitoring of environmental change is comprehensive, accurate, and the ecosystem impacts of change are adequately understood. If environmental conditions were to change, reference points would need to be reset to respond to changes in lower-level ecosystem productivity. However, the linkages of lower-level productivity to fishery production potential, and the potential influences of environmental shifts on this linkage, are not well understood, in the Gulf of Maine as elsewhere. Assumptions inherent in current ecosystem models (e.g., Link and Bundy 2012) may provide inaccurate assessments of new fishery production levels due to, as examples, inadequate accounting of the role of advection of secondary production into the spatial management unit or inadequate accounting of keystone roles of lower trophic level species, such as *Calanus finmarchicus*, which account for much of the trophic lipid budget in the Gulf of Maine and may be particularly vulnerable to continued climate forcing (Johnson et al. 2011). On the positive side, understanding of trophic linkages and the role of advective forcing in the Gulf of Maine is advancing, and modeling tools that could be adapted to inform management are available (e.g., Ji et al. 2009; Record et al. 2010). What is needed is a more formal infrastructure to connect the nonfederal research community to the federal agencies and fisheries management councils where EBM will be executed.

Another area where science could do better to inform management is in the coastal zone. Information is needed for decision making beyond fisheries management, related, for example, to coastal zone land use, aquaculture, contamination of coastal environments, preservation of benthic biodiversity, and control of invasive species. The need is not only for more organized and more comprehensive collection of data, but also for more ecosystem-level research to understand dynamic control of production in bays and estuaries and for more formal infrastructure to match needs of coastal zone managers, with observation and research capacity at research institutions as well as private industry. This latter need is a gap that NERACOOS could fill with additional resources.

The level of funding available to the Gulf of Maine research community is obviously one constraint on its capacity to produce relevant information to management. As alluded to above, another constraint is the need for bridges to connect those in management positions to those in research positions for the transfer of new observational data and methods for analysis and interpretation of results to federal, state-provincial, and community levels, where this information can be used in decision making. This issue has been addressed in previous RARGOM workshops (e.g., Runge and Braasch 2005), which emphasized the need for regional infrastructure that would (1) facilitate regional model evaluation, including skill assessment, evaluation of uncertainty, and model ensemble approaches to predictions; (2) serve to link data analyses, modeling, and prediction capabilities to specific regional management needs; (3) facilitate coordination among government agencies, research institutions, and universities; and (4) develop and demonstrate environmental analysis and forecast products that could be implemented operationally.

What Must We Do?

The final, plenary session of the Gulf of Maine symposium provided an opportunity to synthesize symposium results and perspectives and produce recommendations. A report of this session, with a focus on what must be done to advance EAM in the region, is provided in Cooper et al. (2009). Here, I highlight three recommendations that arose out of discussions specific to the theme session 4 on observations, data collection, analyses, and tools.

The first recommendation is to maintain long time series observations and expand them to cover variables that are presently undersampled, such as coastal benthic and plankton biodiversity (including the microbial plankton). As Fisher and Brown pointed out, ecosystem monitoring is unloved and underfunded (Nisbet 2007), yet the new knowledge and discoveries increase exponentially with length of the time series.

Second, the research community must continue to keep engaged in the development of tools and knowledge to understand not only how the Gulf of Maine is changing, but also the causal mechanisms involved. The effects of climate forcing on coastal communities due to sea level rise, and on ecosystem productivity due to changes in wind, circulation, temperature and salinity fields and acidification of the water column, may occur more rapidly and in completely unexpected ways over the next several decades, and the region would do well to study and be prepared for these changes.

Finally, the community needs to develop and maintain bridges allowing transfer of new research knowledge, understanding, and information-support tools from science to management. One possible mechanism for knowledge and skill transfer in the United States is through the new regional cooperative institute, CINAR (Cooperative Institute for the North Atlantic Region; representing a partnership among the Woods Hole Oceanographic Institution, Rutgers University, the University of Maryland Center for Environmental Science, the University of Maine, and the Gulf of Maine Research Institute). CINAR was established to provide research assistance to NOAA to better understand the physical and biological processes that make possible the forecasting of physical and biological conditions within the region as a tool for effective EBM. Through workshops and individual collaboration with NOAA scientists and managers within this cooperative institute, greater exchange may be achieved. Recommendations arising from the previous RARGOM workshop (Runge and Braasch 2005) merit consideration. Two formal infrastructure improvements have been put forward. A regional modeling center would involve a coordinating entity (e.g., NERACOOS) and distributed output of observations and modeling to desktop computers of researchers and resource managers via standards-based tools. A Gulf of Maine experimental environment forecast center would have as a primary objective the development, testing, and refinement of forecast models that could then be adapted for delivery to decision makers after tailoring of output to end-user needs. Development of new infrastructure to address prospects for global and regional change in the coming decades is warranted, particularly with respect to better exchange between the research community and coastal management needs.

Acknowledgments

The preparation of this chapter was supported in part through award numbers OCE-0709518 and OCE-0815336 from the National Science Foundation.

References

Churchill, J., J. Runge, and C. Chen. 2011. Processes controlling retention of spring-spawned Atlantic cod (*Gadus morhua*) in the western Gulf of Maine and their relationship to an index of recruitment success. Fisheries Oceanography 20:32–46.

Cooper, L. L., R. L. Stephenson, and J. H. Annala, editors 2009. Gulf of Maine symposium—advancing ecosystem research for the future of the gulf: proceedings of a symposium held at St. Andrews, NB, October 5–9, 2009. Canadian Technical Report of Fisheries and Aquatic Sciences 2904.

FAO (Food and Agriculture Organization of the United Nations). 2003. Fisheries management. 2. The ecosystem approach to fisheries. FAO Fisheries, Technical Guidelines for Responsible Fisheries, 4 (Supplement 2), Rome.

Fogarty, M. J., K. D. Friedland, L. Col, R. Gamble, J. Hare, K. Hyde, J. S. Link, S. Lucey, H. Liu, J. Nye, W. J. Overholtz, D. Richardson, B. Rountree, and M. Taylor. 2012. Status of the northeast U.S. Continental Shelf large marine ecosystem: an indicator-based approach. Pages 139–165 *in* R. L. Stephenson, J. H. Annala, J. A. Runge, and M. Hall-Arber, editors. Advancing an ecosystem approach in the Gulf of Maine. American Fisheries Society, Symposium 79, Bethesda, Maryland.

Grant, J. 2010. Coastal communities, participatory research, and far-field effects of aquaculture. Aquaculture and Environmental Interactions 1:85–93.

Greene, C. H., and A. J. Pershing. 2007. Climate drives sea change. Science 315:1084–1085.

Hare, J. A., and J. Kane. 2012. Zooplankton of the Gulf of Maine—a changing perspective. Pages 115–137 *in* R. L. Stephenson, J. H. Annala, J. A. Runge, and M. Hall-Arber, editors. Advancing an ecosystem approach in the Gulf of Maine. American Fisheries Society, Symposium 79, Bethesda, Maryland.

Holliday, M. C., and A. Gautam, editors. 2005. Developing regional marine ecosystem approaches to management. National Oceanic and Atmospheric Administration, National Marine Fisheries Service, Technical Memorandum NMFS-F/SPO-77, Silver Spring, Maryland.

Ji, R., C. Davis, C. Chen, and R. Beardsley. 2008a. Influence of local and external processes on the annual nitrogen cycle and primary productivity on Georges Bank: a 3-D biological-physical modeling study. Journal of Marine Systems 73:31–47.

Ji, R., C. Davis, C. Chen, and R. Beardsley. 2009. Life history traits and spatiotemporal distributional patterns of copepod populations in the Gulf of Maine–Georges Bank region. Marine Ecology Progress Series 384:187–205.

Ji, R., C. Davis, C. Chen, D. Townsend, D. Mountain, and R. Beardsley. 2008b. Modeling the influence of low-salinity water inflow on winter-spring phytoplankton dynamics in the Nova Scotian Shelf–Gulf of Maine region. Journal of Plankton Research 30:1399–1416.

Ji, R., M. Edwards, D. Mackas, J. Runge, and A. Thomas. 2010. Marine plankton phenology and life history in a changing climate: current research and future directions. Journal of Plankton Research 32:1355–1368.

Johnson, C., J. Runge, A. Bucklin, K. A. Curtis, E. Durbin, J. A. Hare, L. S. Incze, J. Link, G. Melvin, T. O'Brien, and L. Van Guelpen. 2011. Biodiversity and ecosystem function in the Gulf of Maine: pattern and role of zooplankton and pelagic nekton. PLoS (Public Library of Science) ONE [online serial] 6(1):e16491. DOI: 10.1371/journal.pone.0016491.

Johnson, C. L., and J. A. Hare. 2012. Zooplankton monitoring in the Gulf of Maine: past, present and future. Pages 205–218 *in* R. L. Stephenson, J. H. Annala, J. A. Runge, and M. Hall-Arber, editors. Advancing an ecosystem approach in the Gulf of Maine. American Fisheries Society, Symposium 79, Bethesda, Maryland.

Kristiansen, T., F. Vikebo, S. Sundby, G. Huse, and O. Fiksen. 2009. Modeling growth of larval cod (*Gadus morhua*) in large-scale seasonal and latitudinal gradients. Deep-Sea Research 59:2001–2011.

Levin, P. S., M. J. Fogarty, S. A. Murawski, and D. Fluharty. 2009. Integrated ecosystem assessments: developing the scientific basis for ecosystem-based management of the ocean. PLoS (Public Library of Science) Biology [online serial] 7(1):e1000014. DOI: 10.1371/journal.pbio.1000014.

Link, J. S., and A. Bundy. 2012. Ecosystem modeling in the Gulf of Maine region: towards an ecosystem approach to fisheries. Pages 281–310 *in* R. L. Stephenson, J. H. Annala, J. A. Runge, and M. Hall-Arber, editors. Advancing an ecosystem approach in the Gulf of Maine. American Fisheries Society, Symposium 79, Bethesda, Maryland.

Mountain, D., and J. Kane. 2010. Major changes to the Georges Bank ecosystem, 1980's to the 1990's. Marine Ecology Progress Series 398:81–91.

Nisbet, E. 2007. Cinderella science. Nature 7:789–790.

Nye, J. A., A. Bundy, N. Shackell, K. D. Friedland, and J. S. Link. 2010. Coherent trends in contiguous survey time-series of major ecological and commercial fish species in the Gulf of Maine ecosystem. ICES Journal of Marine Science 67:26–40.

OECD (Organization for Economic Co-operation and Development). 2003. OECD environmental indicators development, measurement, and use. OECD, Paris.

Overholtz, W. J., L. D. Jacobson, G. D. Melvin, M. Cieri, M. Power, D. Libby, and K. Clark. 2004. Stock assessment of the Gulf of Maine–Georges Bank Atlantic herring complex, 2003. National Oceanic and Atmospheric Administration, National Marine Fisheries Service, Northeast Fisheries Science Center, Reference Document 04-06, Woods Hole, Massachusetts.

Pershing, A. J., C. H. Greene, J. Jossi, L. O'Brien, J. K. T. Brodziak, and B. A. Bailey. 2005. Interdecadal variability in the Gulf of Maine zooplankton community, with potential impacts on fish recruitment. ICES Journal of Marine Science 62:1511–1523.

Record, N. R., A. J. Pershing, J. A. Runge, C. Mayo, B. Monger, and C. Chen. 2010. Improving ecological forecasts of copepod community dynamics using genetic algorithms. Journal of Marine Systems 82:96–110.

Runge, J. A., and E. Braasch. 2005. Modeling needs related to the regional observing system in the Gulf of Maine. Regional Association for Research on the Gulf of Maine, RARGOM Report 05-01, Durham, New Hampshire.

Runge, J. A., A. Kovach, J. Churchill, L. Kerr, J. R. Morrison, R. Beardsley, D. Berlinsky, C. Chen, S. Cadrin, C. Davis, K. Ford, J. H. Grabowski, W. H. Howell, R. Ji, R. Jones, A. Pershing, N. Record, A. Thomas, G. Sherwood, S. Tallack, and D. Townsend. 2010. Understanding climate impacts on recruitment and spatial dynamics of Atlantic cod in the Gulf of Maine: integration of observations and modeling. Progress in Oceanography 87:251–263.

Shackell, N. L., A. Bundy, J. A. Nye and J. S. Link. 2012. Common large-scale responses to climate and fishing across northwest Atlantic ecosystems. ICES Journal of Marine Science 69:151–162.

Sissenwine, M., and S. Murawski. 2004. Moving beyond 'intelligent tinkering': advancing an ecosystem approach to fisheries. Marine Ecology Progress Series 274:291–295.

Smith, P. C., N. R. Pettigrew, P. Yeats, D. W. Townsend, and G. Han. 2012. Regime shift in the Gulf of Maine. Pages 185–203 *in* R. L. Stephenson, J. H. Annala, J. A. Runge, and M. Hall-Arber, editors. Advancing an ecosystem approach in the Gulf of Maine. American Fisheries Society, Symposium 79, Bethesda, Maryland.

Westhead, M. 2012. Tools for integrated policy and management in the Gulf of Maine: a summary of major advancements since 1996. Pages 47–60 *in* R. L. Stephenson, J. H. Annala, J. A. Runge, and M. Hall-Arber, editors. Advancing an ecosystem approach in the Gulf of Maine. American Fisheries Society, Symposium 79, Bethesda, Maryland.

American Fisheries Society Symposium 79:115–137, 2012

Zooplankton of the Gulf of Maine—A Changing Perspective

JONATHAN A. HARE* AND JOSEPH KANE
NOAA Northeast Fisheries Science Center, Narragansett Laboratory
28 Tarzwell Drive, Narragansett, Rhode Island 02818, USA

Abstract.—Numerous studies have examined the dynamics of zooplankton in the Gulf of Maine. Here the authors reanalyze relationships found in these prior studies, using updated data, with the goal of evaluating previously identified zooplankton–environment linkages. These reanalyses support the finding that major changes occurred in the zooplankton community during the late 1980s and again in the late 1990s. Evidence for a broader change in the ecosystem during these periods and mechanisms responsible for changes in the zooplankton are discussed. In general, the results of previous studies are upheld, but it is shown that the relationship between the environmental indicators and zooplankton change through time. This result implies that all data collected in the Gulf of Maine must be considered within a historical context and that the observed environmental–zooplankton linkages are still not well understood. It is possible that changes in the seasonal cycle or shifts in the pressures systems responsible for the North Atlantic oscillation result in nonstationary environmental–zooplankton relations. These results indicate that a mechanistic understanding is required to explain the documented environment–zooplankton linkages rather than correlative explanations. Since the causes of the late-1980s and late-1990s regime shifts are still unclear, future ecosystem-based management in the Gulf of Maine must be supported by continued observation and analysis to identify ecosystem changes soon after they occur. Scenario-driven modeling also is needed to provide guidance as to how the ecosystem will respond to future changes in zooplankton abundance and community structure.

Introduction: Scientific and Historical Context

Continued development of scientific understanding supports the improvement of scientific assessments and advice, which in turn are provided to resource managers. For example, the development of the concept of density-dependent population regulation led to the principle of sustainable yield, which is the current management paradigm for many fisheries around the world (Rosenberg et al. 1993). Similarly, the development of the ecosystem concept lead to the idea of system-level productivity (Ryther 1969) and the use of food chain and food web models to understand ecosystem structure (Polovina 1984). Ecosystem-based approaches to fisheries management seek to bring together the concepts of fisheries and ecosystem science with the goal of sustaining healthy marine ecosystems and the fisheries they support (Link 2002; Pikitch et al. 2004). Much of the history of fisheries and ecosystem science is based on the idea that as we increase our detailed understanding of the state of a population or ecosystem, our ability to provide scientific advice increases (Francis and Hare 1994).

These general ideas are exemplified in plankton studies in the Gulf of Maine ecosystem. Concern over the conservation of haddock *Melanogrammus aeglefinus* in New England during the 1930s led to one of the first fisheries oceanography programs in North America (Anderson 1998). Walford (1938) prefaced the justification for these studies with the following:

* Corresponding author: jon.hare@noaa.gov

> It is a fact of common knowledge to fisherman and marine biologists that the American haddock, like many other animal populations, fluctuate in abundance from year to year. This is a matter of great economic importance, not only because it complicates considerable the task of conservation. It is obviously desirable, therefore to learn as much as possible about these fluctuations and the nature of the elements causing them.

The work of Walford (1938), and Hjort (1914) before him, started a tremendous research effort to understand interanuual variability in fish abundance. Recruitment—the number of young fish that join the adult population and the "birth" term is the classic density-dependent models—was identified as the major causative factor in abundance variability, thereby justifying a century of research to understand the details of recruitment fluctuations, with the explicit or implicit goal of improving managers' ability to maintain sustainable fisheries.

A parallel track is evident in ecosystem studies in the Gulf of Maine. Based in part on the pioneering work of Hjort (1914) and Walford (1938), there was the recognition that conditions during the planktonic life stages of commercially exploited fish could cause recruitment variability. By extension, a more detailed understanding of the planktonic ecosystem could lead to a more complete understanding of recruitment variation. Concurrent with the work of Walford (1938), the ecosystem concept holds that fisheries production in a system is limited by both primary production and the subsequent movement of energy through the ecosystem (Clarke et al. 1946; Ryther 1969). Again, by extension, a more detailed understanding of the planktonic ecosystem could lead to a more complete understanding of total potential fishery yields. These two rationales fueled—directly and indirectly—a century of research into the planktonic ecosystems of the Gulf of Maine.

The planktonic communities of the Gulf of Maine were first described by Bigelow (1926). The zooplankton are dominated by a "*Calanus* community," which includes *Calanus finmarchicus* (copepod), *Pseudocalanus elongates* (copepod), *Metridia lucens* (copepod), *Euchaeta norvegica* (copepod), *Sagitta elegans* (arrow worm), *Themisto* spp. (amphipod termed *Euthemisto* by Bigelow, 1926), euphausiids genera[1], *Limacina retroversa* (pteropod), and *Pleurobrachia pileus* (ctenophore). The phytoplankton are dominated by diatoms and peridinians (dinoflagellates). Bigelow (1926) also described the seasonal progression in these communities and their spatial variability. Since Bigelow (1926), numerous, more detailed descriptions of the planktonic communities, species, populations, and individuals have been completed (e.g., Townsend and Spinrad 1986; Cura 1987; Meise and O'Reilly 1996; Bucklin et al. 1998; Durbin et al. 2003; Johnson et al. 2011). The implicit rationale for many of these studies was that a more complete understanding of the planktonic ecosystem would support improved scientific advice for management (Johnson and Hare 2012; Runge and Jones 2012; both this volume).

The increased understanding of planktonic communities was used to increase the complexity of ecosystem models in the Gulf of Maine. Based on work in the early 20th century, and the developing theory of systems ecology, Clarke et al. (1946) developed a four-box model that included diatoms, zooplankton, the benthos, and fish. A more complex, but structurally similar model with nine compartments was completed by Cohen and Grosslein (1982), which included the planktonic components of primary producers, dissolved inorganic carbon, zooplankton, and bacteria. The system level approach has culminated in complex network models, including those of Link et al. (2008) with 33 compartments and those of Steele et al. (2007) with 14 compartments. These models are evolving from descriptions of the ecosystem to strategic tools to evaluate various forcing effects (e.g., fishing, climate) on ecosystem goods and services (e.g., Overholtz and Link 2009).

This emphasis on improving the detailed understanding of the Gulf of Maine ecosystem had another result: the collection of data over decades. An overview of planktonic sampling in the Gulf of Maine is provided elsewhere in this volume (Johnson and Hare 2012), but consistent

[1] Bigelow (1926) reported eight species of euphausiids from the Gulf of Maine region: ***Thysanoessa inermis***, ***T. longicaudata***, ***T. gregaria***, ***T. raschii***, ***Nematoscelis megalops***, ***Euphausia krohnii***, northern krill ***Meganyctiphanes norvegica***, and ***Thysanopoda acutifrons***.

zooplankton sampling started in the early 1960s with the initiation of a Continuous Plankton Recorder (CPR) transect between Halifax and Portland, Maine/Boston, Massachusetts (Jossi et al. 2003). The Gulf of Maine CPR is operated monthly, and both phytoplankton and zooplankton are enumerated. The Marine Resource Monitoring, Assessment, and Prediction (MARMAP) Program started in the late 1970s (Sibunka and Silverman 1989) and continues today as the Ecosystem Monitoring (EcoMon) Program (Kane 2007). Hydrographic and plankton samples are collected bimonthly over the northeast U.S. continental shelf, including Georges Bank and the Gulf of Maine. The Atlantic Zone Monitoring Program (AZMP) started in 1999 and is ongoing (Therriault et al. 1998; Mitchell et al. 2002). The AZMP conducts broadscale hydrographic and zooplankton surveys along four transects on the Scotian shelf in the spring and fall each year, with some samples collected in the summer on groundfish surveys. In addition, the program collects hydrographic and zooplankton samples twice a month throughout the year at two fixed stations on the inshore central Scotian shelf and in the northwestern coastal Bay of Fundy. The interest in long-term data collection was initially spurred by a desire to provide a baseline of the planktonic components of the ecosystem, as well as describe the seasonal and interannual variability in the system. But after 30–50 years of collection, numerous investigators have used the data to examine multiannual patterns in zooplankton community structure and the environmental forcing that influence community dynamics.

The purpose here is to review prior studies of long-term zooplankton dynamics in the Gulf of Maine, to re-evaluate the various hypotheses and relationships with current data (through 2008), and to update the conceptual model of zooplankton dynamics in the Gulf of Maine. The ultimate goal is to contribute to an ecosystem-based approach to fisheries management in the Gulf of Maine through increasing understanding of the dynamics of zooplankton communities in the ecosystem.

A Multiannual and Decadal Perspective

The first studies examining long-term patterns in zooplankton of the Gulf of Maine focused on *Calanus finmarchicus*—the namesake of Bigelow's "*Calanus* community" and the numerically dominant zooplankton in the system (Bigelow 1926; Fish 1936; Kane 2007). Jossi and Goulet (1993) reported a long-term increase in *C. finmarchicus* in the northwest Atlantic Ocean based on CPR data from 1961 to 1989. This was one of the first examples of long-term change in the pelagic ecosystem of the Gulf of Maine. Greene and Pershing (2000) proposed that the increase in *C. finmarchicus* observed by Jossi and Goulet (1993) was caused by changes in advection in the northwest Atlantic Ocean. These physical changes were linked, in turn, to the North Atlantic oscillation (NAO)—the dominant mode of long-term atmospheric variability in the North Atlantic (Hurrel et al. 2003).

Several studies confirmed the results of Jossi and Goulet (1993) and the link between *C. finmarchicus* and the NAO using the same CPR data set but for longer periods and with different analytical approaches. Conversi et al. (2001) demonstrated significant increases in the NAO and Gulf of Maine sea surface temperature (GOM-SST) over the same time period of increasing *C. finmarchicus*. After de-trending the time series and performing cross-correlation analysis, they found that the NAO was positively correlated with GOM-SST at a 2-year lag (the NAO leading GOM-SST by 2 years), the NAO was positively correlated with *C. finmarchicus* abundance at a 4-year lag (the NAO leading *C. finmarchicus* by 4 years), and GOM-SST was positively correlated with *C. finmarchicus* at a 2-year lag. MERCINA (2001) examined the relation between the NAO and deeper slope water temperature (waters of the shelf in a layer between 150 and 250 m) and found a significant correlation between the NAO and their regional slope water temperature index (RSWT) at a lag of 1 year (the NAO leading the RSWT by 1 year). They also found a significant correlation between *C. finmarchicus* and the RSWT with a 3-year lag. The approximate 2-year lag between sea surface temperature and *C. finmarchicus* was confirmed by Licandro et al. (2001) using the same data but slightly different time series methods. However, Licandro et al. (2001) also found a significant negative relationship between surface temperature and *C. finmarchicus* with a zero lag. The approximate 4-year lag between the NAO and *C.*

finmarchicus abundance was confirmed by Piontkovski et al. (2006). These studies supported the hypothesis that the NAO affects *C. finmarchicus* at an approximately 4-year lag and that temperature was an intermediate indicator of the link between the NAO and *C. finmarchicus* (Figure 1).

The concept of a link between the NAO, ocean temperatures in the Gulf of Maine region, and the abundance of *C. finmarchicus* was further developed in a series of papers (Greene and Pershing 2003; MERCINA 2003, 2004). These studies hypothesize that the NAO influences the distribution of water masses in the northwest Atlantic. These influences have their strongest signal in the slope water in that they effect the distribution of Atlantic temperate slope water (warmer and saltier) and Labrador subarctic slope water (colder and fresher). These two water masses were termed the coupled slope water system, and the RSWT was derived as an indicator of the state of this system. These studies argued that low NAO resulted in a dominance of Labrador subarctic slope water offshore of the North American shelf, a low RSWT, and high *C. finmarchicus* abundance. The 4-year link between the NAO and *C. finmarchicus*, however, was equivocal. MERCINA (2003), in reanalyzing data over two different time periods, found that the correlation between the NAO and *C. finmarchicus* broke down when more recent data were included. Kimmel and Hameed (2008) noted a similar breakdown in the relation between the NAO and *C. finmarchicus* in the northeast Atlantic. These studies suggest that the relationship between the NAO and *C. finmarchicus* varies through time.

Calanus finmarchicus is not the only copepod in Bigelow's "*Calanus* community." In addition to studying *C. finmarchicus*, Licandro et al. (2001) examined the relationship between six other copepod species and Gulf of Maine sea surface temperature. They found that *Oithona* spp. negatively related to sea surface temperature (SST) at a 2.5-year lag (SST leading *Oithona* spp.), opposite of the pattern found for *C. finmarchicus*. They suggested that the inverse relationship between *C. finmarchicus* and *Oithona* spp. may be evidence for shifts in zooplankton community structure from large (*C. finmarchicus*) to small (*Oithona*

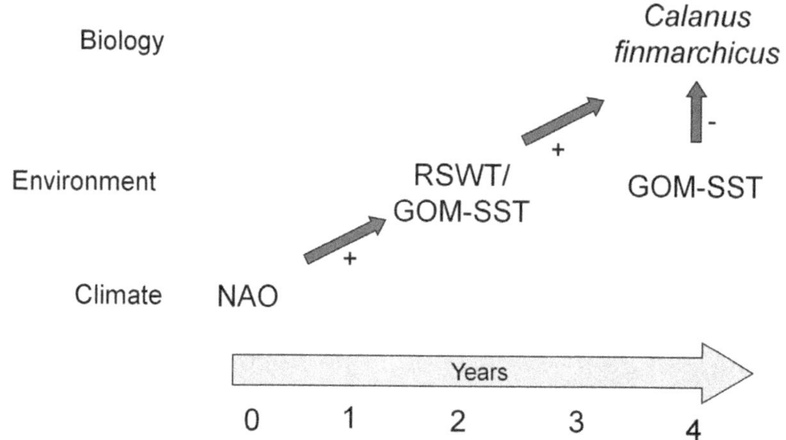

FIGURE 1. Conceptual diagram of climate–environment effects on *Calanus finmarchicus* based on prior work (Conversi et al. 2001; MERCINA 2001; Licandro et al. 2001; Piontkovski et al. 2006). As the North Atlantic oscillation (NAO) fluctuates, the circulation patterns in the northwest Atlantic vary. Specifically, the coupled slope water system is affected, for which the regional slope water temperature index (RSWT) is an indicator. Changes in the coupled slope water system affect *Calanus finmarchicus,* specifically through advection into the Gulf of Maine (GOM) system from the Scotian shelf. Locally sea surface temperature (SST) also affects *C. finmarchicus*; the mechanism of the effect is unclear but could be directly through physiological processes (e.g., metabolism) and population processes (e.g., growth, recruitment) or indirectly through trophic interactions and advective influences. Analyses presented in this study find that this conceptual model is too simple and that all of the linkages vary through time.

spp.) species. Pershing et al. (2005) analyzed the CPR data set with principal component analysis (PCA). They included seven taxa (five overlapping with the study of Licandro et al. [2001]). The first principal component was dominated by *Oithona* spp. and *Centropages typicus*, while the second principal component was dominated by *Calanus finmarchicus*. They identified a shift in community structure in 1989, which was primarily caused by an increase in *Oithona* spp., *Centropages typicus*, *Pseudocalanus* spp., and *Metridia* spp. They identified another community shift in 2001, with a decrease in the copepod species that had increased in 1989. Kane (2007) analyzed a different data set: bongo net collections from Georges Bank. His analysis included 23 taxa and used multidimensional scaling. Similar to Pershing et al. (2005), Kane (2007) found a shift in zooplankton community structure in 1990 and 2002, which resulted primarily from an increase then decrease in small-bodied species (*C. typicus*, *Metridia lucens*, *Temora longicornis*, Thecosomata, *Oithona* spp., Appendicularia, Hyperiidae, and Coelenterata). Kane (2007) also examined the relationship between species abundance and environmental variables. *Calanus finmarchicus* was negatively related with surface temperature, agreeing with the negative zero lag correlation found by Licandro et al. (2001). *Centropages typicus* and *C. hamatus* exhibited the opposite correlation: increasing with increasing temperature. Significant correlations were also found between copepod taxa and salinity; many taxa increased in abundance with decreasing salinity.

To summarize, five main multiannual patterns in zooplankton abundance have been identified in previous studies: (1) a linear increase in *Calanus finmarchicus*, SST, and the NAO over time; (2) a link between the NAO, regional temperature indices, and *C. finmarchicus*, with an approximate 2-year lag at each step (Figure 1); (3) a negative relation between regional temperature indices and *C. finmarchicus* at a 0-year lag (Figure 1); (4) a change over time in the lag relations between the NAO, regional temperature indices, and *C. finmarchicus*; and (5) a community shift in approximately 1990 and 2000 caused by an increase and then a decrease in the abundance of smaller-bodied copepods. This community shift may be linked to salinity variability in the ecosystem. We will re-examine each one of these patterns using updated zooplankton and environmental data through 2008.

Methods

The authors assembled five climate/environmental time series for analysis (Table 1). These series were chosen to represent data used in previous studies. In three cases, the data series were derived from updates of the original source (the NAO, Georges Bank surface temperature [GB-ST], Georges Bank surface salinity [GB-SS]). In two cases, the data series were reconstructed based on descriptions in the original papers (RSWT and GOM-SST). These reconstructed series were compared to the original series for the period of overlap, and in both cases, the agreement was excellent, indicating that the series were successfully reconstructed.

We used two sources of zooplankton data: the CPR data set (Jossi and Goulet 1993; Kane 2009) and the EcoMon data set (Kane 2007). Throughout, we use species names but refer in particular to late copepodid stages for *C. finmarchicus* (CV-CVI), *Centropages typicus* (CIV-CVI), *Oithona* spp. (CIV-CVI), *Metridia lucens* (CIV-CVI), and *Pseudocalanus* spp. (CVI). We use *Calanus* spp. for earlier copepidite stages of the genus (CI-CIV). This nomenclature follows Pershing et al. (2005) and Kane (2007).

The CPR data sets contain zooplankton data from 1961 to the present, collected monthly across the Gulf of Maine from merchant vessels. A spline smoother was fit to the monthly abundance data of each taxa, and anomalies were calculated to remove the average seasonal cycle in abundance (Kane 2007). Monthly anomalies were then averaged in each year to result in annual abundance anomalies. There are several years of missing data in this series (1963, 1974–1977); these values were filled with the average annual anomaly for the entire series. Unlike Pershing et al. (2005), data collected along the entire Gulf of Maine transect were included in the analysis.

The EcoMon data set contains zooplankton data from 1977 to the present, collected approximately bimonthly throughout the Gulf of Maine and Georges Bank region from research

TABLE 1. Description of environmental variables used in this study including the abbreviation used and the source.

Variable	Symbol	Description	Source
North Atlantic oscillation	NAO	Winter (December through March) index of the NAO based on the difference of normalized sea level pressure (SLP) between Lisbon, Portugal, and Stykkisholmur/Reykjavik, Iceland	www.cgd.ucar.edu/cas/jhurrell/Data/naodjfmindex.asc
Gulf of Maine sea surface temperature	GOM-SST	Sea surface temperature extracted from the Extended Reconstructed Sea Surface Temperatures, version 3 (ERSST.v3) data set and averaged over the area 41–45°N and 65–71°W	http://dss.ucar.edu/datasets/ds277.0 /docs/v3bersst.readme
Regional slope water temperature index	RSWT	The dominant mode of variability derived from a principal components analysis of eight slope water temperature anomaly time series. The time series correspond to mean annual slope water temperature anomalies between 150- 250 m in Emerald Basin, Georges Basin, Jordan Basin, Wilkinson Basin, and from four geographic sectors overlying the region's continental slope (Pershing 2001)	Raw data extracted from Fisheries and Oceans Canada Hydrographic Climate Database: www.mar.dfo-mpo.gc.ca/science/ocean/database/doc2006/clim2006app.html

vessels using a 61-cm bongo net (Kane 2007). Similar to the CPR data, a spline smoother was fit to the bimonthly abundance data of each taxa, and anomalies were calculated to remove the average seasonal cycle in abundance (Kane 2007). Monthly anomalies were then averaged in each year to result in annual abundance anomalies. Similar to Kane (2007), data collected over the entire Georges Bank region were included in the analysis.

Linear trends in several of the time series were evaluated with linear regression to update the results of Jossi and Goulet (1993) and Conversi et al. (2001). Time series were also plotted to allow visual inspection. These analyses were performed on the basic time series data described above and in Table 1 (e.g., no transformations, no prewhitening).

Cross-correlation analysis was used to re-evaluate the lag responses between the NAO, regional temperature indices, and *Calanus finmarchicus*, updating the analyses of Conversi et al. (2001), MERCINA (2001), Licandro et al. (2001), and Piontkovski et al. (2006). Prior to analysis, time series were de-trended and prewhitened following Kimmel and Hameed (2008), who showed that both trends and autocorrelation have affected the perceived correlation between *C. finmarchicus* and the NAO. The whit1 function of the Tree Ring Matlab Toolbox was used for prewhitening.[2] Significance was estimated from number of observations in the cross-correlated series with no correction for multiple correlations.

To evaluate the nonstationary of climate–environment–zooplankton relations (as suggested by MERCINA 2003), serial correlations were calculated between de-trended, prewhitened time series. Serial correlation is similar to a moving average but calculates the Pearson correlation coefficient through a time series rather than the average. This approach is used in climate studies to evaluate the stationarity of bivariate correlations through time (Joyce 2002; Vicente-Serrano and López-Moreno 2008). Significance of serial correlation coefficients was estimated as in standard correlation, and no correction is made for multiple comparisons. Only important lags identified in the cross-correlation analysis were examined. Significance was estimated from number of observations in the cross-correlated series, with no correction for multiple correlations.

The multivariate analysis of Pershing et al. (2005) and Kane (2007) were redone using the same statistical approaches and the same data sources, but updated through 2008. The Gulf of Maine CPR data were analyzed with PCA, and the six taxa-stage combinations analyzed by Pershing et al. (2005) were reanalyzed here. The Georges Bank EcoMon data were analyzed with multidimensional scaling (MDS), and the 23 taxa-stage combinations analyzed by Kane (2007) were reanalyzed here. The relations between the dominant mode of zooplankton variability and environmental variables then were analyzed with cross-correlation and serial correlation as above.

[2] www.ltrr.arizona.edu/~dmeko/toolbox.html

Results

Linear Increase in Calanus finmarchicus, Sea Surface Temperature, and North Atlantic Oscillation

Updating the linear regressions of Conversi et al. (2001) resulted in no significant linear trends (Figure 2); however, the slope of all regressions was positive. The disappearance of a significant linear trend with additional data is indicative of oscillating or time-dependent variability, which is consistent with the ideas of Greene and Pershing (2003) and MERCINA (2003, 2004).

Time Lags among North Atlantic Oscillation, Regional Temperature Indices, and Calanus finmarchicus

The most recent 7 years of NAO, RSWT, GOM-SST and *C. finmarchicus* data shows generally near-zero or positive NAO values, generally near zero or positive GOM-SST anomalies, generally positive RSWTs, with negative values in the most recent 2 years, and generally positive anomalies of *C. finmarchicus* abundance, with a negative value in the most recent year (Figure 3).

Updating the cross-correlation analyses with more recent data supports, in part, the results of prior studies (Figure 4). The highest correlation between the NAO and the RSWT is at 2 years, but this correlation is not significant. The highest correlation between the NAO and the GOM-SST is at 3 years, but this correlation also is not significant. There is strong correlation between the RSWT and the GOM-SST at a zero lag. The nonsignificant correlations with additional data may be due to nonstationary in the linkages between the NAO and regional temperatures indices (MERCINA 2003).

With the addition of new data, the relations between *C. finmarchicus* abundance and the NAO, RSWT, and GOM-SST were different than previously reported (Figure 5). Neither the NAO nor RSWT were correlated with *C. finmarchicus* at any positive lags (environment leading *C. finmarchicus*). However, the highest positive correlation with the NAO was at a lag of 4 years. The GOM-SST and *C. finmarchicus* were significant (just) at a lag of 3 years (GOM-SST leading); the highest

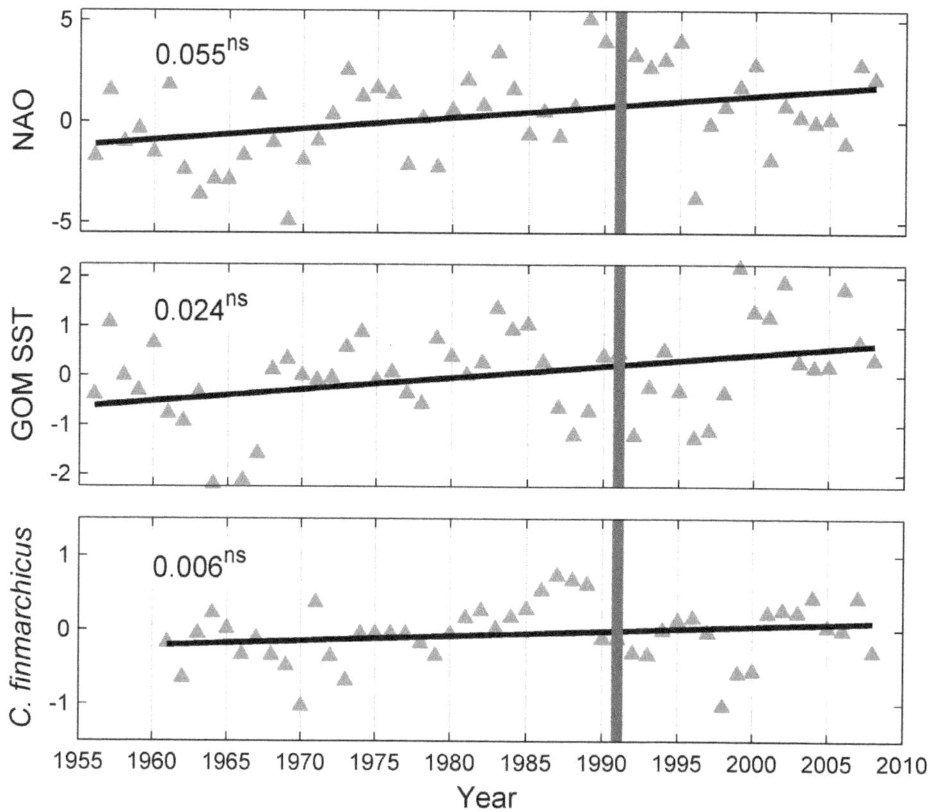

Figure 2. Long-term temporal trends in the North Atlantic oscillation (NAO), Gulf of Maine sea surface temperature (GOM-SST), and *Calanus finmarchicus* abundance. The linear regression is provided as an update of the analysis of Conversi et al. (2001). Their analyses used data through 1991 (indicated by the vertical line). In all three cases, the regressions were not significant. The regression slope and significance values are provided: not significantly different than zero (ns) and significant at $p < 0.05$ (*).

correlation with RSWT was also observed at 3 years. Significant correlation between *C. finmarchicus* and GOM-SST and RSWT were found at a lag of −1 (*C. finmarchicus* leading GOM-SST and RSWT by 1 year). Practically, a biological variable leading a physical variable does not make sense, so interpretation of the significant correlation is difficult.

Nonstationary Relations between North Atlantic Oscillation, Regional Temperature Indices, and Calanus finmarchicus

Serial correlation analysis demonstrated that the relationships between the NAO, RSWT, GOM-SST, and *C. finmarchicus* changed through time (Figure 6A). Before 1980, the NAO and RSWT were not correlated. Starting in the 1980s to approximately 2000, the series are positively correlated. The NAO and GOM-SST are largely not correlated; however, there are three short periods where the correlations were positive: in the 1960s, 1980s, and 2000s. The RSWT and GOM-SST are strongly correlated early in the series before 1980. The correlation decreases for a period from 1980 to 2000 and then increases again near the end of the series. These analyses show that the relationships between the NAO and measures of water temperature in the Gulf of Maine ecosystem change over time. Further, when the correlation between the deep (RSWT) and surface (GOM-SST) temperature indices is strong, the relationship between NAO and

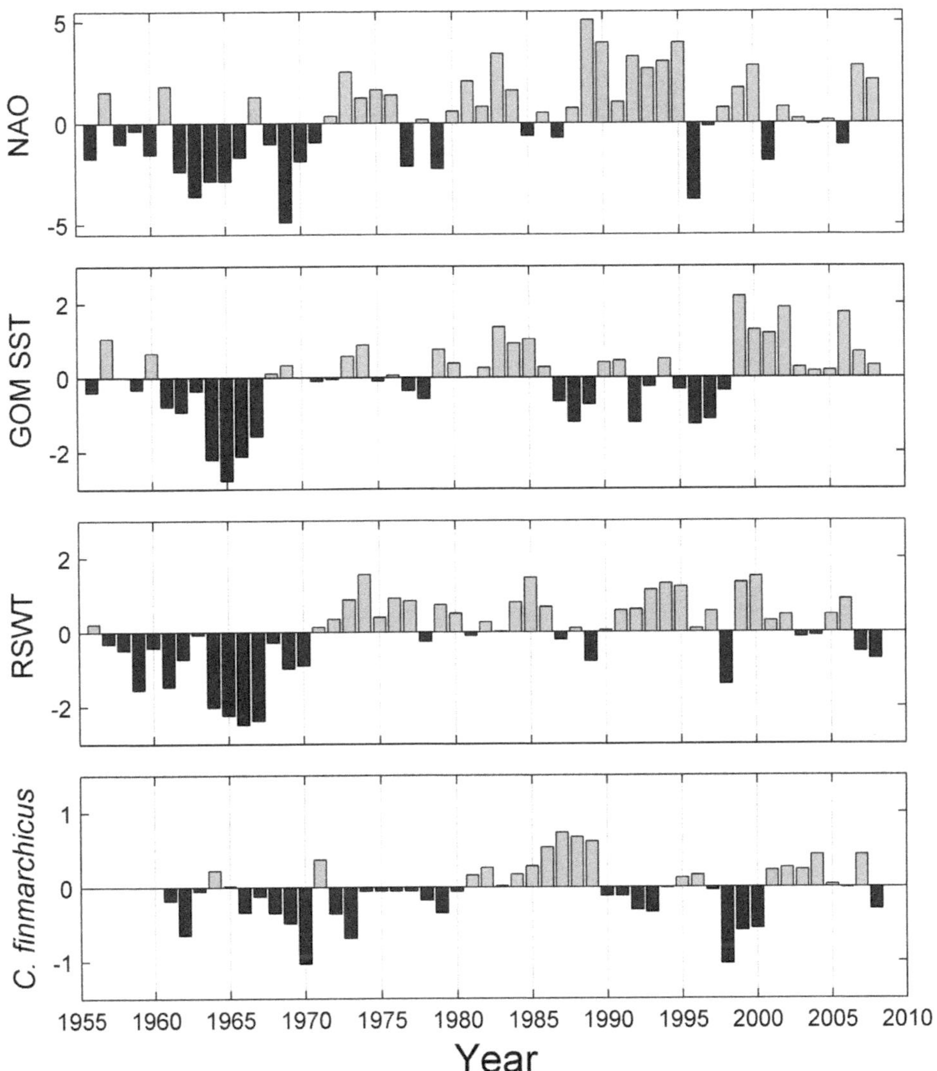

FIGURE 3. Time series of the North Atlantic oscillation (NAO), Gulf of Maine sea surface temperature (GOM-SST), regional slope water temperature index (RSWT), and *Calanus finmarchicus* abundance. Data are plotted as anomalies.

RSWT is weak (Figure 6B). This may suggest an inverse relationship between the strength of local versus remote forcing on the Gulf of Maine ecosystem.

The correlations between *C. finmarchicus* and the climate/environmental indices also change through time (Figure 7). During the 1980s, the relationship between the NAO and *C. finmarchicus* is not evident at either 2- or 4-year lags. *Calanus finmarchicus* and RSWT are positively correlated at a 2-year lag. The correlations between *C. finmarchicus* and GOM-SST shift from positive to nonsignificant at a 2-year lag and from negative to nonsignificant at a 0-year lag. During the 1990s, the relationships change. *Calanus finmarchicus* and the NAO are positively correlated at a 2-year lag, and *C. finmarchicus* and RSWT are positively correlated at a 0-year lag. *Calanus finmarchicus* and GOM-SST are positively correlated at a 2-year lag. Given the lags (2 years) and

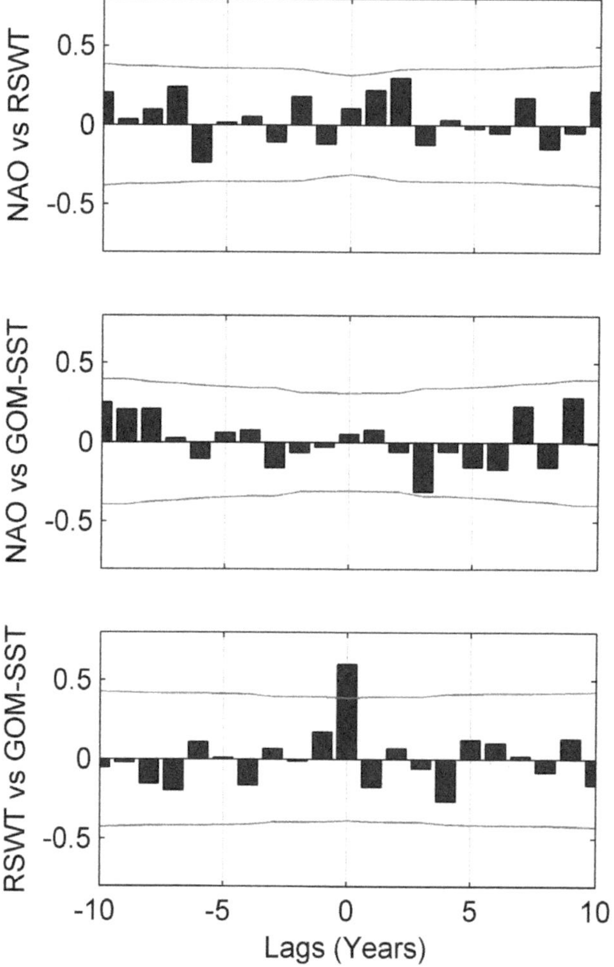

FIGURE 4. Cross-correlation analyses between the North Atlantic oscillation (NAO), Gulf of Maine sea surface temperature (GOM-SST), and the regional slope water temperature index (RSWT). Significance of each correlation ($p < 0.05$) is denoted by thin gray lines and is based on the number of observations included in the correlation at each lag.

a minimum sample size for the serial correlations (10 years), the serial correlations only extend into the early 2000s, but it does appear that the relationships are changing again.

These analyses suggest that the conceptual framework presented in Figure 1 is too simple and that the strength, direction, and lag of the relationships between *C. finmarchicus* and environmental indicators change through time. In the 1980s, there were strong links between the NAO and water temperature and between water temperature and *C. finmarchicus*, each at 2-year lags. However, positive correlations between RSWT and *C. finmarchicus* at a 2-year lag and negative correlations between GOM-SST and *C. finmarchicus* at a 0-year lag were evident, suggesting a tension between the temperature relations with *C. finmarchicus*. The RSWT and GOM-SST are positively correlated at a 0 lag (Figure 4), but the 1980s represents the lowest correlation between the two (Figure 6). This decoupling may represent an increased importance of local effects on GOM-SST (e.g., air temperatures, freshwater inputs, and changing source waters). The role of processes at the 0-year lag affecting *C. finmarchicus* increase in the 1990s, but the direction of

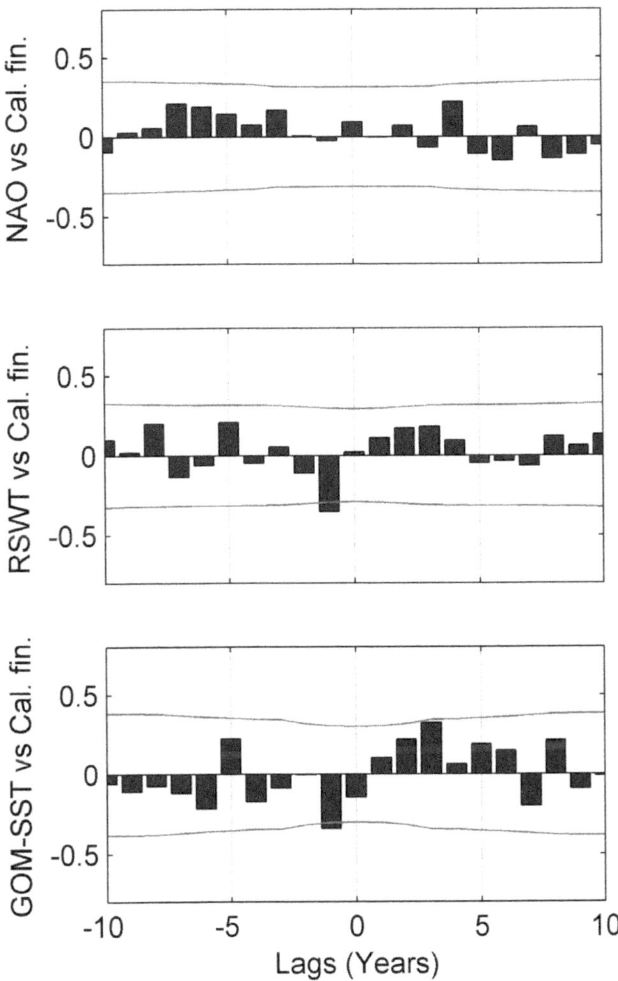

FIGURE 5. Cross-correlation analyses between *Calanus finmarchicus* (Cal. fin.) abundance and the North Atlantic oscillation (NAO), regional slope water temperature index (RSWT), and Gulf of Maine sea surface temperature (GOM-SST). Significance of each correlation ($p < 0.05$) is denoted by thin gray lines and is based on the number of observations included in the correlation at each lag.

effect changes; RSWT is positively related to *C. finmarchicus* (Figure 7). The cause of the switch between negative and positive temperature effects at a 0-year lag is not known and will be addressed in the discussion.

The Zooplankton Community of the Gulf of Maine

Updating the PCA of Pershing et al. (2005) yielded very similar results to the original. The first principal component explained a majority of the annual variability in the seven zooplankton species (69%). *Calanus finmarchicus* was not associated with this component, whereas *Centropages typicus*, *Oithona* spp., *Metridia* spp. and *Pseudocalanus* spp. were associated (Figure 8). The second principal component explained ~15% of the variability; *Calanus finmarchicus*, *Calanus* spp., and euphausiids were most associated with this axis.

The updated multidimensional scaling analysis of Kane (2007) also yielded results similar to the original. Four temporal zooplankton communities can be identified: pre-1980, 1980–1990,

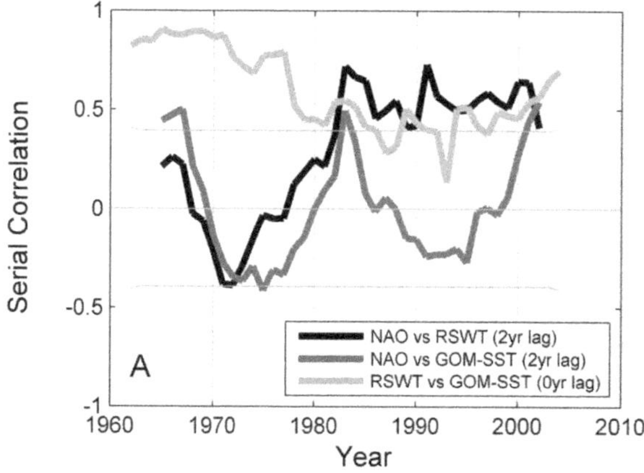

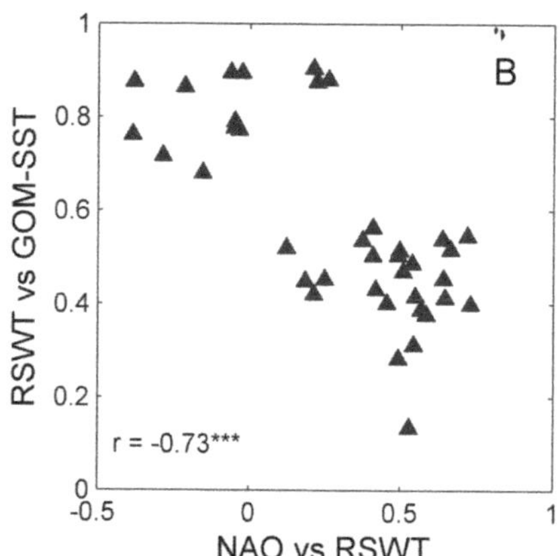

FIGURE 6. (A) Serial correlations between the North Atlantic oscillation (NAO), regional slope water temperature index (RSWT), and Gulf of Maine sea surface temperature (GOM-SST). Correlations with NAO were performed at a lag of 2 years, and correlations between RSWT and GOM-SST were performed at a lag of 0 years. Significance of correlations ($p < 0.05$) is denoted by thin gray lines and is based on the number of observations included in the correlation. (B) Relation between the serial correlations of NAO–RSWT and RSWT–GOM-SST. The correlation between NAO and RSWT is higher when the correlation between RSWT and GOM-SST is lower.

1991–2000, and post-2000 (Figure 9). This same pattern was observed in the updated Pershing analysis, with two important additions. First, in Pershing et al. (2005), the separation between the 1980s and the 2000s was not as pronounced, perhaps a result of the inclusion of fewer taxa in the Pershing et al. (2005) analysis. Second, in the reanalysis of Pershing et al. (2005), the late 1960s/early 1970s remained clearly distinct from other decadal periods, providing support for a distinct zooplankton community pre-1980.

The ordinations of Kane (2007) and Pershing et al. (2005) were similar despite different sampling methods (bongo versus continuous plankton recorder), different areas (Georges Bank versus Gulf of Maine), and different multivariate techniques (MDS versus PCA). The first axis of the MDS was significantly correlated with the first axis of the PCA ($r = 0.81, p < 0.001$), and the second axis of the MDS was significantly correlated with the second axis of the PCA ($r = -0.42, p < 0.05$). Thus, variability in the zooplankton community is spatially consistent across the Gulf of Maine and Georges Bank. Further, the two multivariate axes (whether PCA or MDS) represent the dominant patterns of interannual zooplankton variability across the Gulf of Maine region. The first axis explains a majority of the variability in the zooplankton community and is related to variability in smaller-bodied zooplankton. The second axis is related to *Calanus finmarchicus*.

Based on prior work, the link between these dominant interannual modes of variability and environmental factors is uncertain. Pershing et al. (2005) did not examine the relation between their PCA results and environmental variables. Kane (2007) identified links between individual taxa and environmental variability but did not find any significant relationships between the MDS axes and environmental variables Using the updated analyses of Pershing et al. (2005) and Kane (2007), we evaluated the associations between the primary modes of zooplankton variability in the Gulf of Maine (the first two axes of the MDS analyses: MDS Axis 1 and MDS Axis 2) and a suite of environmental variables, including those related to *C. finmarchicus* by the studies discussed above (the NAO, RWST, and GOM-SST), and those related to individual taxa by Kane (2007; GB-ST and GB-SS; see Table 1). The same analytical approach was used here as was used above for examining the relation between *C. finmarchicus* and environmental variability: cross-correlation

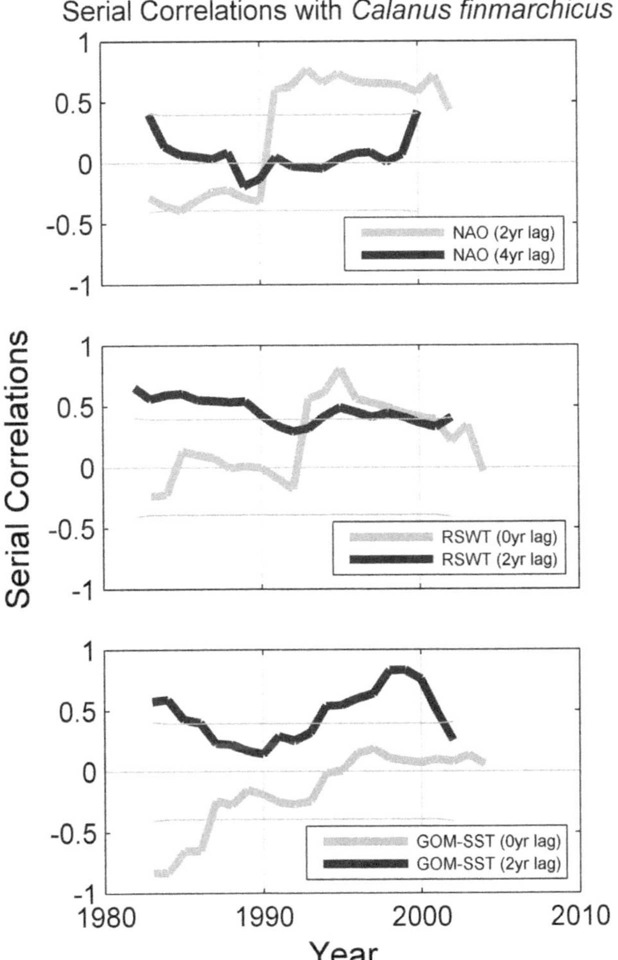

FIGURE 7. Serial correlations between *Calanus finmarchicus* and the North Atlantic oscillation (NAO), regional slope water temperature index (RSWT), and Gulf of Maine sea surface temperature (GOM-SST). The lags used in the serial correlations are provided. Significance of correlations ($p < 0.05$) is denoted by thin gray lines and is based on the number of observations included in the correlation. The time series are truncated to exclude the missing data in the *C. finmarchicus* series.

and serial correlation analyses of de-trended, pre-whitened time series.

Surface temperatures have increased and surface salinities have decreased on Georges Bank (Figure 10). Multiannual variability is also observed with periods of relatively high temperatures (early 2000s) and periods of relatively low salinities (late 1990s and mid-2000s). The time series of MDS Axis 1 clearly shows the shift in 1990 and 2001. The series of MDS Axis 2 shows a shift in 2002. The time series of Pershing et al.'s (2005) PCA axes also exhibit major shifts at these same points (these data are not shown since the MDS and PCA axes are significantly correlated—see above).

There were no clear relations between hydrographic conditions on Georges Bank (GB-ST and GB-SS) and the NAO and RSWT (Table 2). Not surprisingly, GB-ST is highly correlated to GOM-SST at zero lag, and cross-correlations between GB-ST and other variables are very similar to GOM-SST (Figure 4); no significant cross-correlations with the NAO and the highest correlation with RSWT at a zero lag. There were no

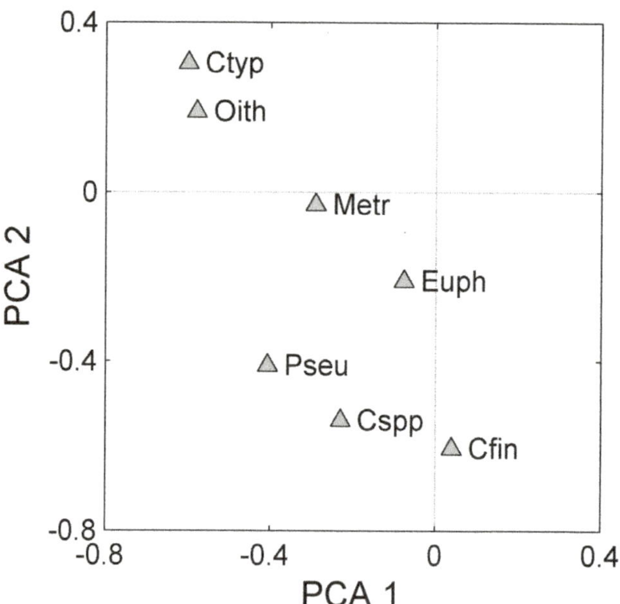

FIGURE 8. Updated principal component analysis (PCA) of Pershing et al. (2005) showing the taxa coefficients for the first two PCA axes. The first axis explains most of the variability (69%) and is primarily composed of *Centropages typicus*, *Oithona* spp., *Metridia* spp., and *Pseudocalanus* spp. The second axis explains less variance (15%) and is primarily composed of *Pseudocalanus* spp., *Calanus* spp., and *Calanus finmarchicus*.

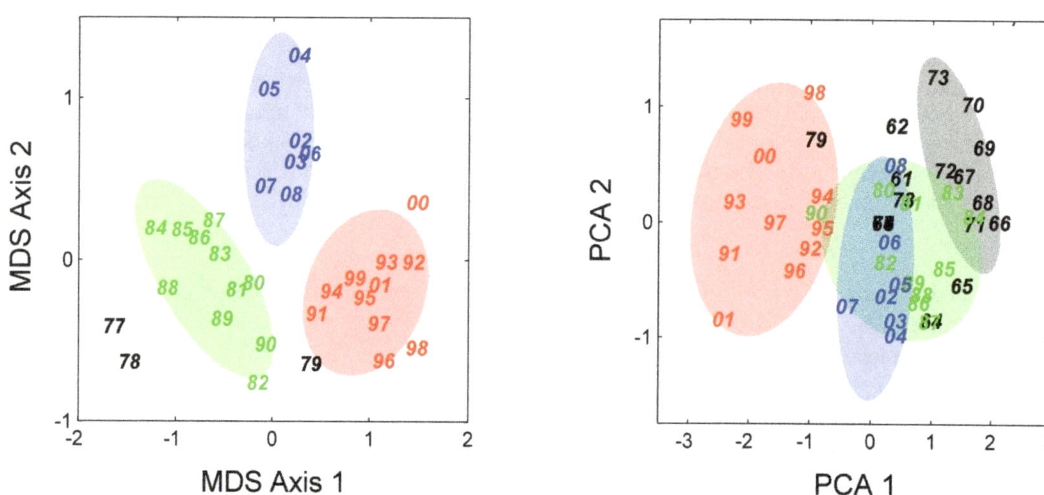

FIGURE 9. Updated multidimensional scaling (MDS) ordination of Kane (2007) and principal component analysis (PCA) of Pershing et al. (2005). Periods of similar years as defined by Kane (2007) are indicated by different colors. The 70% confidence ellipses are shown for each group of years.

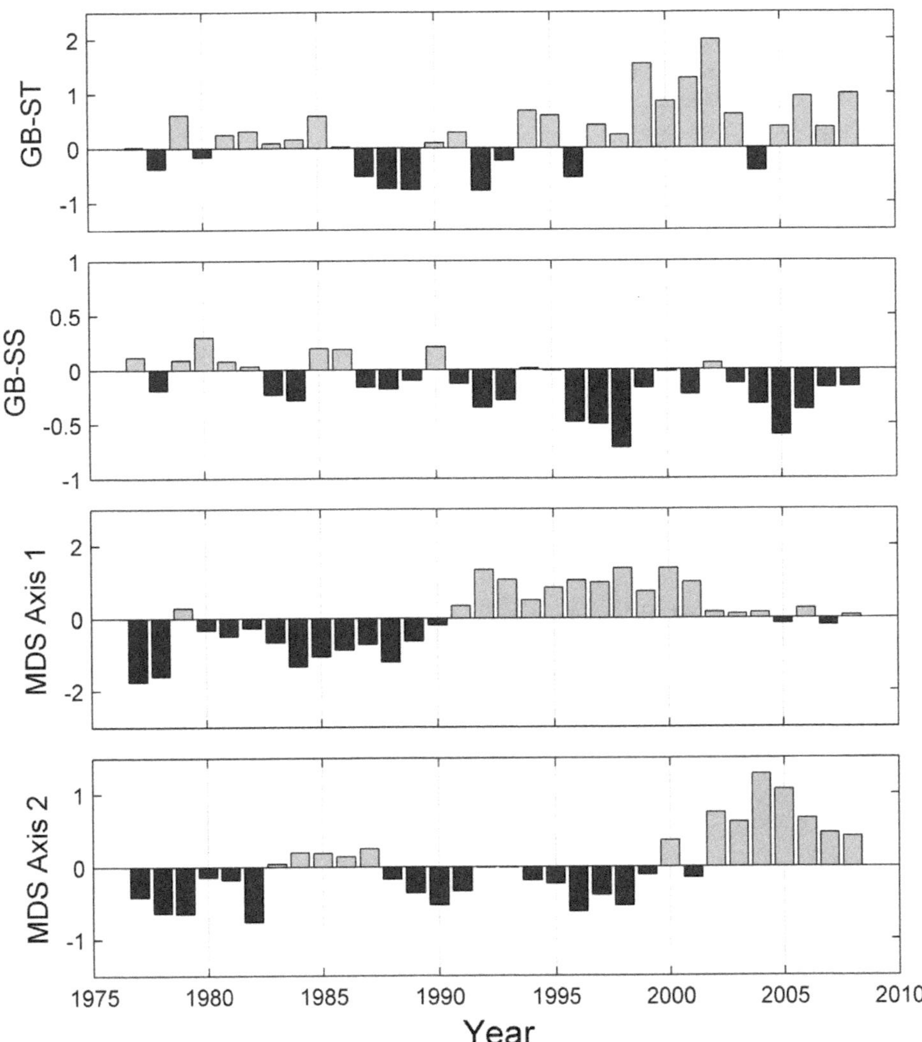

FIGURE 10. Time series of the Georges Bank surface temperature (GB-ST), Georges Bank surface salinity (GB-SS), and the multidimensional scaling (MDS) Axis 1 and MDS Axis 2 (Kane 2007). The Georges Bank salinity and temperature data are anomalies relative to the 1977–1987 period (Mountain 2004).

significant correlations between GB-SS and the NAO, RSWT, or GOM-SST. There is a significant correlation between GB-SS and GB-ST at a 0-year lag. These analyses indicate little linear relation between the NAO, temperature, and salinity in the Gulf of Maine region.

There were also no clear relations between the primary axes of zooplankton variability (MDS Axis 1 and 2) and environmental variables (Table 3). MDS Axis 1 was most highly correlated with the NAO, GOM-SST, and GB-ST at a 3-year lag (environment leading zooplankton), but none of these correlations were significant. MDS Axis 2 also was most highly correlated with the NAO, GOM-SST, and GB-ST but at a 1- or 2-year lag, but again, none of these correlations was significant. Correlations with RSWT and GB-SS were generally low (<0.22), nonsignificant, and greatest at a 0-year lag. These analyses suggest that simple linear relationships are insufficient to explain zooplankton variability in the Gulf of Maine.

TABLE 2. Summary of cross-correlation analyses between Georges Bank surface hydrography and environmental/climate indices. Analyses included Georges Bank surface temperature (GB-ST), Georges Bank surface salinity (GB-SS), North Atlantic oscillation (NAO), regional slope water temperature index (RSWT), and Gulf of Maine sea surface temperature (GOM-SST) (see Table 1). Cross-correlation analyses were performed on de-trended, pre-whitened time series. The maximum correlation, lag of maximum correlation, and significance are provided: not significantly different than zero (ns), and significant at $p < 0.05$ (*).

Variable	GB-SS	GB-ST
NAO	–0.39 (+3 lag) ns	–0.27 (+3 lag) ns
RSWT	0.34 (+0 lag) ns	0.37 (+0 lag) ns
GOM-SST	0.40 (+0 lag) ns	0.75 (+0 lag) *
GB-ST	0.43 (+0 lag) *	

The serial correlations between the multivariate zooplankton axes and environmental indicators demonstrate that these relationships vary over time (Figure 11). MDS Axis 1 was correlated with the NAO at a 3-year lag in the late 1980s and early 1990s. The MDS Axis 1 also was negatively correlated with GOM-SST and GB-ST in the 1980s and with GOM-SST in the later 1990s and early 2000s. This axis was marginally correlated with RSWT at a 2-year lag through the late 1980s and 1990s. Finally, this axis was positively correlated with GB-SS in the middle 1980s and then negatively correlated with GB-SS in the late 1990s and 2000s. The MDS Axis 2 was negatively correlated with the NAO at a 1-year lag in the late 1990s and early 2000s. The MDS Axis 2 also was positively correlated with GOM-SST and GB-ST in the 1980s and with GOM-SST in the late 1990s. This axis was positively correlated with RSWT at a 0-year lag through the mid-1990s. Finally, this axis was negatively correlated with GB-SS at a 0-year lag in the mid-1980s and then negatively correlated with GB-SS in the late 1990s and 2000s. The important result is that these serial correlations show periods of significant and nonsignificant correlations, as well as changes in the sign of correlations from negative to positive and vice versus.

These analyses of zooplankton community structure support the earlier analyses of Pershing et al. (2005) and Kane (2007). Distinct periods of zooplankton community structure are evident. The dominant axis of zooplankton variability (MDS Axis 1) relates to smaller-bodied taxa (e.g., *Centropages typicus* and *Oithona* spp.), and the second axis of zooplankton variability (MDS

TABLE 3. Summary of cross-correlation analyses between the primary axes of zooplankton variability and environmental/climate indices. The axes of zooplankton variability were derived from a multidimensional scaling analysis of zooplankton from Georges Bank (MDS Axis 1 and 2). The environmental/climate indices included North Atlantic oscillation (NAO), regional slope water temperature index (RSWT), Gulf of Maine sea surface temperature (GOM-SST), Georges Bank surface temperature (GB-ST), and Georges Bank surface salinity (GB-SS) (see Table 1). Cross-correlation analyses were performed on de-trended, pre-whitened time series. The maximum correlation, lag of maximum correlation, and significance are provided: not significantly different than zero (ns) and significant at $p < 0.05$ (*)

Variable	MDS Axis 1	MDS Axis 2
NAO	0.29 (+3 lag) ns	–0.36 (+1 lag) ns
RSWT	–0.21 (+2 lag) ns	0.19 (+0 lag) ns
GOM-SST	–0.37 (+3 lag) ns	0.37 (+1 lag) ns
GB-ST	–0.25 (+3 lag) ns	0.30 (+2 lag) ns
GB-SS	–0.15 (0 lag) ns	0.22 (0 lag) ns

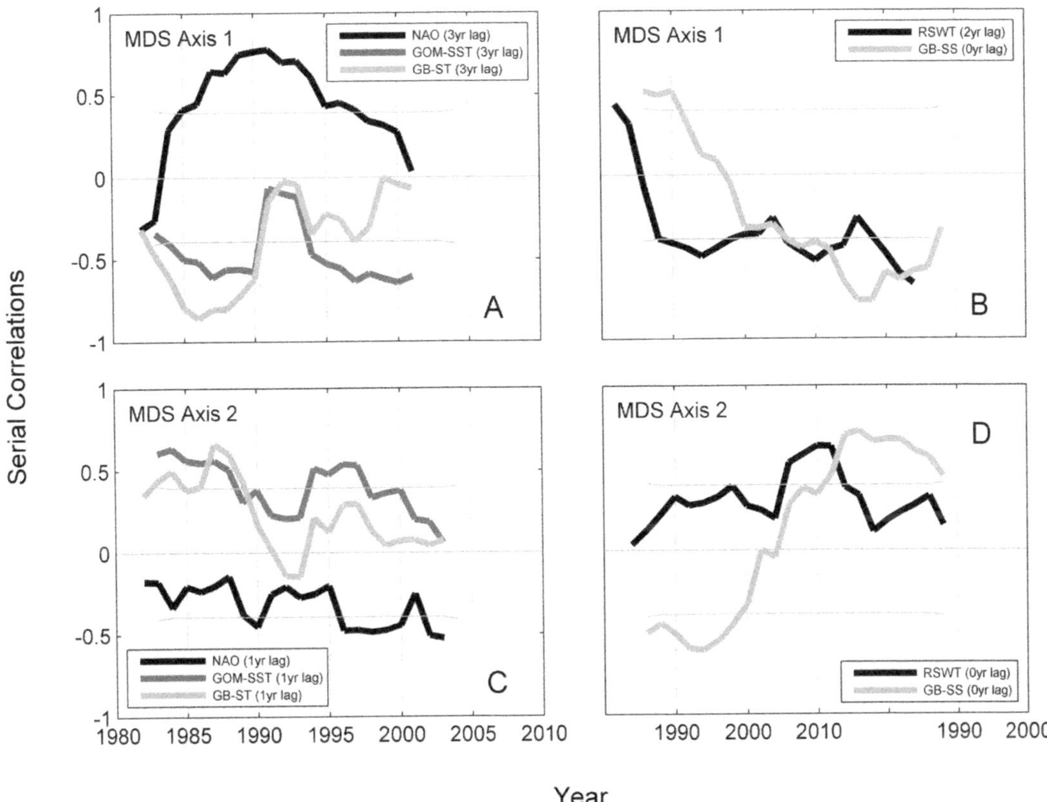

FIGURE 11. Serial correlations between the two zooplankton multidimensional scaling (MDS) axes and environmental variables. (A) The first MDS axis and the North Atlantic oscillation (NAO), Gulf of Maine sea surface temperature (GOM-SST), and Georges Bank surface temperature (GB-ST). Correlations are performed at a 3-year lag (environment leading zooplankton); this lag had the highest correlations in cross-correlation analyses (Table 3). (B) The first MDS axis and regional slope water temperature index (RSWT) and Georges Bank surface salinity (GB-SS). Correlations are performed at a 2-and 0-year lag, respectively (environment leading zooplankton); these lags had the highest correlations in cross-correlation analyses (Table 3). C) The second MDS axis and NAO, GOM-SST, and GB-ST. Correlations are performed at a 1-year lag (environment leading zooplankton); this lag had the highest correlations in cross-correlation analyses (Table 3). (D) The second MDS axis and RSWT and GB-SS. Correlations are performed at a 0-year lag (environment leading zooplankton); this lag had the highest correlation in cross-correlation analyses (Table 3). Environmental variables are described in Table 1. Significance of correlations ($p < 0.05$) is denoted by thin gray lines and is based on the number of observations included in the correlation.

Axis 2) relates to *Calanus finmarchicus*. Importantly, the links between zooplankton community structure and the environment are nonstationary (i.e., dependent on time). This finding explains the equivocal zooplankton–environment links found by Kane (2007), Mountain and Kane (2010), and other researchers studying the Gulf of Maine system and indicates that the conceptual model provided in Figure 1 is too simplistic.

Discussion

Regime Shifts in the Gulf of Maine

Analyses of long-term data sets often show that ecosystems change through time (e.g., regime shifts, deYoung et al. 2008; alternative stable states, Knowlton 2004; Smith et al. 2012, this volume). The idea of "regime shifts" in planktonic communities is best exemplified in the northeast

Pacific (Hare and Mantua 2000; Peterson and Schwing 2003), where changes in the planktonic community are related to changes in the Pacific decadal oscillation—warm and cool phases in the North Pacific lasting 20–30 years. Alternative stable states are similar but traditionally have been described in benthic marine communities and have been explained more through biological interactions than environmental forcing (Schroder et al. 2001). One of the classic examples of alternative stable states comes from benthic communities in the Gulf of Maine (Harris and Tyrrell 2001), with a change from kelp- and red algae-dominated communities in the 1970s to urchin barren's in the 1980s and 1990s to a community composed of several species of benthic algae and the introduced algae *Codium* in the 2000s.

The analyses of Pershing et al. (2005) and Kane (2007) show clear evidence for a change in the zooplankton community structure consistent with the concept of regime shifts. There also is evidence for a 1990 regime shift and the 2000 regime shift in other data from the Gulf of Maine. Townsend et al. (2010) found that nutrients in the Gulf of Maine changed at times consistent with the changes in zooplankton community structure. Nitrate concentrations have decreased and silicate concentrations have increased, and these changes may be related to the relative importance of deepwater inflow into the Gulf of Maine. Such changes in inflow were documented by Smith et al. (2012), with deepwater inflows decreasing and surface inflows increasing, especially along the inner half of the Scotian shelf. These changes in nutrients and circulation are only partially linked to the NAO, similar to the time varying linkages between the NAO and zooplankton reported here.

Changes in the Gulf of Maine ecosystem cooccur with other changes in the larger region. Significant multidecadal shifts have been identified in the diatoms of Narragansett Bay (Borkman and Smayda 2009). Changes in the relative rates of Atlantic cod *Gadus morhua* and haddock recruitment are correlated to changes in zooplankton community structure (Mountain and Kane 2010). Large-scale changes in phytoplankton and zooplankton in the Gulf of Maine could also be expected to affect haddock recruitment (Friedland et al. 2008) and North Atlantic right whale *Eubalaena glacialis* calving rates (Greene and Pershing 2004). Thus, it is important to identify the processes and factors that cause large-scale shifts in the plankton in the Gulf of Maine, that result in the time-varying linkages between the planktonic ecosystem and the environment, and that result in changes in the structure and function of the Gulf of Maine ecosystem.

Recently, Kane (2007), Greene and Pershing (2007), and Mountain and Kane (2010) hypothesized that changes in salinity and high-latitude freshwater input may be an important driver of zooplankton dynamics in the Gulf of Maine. The serial correlations reported here, however, suggest that the relation between zooplankton and salinity has changed sign over the time period, indicating that salinity alone in not related to zooplankton dynamics. This interpretation is susceptible to spurious correlations—the significant tests used in the serial correlations are not corrected for multiple comparisons. This susceptibility also affects studies that relate a biological variable to multiple environmental variables. These analyses must be viewed as exploratory. However, all the serial correlation analyses indicate that the zooplankton–environment linkages change through time, suggesting that simple linear hypotheses are insufficient to explain zooplankton dynamics in the Gulf of Maine.

The Concept of Change

Regime shifts and interdecadal variability have been identified in marine ecosystems across the globe, including the North Pacific (Hare and Mantua 2000; Overland et al. 2008), North Sea (Beaugrand 2004; Weijerman et al. 2005), Scotian shelf (Frank et al. 2005), Black Sea (Oguz and Gilbert 2007), and Baltic Sea (Möllmann et al. 2009). The distinction between regime shifts and interdecadal variability is part semantics and part statistical (Overland et al. 2006; Rudnick and Davis 2006). However, both concepts recognize that marine ecosystems experience multiple years of similar conditions that change to multiyear periods of different conditions (deYoung et al. 2008; Andersen et al. 2009). These changes can be caused by climatic change, climate variability, overfishing, eutrophication, invasive species, or some combination of these factors or others (Harris and Tyrrell 2001; Frank et al. 2005; Oguz and Gilbert 2007; deYoung et al. 2008; Möllmann et al. 2009). Recognizing regime shifts and inter-decadal variability is an important need for both fisheries and ecosystem-based management since system productivity often differs between regimes or multiyear periods.

The results here indicate that simple relations to the NAO or water temperature do not explain multiannual patterns in Gulf of Maine zooplankton. Significant correlations were found between zooplankton variables and the NAO over certain periods of years (e.g., Conversi et al. 2001; Licandro et al. 2001; MERCINA 2001). MERCINA (2004) suggested that the relation between the NAO and *C. finmarchicus* may change through time. Here, this result is substantiated, and it is further shown that the relation between the NAO, other environmental variables, and zooplankton dynamics (more broadly than only C. finmarchicus) vary through time (Figures 7 and 11). Nonstationary relationships between the NAO and environmental variables have been reported in other studies. Joyce (2002) noted a change in the relationship between the NAO and eastern U.S. winter air temperatures, and Vicente-Serrano and López-Moreno (2008) found a change in the relationship between the NAO and European precipitation. Hilmer and Jung (2000) documented a change in the relations between the NAO and Arctic sea ice export.

The number of variables for which nonstationary relationships with the NAO have been documented suggests that the effect of the NAO on the entire North Atlantic varies through time or that additional processes are also affecting the system that act independently of the NAO (Hilmer and Jung 2000; Joyce 2002; Lu and Greatbatch 2002; Vicente-Serrano and López-Moreno 2008). There is evidence that the low pressures cells that define the NAO (the Icelandic low and the Azores high) have shifted eastward, and this may affect the nonstationary relations between the NAO and other variables (see Jung et al. 2003; Cassou et al. 2004; Vicente-Serrano and López-Moreno 2008). Understanding the cause of the nonstationary relations between the NAO and other variables should be a priority for basin-scale climate and ecosystem research. The influence of the eastward shift in the NAO on North Atlantic ecosystems also needs to be studied in more detail. Although environment–ecosystem correlations are useful, the time-varying nature of correlations indicates that mechanistic hypotheses of the effect of the environment on zooplankton dynamics are needed. These mechanistic hypotheses may include advection (MERCINA 2004), changes in seasonality (e.g., Baier and Napp 2003), changes in stratification (Ji et al. 2007), or other processes.

Changes in the relationship between variables through time have been termed phase transitions (Duffy-Anderson et al. 2005). An example of phase transitions is the change in the dominant process in walleye pollock *Theragra chalcogramma* recruitment, consistent with a regime shift in the northeast Pacific. Control of walleye pollock recruitment shifted from environmental effects on larvae to biological effects of juveniles, following an increase in the abundance of predatory flatfishes and Atlantic cod (Bailey 2000; Ciannelli et al. 2005). The presence of regime shifts and phase transitions contributes to the concept of ecosystems as complex adaptive systems, governed by nonlinear dynamics, historical dependency, and multiple outcomes of dynamics (Levin 1998).

Complex adaptive systems present major challenges for ecosystem-based approaches to management because the occurrence of regimes shifts and phase transitions cannot be foreseen (deYoung et al. 2008). Continued observations can detect ecosystem changes, and prior scenario evaluation can provide insight into ecosystem function if a regime shift or phase transition occurs (e.g., Overholtz and Link 2009). Thus ongoing monitoring and strategic modeling need to be seen as components of ecosystem-based management and not as separate activities.

A further implication of ecosystems as adaptive complex systems is that the system can never be fully described, no matter how much detailed research is conducted. The research from Bigelow (1926) to this study need to be viewed in a historical context. Clearly, many aspects of Bigelow's initial work still hold, but the dynamics described by MERCINA (2004) and others are themselves dynamic. Understanding the factors behind these changing dynamics must be a priority for research in the Gulf of Maine. Further, developing a more mechanistic understanding of the effect of the environment on zooplankton dynamics is needed (e.g., Ji et al. 2007). While this modeling research is underway, continued observations must monitor for the next regime shift in zooplankton and the next phase transition in the processes that are related to zooplankton dynamics. These monitoring activities should target both interannual patterns, as well as monthly to seasonal patterns, perhaps by extending the AZMP sampling protocols into the U.S. waters of the Gulf of Maine. Field studies directed by hypotheses could also be used to test specific hypotheses. For example, the coupled slope water hypothesis of MERCINA (2003, 2004) could be tested, based on the extreme negative NAO[3] observed in 2010, with effects in the Gulf of Maine predicted for 2010–2014. Finally, scenarios of changes in zooplankton community structure need to be evaluated in the context of ecosystem-based fisheries management. These activities will prepare us for the next major change in the Gulf of Maine ecosystem, even though the timing and content of this change cannot be predicted. Recognizing that we will not have a complete understanding of the ecosystem in the future, developing science and management to deal with this uncertainty is an important component of ecosystem-based management in the Gulf of Maine ecosystem.

References

Andersen, T., J. Carstensen, E. Hernández-García, and C. M. Duarte. 2009. Ecological thresholds and regime shifts: approaches to identification. Trends in Ecology and Evolution 24:49–57.

Anderson, E. D. 1998. The history of fisheries management and scientific advice: the ICNAF/NAFO history from the end of World War II to the present. Journal of Northwest Atlantic Fishery Science 23:75–94.

Baier, C. T., and J. M. Napp. 2003. Climate-induced variability in *Calanus marshallae* populations. Journal of Plankton Research 25:771–782.

Bailey, K. M. 2000. Shifting control of recruitment of walleye pollock *Theragra chalcogramma* after a major climatic and ecosystem change. Marine Ecology Progress Series 198:215–224.

Beaugrand, G. 2004. The North Sea regime shift: evidence, causes, mechanisms and consequences. Progress in Oceanography 60:245–262.

[3] www.cpc.noaa.gov/data/teledoc/nao_ts.shtml

Bigelow, H. B. 1926. Plankton of the offshore waters of the Gulf of Maine. Bulletin of the Bureau of Fisheries 40:1–509.

Borkman, D. G., and T. Smayda. 2009. Multidecadal (1959–1997) changes in *Skeletonema* abundance and seasonal bloom patterns in Narragansett Bay, Rhode Island, USA. Journal of Sea Research 61:84–94.

Bucklin, A., A. M. Bentley, and S. P. Franzen. 1998. Distribution and relative abundance of *Pseudocalanus moultoni* and *P. newmani* (Copepoda: Calanoida) on Georges Bank using molecular identification of sibling species. Marine Biology 132:97–106.

Cassou, C., L. Terray, J. W. Hurrell, and C. Deser. 2004. North Atlantic winter climate regimes: spatial asymmetry, stationarity with time, and oceanic forcing. Journal of Climate Research 17:1055–1068.

Ciannelli, L., K. M. Bailey, K.-S. Chan, A. Belgrano, and N. C. Stenseth. 2005. Climate change causing phase transitions of walleye pollock (*Theragra chalcogramma*) recruitment dynamics. Proceedings of the Royal Society B 272:1735–1743.

Clarke, G. L., W. T. Edmondson, and W. E. Ricker. 1946. Dynamics of production in a marine area. Ecological Monographs 16:321–337.

Cohen, E. B., and M. D. Grosslein. 1982. An energy budget for Georges Bank. Canadian Special Publication of Fisheries and Aquatic Sciences 59:95–107.

Conversi, A., S. A. Piontkovski, and S. Hameed. 2001. Seasonal and interannual dynamics of *Calanus finmarchicus* in the Gulf of Maine (northeastern US shelf) with reference to the North Atlantic oscillation. Deep-Sea Research Part II 48:519–530.

Cura, J. J. 1987. Phytoplankton. Pages 213–218 *in* R. H. Backus and D. W. Bourne, editors. Georges Bank. The MIT Press, Cambridge, Massachusetts.

deYoung, B., M. Barange, G. Beaugrand, R. Harris, R. I. Perry, M. Scheffer, and F. Werner. 2008. Regime shifts in marine ecosystems: detection, prediction and management. Trends in Ecology and Evolution 23:402–409.

Duffy-Anderson, J. T., K. Bailey, L. Ciannelli, P. Cury, A. Belgrano, and N. C. Stenseth. 2005. Phase transitions in marine fish recruitment processes. Ecological Complexity 2:205–218.

Durbin, E. G., R. Campbell, M. Casas, B. Niehoff, J. Runge, and M. Wagner. 2003. Interannual variation in phytoplankton blooms and zooplankton productivity and abundance in the Gulf of Maine during winter. Marine Ecology Progress Series 254:81–100.

Fish, C. J. 1936. The biology of *Calanus finmarchicus* in the Gulf of Maine and Bay of Fundy. Biological Bulletin 70:118–141.

Francis, R. C., and S. R. Hare. 1994. Decadal-scale regime shifts in the large marine ecosystems of the north-east Pacific: a case for historical science. Fisheries Oceanography 3:279–291.

Frank, K. T., B. Petrie, J. S. Choi, and W. C. Leggett. 2005. Trophic cascades in a formerly cod-dominated ecosystem. Science 308:1621–1623.

Friedland, K. D., J. A. Hare, G. B. Wood, L. A. Col, L. J. Buckley, D., G. Mountain, J. Kane, J. Brodziak, G. R. Lough, and C. H. Pilskaln. 2008. Does the fall phytoplankton bloom control recruitment of Georges Bank haddock, *Melanogrammus aeglefinus*, through parental condition. Canadian Journal of Fisheries and Aquatic Sciences 65:1076–1086.

Greene, C. H., and A. J. Pershing. 2000. The response of *Calanus finmarchicus* populations to climate variability in the northwest Atlantic: basin-scale forcing associated with the North Atlantic oscillation. ICES Journal of Marine Science 57:1536–1544.

Greene, C., and A. J. Pershing. 2003. The flip-side of the North Atlantic oscillation and modal shifts in slope-water circulation patterns. Limnology and Oceanography 48:319–322.

Greene, C., and A. J. Pershing. 2004. Climate and the conservation biology of the North Atlantic right whale: the right whale at the wrong time? Frontiers in Ecology and the Environment 2:29–34.

Greene, C. H., and A. J. Pershing. 2007. Climate drives sea change. Science 315:1084–1085.

Hare, S. R., and N. J. Mantua. 2000. Empirical evidence for North Pacific regime shifts in 1977 and 1989. Progress in Oceanography 47:103–145.

Harris, L. G., and M. C. Tyrrell. 2001. Changing community states in the Gulf of Maine: synergism between invaders, overfishing and climate change. Biological Invasions 3:9–21.

Hilmer, M., and T. Jung. 2000. Evidence for a recent change in the link between the North Atlantic oscillation and Arctic sea ice export. Geophysical Research Letters 27:989–992.

Hjort, J. 1914. Fluctuations in the great fisheries of northern Europe viewed in the light of biological research. Rapports et Procès-Verbaux des Réunions du Conseil 20:1–228.

Hurrel, J. W., Y. Kushnir, G. Ottersen, and M. Visbeck, editors. 2003. The North Atlantic oscillation: climate significance and environmental impact. American Geophysical Union, Geophysical Monograph 134, Washington, D.C.

Ji, R., C. Davis, C. Chen, D. Townsend, D. Mountain, and R. Beardsley. 2007. Influence of ocean freshening on shelf phytoplankton dynamics. Geophysical Research Letters 34:L24607. DOI:10.1029/2007GL032010.

Johnson, C. L., and J. A. Hare, 2012. Zooplankton monitoring in the Gulf of Maine: past, present, and future. Pages 205–218 *in* R. L. Stephenson, J. H. Annala, J. A. Runge, and M. Hall-Arber, editors. Advancing an ecosystem approach in the Gulf of Maine. American Fisheries Society, Symposium 79, Bethesda, Maryland.

Johnson, C. L., J. A. Runge, K. A. Curtis, E. G. Durbin, J. A. Hare, L. S. Incze, J. S. Link, G. D. Melvin, T. D. O'Brien, and L. Van Guelpen. 2011. Biodiversity and ecosystem function in the Gulf of Maine: pattern and role of zooplankton and pelagic nekton. PLoS (Public Library of Science) One [online serial] 6(1):e16491. DOI: 10.1371/journal.pone.0016491.

Jossi, J. W., and J. R. Goulet. 1993. Zooplankton trends: US north-east shelf ecosystem and adjacent regions differ from north-east Atlantic and North Sea. ICES Journal of Marine Science 50:303–313.

Jossi, J. W., A. W. G. John, and D. Sameoto. 2003. Continuous Plankton Recorder sampling off the east coast of North America: history and status. Progress in Oceanography 58:313–325.

Joyce, T. M. 2002. One hundred plus years of wintertime climate variability in the eastern United States. Journal of Climate 15:1076–1086.

Jung, T., M. Hilmer, E. Ruprecht, S. Kleppek, S. K. Gulev, and O. Zolina. 2003. Characteristics of the recent eastward shift of interannual NAO variability. Journal of Climate 16:3371–3382.

Kane, J. 2007. Zooplankton abundance trends on Georges Bank, 1977–2004. ICES Journal of Marine Science 64:909–919.

Kane, J. 2009. A comparison of two zooplankton time series data collected in the Gulf of Maine. Journal of Plankton Research 31:249–259.

Kimmel, D. G., and S. Hameed. 2008. Update on the relationship between the North Atlantic oscillation and *Calanus finmarchicus*. Marine Ecology Progress Series 366:111–117.

Knowlton, N. 2004. Multiple "stable" states and the

conservation of marine ecosystems. Progress in Oceanography 60:387–396.

Levin, S. A. 1998. Ecosystems and the biosphere as complex adaptive systems. Ecosystems 1:431–436.

Licandro, P., A. Conversi, F. Ibanez, and J. Jossi. 2001. Time series analysis of interrupted long-term data set (1961–1991) of zooplankton abundance in Gulf of Maine (northern Atlantic, USA). Oceanologica Acta 24:453–466.

Link, J. S., W. Overholtz, J. O'Reilly, J. Green, D. Dow, D. Palka, C. Legault, J. Vitaliano, V. Guida, M. Fogarty, J. Brodziak, L. Methratta, W. Stockhausen, L. Col, and C Griswold. 2008. The northeast US continental shelf Energy Modeling and Analysis exercise (EMAX): ecological network model development and basic ecosystem metrics. Journal of Marine Systems 74:453–474.

Link, J. S. 2002. What does ecosystem-based fisheries management mean? Fisheries 27:18–21.

Lu, J., and R. J. Greatbatch. 2002. The changing relationship between the NAO and northern hemisphere climate variability. Geophysical Research Letters 29(7):1148. DOI: 10.1029/2001GL014052.

Meise, C. J., and J. E. O'Reilly. 1996. Spatial and seasonal patterns in abundance and age-composition of *Calanus finmarchicus* in the Gulf of Maine and on Georges Bank: 1977–1987. Deep Sea Research Part II 43:1473–1501.

MERCINA (Marine Ecosystem Responses to Climate in the North Atlantic). 2001. Oceanographic responses to climate in the northwest Atlantic. Oceanography 14:76–82.

MERCINA (Marine Ecosystem Responses to Climate in the North Atlantic). 2003. Trans-Atlantic responses of *Calanus finmarchicus* populations to basin-scale forcing associated with the North Atlantic oscillation. Progress in Oceanography 58:301–312.

MERCINA (Marine Ecosystem Responses to Climate in the North Atlantic). 2004. Supply-side ecology response of climate-driven changes in the North Atlantic Ocean. Oceanography 17:60–71.

Mitchell, M. R., G. Harrison, K. Pauley, A. Gagné, G. Maillet, and P. Strain. 2002. Atlantic Zonal Monitoring Program sampling protocol. Canadian Technical Report of Hydrography and Ocean Sciences 223.

Möllmann, C., T. Diekmann, B. Müller-Karulis, G. Kornilovs, M. Plikshs, and P. Axe. 2009. Reorganization of a large marine ecosystem due to atmospheric and anthropogenic pressure: a discontinuous regime shift in the central Baltic Sea. Global Change Biology 15:1377–1393.

Mountain, D. G. 2004. Variability of the water properties in NAFO subareas 5 and 6 during the 1990s. Journal of Northwest Atlantic Fishery Science 34:103–112.

Mountain, D. G., and J. Kane. 2010. Major changes in the Georges Bank ecosystem, 1980s to the 1990s. Marine Ecology Progress Series 398:81–91.

Oguz, T., and D. Gilbert. 2007. Abrupt transitions of the top-down controlled Black Sea pelagic ecosystem during 1960–2000: evidence for regime-shifts under strong fishery exploitation and nutrient enrichment modulated by climate-induced variations. Deep-Sea Research Part I 54:220–242.

Overholtz, W., and J. S. Link. 2009. A simulation model to explore the response of the Gulf of Maine food web to large-scale environmental and ecological changes. Ecological Modelling 220:2491–2502.

Overland, J., S. Rodionov, S. Minobe, and N. Bond. 2008. North Pacific regime shifts: definitions, issues and recent transitions. Progress in Oceanography 77:92–102.

Overland, J. E., D. B. Percival, and H. O. Mqfjeld. 2006. Regime shifts and red noise in the North Pacific. Deep-Sea Research Part I 53:582–588.

Pershing, A. J. 2001. Response of large marine ecosystems to climate variability: patterns, processes, concepts, and methods. Doctoral dissertation. Cornell University, Ithaca, New York

Pershing, A. J., C. H. Greene, J. W. Jossi, L. O'Brien, J. K. T. Brodziak, and B. A. Bailey. 2005. Interdecadal variability in the Gulf of Maine zooplankton community, with potential impacts on fish recruitment. ICES Journal of Marine Science 62:1511–1523.

Peterson, W. T., and F. B. Schwing. 2003. A new climate regime in northeast pacific ecosystems. Geophysical Research Letters 30 (17):1896. DOI: 10.1029/2003GL017528.

Pikitch, E. K., C. Santora, E. A. Babcock, A. Bakun, R. Bonfil, D. O. Conover, P. Dayton, P. Doukakis, D. Fluharty, B. Heneman, E. D. Houde, J. S. Link, P. A. Livingston, M. Mangel, M. K. McAllister, J. Pope, and K. J. Sainsbury. 2004. Ecosystem-based fishery management. Science 305:346–347.

Piontkovski, S. A., T. D. O'Brien, S. F. Umani, E. G. Krupa, T. S. Stuge, K. S. Balymbetov, O. V. Grishaeva, and A. G. Kasymov. 2006. Zooplankton and the North Atlantic oscillation: a basin-scale analysis. Journal of Plankton Research 28:1039–1046.

Polovina, J. 1984. Model of a coral reef ecosystem I. The ECOPATH model and its application to French Frigate Shoals. Marine Biology 3:1–11.

Rosenberg, A. A., M. J. Fogarty, M. P. Sissenwine, J. R. Beddington, and J. G. Shepard. 1993. Achieving sustainable use of renewable resources. Science 262:828–829.

Rudnick, D. L., and R. E. Davis. 2006. Comment on "Regime shifts and red noise in the North Pacific". Deep-Sea Research Part I 53:589–590.

Runge, J. A., and R. J. Jones. 2012. Results of a collaborative project to observe coastal zooplankton and ichthyoplankton abundance and diversity in the western Gulf of Maine: 2003–2008. Pages 345–359 *in* R. L. Stephenson, J. H. Annala, J. A. Runge, and M. Hall-Arber, editors. Advancing an ecosystem approach in the Gulf of Maine. American Fisheries Society, Symposium 79, Bethesda, Maryland.

Ryther, J. H. 1969. Photosynthesis and fish production in the sea. Science 166:72–76.

Schroder, A., L. Persson, and A. De Roos. 2001. Direct experimental evidence for alternative stable states: a review. Oikos 110:3–19.

Sibunka, J. D., and M. J. Silverman. 1989. MARMAP surveys of the continental shelf from Cape Hatteras, North Carolina, to Cape Sable, Nova Scotia (1977–1983). Atlas no. 1, summary of Operations. National Oceanic and Atmospheric Administration, National Marine Fisheries Service, Northeast Fisheries Center, NOAA Technical Memorandum NMFS-F/NEC-68, Woods Hole, Massachusetts.

Smith, P. C., N. R. Pettigrew, P. Yeats, D. W. Townsend, and G. Han. 2012. Regime shift in the Gulf of Maine. Pages 185–203 *in* R. L. Stephenson, J. H. Annala, J. A. Runge, and M. Hall-Arber, editors. Advancing an ecosystem approach in the Gulf of Maine. American Fisheries Society, Symposium 79, Bethesda, Maryland.

Steele, J. H., J. S. Collie, J. J. Bisagni, D. J. Gifford, M. J. Fogarty, J. S. Link, B. K. Sullivan, M. E. Sieracki, A. R. Beet, D. G. Mountain, E. G. Durbin, D. Palka, and W. T. Stockhausen. 2007. Balancing end-to-end budgets of the Georges Bank ecosystem. Progress in Oceanography 74:423–448.

Therriault, J.-C., B. Petrie, P. Pepin, J. Gagnon, D. Gregory, J. Helbig, A. Herman, D. Lefaivre, M. Mitchell, B. Pelchat, J. Runge, and D. Sameoto. 1998. Proposal for a northwest Atlantic zonal monitoring program. Canadian Technical Report of Hydrography and Ocean Sciences 194.

Townsend, D. W., N. D. Rebuck, M. A. Thomas, L. Karp-Boss, and R. M. Gettings. 2010. A changing nutrient regime in the Gulf of Maine. Continental Shelf Research 30:820–832.

Townsend, D. W., and R. W. Spinrad. 1986. Early spring phytoplankton blooms in the Gulf of Maine. Continental Shelf Research 6:515–529.

Vicente-Serrano, S. M., and J. I. López-Moreno. 2008. Nonstationary influence of the North Atlantic oscillation on European precipitation. Journal of Geophysical Research 113: D20120. DOI: 10.1029/2008JD010382.

Walford, L. A. 1938. Effect of currents on distribution and survival of the egs and larvae of the haddock (*Melanogrammus aeglefinus*) on Georges Bank. U.S. Bureau of Fisheries Bulletin 49:1–73.

Weijerman, M., H. Lindeboom, and A. F. Zuur. 2005. Regime shifts in marine ecosystems of the North Sea and Wadden Sea. Marine Ecology Progress Series 298:21–39.

American Fisheries Society Symposium 79:139–165, 2012

Status of the Northeast U.S. Continental Shelf Large Marine Ecosystem: An Indicator-Based Approach

MICHAEL J. FOGARTY*
*Northeast Fisheries Science Center, National Marine Fisheries Service
Woods Hole, Massachusetts 02543, USA*

KEVIN D. FRIEDLAND
*Northeast Fisheries Science Center, National Marine Fisheries Service
Narragansett, Rhode Island 02882, USA*

LAUREL COL AND ROBERT GAMBLE
*Northeast Fisheries Science Center, National Marine Fisheries Service
Woods Hole, Massachusetts 02543, USA*

JONATHAN HARE AND KIMBERLY HYDE
*Northeast Fisheries Science Center, National Marine Fisheries Service
Narragansett, Rhode Island 02882, USA*

JASON S. LINK, SEAN LUCEY, HUI LIU, JANET NYE, AND WILLIAM J. OVERHOLTZ
*Northeast Fisheries Science Center, National Marine Fisheries Service
Woods Hole, Massachusetts 02543, USA*

DAVID RICHARDSON
*Northeast Fisheries Science Center, National Marine Fisheries Service
Narragansett, Rhode Island 02882, USA*

B. ROUNTREE AND MAUREEN TAYLOR
*Northeast Fisheries Science Center, National Marine Fisheries Service
Woods Hole, Massachusetts 02543, USA*

Abstract.—The northeast U.S. Continental Shelf large marine ecosystem (NES LME) has supported important commercial fisheries for several centuries. The NES LME has experienced structural change due to both intensive exploitation and physical forcing in relation to broader climate impacts in the North Atlantic over the past several decades. Here, we examine the combined effects of anthropogenic and environmental factors on the state of the NES LME using a driver-pressure-state-impact-response framework to structure our assessment of patterns of change in this system.

We partitioned both drivers and pressures according to natural and anthropogenic sources. Ecological state variables encompassed a broad spectrum of trophic levels. Impact metrics are based on economic trends in the fisheries. To represent regulatory responses, we trace the history of management actions in this region over the past five decades.

The critical importance of changes in temperature and water column stratification in ecosystem change, in relation to bottom-up forcing, is identified using canonical

* Corresponding author: michael.fogarty@noaa.gov

redundancy analysis. Analysis of anthropogenic pressures indicate a clear effect of fishing pressure, and removals due to fishing, in the dynamics of fish communities in the region, highlighting an important top-down control mechanism. Analysis of zooplankton community dynamics confirms previous indications of a regime-like change in species composition during the 1990s. Observed changes in fish community dynamics appears to be most clearly related to large-scale switches from a demersal to a pelagic fish dominated system and to changes within the demersal fish community itself.

Introduction

A global consensus on the need to adopt a more holistic and integrated approach to management of ocean resources has now emerged (Garcia et al. 2003; Pikitch et al. 2004). In order to assess the status and trends of ecosystems, and to evaluate the impact of different stressors, appropriate indicators must be identified and their overall utility for management validated (Rochet and Trenkel 2003; Trenkel and Rochet 2003; Fulton et al. 2005; Jennings 2005; Rice and Rochet 2005). These indicators should be broadly representative of forcing factors and associated ecosystem states or processes. It is particularly important to understand the relationship between forcing mechanism and effects on ecosystem states to permit the evaluation of management options by manipulating control variables subject to human influence (Jennings 2005).

Indicators can be broadly classified into natural and anthropogenic *drivers*, resulting *pressures*, and ecosystem *states* (OECD 2003). Drivers are factors such as climate and human population size, which, in turn, result in a constellation of pressures exerted on the system. These pressures include human-related impacts, such as removal of living marine resources through harvesting, effects due to shipping, pollution, and habitat loss or degradation. Climate-related pressures include changes in atmospheric and oceanographic processes, directly or indirectly affecting marine life. Indicators of ecosystem state should be chosen that are potentially affected by these drivers and associated pressures. We are further interested in the *impacts* of changes in ecosystem state on the provision of ecosystem goods and services (often measured in social and economic terms), including the resulting effects on human communities, and the response of managers to these impacts. Collectively this driver-pressure-state-impact-response (DPSIR) structure provides a framework for assessing the condition of an ecosystem (including the human dimension). In practice, the pressure-state-response elements have been most intensively investigated, reflecting the relative amount of information available for these classes of indicators (Jennings 2005). The DPSIR framework comprises one component of an integrated ecosystem assessment (IEA) that can be used to complement other modeling and analytical approaches used in development of an IEA (Levin et al. 2008, 2009).

Indicator-based assessments have previously been provided for the northeast U.S. Continental Shelf large marine ecosystem (NES LME; Figure 1), encompassing natural and anthropogenic forcing factors and selected ecosystem state variables (Brodziak and Link 2002; Link et al. 2002). The definition of potential ecosystem-based reference points for state variables (Link 2005) and the evaluation of state indicator performance (Methratta and Link 2006) have also been explored for this system. Here, we expand these efforts to encompass the broader dimensions of the DPSIR approach and draw on a wider compilation of indicators for the NES LME (EcoAP 2009). We examine methods of identifying the most informative metrics and the relationships among drivers, pressures, and ecosystem states to develop an assessment for this region.

Methods

Indicators

We screened a set of indicators from an initial compilation of more than 80 variables describing climate, atmospheric, oceanographic, ecological, and human-related aspects of this system (EcoAP 2009) using the criteria noted above. Our principal goal was to document status and trends for this system. Accordingly, we considered only indicator variables spanning at least 25 years to pro-

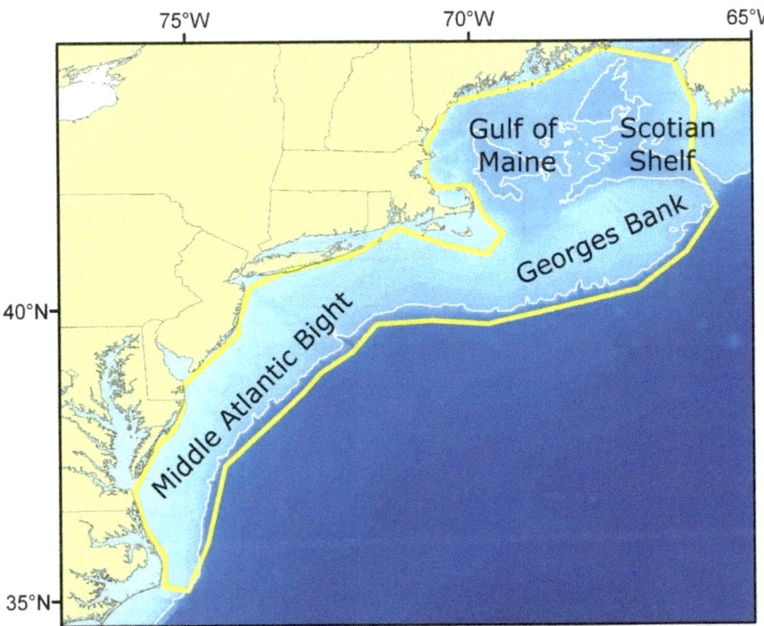

FIGURE 1. Northeast U.S. Continental Shelf large marine ecosystem with major subregions and adjacent Canadian shelf areas identified. Study region delineated in yellow.

vide a near-term historical perspective on change in system dynamics. We focused on interrelationships among indicators classified as drivers, pressures, and ecosystem states. In particular, we wished to identify the most informative indicators from a potentially broad array, on the basis of the demonstrated interactions among drivers and pressures and among pressures and ecosystem state variables. Further details concerning methodology and data sources for the following indicators can be found in EcoAP (2009).

To emphasize trends and general patterns, we illustrate the indicator series as smoothed and normalized variables in the accompanying figures. However, in all statistical analyses, we used the original observations without smoothing. The time span available each indicator available for analysis is provided in the Appendix. Although climate indices and some temperature series are available for more than 100 years, we include only information from 1960 onward in our illustration of trends. For all statistical analyses, we used the longest possible time series available in all cases; in these multivariate analyses, the shortest time series sets the constraint on length of the series to be included.

We selected two macroscale indicators of climate as drivers of change, the North Atlantic oscillation (NAO; Hurrell 1995) and the Atlantic multidecadal oscillation (AMO; Enfield et al. 2001), to represent broadscale physical drivers in the system (Figure 2; Appendix). The NAO is the dominant mode of atmospheric climate variability over the North Atlantic basin (Hurrell 1995) and is known to exert important ecosystem effects (Stenseth et al. 2002). The AMO index is based on spatial patterns in sea surface temperature variability after removing the effects of anthropogenic forcing on temperature, The basin-scale NAO and AMO metrics affect more localized atmospheric and oceanographic conditions, classified as pressure variables in this analysis (Figure 2).

Trends in the population and per capita income for coastal counties, in or adjacent to large metropolitan areas in the Northeast and mid-Atlantic regions were used to characterize human drivers of ecosystem change. Human population and size and wealth have been linked to environmental impacts in coastal systems (Millennium Ecosystem Assessment 2005). These impacts include increases in nitrogen loading and destruction of habitat (including wetland nursery areas)

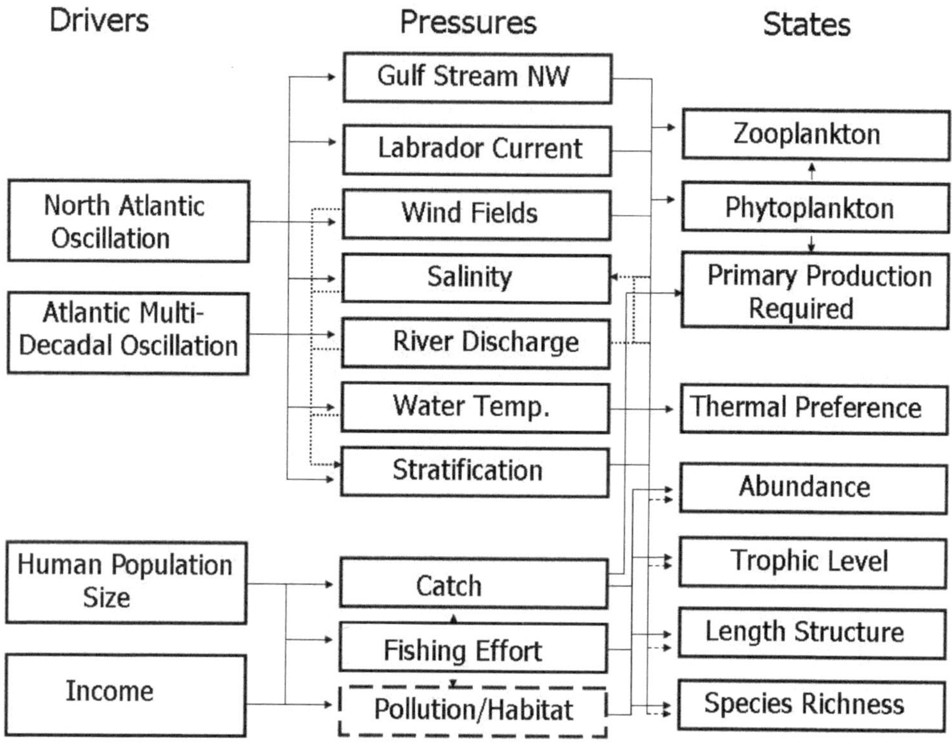

FIGURE 2. Flow diagram for analysis of northeast U.S. Continental Shelf large marine ecosystem indicating physical and anthropogenic drivers and related pressures and classes of ecosystem state variables. Explicit measures of pressures related to pollution and habitat impacts are not available (habitat impacts may be related, however, to fishing activities) Arrows indicate potential relationships among drivers and pressures and among pressures and states. Potential interactions within pressures also indicated (dotted lines). The upper tier of ecosystem state variables relates to phytoplankton and zooplankton metrics. The lower tier of state indicators relates to fish and macroinvertebrate metrics.

as coastal development expands, as well as increased demands on natural resources.

We chose physical pressure variables, known from previous studies to be related to the NAO and the AMO and to be important in this system (Hurrell 1995; Enfield et al. 2001; EcoAP 2009). These large-scale climate drivers affect the physical environment of the NES LME and exert external pressures at its boundaries. These pressures include the Gulf Stream at the southern and offshore boundary, the Labrador Coastal Current at the northern boundary, river discharge at the land boundary, and winds at the atmospheric boundary. In addition to these pressures, climate processes also affect the local physical environment directly, with effects on temperature, salinity, and water column stratification (Figure 2; Appendix). We screened six temperature variables for this system, representing three nearshore coastal station records, the broad-scale extended reconstructed sea surface temperature series (Smith and Reynolds 2003, 2004) and seasonal research vessel survey observations (surface and bottom) for the continental shelf, conducted by the Northeast Fisheries Science Center (NEFSC; Appendix). There are also important relationships among these physical pressure variables. For example, water column stratification is affected by winds, salinity, river discharge, and water temperature (Figure 2).

We used two pressure categories related to exploitation of living marine resources: fishing effort and catch by major species groupings (Figure 2; Table 1; Appendix). The fishing pressure index was the standardized fishing effort for demersal

TABLE 1. Dominant species comprising major taxonomic groups for survey and catch data.

Species group	Species	Scientific name
Principal groundfish	Atlantic cod	*Gadus morhua*
	Haddock	*Melanogrammus aeglefinus*
	Pollock	*Pollachius virens*
	Silver hake	*Merluccius bilinearis*
	Red hake	*Urophycis chuss*
	White hake	*U. tenuis*
	Yellowtail flounder	*Limanda ferruginea*
	Winter flounder	*Pseudopleuronectes americanus*
	American plaice	*Hippoglossoides platessoides*
	Fourspot flounder	*Paralichthys oblongus*
	Witch flounder	*Glyptocephalus cynoglossus*
	Windowpane flounder	*Scophthalmus aquosus*
Other groundfish	Ocean pout	*Zoarces americanus*
	Goosefish	*Lophius americanus*
	Longhorn sculpin	*Myoxocephalus octodecemspinosus*
	Cunner	*Tautogolabrus adspersus*
	Atlantic wolffish	*Anarhichas lupus*
	Acadian redfish	*Sebastes fasciatus*
Elasmobranches	Little skate	*Leucoraja erinacea*
	Winter skate	*L. ocellata*
	Thorny skate	*Amblyraja radiata*
	Barndoor skate	*Dipturus laevis*
	Smooth skate	*Malacoraja senta*
	Spiny dogfish	*Squalus acanthias*
	Smooth dogfish	*Mustelus canis*
Small pelagics	Atlantic mackerel	*Scomber scombrus*
	Butterfish	*Peprilus triacanthus*
	Atlantic herring	*Clupea harengus*
Crustaceans	American lobster	*Homarus americanus*
	Atlantic rock crab	*Cancer irroratus*
	Jonah crab	*C. borealis*
Mollusks	Sea scallop	*Placopecten magellanicus*
	Atlantic surfclam	*Spisula solidissima*
	Ocean quahog	*Arctica islandica*
	Northern shortfin squid	*Illex illicebrosus*
	Longfin inshore squid	*Loligo pealeii*

and pelagic gear types. We indexed the dominant removals by the fishery using catches categorized as groundfish, other finfish, small pelagic fish, elasmobranches, crustaceans, and mollusks. We note that while the removals due to fishing are here characterized as pressures (representing their effect on the population levels of exploited species), catches also represent a major ecosystem provisioning service and are viewed from an entirely different perspective in this context.

Our state variables focused on aggregate or integrative measures of ecological condition for major system components (phytoplankton, zooplankton, and fish and shellfish). Planktonic elements were represented by an index of phytoplankton abundance based on information derived from the NEFSC Continuous Plankton Recorder (CPR) Program (Jossi et al. 2003). This method is size-selective and is best viewed as a measure of the abundance of larger phytoplankton (e.g., diatoms and large dinoflagellates). We also used two measures related to zooplankton derived from the CPR series, total displacement volume and the ratio of small-bodied to large-bodied copepods. For the living marine resource component (principally upper trophic level species), we included aggregate measures of fish and invertebrate abundance (matched to the groupings specified above for catches), as well as factors related to demography, trophodynamics, diversity, and distribution in relation to temperature (Figure 2; Appendix). The latter set of factors included mean trophic level of species represented in the commercial catch and in standardized NEFSC bottom trawl surveys (Smith 2002), mean length of all species in the surveys, survey species richness, and a measure of the thermal preference of species in the survey, weighted by their biomass, to obtain an overall index for the system over time. We also computed the primary production required to support observed levels of catch (Pauly and Christensen 1995)

Our evaluation of impact indicators centered on revenues derived from fishing activities. Both environmental change and the effects of fishing on abundance, demography, and species composition of exploited communities have implications for the amount and value of the catch. In turn, these considerations hold direct implications for fishery-dependent communities. To illustrate these effects on fishing communities, we focus on the effects on changes in groundfish landings and associated revenues for which economic impact information was most readily available. Finally, a brief history of fishery management interventions is provided to illustrate the responses undertaken to address changing resource conditions.

Analysis

For indicators represented by more than one time series, we first extracted the common trend for each variable type to reduce the overall number of indicators considered, while retaining the information derived from multiple sources or locations for these variables. A minimum/maximum autocorrelation factor analysis (MAFA; Solow 1994) was used for this purpose. This method is similar to a principal components analysis but seeks to maximize the first-order autocorrelation in the combined series to emphasize smoothness and trend. In all analyses, the original data were first converted to standard normal deviates and the MAFA applied to the transformed series. This approach was applied to temperature (six series), salinity (two series: surface and bottom), winds (nine series: speed, zonal wind direction, and meridional wind direction for each of three locations), and river discharge (three subregions) to generate single indices for each of these variables to be employed in subsequent statistical analysis.

To examine the relationships among drivers and pressures, and among pressures and ecosystem states, we used a multivariate generalization of multiple regression analysis, canonical redundancy analysis (RDA). Here, redundancy refers to explained variance (Legendre and Legendre 1998). This approach examines the relationship between a matrix of output variables and a set of input or explanatory variables. The ordination of the output variables is constrained to represent a linear combination of the explanatory variables. As part of our screening process to identify the most influential variables, we tested for the effect of each explanatory indicator using a forward stepwise selection procedure and retained only those variables that explained a significant fraction of the variance ($p < 0.05$) in the output variables.

Identification of possible change-points in the system dynamics, reflected in the indicators series, is of particular interest. We tested for evidence of regime-like changes in the compiled driver, pressure, and ecosystem state series using a chronological cluster analysis with temporal contiguity constraints (Legendre and Legendre 1998). This approach ensures that any temporal groupings identified represent sequential time series (Zuur et al. 2007). For this analysis, we set a high level of acceptance ($p < 0.01$) to guard against identifying relatively small changes in system dynamics as significant.

Results and Discussion

Overview of Indicators

Below, we provide additional details on each of the indicators screened for further consideration. Here, we are principally interested in trends in these variables in relation to other factors. We describe trajectories of the drivers, pressures, states, impacts (for the groundfish sector), and responses (represented as history of major management interventions) sequentially in the following overview.

Climate drivers.—The NAO represents the pressure difference between the Icelandic low and the Azores high pressure systems, reflecting the relative strengths of the two systems (Hurrell 1995). When the high pressure system dominates (high NAO), we see increased temperature, precipitation, and westerly winds in the northeastern United States. Conversely, when the low pressure system dominates (low NAO), there is a diminution of westerly winds and precipitation in this region and cooler water temperatures off the northeastern United States (Hurrell 1995). Since the early 1970s, the NAO has primarily been in a positive state (Figure 3a), indicating a dominance of the Azores high pressure system, with associated effects on temperature, wind fields, and precipitation.

The AMO index shows a relatively cool period starting in the early 1960s and extending through the mid-1990s. Since the late 1990s, the AMO has been in a warm phase (Figure 3a). The AMO is

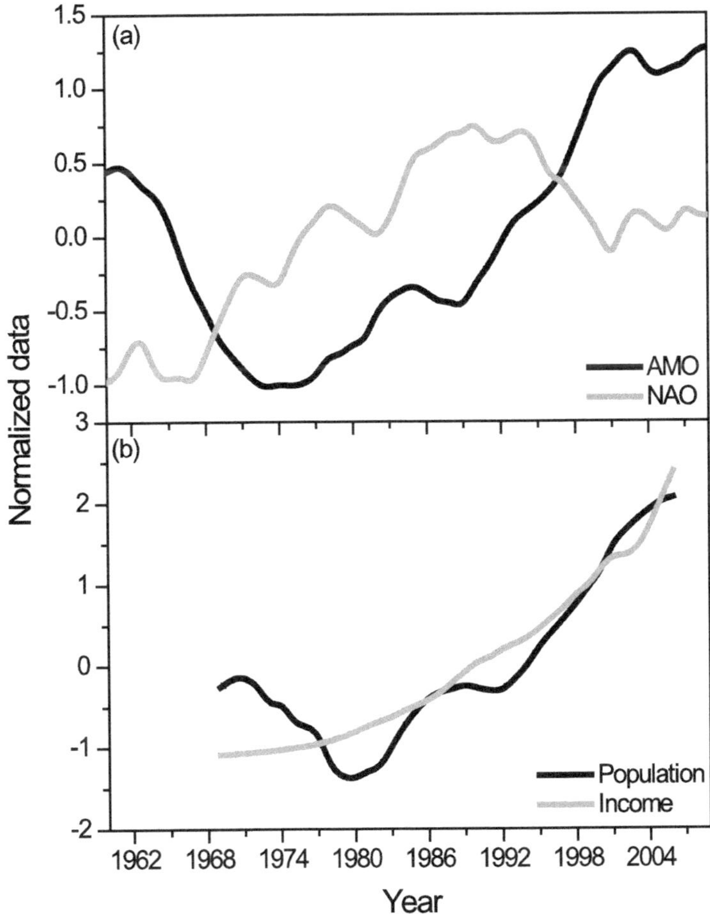

FIGURE 3. Trends in (a) physical and (b) anthropogenic drivers affecting northeast U.S. Continental Shelf large marine ecosystem. AMO = Atlantic multidecadal oscillation; NAO = North Atlantic oscillation.

characterized by warm and cool phases with periods of approximately 20–40 years (Enfield et al. 2001). If past patterns continue to hold, the warm phase will potentially continue for the next several decades. The AMO index is related to air temperatures and rainfall over North America and Europe and is associated with changes in the frequency of droughts in North America and the frequency of severe hurricane events. The AMO has also been related to the North Atlantic branch of the deep thermohaline circulation, which, in turn, is related to the dynamics of the Gulf Stream. Since 1960, the NAO and the AMO have varied inversely (Figure 3a). Steele et al. (2007) had previously inferred that low nitrate levels associated with lower fish productivity prevailed during the 1960s, followed by an increase in the early 1970s. These nutrient levels may have been associated with climatic conditions related to the NAO, particularly the effects of Labrador Current transport.

Anthropogenic drivers.—After declining slightly in the 1970s, the northeast coastal population index increased steadily during 1981–2006 (Figure 3b). Per capita income (expressed in 2000 U.S. dollars) increased steadily from 1969 to 2006, particularly in the New York and Boston metropolitan areas (Figure 3b).

Physical pressures.—All six temperature series examined were consistent in showing a general increase in temperature in the region (Figure 4a–c). As noted above, we extracted the common trend in these series using MAFA and used the combined index for all subsequent analyses. We also examined a measure of available thermal habitat categorized by three temperature bands (>15°C; 5–15°C; and <5°C; EcoAP 2009). This analysis shows an increase in the >15°C band, a decline in the 5–15°C category, and an increase in the <5°C band (Figure 4d). The increase in the latter principally reflects changes in the northern segment of the range and an increase in the influence of the Labrador Current.

Throughout the series, we saw a general increase and then leveling or slight decline of the latitude of the north wall of the Gulf Stream (Figure 5a). The percentage contribution of Labrador Current water in the region declined from 1977 through the late 1980s and then began a reversal to higher levels (Figure 5a). We note an inverse relationship between the Gulf Stream position and Labrador Current indices. The surface- and bottom-water salinities for the region closely mirror the Labrador Current index (Figure 5b). Stratification declined from 1977 to the mid-1980s and then increased sharply (Figure 5c).

Estimates of river discharge throughout coastal regions of the three major subregions of the system show a very strong coherence (Figure 6a), and we constructed a combined index for all subsequent analysis. Wind stress estimates also show a general similarity among widely distributed geographical locations but with weaker coherence than for river discharge. The north–south wind stress component exhibited the weakest geographical coherence (Figure 6d).

Anthropogenic pressures.—Fishing impacts have been identified as one of the dominant human influences on marine systems (Jackson et al. 2001; Lotze et al. 2006; Halpern et al. 2008). Our index of fishing intensity uses standardized fishing effort for both demersal and pelagic fisheries following the transition to extended national jurisdiction in 1976–1977. The initial increase in the domestic fleet activity reflected the availability of low interest loans to build domestic capacity following the exclusion of distant water fleets with the implementation of the 200-mi limit (Figure 7a; Fogarty and Murawski 1998). However, the fishing effort then declined, starting in the mid-1980s, as increasingly restrictive management measures were put in place to rebuild depleted stocks (Figure 7a). The catches for groundfish, other finfish, and small pelagic fishes peaked in the mid-late 1960s and then declined under intense exploitation by foreign and domestic fishing fleets (Figure 7b). In contrast, we observed an initial increase and stabilization of landings of high-value invertebrate resources (Figure 7c) over the past several decades, with a dominance of American lobster Homarus americanus and sea scallop Placopecten magellanicus landings in the total.

Ecosystem states.—We first examined measures of ecosystem state reflecting lower trophic level dynamics (phytoplankton and zooplankton abundance). The CPR color index indicates a general decrease throughout the series (Figure 8a). Based on satellite-derived observations of surface chloro-

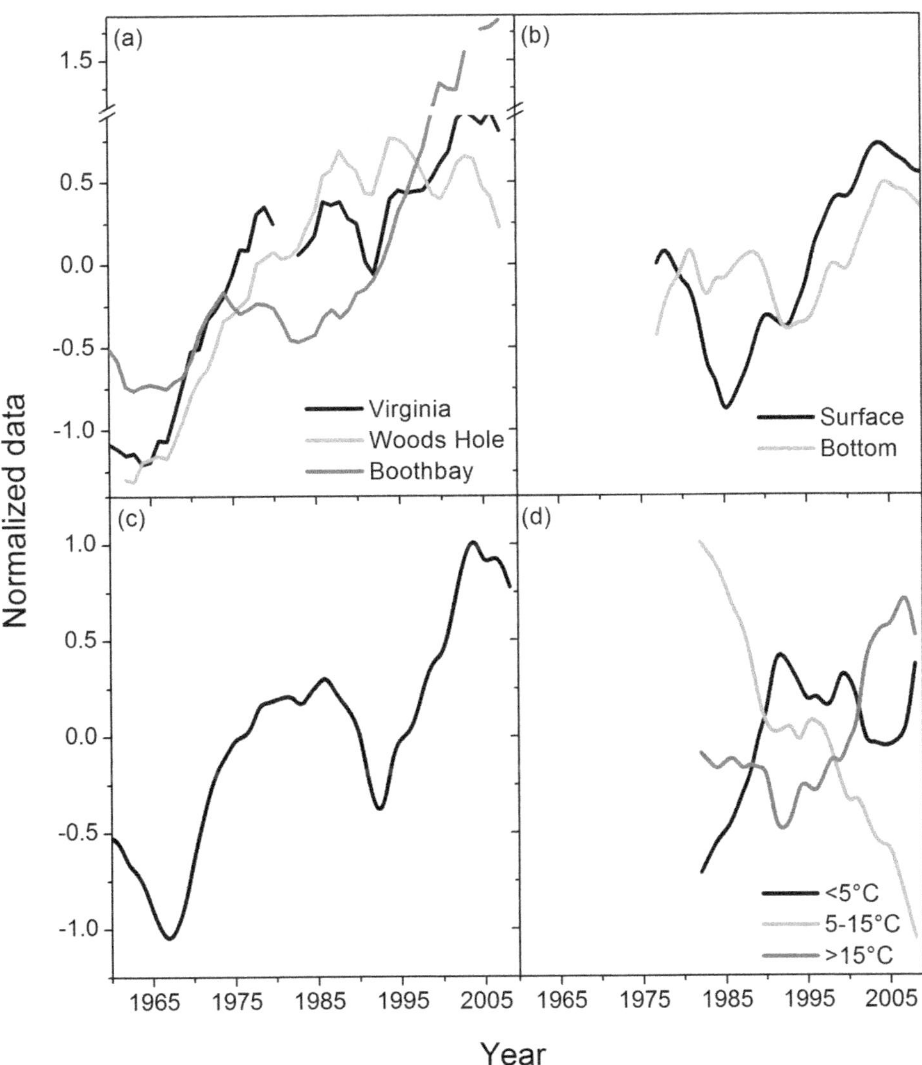

FIGURE 4. Trends in temperature-related variables, including (a) coastal temperatures recorded in Virginia, Woods Hole, Massachusetts and Boothbay Harbor, Maine; (b) surface and bottom temperature recorded during Northeast Fisheries Science Center research vessel surveys; (c) extended reconstructed sea surface temperature series; and (d) thermal habitat estimates for three categories (<5°C, 5–15°C, and >15°C) for northeast U.S. Continental Shelf large marine ecosystem.

phyll for the past decade, no general decline in phytoplankton abundance is evident (recent years have, in fact, been high). This suggests that phytoplankton species composition has changed, with a recent dominance of small-bodied phytoplankton offsetting the apparent decline in net phytoplankton.

Zooplankton biovolume estimates show an initial decline at the start of the available series through the early 1980s, followed by an increase and subsequent stabilization (Figure 8b). Our index of the ratio of small- to large-bodied copepod species shows an increase to a peak in the mid-1990s followed by a decline (Figure 8c). The dominance of small-bodied copepods during the 1990s reflects an increase in species that peak in abundance in autumn. The increase appears to be related, in turn, to an increase in the fall phytoplankton bloom.

Information on longer-term trends in abundance of benthic organisms is available only for se-

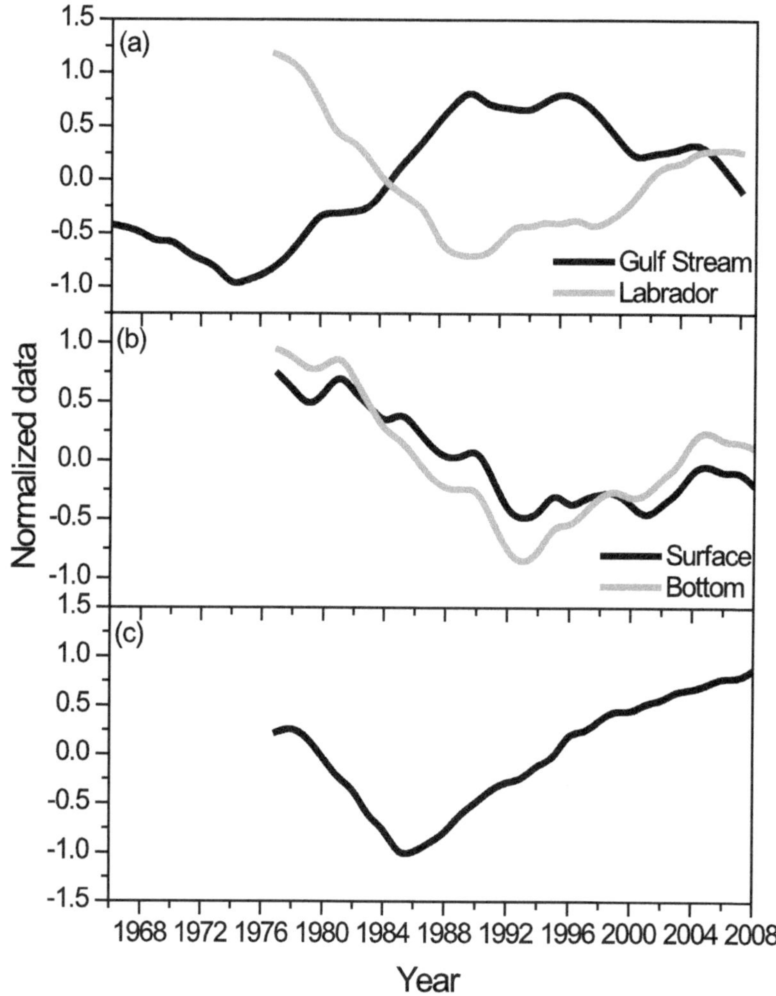

FIGURE 5. Trends in physical pressures, including (a) position of the north wall of the Gulf Stream and the proportion of Labrador Current water entering the Gulf of Maine, (b) surface and bottom salinity recorded during Northeast Fisheries Science Center research vessel surveys, and (c) index of stratification for northeast U.S. Continental Shelf large marine ecosystem.

lected commercially and recreationally important benthic species, based on research vessel surveys. Mollusk and crustacean biomass from surveys show a general increase in abundance over time (Figure 9a). Research vessel surveys also show an increase in elasmobranch and small pelagic species over the past four decades (Figure 9b). In contrast, groundfish species and other finfish species declined under intensive exploitation through the early 1990s and then began to increase again as increasingly restrictive management measures were put in place starting in 1994 (Figure 9c).

An evaluation of the mean trophic level of the catch (all species) shows a marked decline starting in the late 1960s (Figure 10a). However, consideration of the catch of fish species alone shows a more stable pattern, though with a recent decline (Figure 10a). A comparable analysis for fish species obtained in research vessel surveys indicates fluctuations around relatively stable to increasing mean trophic levels for this component of the ecosystem (Figure 10a). These observations point to the dominance of low trophic level invertebrate fisheries in defining the

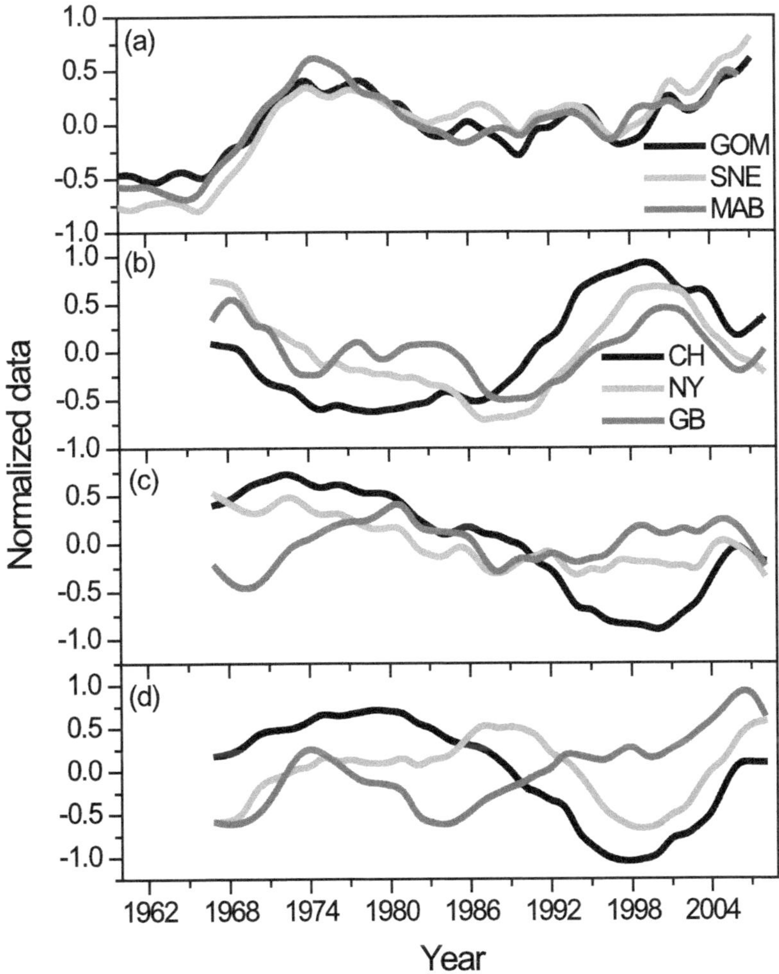

FIGURE 6. Physical pressures reflecting atmospheric processes (precipitation and winds), including (a) river discharge from major river systems in the Gulf of Maine (GoM), southern New England (SNE), and the Middle Atlantic Bight (MAB), (b) annual averages of monthly mean wind stress magnitude at Cape Hatteras (CH), New York (NY), and on Georges Bank (GB), (c) east–west component of wind stress, and (d) north–south component of wind stress. Note: legend for panel (b) applies to panels (c) and (d).

overall trophic level of the catch over the past two decades or more, suggesting a pattern more consistent with fishing through the food web (Essington et al. 2006) than with fishing down the food web (Pauly et al. (1998). The primary production required (PPR) to support the observed level of landings increased through the mid-late 1960s as distant water fleets removed extremely high levels of catches of a broad array of species. The PPR subsequently declined significantly as catches were reduced, and a high fraction of the overall catch was derived from lower trophic level species (Figure 10b). Mean size of species obtained in NEFSC bottom trawl surveys shows a sharp decline to a nadir in the mid-1970s, followed by a slight increase (Figure 10c). The total number of fish species (species richness) observed in spring research vessel surveys has generally increased over the past two decades (Figure 10d). Quite dramatic increases in the fish community composition, weighted by the preferred temperatures of individual species,

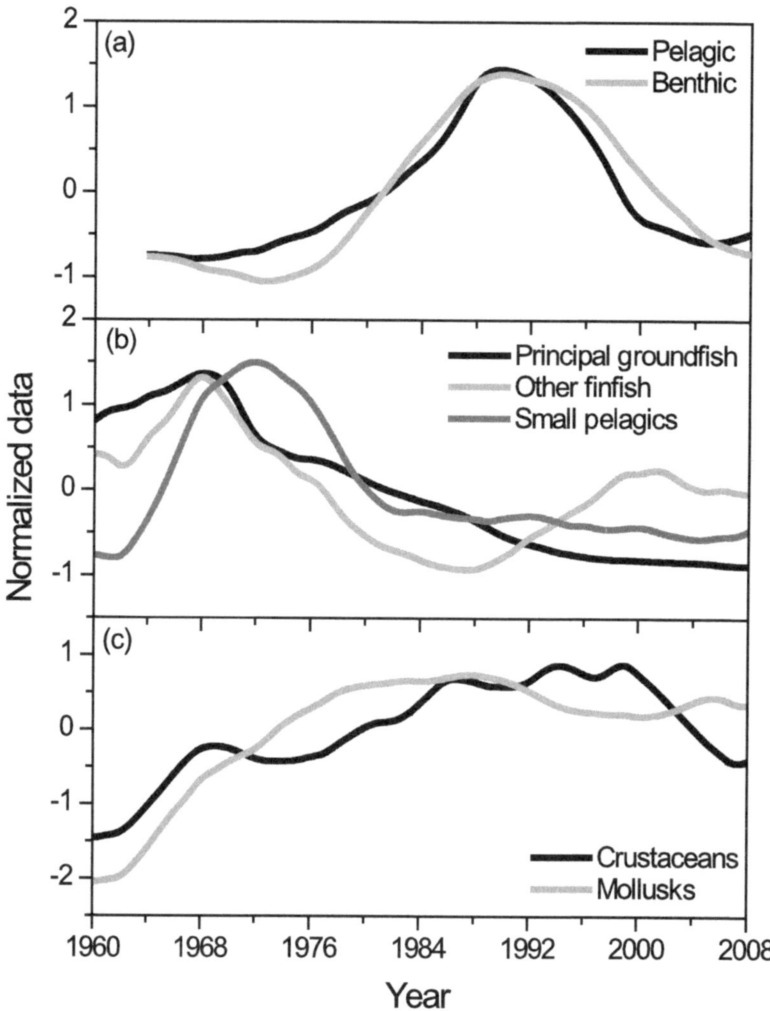

FIGURE 7. Anthropogenic pressures, including (a) standardized fishing effort; (b) landings of principal groundfish, other finfish, and small pelagics; and (c) landings of crustaceans and mollusks for northeast U.S. Continental Shelf large marine ecosystem.

have been observed, reflecting a shift to a more thermophilic species assemblage Figure 10e). Finally, the ratios of pelagic to demersal and elasmobranch to demersal species has fluctuated over time but has generally increased, peaking in the early to mid-1990s (Figure 10f).

Impacts.—Changes in the status of fishery resources exert important impacts on fishers and communities dependent on these resources. Here, we focus on effects on catches and total revenues derived from the groundfish fishery to illustrate these impacts. Total groundfish landings and revenues declined substantially over the past four decades (Figure 11), initially in response to extremely high exploitation rates due to the combined effects of foreign and domestic fishing. Landings declined from 175 kt in 1964 to 79.8 kt in 1976; landings rebounded to 174 kt in 1982 and declined steadily to about 22.7 kt by 2006. Total value averaged about 110 million dollars during 1964–1975, doubled to more than 200 million dollars in 1982, and then plunged to about 57 million dollars in 2006. On a per vessel

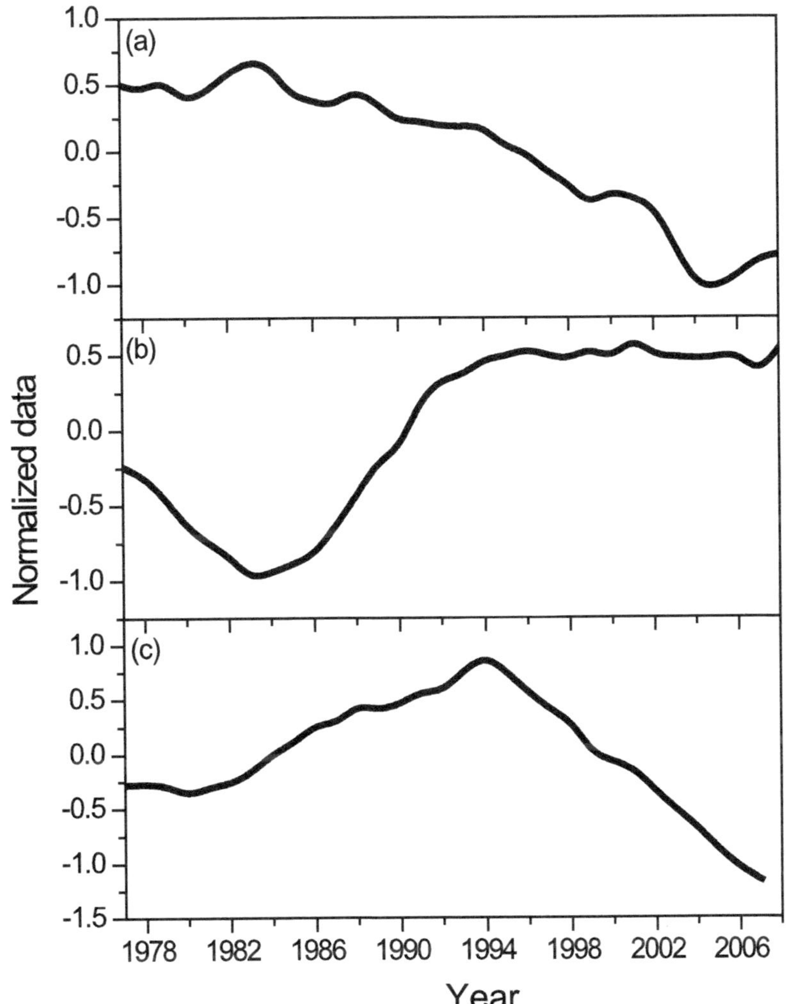

FIGURE 8. Lower trophic level ecosystem state variables, including (a) Continuous Plankton Recorder phytoplankton color index, (b) zooplankton biovolume, and (c) ratio of small to large zooplankton species for northeast U.S. Continental Shelf large marine ecosystem.

basis, however, average revenue actually increased during 2001–2006. Inspection of the smoothed and standardized series of landings and revenue illustrate the sharp decline since the early 1980s. The general decline in these measures can be attributed to overall declines in groundfish catches under intensive exploitation and management restrictions implemented during the past decade, designed to effect recovery of the resource. While landings and revenues have increased in other segments of the broader fishery, notably in the invertebrate fisheries, the decline in the traditional groundfish fishery continues to impact fishing communities. The responsive actions to overharvest are necessary to rebuild the stocks with the aim of developing future economic benefits, but the process is causing severe economic stress in the groundfish fishing community.

Management responses.—As noted above, major management interventions were enacted in the NES LME, with the implementation of extended jurisdiction in 1976–1977. In the following, we focus on major shifts in management

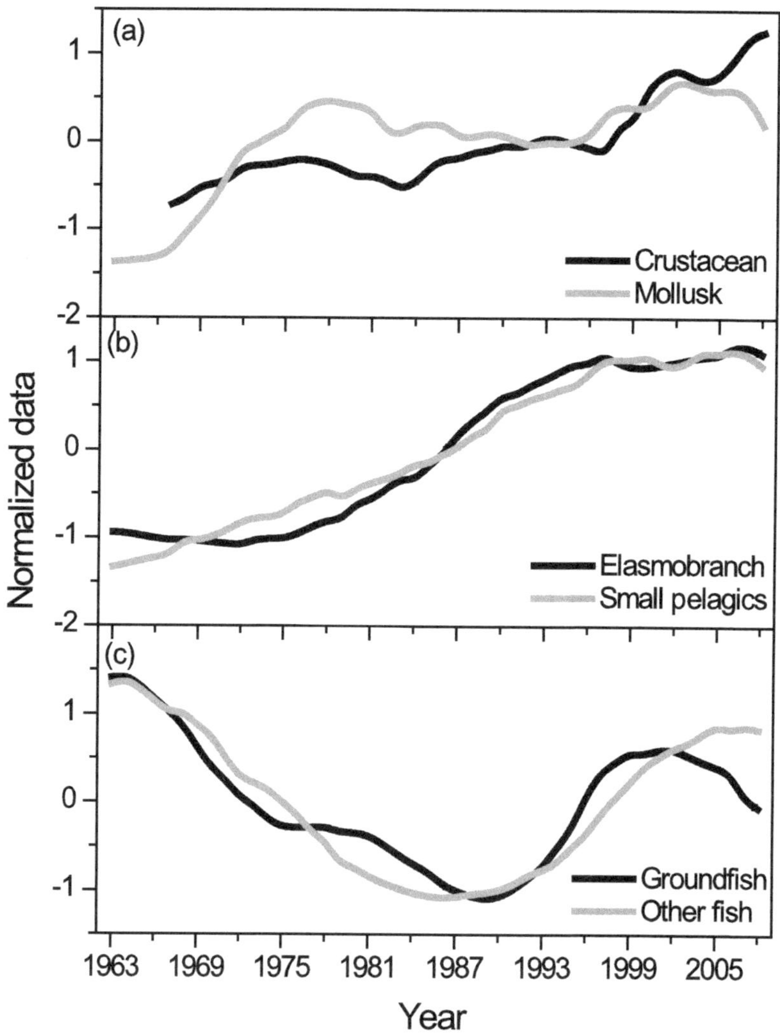

Figure 9. Exploited component ecosystem state variables, including biomass of (a) crustaceans and mollusks (b) elasmobranchs and small pelagics, and (c) and groundfish and other finfish based on NEFSC research vessel surveys for the northeast U.S. Continental Shelf large marine ecosystem.

strategies for dominant resource species that have taken place since then but recognize that many more incremental changes in management have been enacted throughout the fisheries on a near continuous basis. Again, we concentrate on continental shelf species rather than nearshore and estuarine species that are largely managed under state and interstate jurisdiction.

The changes under the Magnuson-Stevens Fishery Management and Conservation Act accompanying the establishment of the 200-mi limit effected large-scale changes in virtually all aspects of offshore demersal, pelagic, and invertebrate fisheries through the exclusion of the distant water fleets. Previous important milestones include the formation of the International Commission for Northwest Atlantic Fisheries (ICNAF) in 1949, which implemented mesh size regulations for otter trawls in 1953 in response to concerns over massive discarding of small fish and apparent declines in abundance of key species, including haddock *Melanogrammus aeglefinus* (Hennemuth and Rockwell 1987). Catch quotas were enacted in 1970–1973 by ICNAF to arrest population

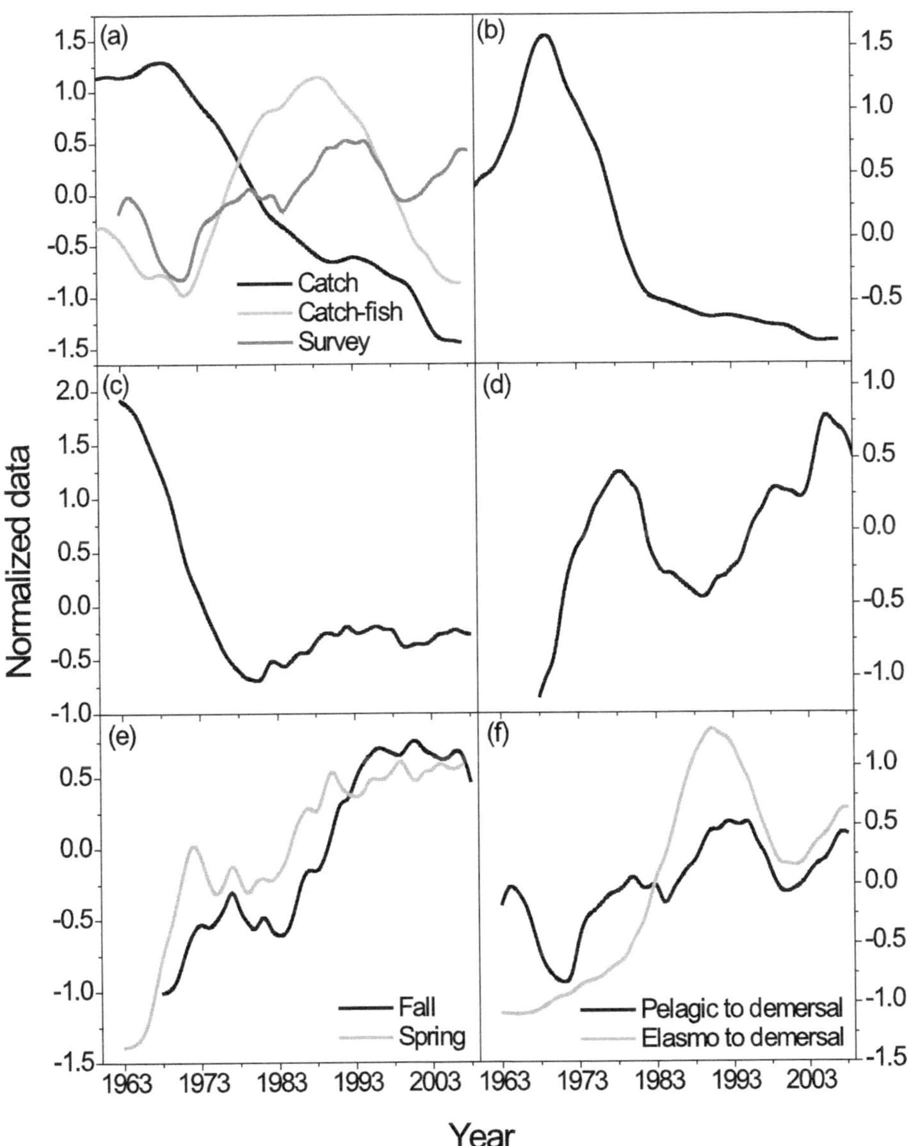

FIGURE 10. Ecosystem state variables, including (a) mean trophic level of the catch (all species), mean trophic level of the catch (fish only), and mean trophic level of sampled species in Northeast Fisheries Science Center (NEFSC) research vessel surveys; (b) primary production required to sustain the catch; (c) mean length of species caught; (d) number of species; (e) mean preferred temperature of the finfish community (°C); and (f) ratio of pelagic to demersal fish and ratio of small elasmobranch to demersal fish for the northeast U.S. Continental Shelf large marine ecosystem. All series except those in panel (a) using catch data are based on in NEFSC research vessel surveys.

declines following intensive foreign fishing and the sequential depletion of key resource species. A "two-tier" quota management system was instituted in 1974, with explicit recognition and allowance for bycatch and discarding practices and interspecific interactions (Hennemuth and Rockwell 1987).

Catch quotas were continued during the initial phases of groundfish management under extended jurisdiction by the New England Fish-

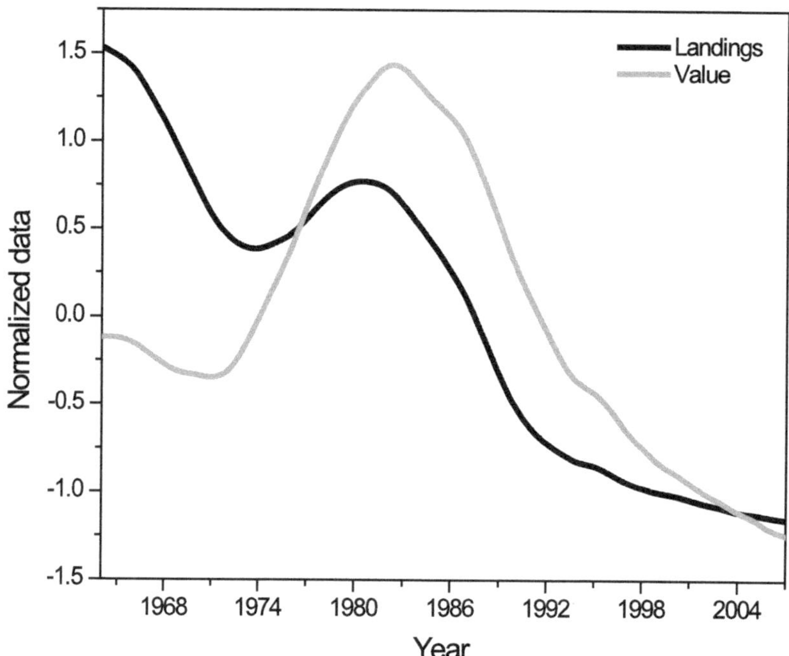

FIGURE 11. Trends in groundfish landings and value as a measure of impacts of changes in ecosystem state.

ery Management Council, providing a nucleus for stock recovery. However, restrictive catch quotas were abandoned in 1982 in the face of widespread quota overruns. Less restrictive measures involving constraints on mesh size, legal size limits for fish, and short-term areal and seasonal closures were adopted in 1982. These qualitative management measures proved ineffective, and groundfish stocks continued to decline until, under court order in 1994, more restrictive management measures were enacted and progressively implemented. These included controls on the allowable number of days at sea, increases in net mesh size, establishment of minimum legal size limits for some species, the establishment of year-round fishery closed areas in which mobile fishing gear was prohibited, the use of "rolling" seasonal closures, and other measures. In 2010, a new sector-based groundfish management plan was instituted in which 17 fishery sectors (effectively cooperatives) were assigned allocations in a catch share system.

The sea scallop fishery currently is the highest revenue fishery in the nation. Fundamental changes in sea scallop management were also implemented in 1994, with the establishment of days-at-sea restrictions, gear modifications, and crew size restrictions. The scallop fishery was also subject to spatial management controls involving exclusion from the fishery closed areas. The closed area restrictions have had an important effect on sea scallop dynamics. Other offshore bivalve fisheries have undergone particularly dramatic regulatory change. The fisheries for Atlantic surfclam *Spisula solidissima* and ocean quohog *Arctica islandica* were among the first managed using an individual transferable quota (ITQ) system in the United States. The ITQ management was established in 1990 for federal waters (3–200 mi from shore).

Although the American lobster fishery is managed under interstate jurisdiction, it has a substantial continental shelf component, and it supports the second most valuable fishery in the NES LME. In 1988, major changes in coast-wide fishery management were implemented, with increases in minimum legal size, trap limitations, escape gap size, and other regulations. Earlier changes in management structures were principally effected through state regulations.

As noted above, the implementation of major management system changes in the groundfish fishery, in particular, is reflected in the chronological cluster analysis of the fishery pressure indicators. These pressures in turn can be related to changes in ecological state variables used to index the condition of living marine resources. The management responses therefore play a demonstrable role in ecosystem status.

Relationships among Drivers, Pressures, and Ecosystem States

An initial evaluation of the relationships among the drivers, pressures, and ecosystem state variables examined in this assessment can be obtained through an examination of the trajectories of the full set of indicators of over time. We have arrayed the smoothed and standardized variables in five panels, representing climate drivers (Figure 12a), physical pressures (Figure 12b), anthropogenic drivers (Figure 12c), fishing pressures variables (Figure 12d), and ecosystem state indicators (Figure 12e), in a color-coded sequence to examine evidence for related patterns of change. The overall increase in the AMO index is clearly (and not unexpectedly) reflected in several temperature-related variables, including the common-trend temperature index, the >15°C thermal habitat indicator, and the stratification index. Similarly, the trajectory of change of the NAO index shows af-

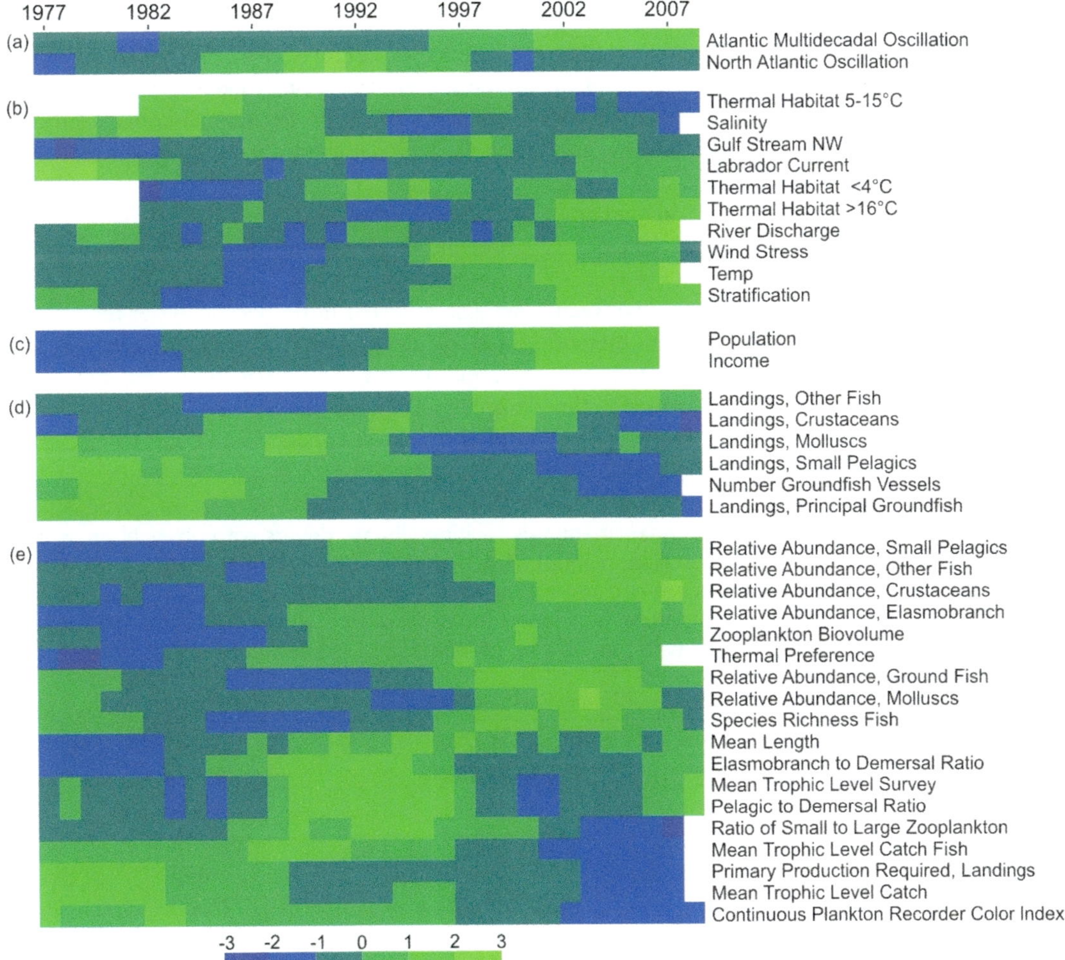

FIGURE 12. Normalized indictors for (a) physical drivers, (b) physical pressures, (c) anthropogenic drivers, (d) anthropogenic pressures, and (e) ecosystem state variables. Green colors indicate high values; blue colors represent low values. Blank rectangles indicate no data for time period.

finities to the position of the north wall of the Gulf Stream. The steadily increasing trend in human drivers (population size and income) is inversely related to landings of groundfish, other finfish, small pelagics, and fishing effort, reflecting both changes in resource status and in management responses designed to effect recovery of these resources.

To quantitatively assess the relationship between the climate drivers and physical pressure variables, we considered the potential of time-delayed effects of climate forcing with lags of up to 4 years for both NAO and AMO using a canonical redundancy analysis. Testing for the effect of these explanatory variables on the output physical pressure variables using a forward selection procedure resulted in the retention of the unlagged AMO and the lag 2 NAO variables in the model (Table 2). This result is consistent with the expectation that NAO effects mediated through far-field forcing from the Arctic would involve an 18- to 24-month delay period in our region (Greene and Pershing 2007). In contrast, the AMO would be expected to reflect an unlagged effect on our combined temperature metric using the first-order MAFA trend line.

We next considered the relationship between anthropogenic drivers and pressures. Both human coastal population size and income were found to explain significant fractions of the variance in the anthropogenic pressures (catch and fishing effort; Table 3). We therefore retained both as useful indicators of human-related drivers of system dynamics. It is clear, however, that the global nature of seafood markets results in a decoupling of demand and local supply. Accordingly, we interpret the human population and income metrics available as simply representing broadscale patterns of human influence in the system as a whole.

Our examination of the relationship between physical pressure variables and lower trophic level dynamics indicated a significant effect of temperature and stratification on the phytoplankton and zooplankton indicators (Table 4). Previous analyses of the change in zooplankton species composition have highlighted the importance of changes in salinity in this region (Mountain and Kane 2010). Changes in temperature and salinity are, of course, important determinants of stratification intensity. Stratification is likely a more proximal indicator of factors affecting phytoplankton and zooplankton dynamics than salinity, per se.

Biomass levels of living marine resource species can be affected by both human-related pressures and changes in the physical environment via recruitment, growth, and survival. These changes are also reflected in the second-order metrics related to demography, relative species composition, thermal preferences, species richness, and mean trophic level used as ecosystem state variables in this analysis. We conducted an RDA using the set

TABLE 2. Results of canonical redundancy analysis relating climate drivers with physical pressures. Drivers include the North Atlantic oscillation index (NAO) and the Atlantic multi-decadal oscillation index (AMO). Lags of up to 4 years are considered for each variable. The pressure variables include winds, salinity, river discharge, temperature, position of the north wall of the Gulf Stream, the percentage contribution of Labrador Current water in the Gulf of Maine, and stratification.

Explanatory variable	*F* statistic	*P*
NAO	1.628	0.150
Lag1 NAO	1.944	0.070
Lag2 NAO	3.043	0.006
Lag3 NAO	1.409	0.218
Lag4 NAO	2.062	0.062
AMO	5.487	0.002
Lag1 AMO	0.764	0.606
Lag2 AMO	0.420	0.874
Lag3 AMO	0.604	0.736
Lag4 AMO	0.988	0.468

TABLE 3. Results of canonical redundancy analysis relating anthropogenic drivers with anthropogenic pressures. Drivers include human population size in the coastal zone and disposable income. The pressure variables include landings of groundfish, other finfish, small pelagics, small elasmobranchs, crustaceans, and mollusks, as well as fishing effort.

Explanatory variable	*F* statistic	*P*
Population size	20.681	0.005
Income	7.664	0.005

of physical and anthropogenic pressure variables to check for immediate linkages between our ecosystem indicators and these pressures. Our suite of pressure indicators can also involve important time lags. Unfortunately, it was not possible to also test for time-delayed effects because of limitations related to the length of the series and the large number of potential time-delayed variables. We found significant relationships between the output variables and removals by the fishery (for groundfish, other finfish, elasmobranchs, and mollusks) and for fishing effort. Removals of small pelagic fish and crustaceans were not significant at the $p < 0.05$ level. No time delays in pressure variables were considered in this analysis. In addition, water temperature and river discharge were found to be significant physical pressure variables in this analysis (Table 5).

Change Point Analysis

An analysis of joint temporal change points in the climate driver series (the NAO and the AMO), using chronological cluster analysis, indicates switches in 1974, 1991, and 1998. In this analysis, we lagged the NAO series by 2 years, based on previous information on time-delayed effects of the NAO in this region (Greene and Pershing 2007) and the results of canonical redundancy analyses described above.

For the anthropogenic driver series, a chronological cluster analysis indicated change points in the joint population and income series in 1986, 1995, and 2001. The identified change points appear to be more directly influenced by inflections in the population series, rather than the income level information, which shows a more strictly monotonic trend.

Consideration of chronological clustering of the set of all physical pressures considered does not reveal a statistically significant change point in the series. We note that the physical pressure series is the shortest of the indicator grouping series examined (starting in 1982 and constrained by the shortest series describing the thermal

TABLE 4. Results of canonical redundancy analysis relating physical pressure variables with lower trophic level ecosystem state variables. The pressure variables include winds, salinity, river discharge, temperature, position of the north wall of the Gulf Stream, the percentage contribution of Labrador Current water in the Gulf of Maine, and stratification. The ecosystem state variables include the Continuous Plankton Recorder color index, zooplankton biovolume, and the ratio of small to large-bodied copepods.

Explanatory variable	*F* statistic	*P*
Winds	0.759	0.735
Salinity	0.375	0.070
River discharge	0.533	0.660
Temperature	8.959	0.005
Gulf Stream north wall	2.039	0.100
Labrador Current	1.720	0.165
Stratification	2.975	0.030

TABLE 5. Results of canonical redundancy analysis relating physical and anthropogenic pressure variables with living marine resource (LMR) ecosystem state variables. The physical pressure variables include winds, salinity, river discharge, temperature, position of the north wall of the Gulf Stream, the percentage contribution of Labrador Current water in the Gulf of Maine, and stratification. The anthropogenic pressure variables include landings of groundfish, other finfish, small pelagics, small elasmobranchs, crustaceans, and mollusks, as well as fishing effort. The LMR ecosystem state variables include biomass of groundfish, other finfish, small pelagics, small elasmobranchs, crustaceans, and mollusks; mean trophic level of species represented in the catch and in standardized Northeast Fisheries Science Center bottom trawl surveys; mean length of all species in the surveys; survey species richness; and a measure of the thermal preference of species in the survey.

Explanatory variable	*F* statistic	*P*
Groundfish landings	12.903	0.002
Other fish landings	7.459	0.002
Pelagic fish landings	1.311	0.256
Elasmobranch landings	3.065	0.002
Crustacean landings	0.852	0.534
Mollusk landings	2.870	0.006
Fishing effort	2.050	0.040
Winds	1.152	0.322
Salinity	1.152	0.278
River discharge	2.058	0.032
Temperature	2.234	0.022
Gulf Stream north wall	0.657	0.702
Labrador Current	1.330	0.258
Stratification	1.086	0.372

habitat metrics). These data constraints, coupled with the diversity of the atmospheric and oceanographic indicators included, may have hampered identification of any change points in the joint series. If the thermal habitat series is excluded, allowing a somewhat longer time series for analysis, we identify a change point in 1983 in the physical pressures.

An analysis of temporal inflection points in the fishing-related pressure variables indicates significant changes in 1977, 1984, and 1994. These change points correspond to the full implementation of extended jurisdiction (the 200-mi limit) and the exclusion of foreign fleets from U.S. waters (1976–1977), the change from total allowable catch-based management to nonquota management for groundfish and the partitioning of Gulf of Maine and Georges Bank into U.S. and Canadian sectors by the World Court (1984), and the initiation of stringent court-ordered groundfish management (1994). These changes in fishing-related pressures can therefore be directly related to responses to system status (see Management Response section below).

Chronological cluster analysis of the phytoplankton and zooplankton series indicates significant inflections in 1990 and 2001. These demarcation points correspond to previously identified regime-shift type changes in the zooplankton community (Kane 2007; Mountain and Kane 2010) and linked to changes in salinity and stratification patterns in this region.

The temporal sequence of upper trophic level state indicators shows a single change point in 1985 for the set of indicators as a whole. Delayed responses to changes in pressures exerted on the upper trophic levels are to be expected, and it is possible that the inflection point detected in 1985 reflects the earlier changes in the structure of the fishery, starting with the implementation of extended jurisdiction and followed by further regulatory change through 1984. The 1985 inflection corresponds to the important switch point in fish community structure reflected in the dominance of small pelagic fish and small elasmobranchs in the system, and inflections in the primary production required to support the catch, mean length in the surveys and the increase in thermophilic species in the system.

The change in relative importance of the groundfish community relative to other components had previously been associated with the decline in the heavily exploited groundfish community (Fogarty and Murawski 1998) and the possibility of a broadly based competitive and predatory release in other components of the fish community. We note that a change point in the physical pressure series was observed in 1983 for the reduced set of environmental factors, excluding the shorter thermal habitat series, possibly reflecting a time-delayed effect on environmental factors.

Conclusions

The U.S. northeast continental shelf has experienced large-scale perturbations over the past several decades due to human activities, notably harvesting (Fogarty and Murawski 1998) and environmental change, including climate-related impacts (Frumhoff et al. 2007; EcoAP 2009). The need for holistic assessments of ecosystem status in relation to natural and anthropogenic forcing is now widely recognized as a critical element in support of ecosystem approaches to fisheries management (Levin et al. 2008, 2009). In particular, linking the pressures related to anthropogenic and natural drivers of ecosystem change to alteration in system status is an essential first step in developing effective management strategies. We have focused on the northeast U.S. Continental Shelf, a designated "large marine ecosystem" as a whole, to provide a broadscale overview of system dynamics. We note, however, that the shelf can be divided into ecologically defined subregions (Anon 2010) and that these subregions are characterized by differences in their species compositions and internal dynamics.

Our analysis indicates important connections between natural and anthropogenic drivers of change and resulting pressures affecting ecosystem states on the northeast continental shelf. Fishing pressure metrics (catches and fishing effort) are not surprisingly shown to be critically important in relation to LMR indicator variables. Although both physical and fishing-related forcing are revealed as important in the dynamics of this system, top-down effects due to fishing appear to be the dominant influence on living marine resources, reflected principally in biomass trends, size composition, and trophic level considerations. In a meta-analysis of 19 fishery ecosystems, including the northeast continental shelf, Link et al. (2010) showed a dominance of anthropogenic forcing in these systems, and in the NES LME in particular. We note that both direct and indirect effects of harvesting appear to be evident in the upper trophic level indicators. Declines in heavily exploited species groups have been accompanied by increases in other groups, possibly reflecting the effects of interspecific interactions.

Although removals of pelagic fish and crustaceans were not significant factors in our RDA, relating anthropogenic pressures to LMR state variables overall, the remaining catch series were highly significant, and retention of all catch variables is recommended for any assessment of the system. As noted earlier, the catch series is critically important in its own right as a measure of an essential ecosystem service. Extending the series to allow examination of time-delayed effects of fishing also has high priority. Changes in management regimes over time are manifest in the time series of fishing-related pressures. Inflection points in these pressure indicators map directly to major management interventions.

The LMR state variables show a major inflection point in the mid-1980s following management responses in the late 1970s and early 1980s. Environmental effects are also evident. Temperature is shown to be important and is reflected in shifts in fish community composition toward a warmer-water assemblage of species (see also Nye et al. 2009).

The lower trophic level processes are dominated by temperature and water column stratification. Transition points identified in the time series for phytoplankton and zooplankton indicators correspond closely to previously identified regime shifts in the zooplankton community and related to changes in salinity. Collectively, the ecosystem state variable RDAs point to the importance of changes in temperature as the dominant environmental pressure. Stratification is also shown to be important and may be particularly useful as an integrative measure because it involves other primary indicators used in this analysis (temperature, salinity, winds, and rivers) to varying degrees.

A full characterization of the status of the NES LME in relation to ecosystem reference points was beyond the scope of this report. Although some recommendations for ecosystem reference points have been made for this system (Link 2005), agreement has not been attempted by managers on choices for reference points at the ecosystem level. As the development of approaches for EBFM is refined and managers are directly engaged in the process, reference points and/or directions will be chosen and a full assessment of ecosystem status will be possible.

Complementary analyses have been undertaken for a nearby continental shelf system on the eastern Scotian shelf (DFO 2003; Choi et al. 2005; Frank et al. 2005). In this system, it appears that overfishing has triggered a cascading effect, which has been amplified and maintained by environmental change even as fishing pressure has been reduced on many ecosystem components. The potential for similar effects in the NES LME should also be considered. Our analysis points to the critical role that exploitation has played in this system and that synergistic effects of environmental forcing, particularly related to temperature, can potentially further alter system structure in ways that are difficult to predict.

Our choice of ecosystem state variables is consistent with previous analyses of the most informative metrics to detect the effects of fishing and environmental forcing on system dynamics. Fulton et al. (2005) found that high priority should be assigned to species groups characterized by fast turnover times (phytoplankton, zooplankton, and bacteria), species subject to commercial fishing pressure, and community- or ecosystem-level metrics. In contrast, indicators based on population-level measures and those that are strongly model-based (e.g., based on network analysis) were found to be less informative. Similarly, Samhouri et al. (2009) reported that indicators of lower trophic level, high productivity groups and complementary indicators such as phytoplankton, zooplanktivorous fish, piscivorous fish, and mean trophic level of the catch collectively performed well as indicators of ecosystem change. It is clear that a suite of carefully chosen indicators is required to capture the full dimensions of change in marine ecosystems, and no single indicator (or small group) can completely fill this role (Murawski 2000; Link et al. 2002; Fulton et al. 2005; Link 2005; Samhouri et al. 2009).

Acknowledgments

We thank the scientists and staff of the Northeast Fisheries Science Center without whom this work would not be possible. For individual contributions, we further express our gratitude to J. Manning, J. Kane, D. Hart, J. Jossi. D. Kitts, E. Thunberg, P. Rago, C. Legault, R. Mayo, L. Jacobson, M. Traver, M. Palmer, S. Wigley, A. Seaver, J. O'Reilly, B. Smith, and T. Ducas. We are grateful for the helpful comments on the manuscript by F. M. Serchuk and two anonymous referees.

References

Anon. 2010. Ecosystem-based fishery management for the northeast U.S. Continental Shelf. NOAA Fisheries Service Fact Sheet FS-2010-02, Silver Spring, Maryland.

Brodziak, J., and J. Link. 2002. Ecosystem-based fishery management: what is it and how can we do it? Bulletin of Marine Science 70:589–611.

Choi, J. S., K. T. Frank, B. D. Petrie, and W. C. Leggett. 2005. Integrated assessment of a large marine ecosystem: a case study of the devolution of the eastern Scotian shelf, Canada. Oceanography and Marine Biology an Annual Review 43:47–67.

DFO (Department of Fisheries and Oceans Canada). 2003. State of the eastern Scotian shelf ecosystem. DFO, Canadian Science Advisory Secretariat Ecosystem Status Report 2003/004, Ottawa.

EcoAP (Ecosystem Assessment Program). 2009. Ecosystem status report for the northeast U.S. continental shelf large marine ecosystem. National Oceanic and Atmospheric Administration, National Marine Fisheries Service, Northeast Fisheries Science Center Reference Document 09-11, Woods Hole, Massachusetts.

Enfield, D. B., A. M. Mestas-Nunez, and P. J. Trimble. 2001. The Atlantic multidecadal oscillation and its relation to rainfall and river flows in the continental US. Geophysical Research Letters 28:2077–2080.

Essington, T. E., A. H. Beaudreau, and J. Wiedenmann. 2006. Fishing through marine food webs. Proceedings of the National Academy of Sciences of the United States of America 103:3171–3175.

Fogarty, M. J., and S. A. Murawski. 1998. Large-scale disturbance and the structure of marine systems: fishery impacts on Georges Bank. Ecological Applications 8:S6–S22.

Frank, K. T., B. Petrie, J. S. Choi, and W. C. Leggett. 2005. Trophic cascades in a formerly cod-dominated ecosystem. Science 308:1621–1623.

Frumhoff, P. C., J. J. McCarthy, J. M. Melillo, S. C. Moser, and D. J. Wuebbles. 2007. Confronting climate change in the U.S. Northeast. A report of the northeast climate impacts assessment. Union of Concerned Scientists, Cambridge, Massachusetts.

Fulton, E. A., A. D. M. Smith, and A. E. Punt. 2005. Which ecological indicators can robustly detect effects of fishing? ICES Journal of Marine Science 62:540–551.

Garcia, S. M., A. Zerbi, C. Aliaume, T. Do Chi, and G. Lasserre. 2003. The ecosystem approach to fisheries: issues, terminology, principles, institutional foundations, implementation and outlook. Food and Agriculture Organization of the United Nations, Fisheries Technical Paper 443, Rome.

Greene, C. H., and A. J. Pershing. 2007. Climate drives sea change. Science 315:1084–1085.

Halpern, B. S., S. Walbridge , K. A. Selkoe, C. V. Kappel, F. Micheli, C. D'Agrosa, J. F. Bruno, K. F. Casey, C. Ebert, H. E. Fox, R. Fujita, D. Heinemann, H. S. Lenihan, E. Madin, M. T. Perry, E. R. Selig, M. Spalding, R. Steneck, and R. Watson. 2008. A global map of human impact on marine ecosystems. Science 319:948–952.

Hennemuth, R. C., and S. Rockwell. 1987. History of conservation and management. Pages 430–446 *in* R. Backus, and D. Bourne, editors. Georges Bank. MIT Press, Cambridge, Massachusetts.

Hurrell, J. W. 1995. Decadal trends in the North Atlantic oscillation: regional temperatures and precipitation. Science 269:676–679.

Jackson, J. B. C., M. X. Kirby, W. H. Berger, K. A. Bjorndal, L. W. Botsford, B. J. Bourque, R. H. Bradbury, R. Cooke, J. Erlandson, J. A. Estes, T. P. Hughes, S. Kidwell, C. B. Lange, H.S. Lenihan, J. M. Pandolfi, C. H. Peterson, R. S. Steneck, M. J. Tegner, and R. R. Warner. 2001. Historical overfishing and the recent collapse of coastal ecosystems. Science 293:629–637.

Jennings, S. 2005. Indicators to support an ecosystem approach to fisheries. Fish and Fisheries 6:212–232.

Jossi, J. W., A. W. G. John, and D. Sameoto. 2003. Continuous plankton recorder sampling off the east coast of North America: history and status. Progress in Oceanography 58:313–325.

Kane, J. 2007. Zooplankton abundance trends on Georges Bank, 1977–2004. ICES Journal of Marine Science 64:909–919.

Legendre, P., and L. Legendre. 1998. Numerical ecology, 2nd edition. Elsevier Press, Amsterdam.

Levin, P. S., M. J. Fogarty, G. C. Matlock, and M. Ernst. 2008. Integrated ecosystem assessments. National Oceanic and Atmospheric Administration, National Marine Fisheries Service, Northwest Fisheries Science Center, Technical Memorandum NMFS-NWFSC-92, Seattle.

Levin, P. S., M. J. Fogarty, S. A. Murawski, and D. Fluharty. 2009. Integrated ecosystem assessments: developing the scientific basis for ecosystem-based management of the ocean. PLoS (Public Library of Science) Biology [online serial] 7(1): e1000014. DOI: 10.1371/journal.pbio.1000014.

Link, J. S. 2005. Translating ecosystem indicators into decision criteria. ICES Journal of Marine Science 62:569–576.

Link, J. S., D. Yemane, L. J. Shannon, M. Coll, Y.-J. Shin, L. Hill, and M. de Fatima Borges. 2010. Relating marine ecosystem indicators to fishing and environmental drivers: an elucidation of contrasting responses. ICES Journal of Marine Science 67:787–795.

Link, J. S., J. K. T. Brodziak, S. F. Edwards, W. J. Overholtz, D. Mountain, J. W. Jossi, T. D. Smith, and M. J. Fogarty. 2002. Marine ecosystem assessment in a fisheries management context. Canadian Journal of Fisheries and Aquatic Sciences 59:1429–1440.

Lotze, H. K., H. S. Lenihan, B. J. Bourque, R. H. Bradbury, R. G. Cooke, M. C. Kay, S. M. Kidwell, M. X. Kirby, C. H. Peterson, and J. B. C. Jackson. 2006. Depletion, degradation, and recovery potential of estuaries and coastal seas. Science 312:1806–1809.

Methratta, E. T., and J. S. Link. 2006. Evaluation of quantitative indicators for marine fish communities. Ecological Indicators 6:575–588.

Millennium Ecosystem Assessment. 2005. Ecosystems and human well-being: synthesis. Island Press, Washington, D.C.

Mountain, D. G., and J. Kane. 2010. Major changes in

the Georges Bank ecosystem, 1980s to the 1990s. Marine Ecology Progress Series 398:81–91.

Murawski, S. A. 2000. Definitions of overfishing from an ecosystem perspective. ICES Journal of Marine Science 57:649–658.

Nye, J. A., J. S. Link, J. A. Hare, and W. J. Overholtz. 2009. Changing spatial distribution of fish stocks in relation to climate and population size on the northeast United States continental shelf. Marine Ecology Progress Series 393:111–129.

OECD (Organisation for Economic Co-operation and Development). 2003. OECD environmental indicators: development, measurement, and use. OECD, Paris.

Pauly, D., and V. Christensen. 1995. Primary production required to sustain global fisheries. Nature 374:255–257.

Pauly, D., V. Christensen, J. Dalsgaard, R. Froese, and R. Torres. 1998. Fishing down marine food webs. Science 279:860–863.

Pikitch, E. K., C. Santora, E. A. Babcock, A. Bakun, R. Bonfil, D. O. Conover, P. Dayton, P. Doukakis, D. Fluharty, B. Heneman, E. D. Houde, J. Link, P. A. Livingston, M. Mangel, M. K. McAllister, J. Pope, and K. J. Sainsbury. 2004. Ecosystem-based fishery management. Science 305:346–347.

Rice, J. C., and M.-J. Rochet. 2005. A framework for selecting a suite of indicators for fisheries management. ICES Journal of Marine Science 62:516–527.

Rochet, M.-J., and V. M. Trenkel. 2003. Which community indicators can measure the impact of fishing? A review and proposals. Canadian Journal of Fisheries and Aquatic Sciences 60:86–99.

Samhouri, J. F., P. S. Levin, and C. J. Harvey. 2009. Quantitative evaluation of marine ecosystem indicator performance using food web models. Ecosystems 12:1283–1298.

Smith, T. D. 2002. The Woods Hole bottom-trawl resource survey: development of fisheries-independent multispecies monitoring. ICES Marine Science Symposia 215:474–482.

Smith, T. M., and R. W. Reynolds. 2003. Extended reconstruction of global sea surface temperatures based on COADS data (1854–1997). Journal of Climate 16:1495–1510.

Smith, T. M., and R. W. Reynolds. 2004. Improved extended reconstruction of SST (1854–1997). Journal of Climate 17:2466–2477.

Solow, A. R. 1994. Detecting changes in the composition of a multispecies community. Biometrics 50:556–565.

Steele, J. H., J. S. Collie, J. J. Bisagni, D. J. Gifford, M. J. Fogarty, J. S. Link, B. K. Sullivan, M. E. Sieracki, A. R. Beet, D. G. Mountain, E. G. Durbin, D. Palka, and W. T. Stockhausen. 2007. Balancing end-to-end budgets of the Georges Bank ecosystem. Progress in Oceanography 74:423–448.

Stenseth, N. C., A. Mysterud, G. Ottersen, J. W. Hurrell, K.-S. Chan, and M. Lima. 2002. Ecological effects of climate fluctuations. Science 297:1292–1296.

Trenkel, V. M., and M.-J. Rochet. 2003. Performance of indicators derived from abundance estimates for detecting the impact of fishing on a community. Canadian Journal of Fisheries and Aquatic Sciences 60:67–85.

Zuur, A. F., E. N. Ieno, and G. M. Smith. 2007. Analyzing ecological data. Springer Science, New York.

Appendix

APPENDIX TABLE 1. List of indicators used, time period covered and sources for analyses described in this paper. SST = sea surface temperature; NEFSC = Northeast Fisheries Science Center; GoM = Gulf of Maine; NAFO = Northwest Atlantic Fisheries Organization.

Indicators	Time period	Source
Physical drivers		
North Atlantic oscillation	1960–2007	www.cgd.ucar.edu/cas/jhurrell/Data/naodjfmindex.asc
Atlantic multi-decadal oscillation	1960–2007	www.cdc.noaa.gov/Correlation/amon.us.long.data
Human drivers		
Population	1968–2006	www.bea.gov
Income	1968–2006	www.bea.gov
Physical pressures		
Temperature		
Extended reconstructed SST	1960–2007	www.cdc.noaa.gov/data/gridded/data.noaa.ersst.html
Coastal temperature, Virginia	1960–2007	ftp://ftp.nefsc.noaa.gov/pub/hydro/spool_hydro/
Coastal temperature, Woods Hole	1960–2007	NEFSC Archives
Costal temperature, Boothbay Harbor	1960–2007	Maine Department of Marine Resources
Survey SST	1977–2007	NEFSC Survey Data Base
Survey bottom sea temperature	1977–2007	NEFSC Survey Data Base
Thermal habitat <4°C	1982–2007	Ecosystem Assessment Program
Thermal habitat >5°C and <15°C	1982–2007	Ecosystem Assessment Program
Thermal habitat >16°C	1982–2007	Ecosystem Assessment Program
River discharge		
River flow—Gulf of Maine	1960–2007	http://waterdata.usgs.gov/nwis
River flow—Middle Atlantic Bight	1960–2007	http://waterdata.usgs.gov/nwis
River flow—southern New England	1960–2007	http://waterdata.usgs.gov/nwis
Wind fields		
Wind stress, Cape Hatteras	1967–2007	http://waterdata.usgs.gov/nwis
Wind stress, New York	1967–2007	http://waterdata.usgs.gov/nwis
Wind stress, Georges Bank	1967–2007	http://waterdata.usgs.gov/nwis
Wind stress east–west, Cape Hatteras	1967–2007	http://waterdata.usgs.gov/nwis
Wind stress east–west, New York	1967–2007	http://waterdata.usgs.gov/nwis
Wind stress east–west, Georges Bank	1967–2007	http://waterdata.usgs.gov/nwis
Wind stress north–south, Cape Hatteras	1967–2007	http://waterdata.usgs.gov/nwis
Wind stress north–south, New York	1967–2007	http://waterdata.usgs.gov/nwis
Wind stress north–south, Georges Bank	1967–2007	http://waterdata.usgs.gov/nwis
Other		
Stratification	1977–2007	www.nefsc.noaa.gov/publications/crd/crd0408/index.htm
Survey surface salinity	1977–2007	NEFSC Database System Bottom Trawl Survey Database

Appendix Table 1. Continued.

Indicators	Time period	Source
Other (continued)		
Survey bottom salinity	1977–2007	NEFSC Database System Bottom Trawl Survey Database
Gulf Stream location	1977–2007	http://web.pml.ac.uk/gulfstream/Web2008%20(2).pdf
% Labrador-Subarctic slope water in GoM	1977–2007	ftp://ftp.nefsc.noaa.gov/pub/hydro/spool_hydro/
Human pressures (fishery removals)		
Number groundfish vessels	1965–2008	NEFSC Commercial Fisheries Database System
Landings, principal groundfish	1960–2008	NEFSC Commercial Fisheries Database System & www.nafo.int
Landings, other fish	1960–2008	NEFSC Commercial Fisheries Database System & www.nafo.int
Landings, small pelagics	1960–2008	NEFSC Commercial Fisheries Database System & www.nafo.int
Landings, crustaceans	1960–2008	NEFSC Commercial Fisheries Database System & www.nafo.int
Landings, mollusks	1960–2008	NEFSC Commercial Fisheries Database System & www.nafo.int
Ecosystem state variables		
Plankton		
Continuous Plankton Recorder Color Index	1977–2007	NEFSC Plankton Database System
Zooplankton ecosystem biovolume	1977–2007	NEFSC Plankton Database System
Ratio of small to large zooplankton	1977–2007	NEFSC Plankton Database System
Nekton/ben\thos		
Relative abundance, crustaceans	1965–2007	NEFSC Database System Bottom Trawl Survey Database
Relative abundance, elasmobranch	1963–2007	NEFSC Database System Bottom Trawl Survey Database
Relative abundance, groundfish	1963–2007	NEFSC Database System Bottom Trawl Survey Database
Relative abundance, mollusks	1963–2007	NEFSC Database System Bottom Trawl Survey Database
Relative abundance, other fish	1963–2007	NEFSC Database System Bottom Trawl Survey Database
Relative abundance, small pelagics	1963–2007	NEFSC Database System Bottom Trawl Survey Database
Relative abundance, all species	1963–2007	NEFSC Database System Bottom Trawl Survey Database
Demography/trophic level		
Mean trophic level catch	1960–2007	www.fishbase.org and NAFO database
Mean trophic level survey	1963–2007	www.fishbase.org and NAFO database

Appendix Table 1. Continued.

Indicators	Time period	Source
Demographic/trophic level (continued)		
Primary production required, landings	1960–2007	www.fishbase.org and NEFSC Bottom Trawl Survey Database
Mean length	1963–2007	NEFSC Database System Bottom Trawl Survey Database
Community composition		
Thermal preference	1963–2007	www.fishbase.org and NEFSC Bottom Trawl Survey Database
Pelagic to demersal ratio	1963–2007	NEFSC Bottom Trawl Survey Database
Elasmobranch to demersal groundfish ratio	1963–2007	NEFSC Bottom Trawl Survey Database
Impacts		
Groundfish fishery revenue	1965–2006	NEFSC Commercial Fisheries Database System

American Fisheries Society Symposium 79:167–183, 2012

Spatial Patterns of Subtidal Benthic Invertebrates and Environmental Factors in the Nearshore Gulf of Maine

Stephen S. Hale*

Atlantic Ecology Division, National Health and Environmental Effects Research Laboratory
Office of Research and Development, U.S. Environmental Protection Agency
Narragansett, Rhode Island 02882, USA

Abstract.—Spatial patterns of subtidal benthic invertebrates and physical-chemical variables in the nearshore Gulf of Maine (Acadian biogeographic province) were studied to provide information to calibrate benthic indices of ecological condition, determine physical-chemical factors affecting species distributions, and compare recent data with historical biogeographic studies. Knowledge of the distribution of species and how they are affected by biotic, environmental, and anthropogenic factors is essential to the pursuit of ecosystem-based management. Five years (2000–2004) of data from 268 reference stations of the National Coastal Assessment were used. Multidimensional scaling done on Bray-Curtis similarity matrices of species' relative abundance (367 species) showed faunal transitions around Cape Ann and Cape Elizabeth, with a weaker transition around Penobscot Bay. The southernmost area shared 41% of its species with the northernmost area. An ordination of environmental data (temperature, salinity, sediment percent silt-clay, depth) correlated well with the ordination of benthic relative abundance data ($R = 0.75, p < 0.03$). Temperature was the most important factor affecting broad species distribution patterns, followed by salinity. A multivariate regression tree first split the fauna at a temperature of 16°C. Species richness increased with increasing salinity but showed no relationship with latitude or percent silt-clay. Accuracy of benthic indices for the nearshore Gulf of Maine might be improved by taking biogeographical differences among subregions into account. These results provide a foundation for ecosystem-based management, valuation of ecosystem services, conservation, and ocean spatial planning.

Introduction

Knowledge of how species and communities are distributed in space and time in relation to biotic, abiotic, and anthropogenic factors is fundamental to ecosystem-based management, conservation, and ocean planning. These data are necessary for the detection of change caused by long-term and broadscale phenomena such as climate change, species invasions, and ocean acidification. Traditional biogeographical studies were carried out to delineate and explain taxon distributions, understand evolution and history of fauna, and infer historical climate and continental movement (Hedgepeth 1957; Briggs 1974; Franz and Merrill 1980). Recent biogeographical techniques and studies have examined effects of global climate change and associated species range shifts (Beaugrand et al. 2002; Kirby et al. 2008); species invasions, fisheries exploitation, and other long-term changes in faunal assemblages (Buchsbaum and Powell 2008); and chemical contaminants and eutrophication (Llansó et al. 2002). They have been used to set boundaries for ecosystem modeling (Campbell 1987), make model predictions of species' distributions (Elith et al. 2006), and form the basis for marine biodiversity and con-

* Corresponding author: hale.stephen@epa.gov

servation (Lourie and Vincent 2004; Cook and Auster 2007; Spalding et al. 2007). Biogeographic information is needed to measure the provision of ecosystem services and move toward ecosystem-based management (Tsontos and Kiefer 2003; McLeod and Leslie 2009).

The present study in the estuaries and near-shore coastal Gulf of Maine (Acadian biogeographic province; Figure 1) relates to development of a benthic index of environmental condition (Hale and Heltshe 2008). Biogeographic information is needed to calibrate benthic indices that the National Coastal Assessment (NCA) uses as ecological indicators of effects of pollution on health of bottom communities in broad biogeographical provinces (e.g., Van Dolah et al. 1999; Paul et al. 2001). Species distribution patterns depend on temperature, food, shelter, salinity, sediment grain size, oxygen levels, predators, and other factors (Briggs 1995); however, these distributions are altered by anthropogenic stressors, including chemical contaminants and eutrophication. For benthic indices to work, it is necessary to understand the distribution and abundances of species in the absence of human influences (Llansó et al. 2002). Biogeographic information is also needed

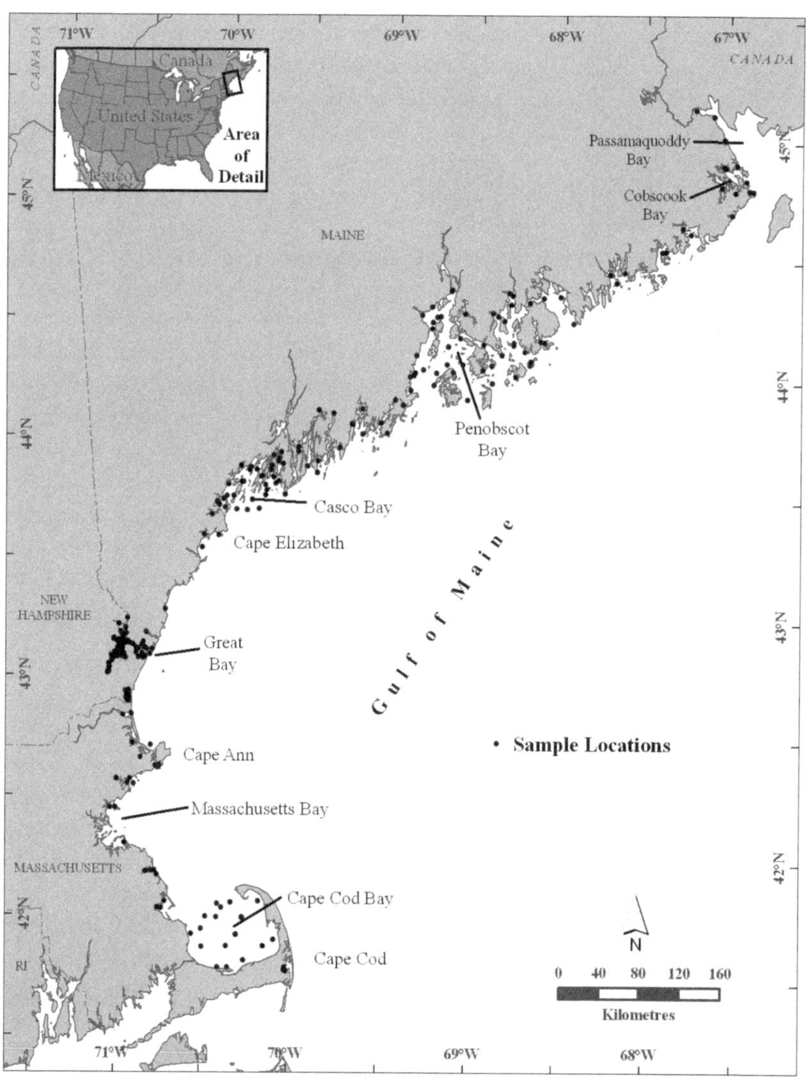

FIGURE 1. Gulf of Maine showing National Coastal Assessment stations used in analyses.

to determine whether there are subregions of faunal assemblages within the Acadian biogeographic province strong enough to justify adjustments to the benthic index. Can an index that works in Cape Cod Bay also work in Passamaquoddy Bay? For example, the benthic index of Weisberg et al. (1997), developed specifically for Chesapeake Bay, has a higher classification accuracy in that bay than does the benthic index developed by Paul et al. (2001) for the larger Virginian biogeographic province (Ranasinghe et al. 2002; Llansó et al. 2009). To support ecological syntheses just north of the study area, Brunel et al. (1998) divided the estuary and Gulf of St. Lawrence into 20 subregions, based on physiographic, oceanographic, biogeographic, and bathymetric criteria.

Present-day geography, oceanographic conditions, movement of water masses, temperature, and freshwater from rivers strongly influence species distribution patterns. Faunal groupings in the study area are also a result of environmental conditions in the Pleistocene and Holocene (Franz and Merrill 1980). This area was repopulated after the retreat of Wisconsin era glaciers, which reached their maxima around 20,000 years ago. Unglaciated refugia existed in Nova Scotia and Newfoundland (Wares 2002). Many present-day species are of trans-Arctic or European origin (Franz and Merrill 1980). The cold water of the Labrador Current flowing out of the Labrador Sea and then southwest affects temperature, salinity, upwelling, productivity, and transport of larvae; it has, in large part, determined the nature of the fauna from Newfoundland to Cape Hatteras (Briggs 1995). Labrador Current and St. Lawrence River water enters the Gulf of Maine over the Scotian shelf and circulates counterclockwise through the Bay of Fundy and southwest along the coast of Maine, where it is known as the Maine Coastal Current (Pettigrew et al. 2005). Part of the eastern Maine Coastal Current veers offshore near Penobscot Bay; the rest continues south along the coast as the western Maine Coastal Current, picking up fresher and warmer water from rivers like the Penobscot, Kennebec-Androscoggin, and, further south, the Merrimac. The western current is shallower, warmer, more influenced by rivers, and more variable than the eastern (Pettigrew et al. 2005; Tilburg et al. 2011). The current exits the Gulf of Maine over the Nantucket shoals.

Shallow-water marine invertebrates are typically restricted in their latitudinal ranges along continental coasts, primarily by water temperature, but also by physical and biological barriers (Calder 1992; Briggs 1995). Faunal discontinuities are often found near capes where water masses with different hydrographic characteristics converge and isotherms get compressed. Summer temperature maximums establish southern limits and winter temperature minimums establish northern limits (Hazel 1970). However, at the scale of the nearshore Gulf of Maine, other environmental factors, such as salinity and sediment grain size, can have a strong effect (Engle and Summers 1999; Van Dolah et al. 1999; Llansó et al. 2002). Salinity is the major factor influencing species composition of northeastern estuaries such as Chesapeake Bay (Llansó et al. 2002) and the Hudson-Raritan estuary, but not Boston Harbor with its strong tidal mixing (Gallagher and Keay 1998).

The Acadian biogeographical province (Cape Cod north to Nova Scotia; Figure 1), named by Dana (1853, cited by Hazel 1970), has been the subject of numerous classic biogeographic studies (reviewed by Hazel 1970). Roman et al. (2000) described the general ecological characteristics of estuaries in the region. The glacial history of Maine left a predominance of sand beaches and salt marsh in southern Maine and a rocky coast with mud and coarse sand beaches in northern Maine (Larsen and Doggett 1990). Cape Cod has traditionally been considered the boundary between the Acadian biogeographic province to the north and the Virginian biogeographic province to the south, based on evidence from several faunal and floral groups (Hayden and Dolan 1976; Briggs 1995; Spalding et al. 2007). The boundary is dynamic and leaky (Longhurst 1998)—many warmwater invertebrates extend northward into the Gulf of Maine and coldwater invertebrates extend south of Cape Cod to Long Island Sound and beyond (Bousfield and Laubitz 1972). A warm pocket of water that develops from Cape Cod Bay to Ipswich Bay, Massachusetts during July to September allows Virginian province species to occupy this region (Hazel 1970; Campbell 1987). Additionally, Larsen (2004) described pockets of warmwater refugia in mid-coast Maine for Virginian province species that were isolated in the heads of estuaries by sea level rise and an increase in tidal amplitude after the last glaciation. Most summers, these areas are warm enough to allow successful reproduction.

Earlier biogeographic studies disagree about boundaries and even about the existence of the Acadian and Virginian biogeographical provinces (reviewed by Hazel 1970). Watling (1979), using distributional data of amphipods, ostracods, mollusks, and polychaetes, suggested two biogeographic provinces in the Gulf of Maine: Nova Scotian, extending south to the northern edge of Georges Bank, and Transhatteran, including the estuarine and nearshore shallow waters from Penobscot Bay south to northern Florida. Franz and Merrill (1980) proposed three overlapping molluscan groups: Transhatteran group, comprised predominantly of shallow shelf and estuarine species distributed from south of Cape Hatteras north to Cape Cod and endemic to the American Atlantic coast; a Boreal group ranging from Cape Hatteras to Labrador and containing both endemic and amphiatlantic (occurring on both sides of the North Atlantic) species; and an Arctic-Boreal group made up of predominantly panarctic-boreal species. Cape Cod is a major environmental barrier to the southward distribution of arctic-boreal species and the northward distribution of Transhatteran species. Pocklington and Tremblay (1987), working with polychaetes, identified an Acadian faunal zone ranging from St. Lawrence Bay to intermediate depths of Georges Bank where it lies offshore of the Virginian faunal zone. Faunal transitions within the Gulf of Maine have been suggested around Cape Ann (42.6°N), the mouth of the Saco River (43.5°N), Penobscot Bay (44°N), and the Schoodic Peninsula-Jonesport area (44.3°N) (Bousfield and Thomas 1975; Watling 1979; Campbell 1987; Larsen and Doggett 1990; Trott 2007).

Some of the incongruity of earlier studies arises from the use of different taxa (often limited to class level or below) and different habitats (Franz and Merrill 1980). The broadscale data from NCA provided an opportunity to address some of these limitations with all taxa from subtidal macrobenthic communities. The NCA data were collected for a survey of ecological condition but are well suited for biogeographical studies because the stations were randomly located on a hexagonal grid and represent wide ranges of environmental factors. These data will be useful in assessing species range shifts resulting from global climate change (Sorte et al. 2010).

The main purpose of this study was to provide biogeographic information needed to calibrate the Acadian province benthic index (Hale and Heltshe 2008). A secondary goal was to evaluate classical biogeographic studies done along the Gulf of Maine coast to provide information needed for ecosystem-based management. The study focused on the U.S. part of the Gulf of Maine (41.7°N to 45.5°N), in the Acadian biogeographic province at the northern end of the U.S. Atlantic coast (Figure 1).

Methods

During 2000–2006, as part of the U.S. Environmental Protection Agency's NCA, subtidal estuarine and nearshore coastal stations in Maine, New Hampshire, and Massachusetts were sampled to assess ecological condition (National Coastal Assessment Web site, www.epa.gov/emap/nca). The sample design randomly allocated stations on a hexagonal grid, and a standardized protocol was used for sample collection and analysis. Samples were taken during a summer index period (July through September) when it was assumed that stresses (e.g., hypoxia, temperature) to the benthic community were highest. Benthic macroinvertebrate assemblages were sampled using a 0.04-m^2 Young-modified Van Veen grab; the samples were sieved with a 0.5-mm mesh. Concurrent samples were taken for sediment percent silt-clay, chemical contaminants, and toxicity, as well as physical-chemical properties of the bottom water (National Coastal Assessment Web site, www.epa.gov/emap/nca). Five years (2000–2004) of NCA data from Cape Cod north to the United States–Canada border (339 stations) were used for the analyses in this study. Only organisms identified to species level were used.

Because the objective was to determine species distributions in the absence of pollution (Engle and Summers 2000; Llansó et al. 2002), degraded stations (as determined by levels of sediment contaminants, sediment total organic carbon, dissolved oxygen in the bottom layer of the water column, and a sediment toxicity test; Hale and Heltshe 2008) were removed, leaving 268 stations with 367 species (Figure 1). The large tidal range in the Gulf of Maine (mean of 2.7 m, and up to 5.8 m; Roman et al. 2000) made it difficult to sample shallow inshore stations near low tide, when salinity would be lowest due to river flow. As a result, the data set mainly comprised stations of salinity ≥18 (oligohaline-1%, mesohaline-4%, polyhaline-27%, euhaline-68%). Classification of these stations into groups was necessary to reduce the dimensionality to a manageable size for analysis. Categories of geographic areas, latitude, and temperature were used. Because there were only 11 oligohaline or mesohaline stations in the data set after dropping degraded ones, the stations were not categorized by salinity classes (dropping degraded stations took a greater toll on lower salinity stations, which tended to be in harbors and upper reaches of estuaries, than on higher salinity ones). The influence of salinity and sediment grain size at the next larger geographic scale (Delaware Bay to Passamaquoddy Bay) was examined by Hale (2010). For each category of each abiotic variable, each species' abundance as a proportion of the total abundance of all species in that category was calculated. Proportions were used to better account for uneven abundance of individuals among categories. Henceforth, this is referred to as species' relative abundance. Fourth-root transformation was used to down-weight the influence of the most abundant fauna without giving too much weight to rare species (Clarke and Warwick 2001). Then nonmetric, multidimensional scalings (MDS) on Bray-Curtis similarity matrices were run to determine how assemblages grouped together. The analysis of similarity test of Clarke et al. (2008) determined the statistical significance of separation between groups, with the null hypothesis that the pattern was random. The SAS software (SAS 2003) was used for data manipulation and to calculate descriptive statistics; PRIMER version 6 (Clarke and Warwick 2001; Clarke and Gorley 2006) was used for all other statistical analyses.

First, the study area was divided into seven smaller geographic areas: Cape Cod Bay, Massachusetts Bay, Cape Ann to Cape Elizabeth, Great Bay, Casco Bay, mid-coast Maine to Penobscot Bay, and northeast of Penobscot Bay (Figure 1). Then an MDS was run on species' relative abundance in each subarea. Next, stations were grouped into categories of temperature categories taken from Bousfield and Thomas (1975): <12°C, 12–15°C, 15–18°C, and >18°C. An MDS was run by those categories. Next, stations were grouped into seven 0.5° bands of latitude. The bands were labeled with their southernmost border (e.g., the 42.5° band included all stations from latitude 42.50° to 42.99°). The 41.5° band included only stations north of the Cape Cod elbow, and the 45.0° band, with only three stations, was dropped. Then, an MDS was run on species' relative abundance in the latitude bands. The SIMPER program of PRIMER was run on the groups identified by the MDS to find those species that contributed the most to the Bray-Curtis similarity of each group and also those that led to the dissimilarity between groups. Another MDS, using Euclidean distances, was run on normalized means of the physical-chemical environmental variables (depth, temperature, salinity, sediment percent silt-clay, dissolved oxygen, and sediment total organic carbon) of the seven latitude bands to compare the pattern of environmental variables with the pattern of species' relative abundance data. The BEST procedure of PRIMER was used to find the combination of environmental variables that best "explained" the observed patterns of species' relative abundance data under the null hypothesis of a random relationship between the two ordinations. Last, PRIMER's LINKTREE procedure, a multivariate classification and regression tree, was used to find what threshold level of each environmental variable was responsible for splitting one species group from another, and the SIMPROF procedure was used to test for statistical significance.

Results

Polychaetes made up 61% of the species, crustaceans 23%, bivalves 12%, gastropods 3%, and echinoderms 0.5%. Species richness (number of species per 0.04-m^2 grab) generally increased with salinity (r = 0.3, p < 0.0001), although euhaline stations showed a wide range (Figure 2). Depth (r = 0.51, p < 0.0001) and temperature (r = –0.49, p < 0.0001) were both correlated with the number of species; in part, this was an artifact of low salinity stations being found only in shallower, warmer areas. Species richness showed no strong patterns

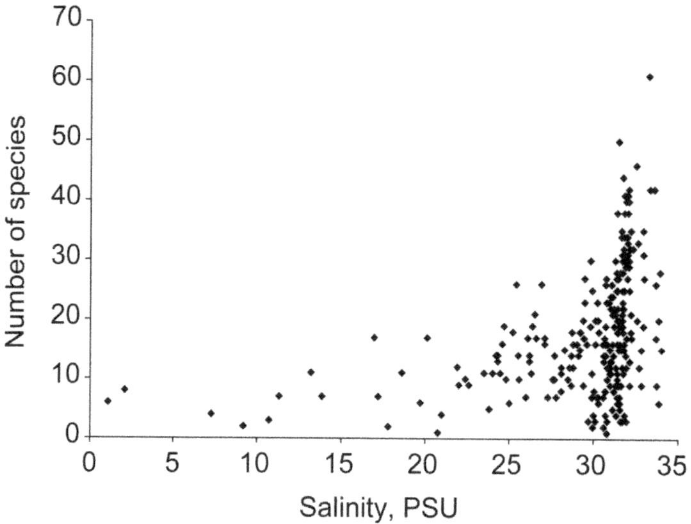

FIGURE 2. Number of species in 0.04-m² grab as a function of salinity. PSU = practical salinity units.

with either latitude (p = 0.11) or percent silt-clay (p = 0.07). A species accumulation curve running from south to north (Figure 3) shows a steady accumulation of additional species, without major increases in slope that would indicate a dramatic addition of new species (e.g., L41.5 has ~160 species out of 367; L42 added ~40 more not seen in L41.5, and so on).

On the MDS plots of multivariate species data (367 species), the closer two points are, the greater the similarity of the faunal assemblages. The MDS of the seven geographic areas (Figure 4) showed three groups of significantly different (p < 0.02) benthic assemblages: (1) Massachusetts Bay to Cape Elizabeth, (2) Great Bay, and (3) Cape Elizabeth to Passamaquoddy Bay. Cape Cod Bay

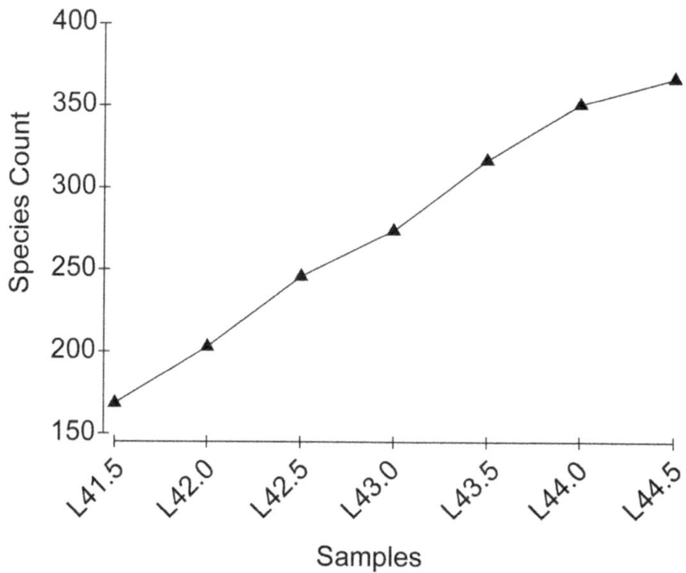

FIGURE 3. Species accumulation curve. Labels represent half-degree latitude bands where the number shown is the southernmost boundary of the band.

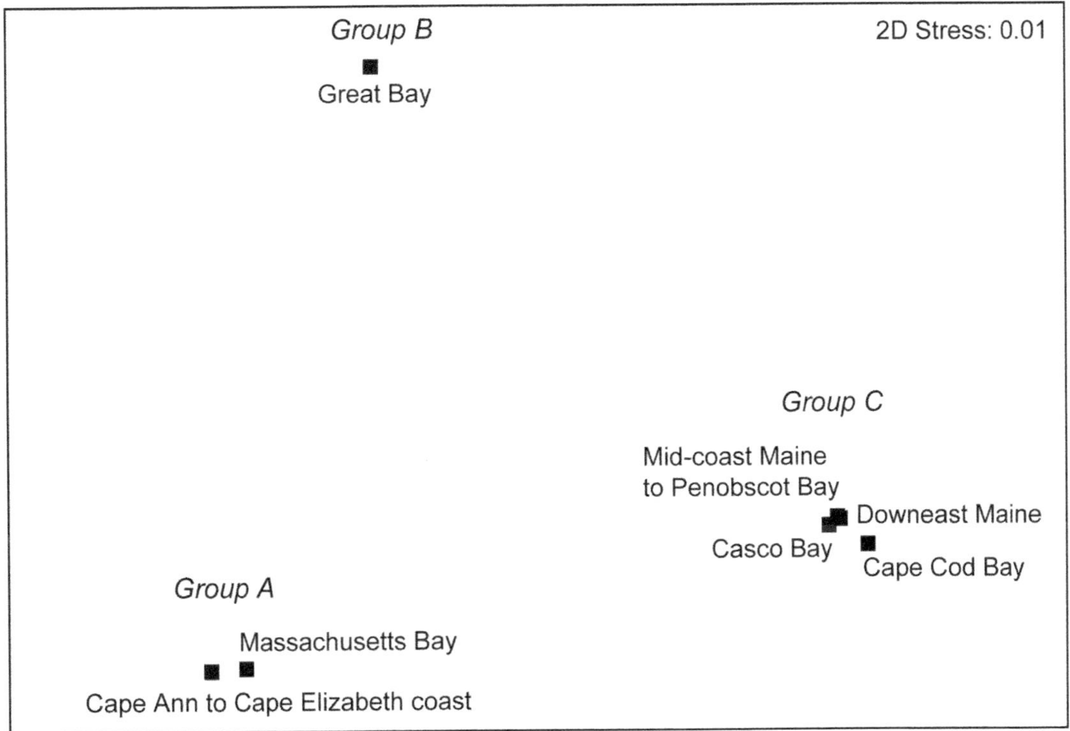

FIGURE 4. Multidimensional scaling of benthic species' relative abundance data by geographic area showing three groups of benthic assemblages: (A) Massachusetts Bay to Cape Elizabeth, (B) Great Bay, and (C) Cape Elizabeth to Passamaquoddy Bay.

appears with the northern group, mainly because several of the stations in Cape Cod Bay were in the center of the bay (deeper, colder, less saline), rather than in estuaries (Figure 1). The separation of species' relative abundance in the MDS by temperature categories showed that the categories defined by Bousfield and Thomas (1975) make ecological sense (Figure 5).

The MDS by latitude bands (Figure 6) showed three groups; the null hypothesis of no differences among the groups was strongly rejected (global $R = 0.90, p < 0.01$). In this analysis, Great Bay dominated the area between Cape Ann and Cape Elizabeth. The Bray-Curtis similarity within groups was at least 54.0. Each group had about 40% of its species in common with each of the other two groups (Table 1). Species that contributed the most to within-group Bray-Curtis similarity can be considered typical of that group; the top five were polychaetes (Table 2). On the other hand, species that contributed the most to group dissimilarity can be used to help decide whether faunal transitions occur in the Acadian biogeographical province; again, most of the top five were polychaetes, except for the ostracod *Parasterope pollex*, the amphipod *Protohaustorius wigleyi*, and the amethyst gemclam *Gemma gemma* (Table 3).

The means and standard deviations for environmental variables of different latitude bands showed the environmental gradients of the Acadian biogeographical province that influenced species distribution patterns (Table 4). The MDS on these environmental variables showed a nonrandom distribution (Figure 7), with a pattern somewhat similar to that for the species' relative abundances. These physical-chemical groups are statistically separate (global $R = 0.52, p < 0.04$), although not as strongly as with the species ordination. Even more so than with the biotic data, the deeper, colder Cape Cod Bay stations (L41.5 band) fell close to the bands north of Cape Elizabeth. The BEST procedure's null hypothesis that the pattern match of the environmental variable

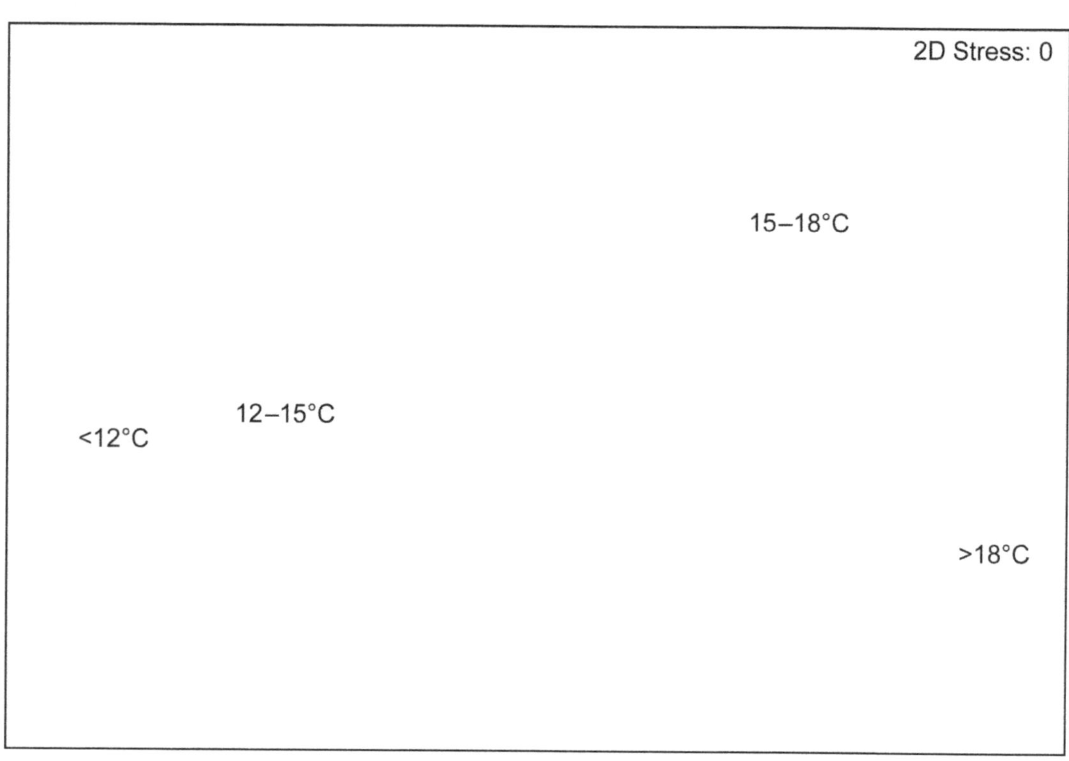

FIGURE 5. Multidimensional scaling of benthic species' relative abundance data by bottom temperature.

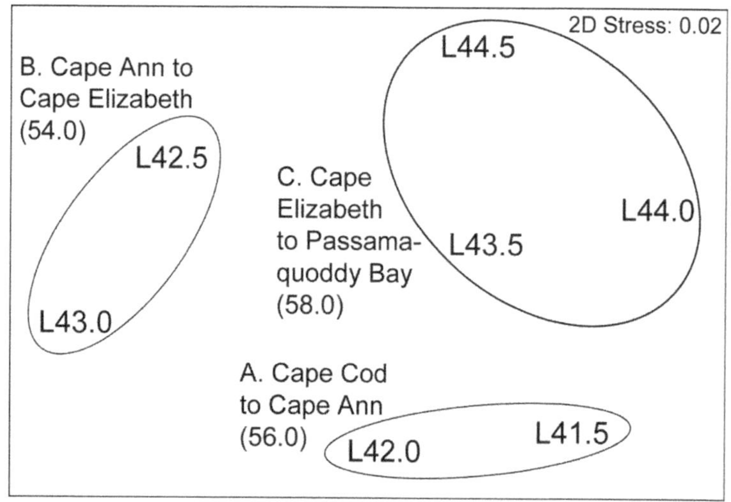

FIGURE 6. Multidimensional scaling of benthic species' relative abundance data by latitude band showing three groups. Labels represent half-degree latitude bands where the number shown (e.g., L41.5) is the southernmost boundary of the band. Numbers in parentheses are the Bray-Curtis similarities for the points within the hand-drawn circles.

TABLE 1. Percent of benthic species in common among latitudinal groups.

Group	A. Cape Cod to Cape Ann	B. Cape Ann to Cape Elizabeth	C. Cape Elizabeth to Passamaquoddy Bay
A. Cape Cod to Cape Ann	X		
B. Cape Ann to Cape Elizabeth	38	X	
C. Cape Elizabeth to Passamaquoddy Bay	41	42	X

ordination and the benthic relative abundance ordination was due to chance was strongly rejected (Spearman rank correlation = 0.75, $p < 0.03$). The strongest single factor was temperature ($r = 0.65$), followed by salinity ($r = 0.56$).

The linkage tree shows what levels of environmental variables (temperature, salinity, and percent silt-clay) account for splits in faunal structure (Figure 8). The first separation (split I) of the biological data by LINKTREE was at a summer bottom temperature of around 16°C ($R = 0.80$, $p < 0.05$). Percent silt-clay (at split II) became an important factor ($R = 0.75$, $p < 0.05$) after temperature was taken into account. Salinity was not a factor in this data set with low numbers of oligohaline and mesohaline stations. Each terminal node of species data are defined by a set of environmental characteristics (e.g., L43.5–L44.5 was characterized by temperature <16.2°C and silt-clay >48.9)%. The node in the upper left of Figure 8 reflects a warmer Great Bay. The node in the lower left reflects the sandier stations in the southern part of the Gulf of Maine. The terminal nodes on this grouping that was constrained by

TABLE 2. Top five species contributing to similarity within each latitudinal group.

A. Cape Cod to Cape Ann
Average Bray-Curtis similarity: 56.1
Acmira catherinae
Euchone incolor
Mediomastus californiensis
Tharyx acutus
Ninoe nigripes

B. Cape Ann to Cape Elizabeth
Average Bray-Curtis similarity: 54.1
Streblospio benedicti
Polydora cornuta
Pygospio elegans
Tharyx acutus
Nereis diversicolor

C. Cape Elizabeth to Passamaquoddy Bay
Average Bray-Curtis similarity: 58.0
Acmira catherinae
Streblospio benedicti
Prionospio steenstrupi
Polydora cornuta
Exogone hebes

Table 3. Top five species contributing to dissimilarity between latitudinal groups. Bray-Curtis dissimilarity in parentheses.

Group	A. Cape Cod to Cape Ann	B. Cape Cod to Cape Elizabeth
B. Cape Ann to Cape Elizabeth	(55.4) *Euchone incolor* *Terebellides stroemi* *Aricidea quadrilobata* *Streblospio benedicti* *Cossura soyeri*	
C. Cape Elizabeth to Passamaquoddy Bay	(49.2) *Spio limicola* *Aricidea suecica* *Parasterope pollex* *Protohaustorius wigleyi*	(55.3) *Cossura soyeri* *Aricidea suecica* *Nereis diversicolor* *Marenzelleria viridis*

environmental factors generally follows the pattern of the unconstrained species' relative abundance MDS shown in Figure 3. This indicates that the environmental factors measured provided an adequate explanation for the biological grouping observed (De'ath 2002).

Discussion

Benthic Faunal Transition Areas

This study indicated three faunal groups with transitions in the vicinity of Cape Ann (42.5°N) and Cape Elizabeth (43.5°N). These are not hard boundaries; rather, there is extensive overlapping of species (Pocklington and Tremblay 1987) and year-to-year variation, as noted by Gabriel (1992) and Mahon et al. (1998) for demersal fishes in the area. For 2000–2001, the most northerly transition area was at Penobscot Bay (44°N), rather than Cape Elizabeth. The lower salinity, higher temperature, and shallower Great Bay dominated the Cape Ann to Cape Elizabeth sampling area. In part, the transitions based on latitude bands (Figure 6) represent a difference between the faunal assemblages of Great Bay from those to the south and north. In the MDS by geographic area (Figure 4), Great Bay stood apart from the other areas. Also, in that analysis, the deep and cold Cape Cod Bay stations in the central part of that bay grouped with those from north of Cape Elizabeth. However, the MDS by latitude band showed the Cape Cod Bay faunal assemblage more similar to the adjacent one to the north. These differences are likely due to the homogenizing effect of latitude bands. The latitude band approach is useful for modeling and examining regional eco-

Table 4. Mean (SD) for environmental variables by latitude band. Labels represent half-degree latitude bands where the number shown is the southernmost boundary of the band.

Latitude band (°N)	Number stations	Depth (m)	Temp (°C)	Salinity	Silt-clay (%)
44.5	17	9.7 (6.5)	12.9 (1.2)	32.6 (1.1)	48.9 (32.8)
44.0	51	21.4 (16.0)	12.1 (2.7)	31.5 (1.3)	63.7 (32.3)
43.5	49	13.8 (12.9)	13.3 (3.2)	31.2 (1.4)	59.9 (29.5)
43.0	86	1.6 (1.7)	19.4 (3.6)	25.9 (6.3)	50.8 (29.4)
42.5	28	3.6 (3.3)	16.3 (3.9)	29.3 (5.0)	18.4 (22.3)
42.0	14	9.7 (16.7)	16.2 (5.7)	27.4 (6.5)	21.2 (26.8)
41.5	20	22.0 (14.5)	10.9 (5.1)	31.5 (0.5)	41.4 (33.1)

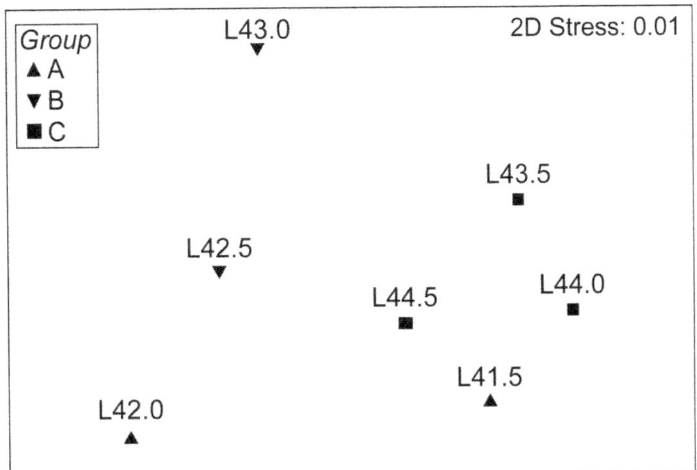

FIGURE 7. Multidimensional scaling of environmental variables. Labels represent half-degree latitude bands where the number shown is the southernmost boundary of the band. The three groups identified in Figure 6 are indicated. Group A—Cape Cod to Cape Ann; Group B—Cape Ann to Cape Elizabeth; Group C—Cape Elizabeth to Passamaquoddy Bay.

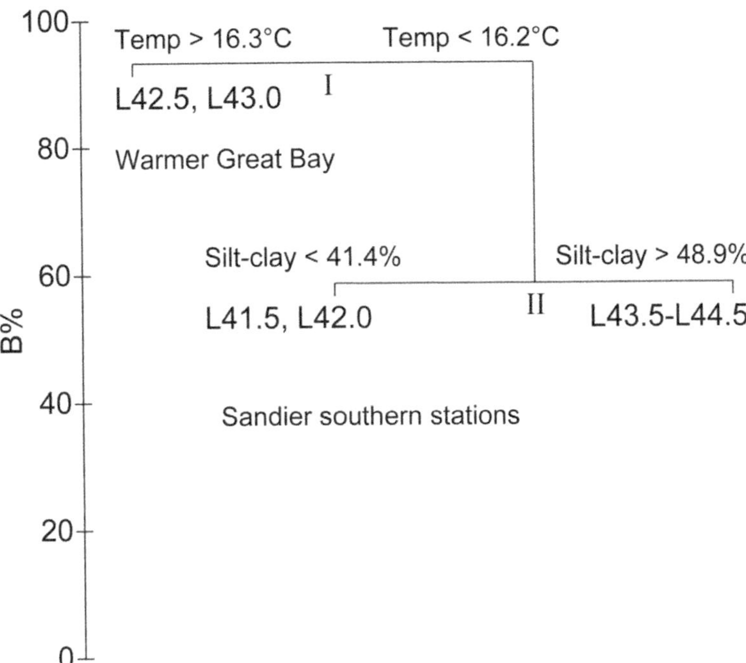

FIGURE 8. Multivariate regression tree (LINKTREE) splits of benthic community half-degree latitude bands. Labels represent half-degree latitude bands where the number shown is the southernmost boundary of the band. B% is a measure of group differences; points with lower values would tend to group more closely together on the multidimensional scaling plot.

logical relationships (Engle and Summers 1999) while the geographic subarea approach might be more useful for specific management purposes.

A transition in the Cobscook Bay area—which Trott (2009) called a biological hotspot—was suggested in that the latitude band with Cobscook and Passamaquoddy bays (L44.5) had the highest mean number of species per station; however, more stations further north are needed to verify this. In general, stations south of Cape Elizabeth were warmer and sandier; those north of Cape Elizabeth were colder, muddier, and deeper. The Transhatteran province proposed by Watling (1979), extending up to Penobscot Bay, was supported by evidence from oligohaline and mesohaline benthic fauna in estuarine and nearshore waters (Hale 2010).

The transition areas found generally agreed with previous studies. The Cape Elizabeth area is further south than the Saco River area (43.5°N) noted by some of these studies, putting Casco Bay in the group to the north, but this transition appears to vary year to year. Campbell (1987) divided the Gulf of Maine into several ecological subsystems and set boundaries at Cape Ann, the mouth of the Saco River (just south of Cape Elizabeth), Pemaquid Point (southwestern edge of Penobscot Bay), and Schoodic Point. Bousfield and Laubitz (1972), on the basis of surface water characteristics and littoral zone invertebrates, divided the Acadian province into two zoogeographical subregions: the area north of Penobscot Bay by its uniformly cold (<12°C) summer surface temperatures and relatively high salinities, in which boreal and subarctic faunal elements dominate; and the area south of Penobscot Bay, containing pockets of warm brackish waters in which Virginian faunal elements dominate. Later, Bousfield and Thomas (1975) mapped summer surface water temperatures and proposed four faunal zones: a <12°C subarctic zone in eastern Maine and western New Brunswick, a 12–15°C boreal region in central Maine; a 15–18°C cool temperate zone extending from southern Maine to Massachusetts Bay, and a >18°C temperate or Virginian zone south of Cape Cod and in isolated shallow embayments in the Gulf of Maine. These zones generally coincide with those of the present study; the 16°C split point of the multivariate regression tree (Figure 8) aligns with their 15°C point in southern Maine. Bottom temperatures greater than 18°C were found in Great Bay. The MDS by temperature (Figure 5) showed these categories have ecological meaning with regard to the benthos; Brunel et al. (1998) found the same categories useful in the Gulf of St. Lawrence.

Larsen and Doggett (1990) reviewed the evidence for zoogeographic subregions within the Gulf of Maine and agreed with Bousfield and Thomas' (1975) four faunistic zones. Larsen and Doggett's (1990) results from a study of macrofauna from eight sand beaches along the Maine coast showed transitions at the Sheepscot River (43.7°N, just north of Casco Bay) and east of Penobscot Bay around Schoodic Peninsula (44.3°N). The northern limit of many species, including some characteristic sand beach amphipods of the family Haustoriidae, is in central Maine; species with subarctic affinities are represented in northern Maine (Bousfield and Laubitz 1972; Bousfield and Thomas 1975). On the other hand, Larsen and Doggett's (1991) survey of macrofauna at five intertidal mud flats along the coast of Maine (only one site northeast of Penobscot Bay) showed little evidence of geographic faunal transitions, as all the communities were dominated by a few widely distributed species. Trott (2007) found distinct groups of intertidal invertebrates north and south of Penobscot Bay, supporting the notion that Penobscot Bay represents a faunal transition area. He attributed this to differences between the colder eastern Maine Coastal Current (EMCC) and the warmer, fresher surface current of the western Maine Coastal Current (WMCC).

Two studies have suggested individual species range limits in the Schoodic Point-Jonesport area, near the change from the EMCC to the WMCC. Rawson et al. (2001) found that the southward extension of the foolish mussel *Mytilus trossulus* declined dramatically around Little Machias Bay. East of Jonesport, the EMCC remains close to the shore, keeping nearshore temperatures relatively cool compared to those west of Jonesport. However, a satellite image of summer sea surface temperature shows, to the east of Jonesport, warmer waters in estuaries and bays inshore of the EMCC (Rawson et al. 2001). Stephenson et al. (2009) hypothesized that the northward invasion of the Asian shore crab *Hemigrapsus sanguineus* into the Gulf of Maine is limited by summer temperatures and that its expansion stalled when it reached the terminus of the EMCC. The crab was established southwest of Casco Bay (mean summer surface temperature >16°C, isolated in the Casco Bay to Schoodic Peninsula area (12.5–16°C) and not detected in eastern Maine (<12°C). Although these two studies of individual species suggest a transition in the area where the EMCC becomes the WMCC, the present study, at the level of benthic communities with up to 61 species per station, each with its own geographic range, did not show this area as a major faunal transition zone. In summary, the evidence to date for distinct nearshore faunal assemblages is varied and there is extensive overlap, but it seems reasonable to conclude that faunal transitions of varying degrees occur around Cape Cod, Cape Ann, the Cape Elizabeth to Saco River area, the Penobscot Bay to Jonesport area, and possibly the high biodiversity Cobscook Bay-Passamaquoddy Bay area. Study of other taxa could help clarify these transition zones. Knowledge of these subregions is important in applying conservation strategies and ecosystem-based management. Management will be improved by a better understanding of how environmental heterogeneity affects ecosystem structure and function. Recognizing faunal assemblages at finer scales than biogeographic provinces would be the most conservative approach for planning a system of representative marine protected areas (Cook and Auster 2007).

Physical-Chemical Habitat Factors Defining Groups

As many biogeographic studies have shown (Engle and Summers 1999; O'Hara and Poore 2000; Blanchette et al. 2008), temperature is the dominant factor affecting broadscale species distribution patterns of marine benthic species, with salinity and sediment grain size affecting smaller scales. In the Gulf of Mexico and southeastern United States, Engle and Summers (1999) found that within 1° latitude bands, salinity was the most important factor affecting benthic species distribution; sediment grain size (and sometimes depth) were secondary factors. A study within Massachusetts Bay (NCCOS 2006) found zones of benthic species driven by sediment grain size, with salinity and depth as secondary factors. Llansó et al. (2002), working with benthic invertebrates along the mid-Atlantic U.S. coast, stated that sediment composition mattered at local scales, but this finding was restricted to polyhaline sites. In the present study, the northward shift of the faunal transition zone from Cape Elizabeth to Penobscot Bay for 2000–2001 may be related to interannual variations in the strength of the WMCC (Pettigrew et al. 2005). A biogeographic study at the next larger scale (Delaware Bay to Passamaquoddy Bay; Hale 2010) showed that while temperature was the most important factor for polyhaline and euhaline stations (salinity >18), at low salinity stations, salinity was more important than latitude or temperature; it even overrode the effect of Cape Cod as a biogeographical boundary. Van Dolah et al. (1999), in Carolinian biogeographic province estuaries, and Llansó et al. (2002), in estuaries of the southern Virginian biogeographic province, also found that lower salinity stations were less geographically distinct compared with polyhaline and euhaline ones.

Conclusions

Depending on the research or management purpose, the three faunal groups identified by this study could be used to further calibrate the Acadian province benthic index of Hale and Heltshe (2008), or others, or to develop specific indices for subregions of the Gulf of Maine. The three metrics of that index (Shannon-Wiener diversity, a measure of species pollution tolerance, and percent capitellid polychaetes) might be expected to respond the same way in different subregions, but the coefficients might change or other metrics become more efficacious at detecting degraded stations. The purpose, accuracy desired, and cost of sampling must be considered. With more low-salinity stations, it could be useful to apply a salinity correction to the Shannon-Wiener diversity measure. The benthic index supports the goals of ecosystem-based management for the Gulf of Maine, which has to consider the full range of biotic and habitat diversity in order to make effective ecosystem-based decisions. Different management actions may be required for different biogeographic subregions of the gulf.

This study sampled only subtidal, soft-bottom, benthic macroinvertebrates. More oligohaline and mesohaline stations are needed to clarify the effect of salinity on biogeographical patterns. The NCA data were collected in the summer, and how strong the patterns noted would be in other seasons is unknown. Future needs of an ecosystem approach for the Gulf of Maine will require biogeographic information for all taxa in all habitats across the entire Gulf. Current biogeographical studies, which benefit from better large-scale monitoring programs based on multiple taxa, better availability of both broad- and fine-scale physical-chemical data, new taxonomic techniques such as DNA bar coding, and use of ecoinformatics to bring together data from many taxa (Hale and Hollister 2009), will markedly improve understanding of how and why species are distributed in space and time. Databases such as the Ocean Biogeographic Information System (Ocean Biogeographic Information System Web site, www.iobis.org) that enable NCA data to be overlaid with those from numerous other studies could help clarify faunal transition zones.

In addition to updating classic biogeographic studies, and providing necessary information to develop and calibrate ecological indicators of environmental condition, results of the present study may help address broadscale and long-term issues such as climate change, species invasions, and species range shifts. The results provide a foundation for ecosystem-based management, conservation, quantifying ecosystem services, and ocean spatial planning. They will serve as a baseline for predicting future biogeographic shifts in benthic assemblages as a result of global climate change (Engle and Summers 1999; O'Hara and Poore 2000; Hiscock et al. 2004; Trott 2007), changes in nutrient loading, and other pressures (Buchsbaum and Powell 2008). Given the importance of temperature and salinity in affecting the spatial patterns of benthic fauna, it is likely that the patterns identified in this study will shift as predicted river and ocean temperatures continue to rise, precipitation patterns in watersheds of the study area become more variable, and oceanic source water into the Gulf of Maine is altered (Hayhoe et al. 2008; Smith et al. 2012, this volume). The rise of sea surface temperature in the North Atlantic is changing the pelagic community, which in turn affects recruitment of many benthic invertebrates (Kirby et al. 2008). The northward 10°latitude range shift for copepods in the North Atlantic from 1960 to 1999 could have serious consequences for marine fisheries (Beaugrand et al. 2002). Benthic communities will be affected by variation in food supply as the timing of phytoplankton blooms changes and by shifts in the distribution and abundance of their demersal fish predators (Nye et al. 2009). Range shifts of marine species in response to global climate change have become widespread (Sorte et al. 2010). We can speculate that at some point, the Virginian province fauna in refugia in mid-coast Maine discussed by Larsen (2004), after a separation of several thousand years, may rejoin the Virginian province faunal group that will continue to penetrate from the south. Application of robust ecosystem-based management in the Gulf of Maine will require, in addition to physical-chemical monitoring, continued and improved monitoring of several taxonomic groups in varied habitats.

Acknowledgments

Thanks to the Maine, New Hampshire, and Massachusetts field crews who collected the data. Discussions with P. Larsen of the Bigelow Oceanographic Laboratory and T. Trott of Suffolk University helped elucidate subregions of the Acadian biogeographic province. J. Heltshe, H. Buffum, C. Audette, M. Charpentier, and P. Decastro provided valuable assistance in preparing the manuscript. Reviews by D. Campbell, C. Strobel, M. Cantwell, and two anonymous reviewers led to many improvements. Although the research described in this article has been funded wholly by the U.S. Environmental Protection Agency, it has not been subjected to agency review. Therefore, it does not necessarily reflect the views of the agency. This is contribution number AED-10–003 of the U.S. Environmental Protection Agency, Office of Research and Development, National Health and Environmental Effects Research Laboratory, Atlantic Ecology Division, Narragansett, Rhode Island. Mention of trade names or commercial products does not constitute endorsement or recommendation for use.

References

Beaugrand, G., P. C. Reid, F. Ibañez, J. A. Lindley, and M. Edwards. 2002. Reorganization of North Atlantic marine copepod biodiversity and climate. Science 296:1692–1694.

Blanchette, C. A., C. M. Miner, P. T. Raimondi, D. Lohse, K. E. K. Heady, and B. R. Broitman. 2008. Biogeographical patterns of rocky intertidal communities along the Pacific coast of North America. Journal of Biogeography 35:1593–1607.

Bousfield, E. L., and D. R. Laubitz. 1972. Station lists and new distributional records of littoral marine invertebrates of the Canadian Atlantic and New England regions. National Museums of Canada, National Museum of Natural Sciences, Publications in Biological Oceanography no. 5, Ottawa.

Bousfield, E. L., and M. L. H. Thomas. 1975. Postglacial changes in distribution of littoral marine invertebrates in the Canadian Atlantic region. Proceedings of the Nova Scotian Institute of Science 27(Supplement 3):47–60.

Briggs, J. C. 1974. Marine zoogeography. McGraw-Hill, New York.

Briggs, J. C. 1995. Global biogeography. Elsevier, Amsterdam.

Brunel, P., L. Bosse, and G. Lamarche. 1998. Catalogue of the marine invertebrates of the estuary and Gulf of Saint Lawrence. Canadian Special Publication of Fisheries and Aquatic Sciences 126.

Buchsbaum, R., and J. C. Powell. 2008. Symposium review: long-term shifts in faunal assemblages in eastern North American estuaries: a review of a workshop held at the biennial meeting of the Coastal and Estuarine Research Federation (CERF), November 2007, Providence, Rhode Island. Reviews in Fish Biology and Fisheries 18:447–450.

Calder, D. R. 1992. Similarity analysis of hydroid assemblages along a latitudinal gradient in the western North Atlantic. Canadian Journal of Zoology 70:1078–1085.

Campbell, D. E. 1987. System ecology of the Gulf of Maine. Maine Department of Marine Resources, West Boothbay Harbor.

Clarke, K. R., and R. N. Gorley. 2006. PRIMER v6: user manual/tutorial. Primer-E Ltd, Plymouth Marine Laboratory, Plymouth, UK.

Clarke, K. R., P. J. Somerfield, and R. N. Gorley. 2008. Testing of null hypotheses in exploratory community analyses: similarity profiles and biota-environment linkage. Journal of Experimental Marine Biology and Ecology 366:56–69.

Clarke, K. R., and R. M. Warwick. 2001. Change in marine communities: an approach to statistical analysis and interpretation, 2nd edition. Primer-E Ltd, Plymouth Marine Laboratory, Plymouth, UK.

Cook, R. R., and P. J. Auster. 2007. A bioregional classification for the continental shelf of northeastern North America for conservation analysis and planning based on representation. National Oceanic and Atmospheric Administration, Marine Sanctuaries Conservation Series MNSP-07-03, Silver Spring, Maryland.

De'ath, G. 2002. Multivariate regression trees: a new technique for modeling species-environment relationships. Ecology 83:1105–1117.

Elith, J., C. H. Graham, R. P. Anderson, M. Dudik, S. Ferrier, A. Guisan, R. J. Hijmans, F. Huettmann, J. R. Leathwick, A. Lehmann, J. Li, L. G. Lohmann, B. A. Loiselle, G. Manion, C. Moritz, M. Nakamura, Y. Nakazawa, J. M. Overton, A. T. Peterson, S. J. Phillips, K. Richardson, R. Scachetti-Pereira, R. E. Schapire, J. Soberón, S. Williams, M. S. Wisz, and N. E. Zimmermann. 2006. Novel methods improve prediction of species' distributions from occurrence data. Ecography 29:129–151.

Engle, V. D., and J. K. Summers. 1999. Latitudinal gradients in benthic community composition in western Atlantic estuaries. Journal of Biogeography 26:1007–1023.

Engle, V. D., and J. K. Summers. 2000. Biogeography of benthic macroinvertebrates in estuaries along the Gulf of Mexico and western Atlantic coasts. Hydrobiologia 436:17–33.

Franz, D. R., and A. S. Merrill. 1980. The origins and determinants of distribution of molluscan faunal groups on the shallow continental shelf of the northwest Atlantic. Malacologia 19:227–248.

Gabriel, W. L. 1992. Persistence of demersal fish assemblages between Cape Hatteras and Nova Scotia, northwest Atlantic. Journal of Northwest Atlantic Fisheries Science 14:29–46.

Gallagher, E. D., and K. E. Keay. 1998. Organism-sediment-contaminant interactions in Boston Harbor. Pages 89–132 *in* K. D. Stolzenbach and E. E. Adams, editors. Contaminated sediments in Boston Harbor. MIT Sea Grant Publication 98-1, Boston.

Hale, S. S. 2010. Biogeographical patterns of marine benthic macroinvertebrates along the Atlantic coast of the northeastern USA. Estuaries and Coasts 33:1039–1053.

Hale, S. S., and J. F. Heltshe. 2008. Signals from the benthos: development and evaluation of a benthic index for the nearshore Gulf of Maine. Ecological Indicators 8:338–350.

Hale, S. S., and J. W. Hollister. 2009. Beyond data management: how ecoinformatics can benefit environmental monitoring programs. Environmental Monitoring and Assessment 150:227–235.

Hayden, B. P., and R. Dolan. 1976. Coastal marine fauna and marine climates of the Americas. Journal of Biogeography 3:71–81.

Hayhoe, K., C. Wake, B. Anderson, X.-Z. Liang, E. Maurer, J. Zhu, J. Bradbury, A. DeGaetano, A. Herter, and D. Wuebbles. 2008. Regional climate change projections for the Northeast. U.S. Mitigation and Adaptation Strategies for Global Change 13:425–436.

Hazel, J. E. 1970. Atlantic continental shelf and slope of the United States—ostracode zoogeography in the southern Nova Scotian and northern Virginian faunal provinces. Geological Survey Professional Paper 529-E. U.S. Government Printing Office, Washington, D.C.

Hedgepeth, J. W. 1957. Marine biogeography. Pages 359–382 *in* J. W. Hedgepeth, editor. Treatise on marine ecology and paleoecology. The Geological Society of America, Memoir 67, Boulder, Colorado.

Hiscock, K., A. Southward, I. Titley, and S. Hawkins. 2004. Effects of changing temperature on benthic marine life in Britain and Ireland. Aquatic Conservation: Marine and Freshwater Ecosystems 14:333–362.

Kirby, R. R., G. Beaugrand, and J. A. Lindley. 2008. Climate-induced effects on the meroplankton and the benthic-pelagic ecology of the North Sea. Limnology and Oceanography 53:1805–1815.

Larsen, P. F. 2004. Notes on the environmental setting and biodiversity of Cobscook Bay, Maine: a boreal, macrotidal estuary. Pages 13–22 *in* P. F. Larsen, editor. Ecosystem modeling in Cobscook Bay, Maine: a boreal, macrotidal estuary. Northeastern Naturalist 11(Special Issue 2).

Larsen, P. F., and L. F. Doggett. 1990. Sand beach macrofauna of the Gulf of Maine with inference on the role of oceanic fronts in determining community composition. Journal of Coastal Research 6:913–926.

Larsen, P. F., and L. F. Doggett. 1991. The macroinvertebrates associated with mud flats of the Gulf of Maine. Journal of Coastal Research 7:365–375.

Llansó, R. J., L. C. Scott, D. M. Dauer, J. L. Hyland, and D. E. Russell. 2002. An estuarine benthic index of biotic integrity for the mid-Atlantic region of the United States. I. Classification of assemblages and habitat definition. Estuaries 25:1219–1230.

Llansó, R. J., J. H. Vølstad, D. M. Dauer, and J. R. Dew. 2009. Assessing benthic community condition in Chesapeake Bay: does the use of different benthic indices matter? Environmental Monitoring and Assessment 150:119–127.

Longhurst, A. 1998. Ecological geography of the sea. Academic Press, San Diego, California.

Lourie, S. A., and A. C. Vincent. 2004. Using biogeography to help set priorities in marine conservation. Conservation Biology 18:1004–1020.

Mahon, R., S. K. Brown, K. C. T. Zwanenburg, D. B. Atkinson, K. R. Buja, L. Claflin, G. D. Howell, M. E. Monaco, R. N. O'Boyle, and M. Sinclair. 1998. Assemblages and biogeography of demersal fishes of the east coast of North America. Canadian Journal of Fisheries and Aquatic Sciences 55:1704–1738.

McLeod, K., and H. Leslie, editors. 2009. Ecosystem-based management for the oceans. Island Press, Washington, D.C.

NCCOS (NOAA National Centers for Coastal Ocean Science). 2006. An ecological characterization of the Stellwagen Bank National Marine Sanctuary eegion: oceanographic, biogeographic, and contaminants assessment. National Oceanic and Atmospheric Administration, National Ocean Service, Center for Coastal Monitoring and Assessment, NOAA Technical Memorandum NOS NCCOS 45, Silver Spring, Maryland.

Nye, J. A., J. S. Link, J. A. Hare, and W. J. Overholtz. 2009. Changing spatial distributions of fish stocks in relation to climate and population size on the northeast United States continental shelf. Marine Ecology Progress Series 393:111–129.

O'Hara, T. D., and G. C. B. Poore. 2000. Patterns of distribution for southern Australian marine echinoderms and decapods. Journal of Biogeography 27:1321–1335.

Paul, J .F., K. J. Scott, D. E. Campbell, J. H. Gentile, C. S. Strobel, R. M. Valente, S .B. Weisberg, A. F. Holland, and J. A. Ranasinghe. 2001. Developing and applying a benthic index of estuarine condition for the Virginian biogeographic province. Ecological Indicators 1:83–99.

Pettigrew, N .R., J. H. Churchill, C. D. Janzen, L. J. Mangum, R. P. Signell, A. C. Thomas, D. W. Townsend, J. P. Wallinga, and H. Xue. 2005. The kinematic and hydrographic structure of the

Gulf of Maine Coastal Current. Deep-Sea Research Part II 52:2369–2391.

Pocklington, P., and M. J. Tremblay. 1987. Faunal zones in the northwestern Atlantic based on polychaete distribution. Canadian Journal of Zoology 65:391–402.

Ranasinghe, J. A., J. B. Frithsen, F. W. Kutz, J. F. Paul, D. E. Russell, R. A. Batiuk, J. L. Hyland, J. Scott, and D. M. Dauer. 2002. Application of two indices of benthic community condition in Chesapeake Bay. Environmetrics 13:499–511.

Rawson, P. D., S. Hayhurst, and B. Vanscoyoc. 2001. Species composition of blue mussel populations in the northeastern Gulf of Maine. Journal of Shellfish Research 20:31–38.

Roman, C. T., N. Jaworski, F. T. Short, S. Findlay, and R. S. Warren. 2000. Estuaries of the northeastern United States: habitat and land use signatures. Estuaries 23:743–764.

SAS. 2003. SAS, version 9. SAS Institute, Cary, North Carolina.

Smith, P. C., N. R. Pettigrew, P. Yeats, D. W. Townsend, and G. Han. 2012. Regime shift in the Gulf of Maine. Pages 185–203 *in* R. L. Stephenson, J. H. Annala, J. A. Runge, and M. Hall-Arber, editors. Advancing an ecosystem approach in the Gulf of Maine. American Fisheries Society, Symposium 79, Bethesda, Maryland.

Sorte, C. J. B., S. L. Williams, and J. T. Carlton. 2010. Marine range shifts and species introductions: comparative spread rates and community impacts. Global Ecology and Biogeography 19:303–316.

Spalding, M. D., H. E. Fox, G .R. Allen, N. Davidson, Z. A. Ferdaña, M. Finlayson, B. S. Halpern, M. A. Jorge, A. Lombana, S. A. Lourie, K. D. Martin, E. McManus, J. Molnar, C. A. Recchia, and J. Robertson. 2007. Marine ecoregions of the world: a bioregionalization of coastal and shelf areas. BioScience 57:573–583.

Stephenson, E. H., R. S. Steneck, and R. H. Seeley. 2009. Possible temperature limits to range expansion of non-native Asian shore crabs. Journal of Experimental Marine Biology and Ecology 375:21–31.

Tilburg, C. E., S. Gill, S. I. Zeeman, A. Carlson, T. Arienti, J. Eickhorst, and P. O. Yund. 2011. Characteristics of a shallow river plume: observations from the Saco River Coastal Observing System. Estuaries and Coasts 34:785–799.

Trott, T. J. 2007. Zoogeography and changes in macroinvertebrate community diversity of rocky intertidal habitats on the Maine coast. Challenges in environmental management in the Bay of Fundy–Gulf of Maine. Pages 54–73 *in* G. W. Pohle, P. G. Wells, and S. J. Rolston, editors. Proceedings of the 7th Bay of Fundy Science Workshop, St. Andrews, New Brunswick 24–27 October 2006. Bay of Fundy Ecosystem Partnership, Report No. 3, Wolfville, Nova Scotia.

Tsontos, V. M., and D. A. Kiefer. 2003. The Gulf of Maine biogeographical information system project: developing a spatial data management framework in support of OBIS. Oceanologica Acta 25:199–206.

Van Dolah, R. F., J. F. Hyland, A. F. Holland, J. S. Rosen, and T. R Snoots. 1999. A benthic index of biological integrity for assessing habitat quality in estuaries of the southeastern United States. Marine Environmental Research 48:269–283.

Wares, J. P. 2002. Community genetics in the northwestern Atlantic intertidal. Molecular Ecology 11:1131–1144.

Watling, L. 1979. Zoogeographic affinities of northeastern North American gammaridean Amphipoda. Bulletin of the Biological Society of Washington 3:256–282.

Weisberg, S. B., J. A. Ranasinghe, D. M. Dauer, L. Schaffner, and R. J. Diaz. 1997. An estuarine benthic index of biotic integrity. Estuaries 20:149–158.

American Fisheries Society Symposium 79:185–203, 2012

Regime Shift in the Gulf of Maine

Peter C. Smith*
Fisheries and Oceans Canada, Bedford Institute of Oceanography
Post Office Box 1006, 1 Challenger Drive, Dartmouth, Nova Scotia B2Y 4A2, Canada

Neal R. Pettigrew
University of Maine, Orono, Maine 04469, USA

Philip Yeats
Fisheries and Oceans Canada, Bedford Institute of Oceanography
Post Office Box 1006, Dartmouth, Nova Scotia B2Y 4A2, Canada

David W. Townsend
School of Marine Sciences, University of Maine
Orono, Maine 04469, USA

Guoqi Han
Northwest Atlantic Fisheries Center
Post Office Box 3667, St. John's, Newfoundland A1C 5X1, Canada

Abstract.—Conventional wisdom, based on observations spanning two and a half decades (1975–2000), asserts that inflow to the Gulf of Maine (GoM) occurs primarily in two areas: inshore on the Scotian Shelf off Cape Sable, Nova Scotia and on the eastern side of the Northeast Channel (NEC). In particular, the monthly mean currents in the eastern NEC have shown persistent inflow at all depths and in all seasons, except for the occasional, but brief, reversals near the bottom (~200 m). Conversely, the flow on the western side of the NEC is normally directed out of the gulf in the surface layer and at mid-depth, consistent with the clockwise gyre over Georges Bank, but those currents do show relatively frequent reversals to inflow in the deeper layers (150–200 m), in sympathy with the flow on the eastern side. At some point between the year 2000, when the last Bedford Institute of Oceanography (BIO)/U.S. GLOBEC mooring was removed from the eastern NEC, and 2004, when a new mooring was placed there as part of the U.S. ocean observing array, a transformation occurred. The recent data, collected from a representative location in the eastern NEC, show a strongly seasonal current signal marked by persistent periods of outflow in the deep layers (>100 m), particularly in winter. This observation was first reported by Pettigrew et al. (2008), where the outflow currents occasionally extend to the surface layers as well, most notably in the winters of 2004–2005 and 2006–2007. Additional data and analyses reported here suggest that this new mode of behavior in the NEC currents could have important consequences for the GoM ecosystem. Possible causes for this "regime shift" in the NEC circulation and implications for the GoM deepwater nutrient fields and ecosystem are discussed.

* Corresponding author: peter.smith@dfo-mpo.gc.ca

Introduction

Background

Studies of the circulation in the GoM during the 1970s and 1980s suggest that the primary inflows were located in the nearshore region of the Scotian Shelf off Cape Sable (CS; Smith 1983, 1989) and in the deep and shallow layers on the eastern side of the Northeast Channel (NEC; Ramp et al. 1985; Smith et al. 2001; see Figure 1). These results, along with other historical hydrographic surveys (e.g., Bigelow 1927; Brooks 1985) led to the conception of a three-component mechanism for the exchange of shelf and offshore waters in the NEC (Brooks 1992). According to this model, the bottom layer consists of warm, saline, nutrient-rich slope water, which hugs the eastern side of the NEC as it moves into Georges basin but intermittently extends across the channel, causing deep inflow there as well. On average, the top of the deep inflow layer (as indicated by the 34 g/kg isohaline contour) generally lies between 75 and 100 m on the eastern side of the NEC and descends to 125–150 m on the west. Occasional reversals of the deep inflow appear to be associated with wind forcing and/or events in the slope water (e.g., impingement of Gulf Stream warm core rings; Bisagni and Smith 1998; Smith et al. 2001) and can have significant impacts on both the heat and salt budgets, as well as plankton bloom timing and dynamics (Townsend and

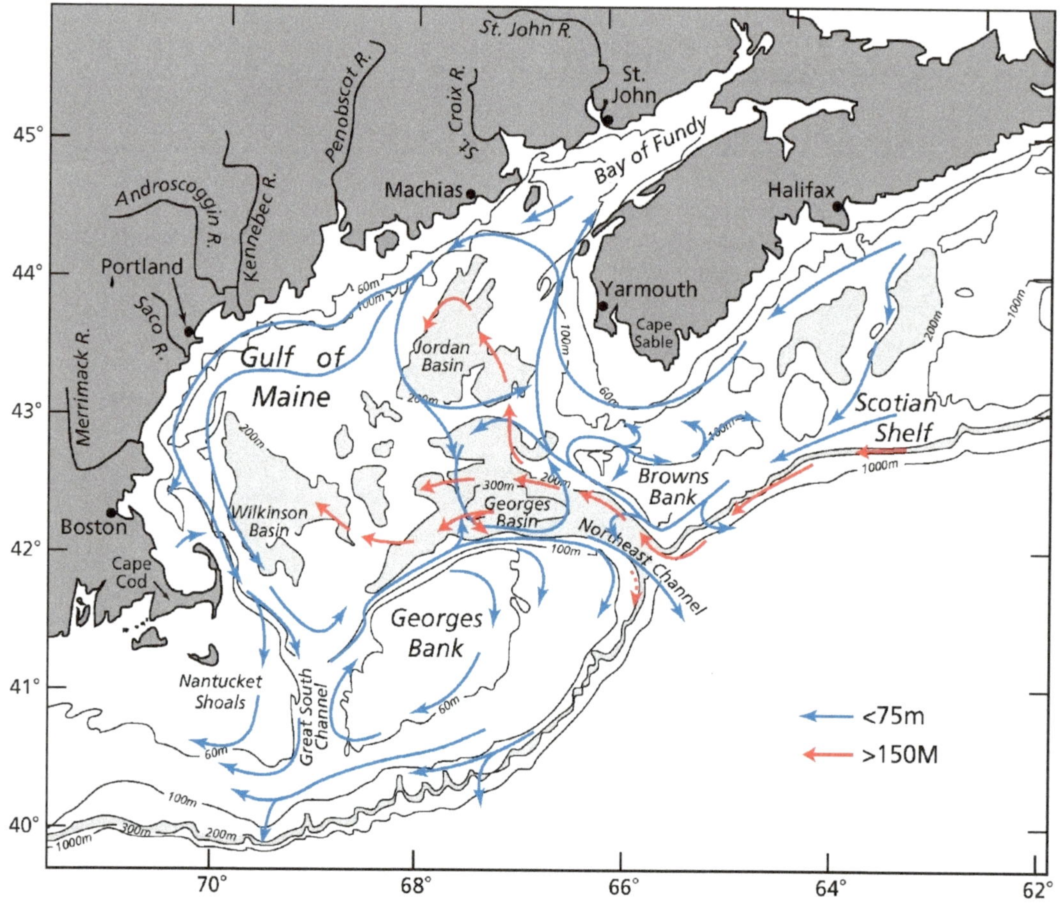

FIGURE 1. Schematic diagram of surface layer (blue arrows) and deep layer (red arrows) circulation in the Gulf of Maine during the stratified season (courtesy of Butman and Beardsley 1992).

Spinrad 1986) of the gulf. Above the slope water on the east, cool, fresh Scotian Shelf water moves into the Gulf of Maine (GoM), while on the western side, the flow is persistently outward, consisting primarily of Maine surface and intermediate water (Brooks 1992). The pattern of axial currents near the sill of the NEC is revealed by the average from a set of vessel-mounted Acoustic Doppler Current Profiler (ADCP) transects collected over a full M2 tidal cycle in June 1994 (Figure 2). During this period, the inflow currents were confined roughly to the eastern half of the NEC, which appears to be the prevailing pattern according to U.S. GLOBEC average currents at selected depths from 1993 to 1995 (Figure 3). Axial currents deeper than 125 m on the western side of the NEC were highly intermittent, with an overall mean of roughly zero. During the 1990s, the U.S. GLOBEC Georges Bank Program sponsored a wide variety of comprehensive biophysical observations in the GoM with a natural focus on Georges Bank. Among these were current meter and hydrographic measurements in the NEC designed to monitor exchanges between "upstream" (Scotian Shelf) and "offshore" (slope and Gulf Stream) waters and those of the GoM (Figure 3). In the period 1993–1996, Smith et al. (2001) noted a distinct change in the interannual variability of the inflow regime from enhanced warm, saline deep (>75 m) inflow in the eastern NEC to a time of predominant cold, fresh inflow episodes in the shallow layers off CS and in the NEC during 1995–1996. Moreover, the transport anomalies at CS and in the NEC were found to be generally out of phase, that is, increased inflow at CS was associated with reduced inflow in the deep NEC and cooler, fresher conditions at both sites and, conversely, increased flow of deep NEC water was associated with reduced inflow at CS and warmer and saltier conditions.

Smith et al. (2001) used a simple box model to estimate the influence of the observed changes in boundary fluxes on the mass and salt balances within the GoM. Comparing the last (April 1995–September 1996) to the first (October 1993–March 1995) half of the measurements, they found that the volumetric flow rate through the GoM had increased by 10^5 m^3/s (roughly 17%) and that enhanced freshwater fluxes in the surface layer had produced a decrease of 0.73 in the salinity of the outflow waters. Furthermore, the estimated CS

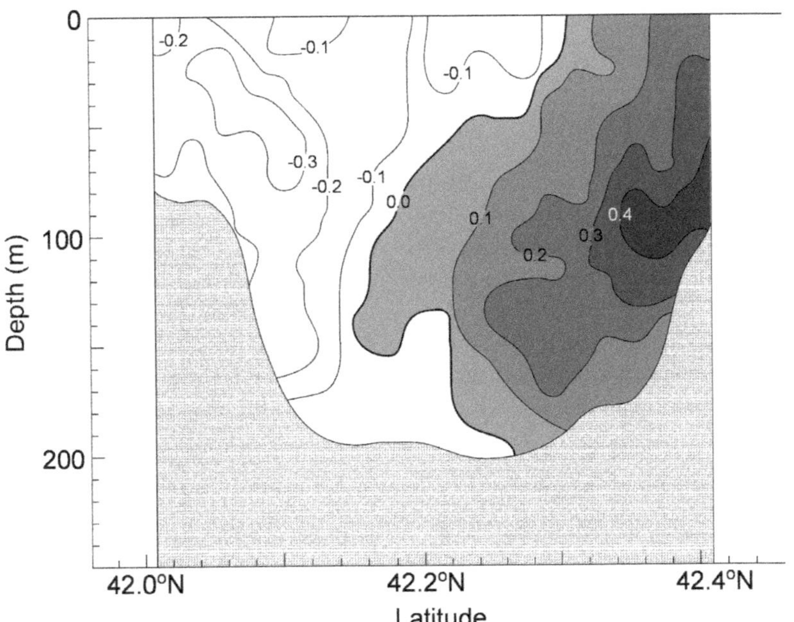

FIGURE 2. Mean axial (305°T) current component derived by averaging data over a series of vessel-mounted Acoustic Doppler Current Profiler (ADCP) transects near the sill of the Northeast Current during June 1994 (inflow: shaded gray, outflow: white).

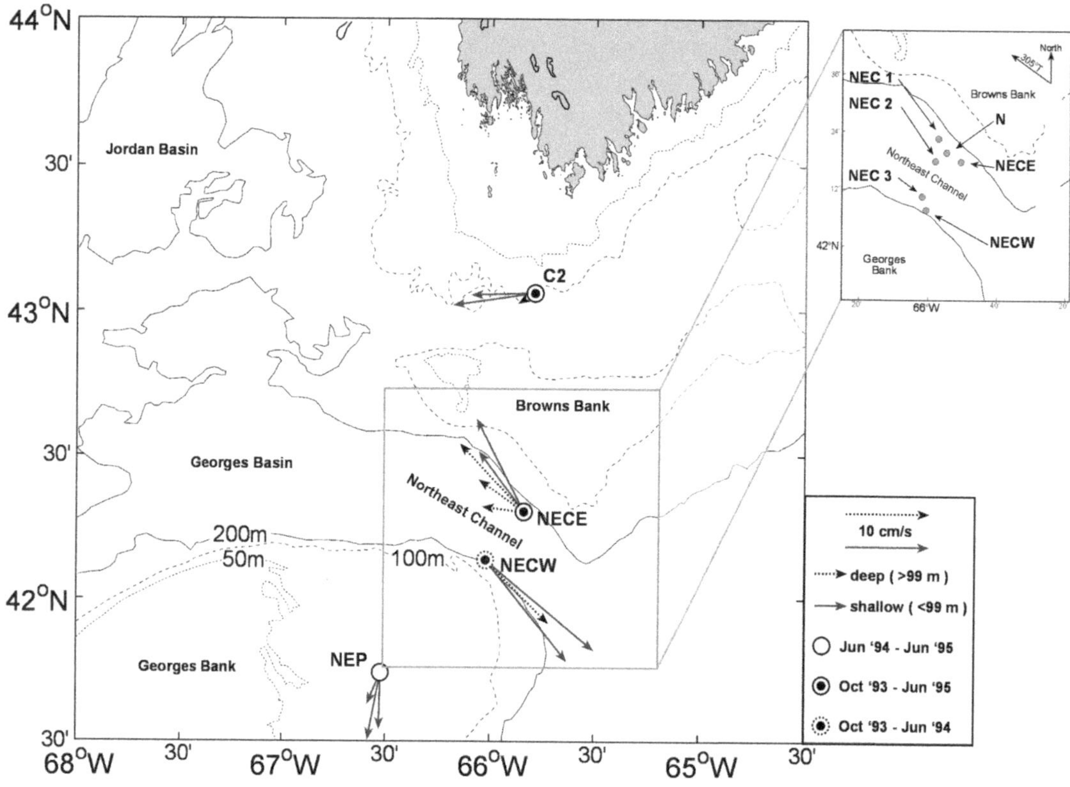

FIGURE 3. Long-term average currents collected during October 1993 to June 1995. Eastern (NECE) and western (NECW) moorings in the Northeast Channel (NEC) lie on the 200-m isobath near the sill. The xa Sable mooring (C2) lies on the 110-m isobath inshore. Inset—Locations of historical moorings in Northeast Channel, including NEC 1–3: 1977–1978 (Ramp et al. 1985), NECE–NECW: 1993–1999, GLOBEC moorings sites (Smith et al. 2001), and Buoy N: 2004–2008 NERACOOS (Northeast Regional Association for Coastal Ocean Observing Systems) mooring site (Pettigrew et al. 2008).

transport (297×10^3 m^3/s) was roughly twice that calculated by Smith (1983) for the 1979–1980 period (140×10^3 m^3/s), whereas the deep NEC transport (143×10^3 m^3/s) was slightly more than half that found by Ramp et al. (1985) for the 1977–1979 period (260×10^3 m^3/s). Nevertheless, the estimated average volumetric flow rate through the system over this period (563×10^3 m^3/s) is close to the climatological mean value of Loder et al. (1998) (500×10^3 m^3/s). For purposes of this paper, that level of interannual variability in boundary mass fluxes will be considered "normal."

More recently, scientists from the University of Maine have maintained (first on behalf of GoMOOS [Gulf of Maine Ocean Observing System]; now for NERACOOS [Northeast Regional Association for Coastal Ocean Observing Systems]) a single mooring on the eastern side of the NEC, in close proximity to both the GLOBEC and Ramp mooring sites (Figure 3 inset). Among other instrumentation, this mooring contains a long-range ADCP for monitoring currents over the water depths from 8 to 202 m and a surface current meter at 2 m. Recent hydrographic and nutrient data from an NEC transect near the sill suggest that the elements of the three-component exchange process are still in place (Figure 4), although aliasing by the strong tidal currents may obscure the true average properties. Here, we report results from this new mooring in comparison with results from the previous studies. We calculate boundary fluxes and annual cycles of flow in the NEC and estimate with a box model changes in mass balance. Finally, we discuss possible causes and ecosystem-level impli-

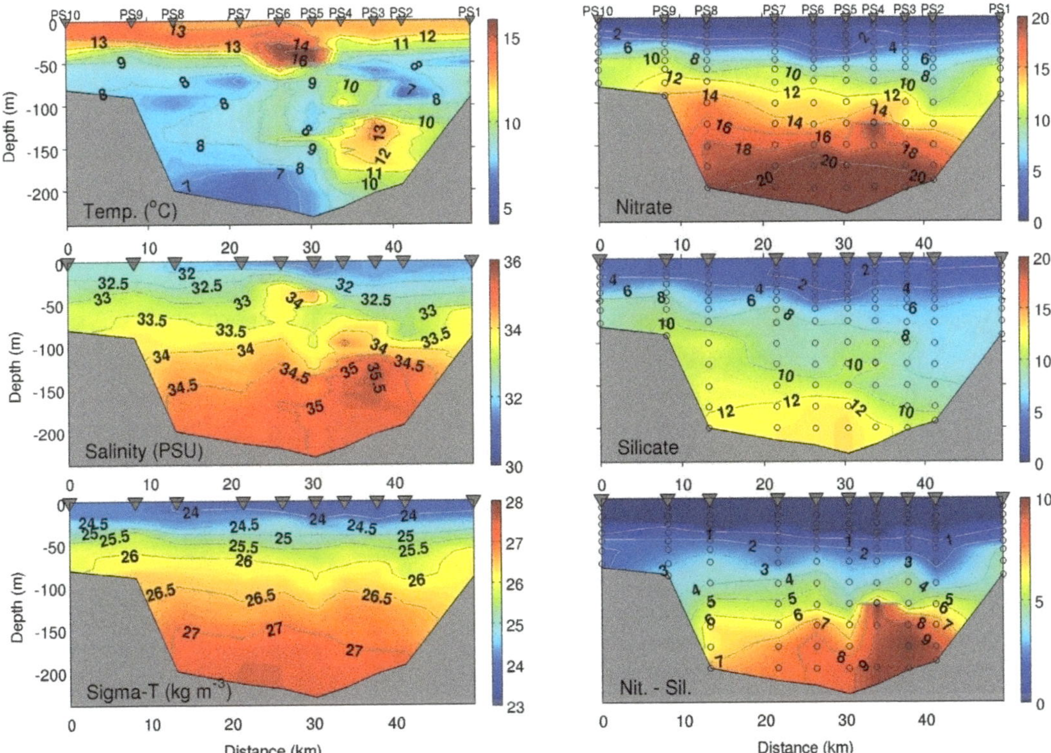

FIGURE 4. Hydrographic properties: temperature, salinity, and sigma-t (left); nutrients: nitrate, silicate, and nitrate residual (right) from a Northeast Channel (NEC) transect near the sill on October 12, 2008. This section runs roughly between the 100-m isobaths on Browns and Georges Banks, passing through the historical eastern and western NEC mooring sites (Figure 3).

cations of the new mode of behavior in the NEC currents.

Regime Shift/Mass Balance Changes

Regime Shift in Circulation

Between its deployment in early June 2004 and recovery in July 2008, NERACOOS Buoy N has achieved 94% data return. A contour plot of the monthly mean axial (305°T) current component at that site for the period July 2004–July 2008 is compared to a similar figure for the GLOBEC mooring eastern NEC during July 1993–July 1997 in Figure 5. The differences are striking. In the GLOBEC data, persistent inflow occurs over the full depth of the water column, with only occasional, weak reversals at depth on a seasonal timescale (note that seasonal reversals appear to be more common in the latter portion of the GLOBEC records). On the other hand, the Buoy N data show frequent, sustained outflows, especially at depth, with occasional sustained outflows at all depths, such as in the winters of 2004–2005 and 2006–2007, as first reported by Pettigrew et al. (2008). The phases of the seasonal cycles appear to be similar in both records, but the recent data are definitely biased toward outflow.

In order to quantify these differences, we intercompared the axial current components within the layers associated with the three-component exchange model, assuming that the average depths of the layers are shallow inflow on the east (0–75 m), deep inflow on the east (75–200 m), and outflow on the west (0–125 m). Dealing first with the depth-averaged currents, Figure 6 shows monthly mean inflows on the eastern side of the NEC (NECE) and on the western side (NECW, for the

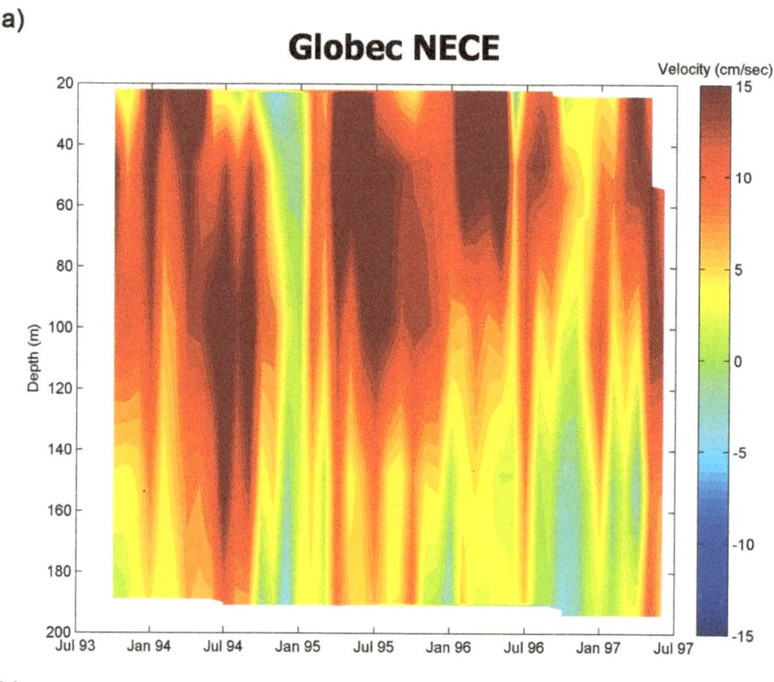

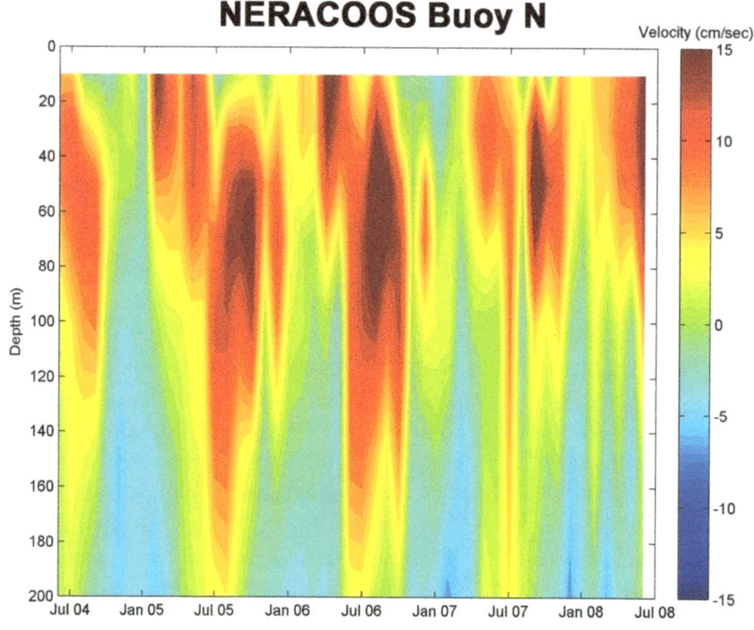

FIGURE 5. Comparison of 4 years of axial (305°T) current component data from (a) GLOBEC NECE (eastern Northeast Channel) mooring (July 1993–June 1997), and (b) NERACOOS (Northeast Regional Association for Coastal Ocean Observing Systems) Buoy N (July 2004–June 2008). GLOBEC estimates are based on currents measured at 20, 50, 100, 150, and 200 m (as indicated by symbols on the left axis), whereas NERACOOS Acoustic Doppler Current Profiler (ADCP) data have a vertical resolution of 20 m.

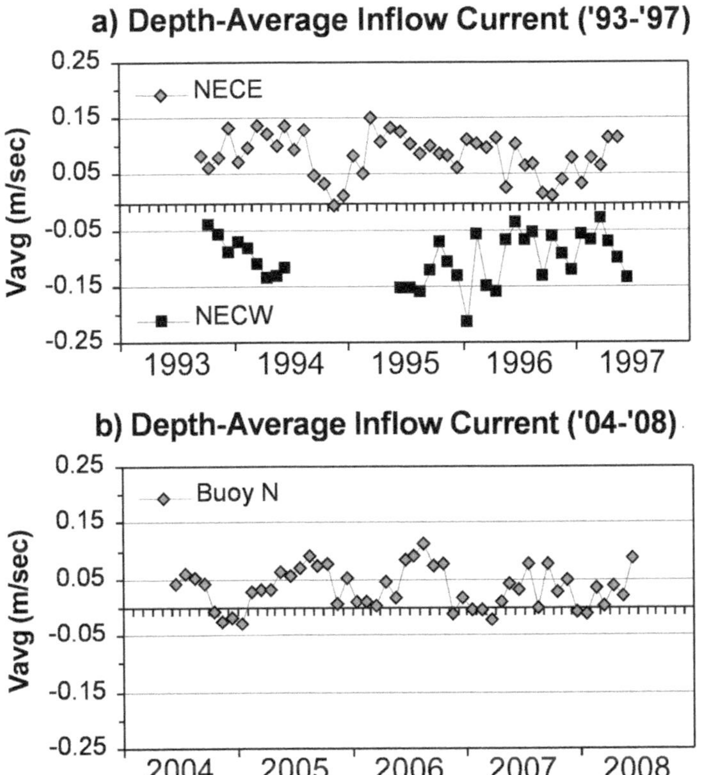

FIGURE 6. Monthly mean inflow currents (305°T) during the 1990s (a) and 2000s (b).

1990s only). Except for a single month in 1994, the depth-averaged inflow current at NECE during the 1990s was positive, with magnitudes of 8–10 cm/s. Outflow current magnitudes at NECW during the same period were of the same order (i.e., −8 to −10 cm/s). On the other hand, the net depth-averaged inflow at Buoy N during 2004–2008 fluctuated, exhibiting summer maxima but overall weaker inflow (~3–4 cm/s) with frequent seasonal reversals to net outflow conditions in winter.

Observed inflow/outflow currents on the east and west for the 1993–1997 (Figure 7a, 7b) and 1998–1999 (not shown) periods are similar. On the east, currents show generally stronger inflow in the surface layer, with some synchronous features in both layers (e.g., quiescent period in November 1994). However, occasional bursts of surface layer inflow create vertical shear of order 10 cm/s between layers. As noted earlier, outflow currents on the western side appear to penetrate to roughly 125 m, below which the flow is variably in and out of the gulf. Furthermore, a vertical shear of order 10–20 cm/s between layers persists on the west side of the NEC. Similar features are found in the current structures from the 1998–1999 period (not shown). Finally, the currents on the eastern NEC at Buoy N during 2004–2008 show pulses of summer inflow, alternating with extended periods of winter outflow in the bottom layer, while the surface layer inflow maintains a weak shear of order 5 cm/s (Figure 7c). More specifically, at this site, the net inflow below a depth of 125 m is effectively zero while the average inflow above that layer is 5.8 cm/s. Clearly, the character of the flow regime in the NEC has changed dramatically between the two periods analyzed here (the 1990s and 2004–2008).

Boundary Fluxes-Annual Cycles

To shed light on the character of the boundary mass flux in the NEC, methods used to analyze the GLOBEC Phase I observations (October 1993–September 1996) were extended to include GLOBEC data from September 1996–June 1997

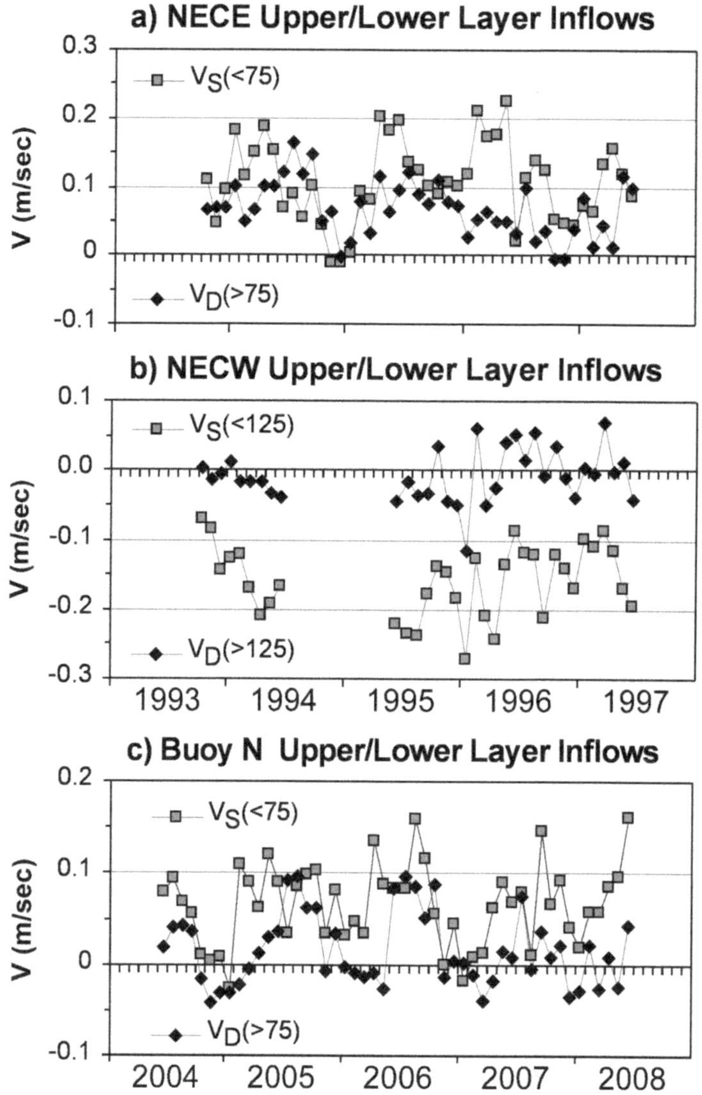

FIGURE 7. Two-layer inflow currents (305°T) during the 1990s on the eastern (a) and western (b) sides of the Northeast Channel (NEC) and (c) during 2004–2008 at Buoy N. The interface between the two-layers is assumed to descend 50 m from east to west on average, based on hydrographic and other data.

and November 1998–September 1999, as well as the NERACOOS data from June 2004–June 2008. Procedures for calculating the mass flux at the NEC mooring sites, detailed in Smith et al. (2001), were applied to calculate monthly mean differential flux indices,

$$q = \int_{z_1}^{z_2} V dz \cong \sum_{i=1}^{N} V_i \Delta z_i \quad , \tag{1}$$

where V represents the layer-average axial component (305°T) of the inflow current and Δz_i are discrete intervals within $z_1 \le z \le z_2$ over which the measured inflow/outflow currents are averaged. Consistent with the three-component exchange process, three layer-average differential mass fluxes were computed, corresponding to

$$\begin{aligned}(z_1, z_2) &= (-H, 0)\,\text{depth-average}, H \cong 200\text{ m}\\ &= (-H, -75\text{ m})\,\text{deep layer, and}\\ &= (-75\text{ m}, 0)\,\text{surface layer}\end{aligned}$$

These differential flux quantities, q, are measurable at each of the mooring sites and are considered to be indices of the inflow/outflow volumetric transport in the sense that multiplication by the appropriate along-boundary dimension and suitably offset, they provide an estimate of the total transport in that layer (see next subsection).

Furthermore, the seasonal variations of the inflow currents (Figure 8) and fluxes were quantified using least-squares multiple regression with $n-3$ degrees of freedom, where n is the number of monthly data points (Smith et al. 2001). The estimator was of the form

$$Y_v = B_0 + B_1 \cos\left[\pi\left(t_v - \Phi_1\right)/6\right] + \varepsilon \tag{2}$$

where t_v is the month number (=0.5 for January, $v = 1$) and ε is the residual. Table 1 compares the statistically significant annual cycles, and their associated 95% confidence intervals (Δ,δ) (see Smith et al. 2001), for two sets of GLOBEC observations (1993–1996, 1993–1999) and the NERACOOS data (2004–2008). Depth average, surface-layer average, and deep-layer average quantities are denoted by q_T, q_S, and q_D, respectively, and coincident estimates for the annual cycle of the depth-average Scotian Shelf inflow off Cape Sable, Nova Scotia (q_2) are included for comparison.

Several features of the annual cycle regressions serve to quantify both similarities and differences between the GLOBEC and NERACOOS data sets. The major difference, as suggested earlier, lies in the mean fluxes. For instance, the NERACOOS depth-average mean inflow (B_0) is less than 40% of those from the GLOBEC eras, and the annual cycle provides a weak winter reversal (Figure 8b). This reversal is absent from the surface layer (Figure 8d) but pronounced in the deep layer (Figure 8f). The amplitudes of the annual cycles for both periods, though different, are statistically equivalent within the error estimates, for both full-depth- and layer-average fluxes. The phases of the annual cycle of the NERACOOS inflows appear to lag those of the GLOBEC period by the order of 1–2 months, though only one of those differences exceeds the error bounds. Note that the peak of the annual CS inflow cycle (January) is fully out of phase with the deep NEC inflow.

Box Model: Estimates of the Changing Mass Balance

Smith et al. (2001) showed that because of coherent relationships at low frequencies among current measurements in the NEC and across the inner Scotian Shelf off CS, the net volumetric inflow in those two locations could be indexed to the low-frequency (detided monthly mean) differential transports measured at the mooring sites NECE and C2 (see Figure 3). Assuming these relationships hold throughout both the GLOBEC and NERACOOS observation periods, the net inflows at CS and in the NEC shallow and deep layers may be indexed to the differential transports at C2 and NECE as

$$Q_{CS}\left(m^3/s\right) = 1.68 q_2 W_{CS} + 45\times10^3, \quad W_{CS} = 20\text{ km} \tag{3a}$$

$$Q_S\left(m^3/s\right) = 1.75 q_S W_{NEC} - 112\times10^3, \quad W_{NEC} = 15\text{ km} \tag{3b}$$

$$Q_D\left(m^3/s\right) = 1.75 q_D W_{NEC} - 112\times10^3, \quad W_{NEC} = 15\text{ km} \tag{3c}$$

where W_{CS} and W_{NEC} are the effective widths of the inflow current fields, derived from regressions of total transport through a particular segment of the boundary cross section against observations at a particular point on that boundary. Direct evidence for the surface layer inflow relationship is questionable due to the lack of near-surface current data, so the deep NECE regression has been applied following Smith et al. (2001).

The box model conserving mass and salt (Figure 9) formulated by Smith et al. (2001) features the major inflows at CS (given by equation 3a) and in the NEC (equations 3b, 3c), plus small contributions from river runoff into the Gulf (Q_R~3 × 10^3 m^3/s) and the difference between precipitation and evaporation (Q_{P-E} ~ 10^3 m^3/s). The two sinks of mass for the model are the longshore flux, Q_N, on the New England Shelf, and an offshore flux, αQ_N, which are assumed to remain in fixed ratio over the long term.[1] Assuming the river and $P-E$ contributions are negligible, conservation of mass requires:

[1] The basis for this split of the two sinks is represented in Figure 1 as the distinction between the flow that remains inside the shelf break (~200 m isobath) as it exits the region (e.g., through Great South Channel or around the Northeast Peak) and that which crosses the shelf break to the offshore region. This distinction is not critical to our results and conclusions in any way.

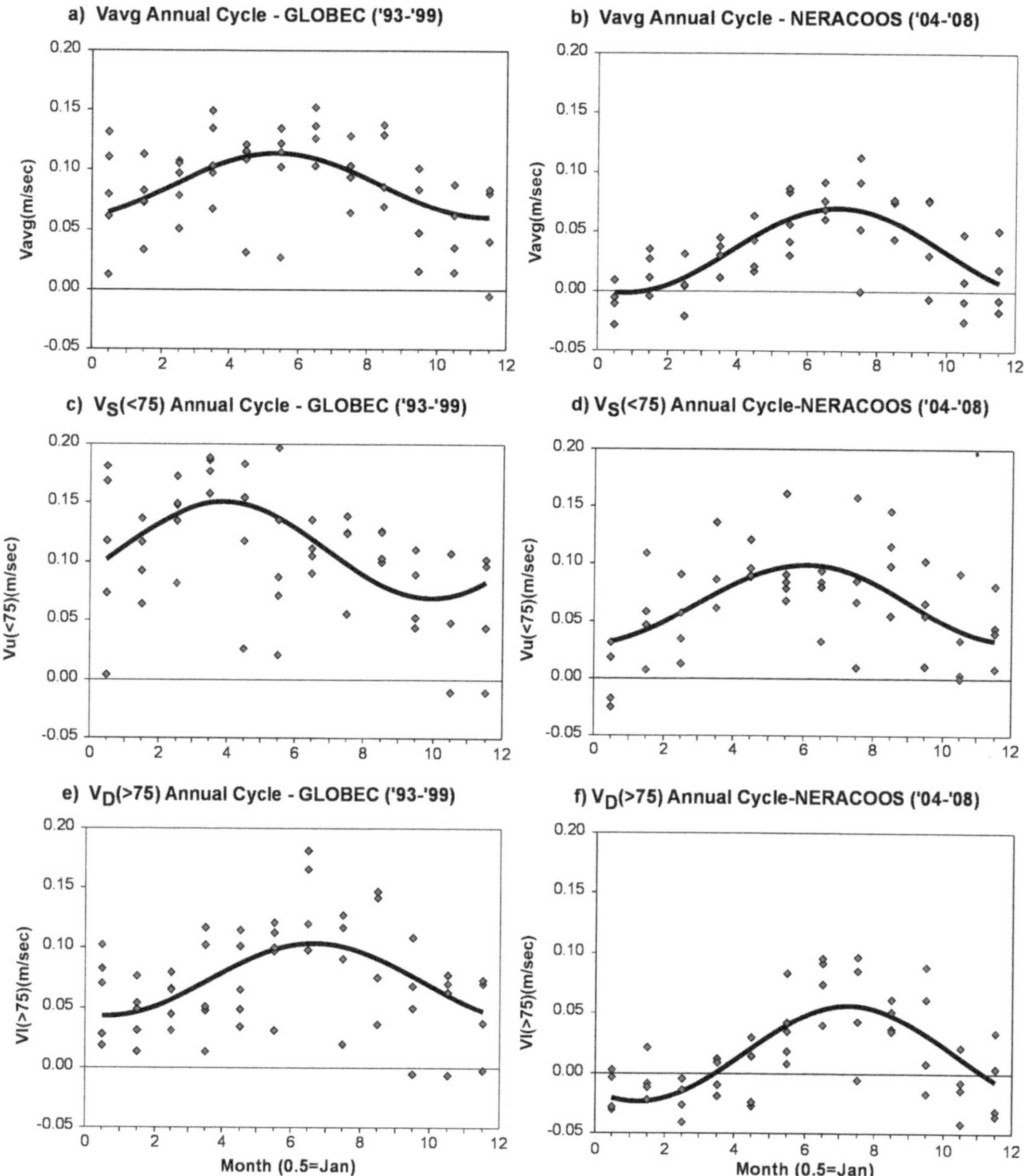

FIGURE 8. Annual cycles for depth (a, b), surface (c, d) and deep-layer-average (e, f) GLOBEC inflow (1993–1999, left column) and inflow currents in the Northeast Channel for NERACOOS (Northeast Regional Association for Coastal Ocean Observing Systems; 2004–2008, right column) data sets.

$$(1+\alpha)Q_N = Q_{CS} + Q_S + Q_D \quad (4)$$

To define the differential mass transport at a given location, Smith et al. (2001) chose periods when moored observations were widely available for the inflow as a whole and correlated the relatively well-sampled cross-sectional mass flux estimates to the differential fluxes for the index station (e.g., C2) at low frequencies. The correlation structures in the full data sets indicate that the currents at these locations are the primary determinants of the net mass fluxes into the Gulf of Maine. Also,

Table 1. Layer-average annual cycles for differential transports. q_T, q_S, and q_D are the differential transport indices for depth average, surface-, and deep-layer average inflow in the NEC, respectively. q_2 is the depth-average off Cape Sable, Nova Scotia.

Flux	Quantity	GLOBEC 1993–1996	GLOBEC 1993–1999	NERACOOS 2004–2008
q_T (m²/s)	n	36	54	49
Mean	B_0 (m²/s)	18.5	17.6	7.0
Amplitude	$B_1 \pm \Delta_1$ (m²/s)	5.4±3.2	5.4±2.6	7.2±2.0
Phase	$\Phi_1 \pm \delta_1$ (mon.)	5.2±1.1	5.4±0.9	6.9±0.6
SE	σ_ε (m²/s)	6.7	6.8	5.2
q_S (m²/s)				
Mean	B_0 (m²/s)	8.8	8.4	4.9
Amplitude	$B_1 \pm \Delta_1$ (m²/s)	3.5±1.9	3.1±1.4	2.6±1.1
Phase	$\Phi_1 \pm \delta_1$ (mon.)	4.1±1.1	3.9±0.9	6.0±0.9
SE	σ_ε (m²/s)	3.9	3.7	2.8
q_D (m²/s)				
Mean	B_0 (m²/s)	9.7	9.0	2.0
Amplitude	$B_1 \pm \Delta_1$ (m²/s)	3.2±2.2	3.9±1.9	5.0±1.5
Phase	$\Phi_1 \pm \delta_1$ (mon.)	6.5±1.3	6.6±0.9	7.3±0.6
SE	σ_ε (m²/s)	4.6	4.7	3.6
q_2 (m²/s)				
Mean	B_0 (m²/s)	7.5		
Amplitude	$B_1 \pm \Delta_1$ (m²/s)	3.1±1.8		
Phase	$\Phi_1 \pm \delta_1$ (mon.)	0.4±1.2		
SE	σ_ε (m²/s)	3.8		

during GLOBEC periods when measurements were available from both the NEC and Cape Sable, a robust inverse relationship was noted between the low-frequency (detided monthly mean) deep (>75m) differential transport on the eastern side of the NEC and that of the full-depth inflow off Cape Sable on the 100-m isobath (mooring C2). Hence, to estimate the differential transport at Cape Sable for the late GLOBEC (1998–1999) and NERACOOS (2004–2008) periods, when direct observations were not available, we make use of this inverse relationship between the observed anomalies to estimate the Scotian Shelf inflow at Cape Sable. Specifically, if we define the differential transport anomaly as

$$q'(t) = q(t) - B_0 - B_1 \cos\left[\pi(t - \Phi_1)/6\right], \quad (5)$$

then the early GLOBEC data reveal a significant inverse correlation of the form (see also Figure 10)

$$q_2'(t) = -0.474 q_D' + 0.070. \quad r^2 = 0.48 \quad (6)$$

Using the estimates of q_2' from the inverse regression (6), along with the observed values of shallow- and deep-layer NEC differential transports for the late GLOBEC and NERACOOS periods, the mass balances for all three periods may be calculated and compared, using the formulae (3a–c) and assuming that river and atmospheric inputs are negligible.

The results of the box model analysis (Table 2) indicate, first of all, that the transports during the latter half of the 1990s are virtually identical to those of the earlier period. Moreover, the estimated total volumetric fluxes off CS and in the NEC are reasonably consistent with earlier estimates of 140 and 260 m³/s, respectively, from direct observations by Smith (1983) and Ramp et al. (1985). However, there has been a drastic

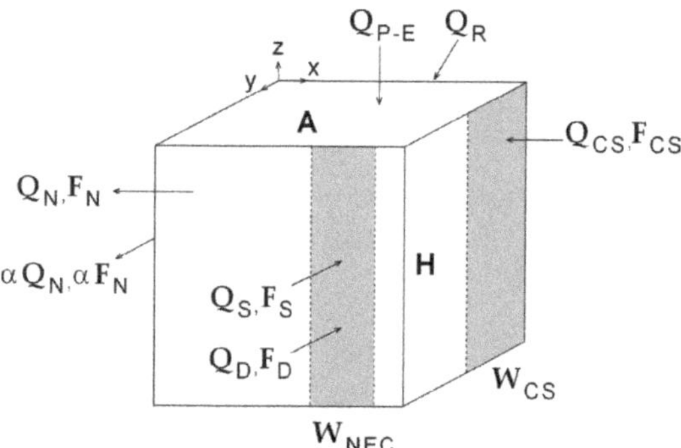

FIGURE 9. Box model for volumetric (Q in m^3/s) and freshwater (F in g kg^{-1} m^3/s) transports in the Gulf of Maine (Smith et al. 2001). Inflow volumetric transport components Q_{CS}, Q_S, and Q_D enter via the Scotian Shelf off Cape Sable (CS) and shallow (S) and deep (D) NEC (Northeast Channel), respectively, via sections of width W_{CS} and W_{NEC}. River discharge (Q_R) and precipitation-evaporation (Q_{P-E}) enter from land and surface boundaries, and Q_N and αQ_N exit via the New England shelf and offshore. A and H are the estimated surface area and mean depth.

change in the mass balance of the gulf between the GLOBEC years of the 1990s and the more recent NERACOOS era, 2004–2008. Specifically, the effective reversal of the net deep inflow transport, due to linkages to outflow on the western side of the NEC, is consistent with the observed pattern of sustained outflow at depth in the NERACOOS data. Moreover, the surface inflow in the NEC, while still positive, is considerably reduced from that in the GLOBEC period. Also, by virtue of the inverse correlation with deep inflow, the Scotian Shelf inflow at CS appears to have increased by 40%. Overall, the results indicate that the mass flux through the gulf is reduced by roughly 30% in recent years. According to Table 2, this reduction is entirely at the expense of the NEC inflow, especially the deep component.

These results must be considered tentative at this point since they are based on many assumptions. Not the least of these is the inferred inverse correlation between the differential flux anomalies at CS and the deep NEC. This idea is not new however. Brooks (1992) pointed out such a relationship between the deep inflow observations of Ramp et al. (1985) and the freshwater flux into the gulf, both from the Scotian Shelf and from local rivers, the former being the dominant source. According to this hypothesis, in "wet" years, high freshwater input by the Scotian Shelf water, derived in part from the St. Lawrence River, should delay or retard deep NEC inflow by imposing an adverse, barotropic pressure gradient. Conversely, in "dry" years, a weaker barotropic gradient should allow deep injection earlier. It is interesting to note that the increase in the phase lag of the annual cycle of deep inflow during the NERACOOS versus GLOBEC period (Table 1), though not statistically significant, is consistent with this hypothesis. At a more basic level, the 6-month lag between the CS and deep NEC differential transports (Table 1) reflects this inverse relationship between the annual cycles themselves.

Discussion

Possible Causes

As indicated by the box model results, among the possible causes for this drastic change in the magnitude and character of the NEC inflow currents is a substantial increase of the fresher coastal inflows from the Scotian Shelf at CS. In a geostrophic sense, this would require an adverse (positive onshore) barotropic pressure gradient that would, in turn, oppose the deep inflow. In any case, one likely source of enhanced coastal inflow would be an increase in the freshwater input to the Nova Scotian Current by its major sources in the Gulf of St. Lawrence and Labrador Sea. Freshwater input to the Gulf of St. Lawrence is monitored by an index known as RIVSUM, which represents the combined discharges of the St. Lawrence, Ottawa, and Saguenay rivers, the main contributors to the gulf (Sutcliffe et al. 1976). However, records of the RIVSUM index over the past 50 years reveal no major changes in freshwater input to the Gulf of St. Lawrence over the past two decades. In fact, the average discharge in the early 1990s appears to exceed that in the 2000s. On the other hand, according to Loder et al. (1998), RIVSUM accounts for slightly greater than 40% of the freshwater exiting the Cabot Strait and roughly 30% of that carried by the Nova Scotian Current off Halifax. The remainder of the freshwater transport off Halifax is derived from the in-

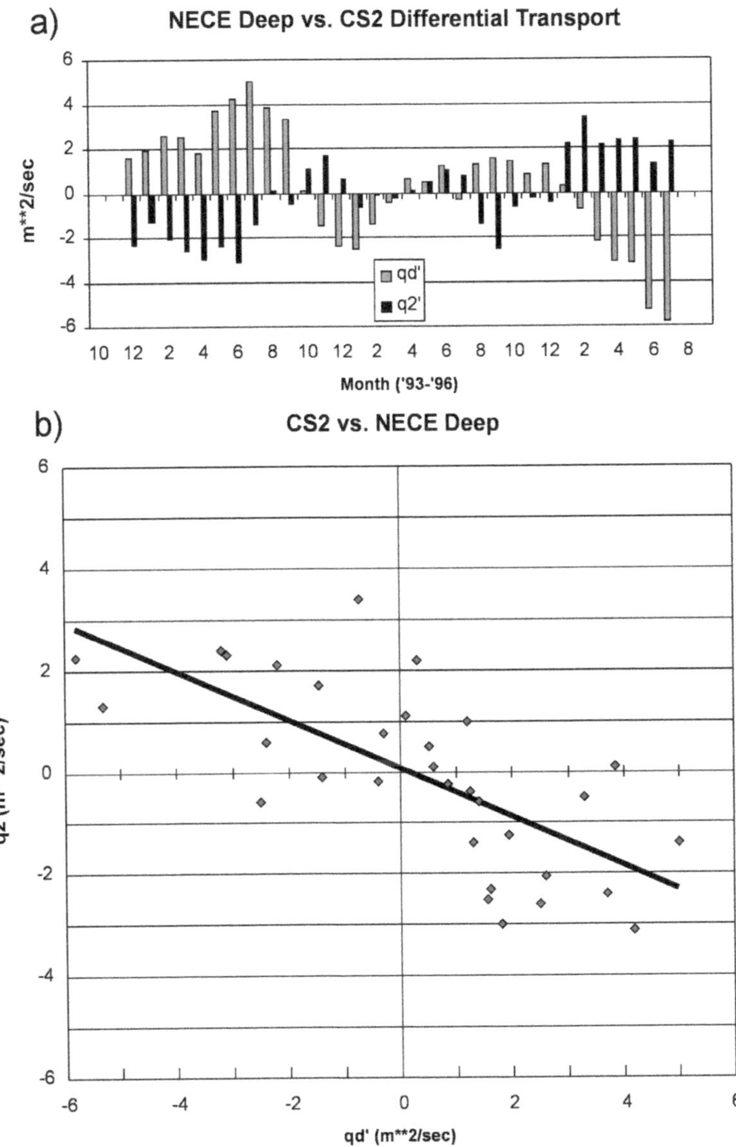

FIGURE 10. Time series (a) and regression plot (b) showing inverse correlation between differential transport anomalies at Cape Sable (C2 on the Scotian Shelf) and in the deep eastern Northeast Channel (NECE; see Figure 3).

TABLE 2. Box model transport estimates.

Mass flux component	GLOBEC I 1993–1996	GLOBEC III 1998–1999	NERACOOS 2004–2008	CHANGE: 2000s vs. 1990s
Q_{CS} (m^3/s)	297	301	420	+40%
Q_S (m^3/s)	120	121	16	–87%
Q_D (m^3/s)	144	140	–60	–141%
$(1 + \alpha)Q_N$ (m^3/s)	560	562	380	–32%

shore branches of the Labrador Current that pass through the Strait of Belle Isle and the Avalon Channel. Thus, RIVSUM is, at best, a weak indicator of the freshwater transport on the Scotian Shelf. Similar arguments can be made for the mass balance in the Gulf of St. Lawrence, prominent sources for which lie in the inshore branches of the Labrador Current.

Another approach to this problem might be to look at trends in sea surface height (SSH) anomalies measured by satellite-borne altimeters, which pass over the region regularly. The trend in sea-level height anomaly, derived from multi-mission satellite altimetry for the period 1992–2008, shows a contrast of the sea-level trend between the Scotian Shelf–Gulf of Maine (rise) and the offshore slope water region (fall) over the past 15 years. In fact, according to a global picture (Wunsch et al. 2007), this trend extends eastward to at least the tail of the Grand Banks, geostrophically favoring an accelerated westward flow over that time period. Considering just the contrasting trends between the Yarmouth area and nearby slope water, the order of magnitude of the onshore–offshore height difference developed after just one decade would be the order of 13 cm. In conjunction with the satellite data, consideration should be given to the longer coastal sea level records on the Scotian Shelf from Halifax and Yarmouth, Nova Scotia. Figure 11a, 11b show the

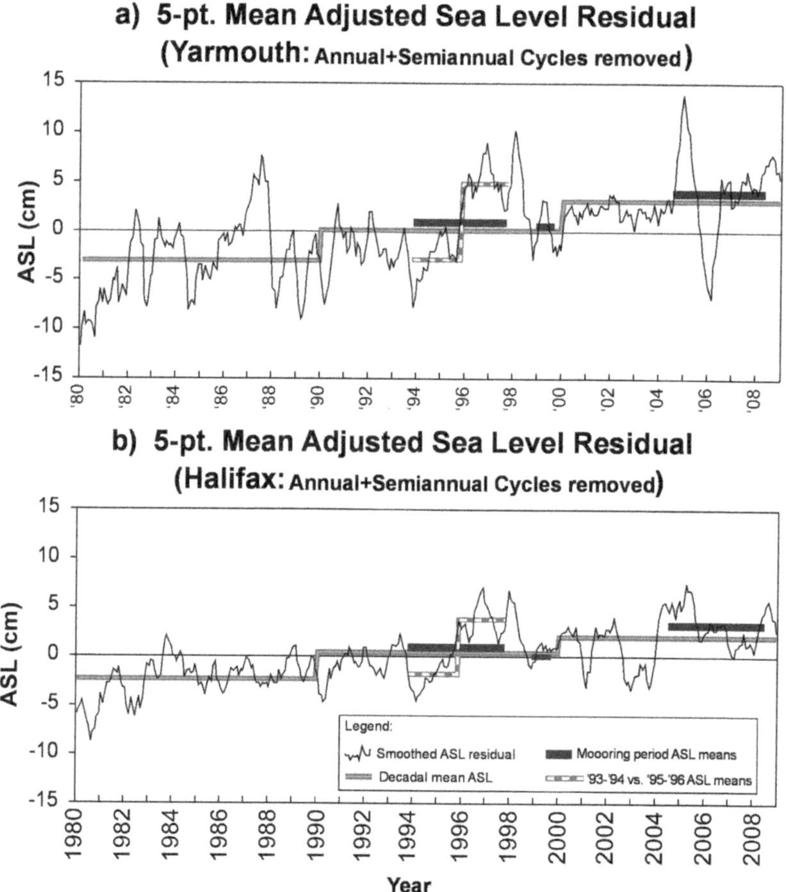

FIGURE 11. Five-point mean adjusted sea level anomaly (1980–2008) records from (a) Yarmouth, Nova Scotia and (b) Halifax, Nova Scotia. Tide gauge records have been adjusted for barometric pressure and annual and semiannual signals have been removed by least-squares regression. Decadal (purple) and mooring period (yellow) means are indicated, as well as 2-year means for the first and last halves of the 1993–1997 GLOBEC period.

nearly 30-year time series (1980–2008) of monthly mean adjusted sea level (ASL) anomalies (corrected for the inverse barometer effect, annual and semiannual cycles are removed by least-squares regression, and 5-point running mean is applied). Ten-year means of these data show a clear positive trend in mean relative sea level of order +2–3 cm per decade, in spite of strong interannual variability. Though a significant fraction of this change may be attributed to postglacial rebound (Peltier 2004), the geocentric altimetry height measurements support the contention that there has been some eustatic sea level rise at both Yarmouth and Halifax over the past two decades. The ASL averages over the various mooring periods are also consistent with the decadal averages, but within a given mooring period, shorter-term averages can produce trends an order of magnitude different (e.g., 1993–1994 versus 1996–1997 averages suggest an ~8 cm increase over 2 years or roughly 40 cm per decade). The shorter-term differences were thought and explained to be related to the Gulf Stream north–south movement (Han 2002, 2007).

Seasonal maps of the SSH derived from multi-mission altimeter data capture some of the synoptic features related to coastal ASL and the NEC current records. In particular, the dominant outflow currents over the full depth of the water column at Buoy N in the fall/winter of 2004/2005 are coincident with a strong, positive SSH anomaly (>10 cm) observed near Yarmouth during the same season (Figure 12d). Continuing in 2005, the Yarmouth anomaly dropped to roughly +6 cm in winter (Figure 12e), then to ~+2 cm in spring (Figure 12f), and became negative by the summer (Figure 12g). These features are all qualitatively consistent with the rise and fall of adjusted sea level in the tide gauge record (Figure 11a).

Thus, it appears that the inverse relationship demonstrated by NEC deep inflow versus Scotian Shelf inflow and the accompanying offshore pressure gradient reflected in Yarmouth SSH is a robust feature of the Gulf of Maine circulation at all timescales from seasonal, to annual, to interannual. A viable hypothesis, therefore, is that a recent rise in sea level off southwest Nova Scotia is associated with both an increased mass flux into the GoM from the Scotian Shelf and an adverse pressure gradient in the NEC, which opposes and reduces the deep inflow of nutrient-rich slope water to the gulf. This increased mass flux may also be exerting its effects on the circulation on the Scotian Shelf and might have its origin further upstream in the Labrador Sea.

Potential Consequences

One of the primary consequences of the regime change in the NEC circulation is alteration in the nutrient supply to the inner GoM and its impact on the ecosystem. The shift in the balance between deep inflows in the NEC and shallower inflows from the inshore Scotian Shelf will decrease the net input of nutrients to the GoM because nutrient concentrations invariably increase with depth and also change the relative inputs of nitrate and silicate because silicate shows less gradient with depth (see Figure 4). Townsend et al. (2010) have shown, based on five decades of nutrient data compiled for the deep waters (>100 m) of the eastern gulf (Rebuck et al. 2009), that nitrate concentrations have been steadily declining since the 1970s, whereas the average silicate concentration has remained roughly constant, leading to a decrease in the traditionally positive residual nitrate (nitrate concentration minus silicate concentration). Less residual nitrate will favor production of diatoms. Similar temporal trends in nutrient concentrations since the 1970s have been observed in Scotian Shelf waters at depths greater than 60 m (Yeats et al. 2011). Since the early 1990s, a decrease in residual nitrate has been noted on the Scotian Shelf (60–100 m, Figure 13a) and in the Gulf of St. Lawrence at virtually all depths (Figure 13b). These conditions in the shelf waters are thought to reflect nitrate losses from sediment denitrification and silicate increases from river input and in situ regeneration (Townsend et al. 2010). However, at the same time, the opposite trend has been observed in the Labrador Sea, where silicate concentrations appear to be decreasing more rapidly than nitrate, leading to an increase in residual nitrate (Figure 13c). This suggests that the changing nutrient ratios in the inshore inputs to the GoM are more closely tied to biogeochemical changes taking place between the Labrador Sea and the Gulf of Maine (e.g., along a shelf-bound route of the coastal currents) than to changes in upstream nutrient sources. Thus, the regime shift in GoM water transport and bio-

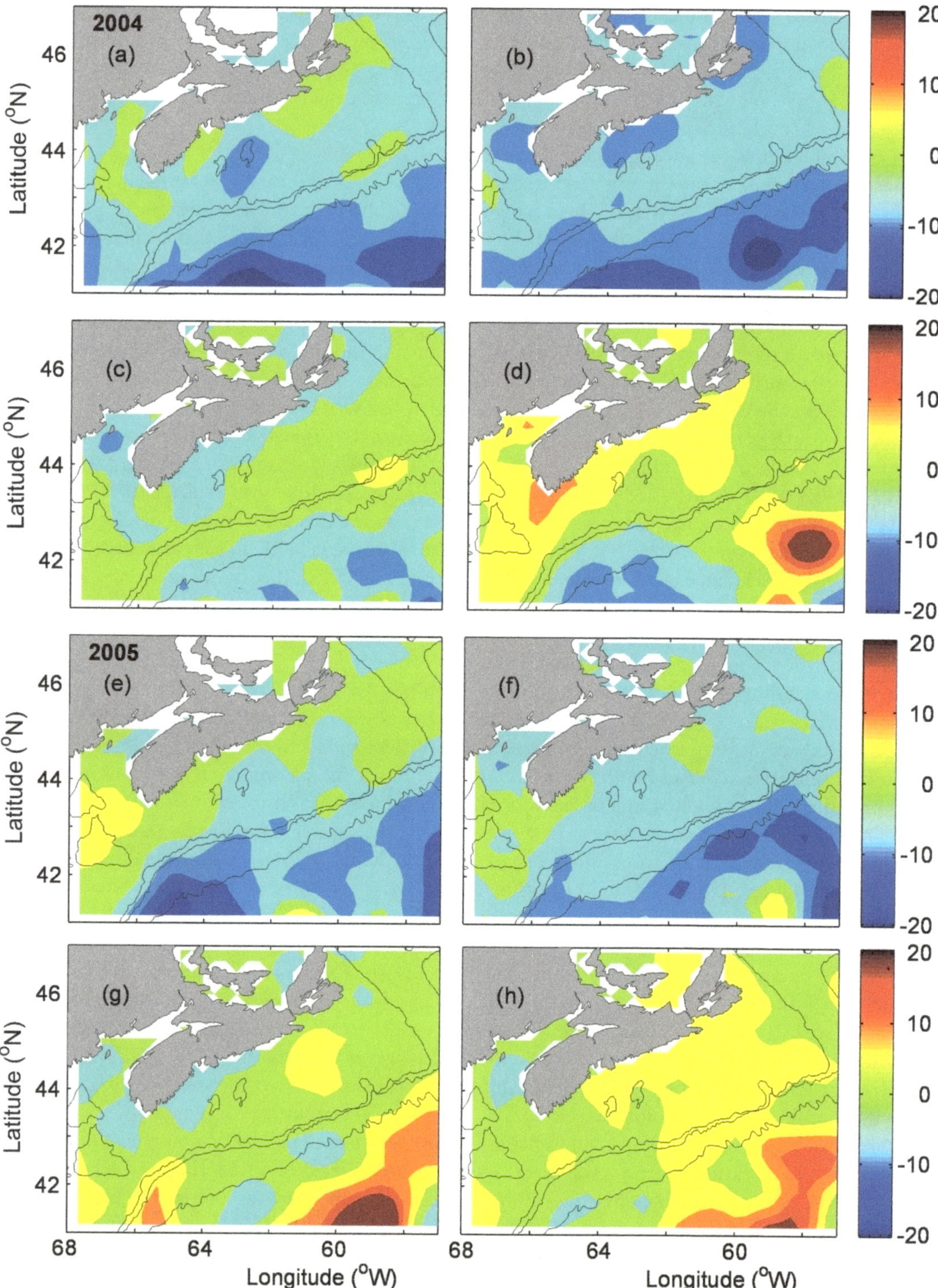

FIGURE 12. Maps of sea surface height anomalies (in cm) for 2004–2005 created from satellite remote-sensing data: (a, e) winter, (b, f) spring, (c, g) summer, (d, h) fall.

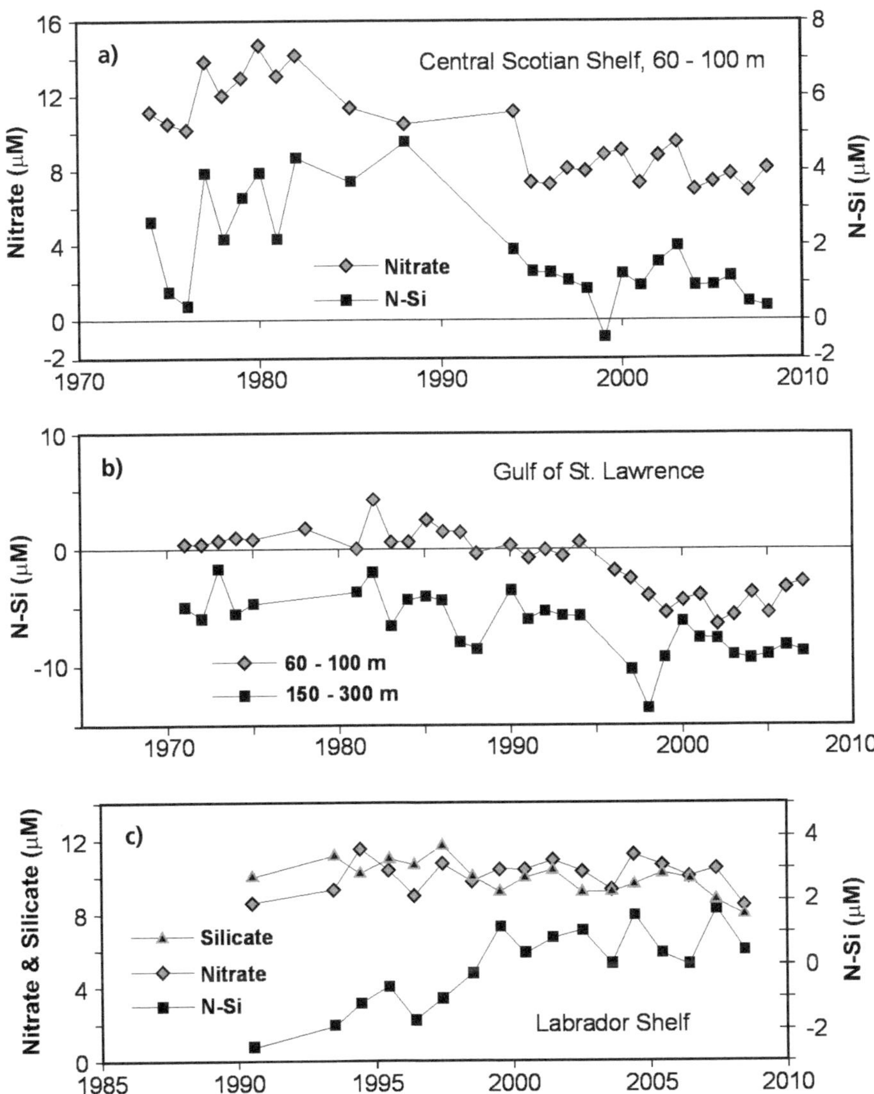

FIGURE 13. Time series for (a) nitrate and residual nitrate based on annual average concentrations for all data for the 60–100 m depth range on the central Scotian Shelf (62–64 W, coastline to edge of continental shelf); (b) residual nitrate for middepth and deep samples in the Gulf of St. Lawrence (Cabot Strait in east to Pointe des Monts in west and Belle Isle Strait in north); and (c) nitrate, silicate, and residual nitrate in the 60–200-m-depth range for the Labrador Shelf stations (53.5 N to 55.0 N) on the annual surveys along the WOCE AR7W section from Labrador to Greenland.

geochemical process in upstream coastal waters combine to generate the observed temporal trends (Townsend et al. 2010) in nutrient concentrations in eastern GoM waters.

With regard to the biological impacts at lower trophic levels of the GoM ecosystem, laboratory and field studies suggest that a competitive interaction exists in the bloom dynamics of diatoms and the toxic dinoflagellate *Alexandrium fundyense* (Sathyendranath et al. 2004; Townsend et al. 2010). The result is an inverse relation in their areal and temporal distributions through some form of reverse alleopathic interference, which appears to have a basis in the nutrient regime (Townsend et al. 2005). Local nutrient concentrations affect phytoplankton growth, with high silicate favoring diatoms and nitrates favoring dinoflagellates. Thus, the changing balance of these two nutrients presently favors the diatoms and may be altering the structure of the planktonic ecosystem with respect to the relative abundances of these two species (Townsend et al. 2010).

Furthermore, the observed oceanographic regime change between the 1990s and late 2000s is reflected by changes in the GoM zooplankton community (Hare and Kane 2012, this volume) deduced from nearly five decades of observations. Their multidecadal analysis, based on earlier studies, reveals distinct changes in the composition of the community between the 1990s, when small-bodied taxa prevailed, and the early 2000s, when the dominance reverted to a group of larger species, including *Calanus finmarchicus*. Linking these changes to environmental variability is problematic, however, because of nonstationarity of the environment–zooplankton linkages.

In contrast to Hare and Kane (2012), observations of the coastal plankton communities in the western GoM (Runge and Jones 2012, this volume) suggest that the period between 2003 and 2005 is marked by a sharp *decline* in *C. finmarchicus* abundance on Jeffrey's Ledge, accompanied by reduced surface salinities. These authors attribute the drop in salinity to either an increase in freshwater transport by the Scotian Shelf inflow (carried to Jeffrey's Ledge by the eastern and western Maine coastal currents) and/or the local input of freshwater by Maine rivers. They also hypothesize that the freshwater is associated with a change in physical forcing of the Jeffrey's Ledge circulation, affecting advection of *C. finmarchicus* onto, or out of, the coastal shelf. If the origin of the freshwater is on the Scotian Shelf, fewer *C. finmarchicus* are thought to be transported by the Maine coastal currents.

Conclusions

Next Steps

Between years 2000 and 2004, a profound change has taken place in the Gulf of Maine circulation and ecosystem. The traditional influx of nutrient-rich slope water has been considerably reduced relative to historical levels, leading to changes in the mass, heat, salt, and nutrient budgets for the gulf system. A simple box model has been used to quantify some of the changes in the mass balance implied by the GLOBEC (1993–1999) and more recent NERACOOS (2004–2008) observations. In accordance with historical observations, the box model predicts that the regime shift would be associated with an increased influx of coastal waters from the Scotian Shelf. Such behavior is consistent with a hypothesis of Brooks (1992) that claims that excess Scotian Shelf water inflow sets up an adverse pressure gradient that limits the deep inflow in the NEC.

An independent test of this hypothesis ought to be available when recent measurements of the CS inflow currents have been processed. If a strong increase of the CS inflow, carrying significant amounts of coastal current waters from the Labrador Sea and Gulf of St. Lawrence, is confirmed, further significant impacts on the Gulf of Maine nutrient regime and ecosystem should be anticipated.

Acknowledgments

The authors would like to thank the GLOBEC and NERACOOS program sponsors for supporting their field studies and analysis in Northeast Channel and the western Scotian Shelf. We also acknowledge very useful and inspiring comments from Brian Petrie and Charles Hannah. We are also grateful to David Brickman and Adam Drozdowski for their supporting model calculations in the Gulf of Maine. Editorial suggestions from Dr. R. Stephenson, Dr. J. Runge, and D. Lehman are also appreciated.

References

Bigelow, H. B. 1927. Physical oceanography of the Gulf of Maine. Fisheries Bulletin 40:511–1027.

Bisagni, J. J., and P. C. Smith. 1998. Eddy-induced flow of Scotian Shelf water across Northeast Channel, Gulf of Maine. Continental Shelf Research 18:515–539.

Brooks, D. A. 1985. Vernal circulation in the Gulf of Maine. Journal of Geophysical Research 90(C5):4687–4705.

Brooks, D. A. 1992. A brief overview of the physical oceanography of the Gulf of Maine. Pages 51–74 *in* J. Wiggin and C. N. K. Mooers, editors. Proceedings of the Gulf of Maine scientific workshop. Urban Harbors Institute, University of Massachusetts, Boston.

Butman, B., and R. C. Beardsley. 1992. Science in the Gulf of Maine: directions for the 1990s. Pages 23–38 *in* J. Wiggin and C. N. K. Mooers, editors. Proceedings of the Gulf of Maine scientific workshop. Urban Harbors Institute, University of Massachusetts, Boston.

Han, G. 2002. Interannual sea level variations in the Scotia-Maine region in the 1990s. Canadian Journal of Remote Sensing 28:581–587.

Han, G. 2007. Satellite observations of seasonal and interannual changes of sea level and currents over the Scotian Slope. Journal of Physical Oceanography 37:1051–1065.

Hare, J. A., and J. Kane. 2012. Zooplankton of the Gulf of Maine—a changing perspective. Pages 115–137 *in* R. L. Stephenson, J. H. Annala, J. A. Runge, and M. Hall-Arber, editors. Advancing an ecosystem approach in the Gulf of Maine. American Fisheries Society, Symposium 79, Bethesda, Maryland.

Loder, J. W., B. Petrie, and G. Gawarkiewicz. 1998. The coastal ocean of northeastern North America: a large-scale view. Pages 3–27 *in* A. Robinson and K. Brink, editors. The sea, volume 11. Wiley, New York.

Peltier, W. R. 2004. Global glacial isostasy and the surface of the ice-age earth: the ICE-5G(VM2) model and GRACE. Annual Review of Earth Planet Science 32:111–149.

Pettigrew, N. R., C. S. Roesler, F. Neville, and H. E. Deese. 2008. An operational real-time ocean sensor network in the Gulf of Maine. Pages 213–238 *in* S. Nittel, A. Labrinidis and A. Stefanidis, editors. GeoSensors networks, GSN 2006. Springer, LNCS 4540, New York.

Ramp, S. R., R. J. Schlitz, W. R. Wright. 1985. The deep flow through Northeast Channel, Gulf of Maine. Journal of Physical Oceanography 15:1790–1808.

Rebuck, N., D. W. Townsend, and M. A. Thomas. 2009. Gulf of Maine region nutrient and hydrographic database. Available: http://grampus.umeoce.maine.edu/nutrients. (November 2011).

Runge, J. A., and R. J. Jones. 2012. Results of a collaborative project to observe coastal zooplankton and ichthyoplankton abundance and diversity in the western Gulf of Maine: 2003–2008. Pages 345–359 *in* R. L. Stephenson, J. H. Annala, J. A. Runge, and M. Hall-Arber, editors. Advancing an ecosystem approach in the Gulf of Maine. American Fisheries Society, Symposium 79, Bethesda, Maryland.

Sathyendranath, S., L. Watts, E. Devred, T. Platt, C. Caverhill, and H. Maass. 2004. Discrimination of diatoms from other phytoplankton using ocean-colour data. Marine Ecology Progress Series 272:59–68.

Smith, P. C. 1983. The mean and seasonal circulation off southwest Nova Scotia. Journal of Physical Oceanography 13:1034–1054.

Smith, P. C. 1989. Seasonal and interannual variability of current, temperature and salinity off southwest Nova Scotia. Canadian Journal of Fisheries and Aquatic Sciences 46(Supplement 1):4–20.

Smith, P. C., R. W. Houghton, R. G. Fairbanks, and D. G. Mountain. 2001. Interannual variability of boundary fluxes and water mass properties in the Gulf of Maine and on Georges Bank: 1993–1997. Deep-Sea Research Part II 48:37–70.

Sutcliffe, W. H. Jr., R. H. Loucks, and K. F. Drinkwater. 1976. Coastal circulation and physical oceanography of the Scotian Shelf and Gulf of Maine. Journal of the Fisheries Research Board of Canada 33:98–115.

Townsend, D. W., N. R. Pettigrew, and A. C. Thomas. 2005. On the nature of Alexandrium fundyense blooms in the Gulf of Maine. Deep-Sea Research Part II 52:2603–2630.

Townsend, D.W., N.D. Rebuck, M.A. Thomas, L. Karp-Boss, R.M. Gettings. 2010. A changing nutrient regime in the Gulf of Maine. Continental Shelf Research 30:820–832.

Townsend, D. W., and R. W. Spinrad. 1986. Early spring phytoplankton blooms in the Gulf of Maine. Continental Shelf Research 6:515–529.

Wunsch, C., R. Ponte, and P. Heimbach. 2007. Decadal trends in global sealevel patterns. Journal of Climate 20:5889–5911.

Yeats, P., S. Ryan, and G. Harrison. 2011. Temporal trends in nutrient and oxygen concentrations in the Labrador Sea and on the Scotian Shelf. AZMP Bulletin 9:23–27.

American Fisheries Society Symposium 79:205–218, 2012

Zooplankton Monitoring in the Gulf of Maine: Past, Present, and Future

CATHERINE L. JOHNSON*
Fisheries and Oceans Canada, Bedford Institute of Oceanography
Post Office Box 1006, Dartmouth, Nova Scotia, B2Y 4A2, Canada

JONATHAN A. HARE
NOAA, NMFS, Northeast Fisheries Science Center
28 Tarzwell Drive, Narragansett, Rhode Island, 02882, USA

Abstract.—Zooplankton communities perform a critical role as secondary producers in marine ecosystems. They are vulnerable to climate-induced changes in the marine environment, including temperature, stratification, and circulation, but the effects of these changes are difficult to discern without sustained ocean monitoring. The physical, chemical, and biological environment of the Gulf of Maine, including Georges Bank, is strongly influenced by inflow from the Scotian Shelf and through the Northeast Channel, and thus observations both in the Gulf of Maine and in upstream regions are necessary to understand plankton variability and change in the Gulf of Maine. Large-scale, quasi synoptic plankton surveys have been performed in the Gulf of Maine since Bigelow's work at the beginning of the 20th century. More recently, ongoing plankton monitoring efforts include Continuous Plankton Recorder sampling in the Gulf of Maine and on the Scotian Shelf, U.S. National Marine Fisheries Service's MARMAP (Marine Resources Monitoring, Assessment, and Prediction) and EcoMon (Ecosystem Monitoring) programs sampling the northeast U.S. Continental Shelf, including the Gulf of Maine, and Fisheries and Oceans Canada's Atlantic Zone Monitoring Program on the Scotian Shelf and in the eastern Gulf of Maine. Here, we review and compare past and ongoing zooplankton monitoring programs in the Gulf of Maine region, including Georges Bank and the western Scotian Shelf, to facilitate retrospective analysis and broadscale synthesis of zooplankton dynamics in the Gulf of Maine. Additional sustained sampling at greater-than-monthly frequency at selected sites in the Gulf of Maine would be necessary to detect changes in phenology (i.e. seasonal timing of biological events). Sustained zooplankton sampling in critical nearshore fish habitats and in key feeding areas for upper trophic level organisms, such as marine mammals and seabirds, would yield significant insights into their dynamics. The ecosystem dynamics of the Gulf of Maine are strongly influenced by large-scale forcing and variability in upstream inflow. Improved coordination of sampling and data analysis among monitoring programs, effective data management, and use of multiple modeling approaches will all enhance the mechanistic understanding of the structure and function of the Gulf of Maine pelagic ecosystem.

Introduction

Zooplankton species are important lower trophic level members of marine ecosystems. They are a good indicator of ecosystem status since their populations respond relatively quickly to environmental variability and they are usually not fished (Mackas and Beaugrand 2009). To serve as indicators, zooplankton populations must be quanti-

* Corresponding author: catherine.johnson@dfo-mpo.gc.ca

fied, and this presents several challenges. First, their abundance and community composition is spatially and temporally variable requiring sampling over space and time. Second, quantitative assessment of abundance, biomass, and community composition is sensitive to sampling methodology, including the seasonal timing, location, and depth of sampling, as well as gear type and mesh size. Therefore, consistent, repeated sampling is essential to identify changes in zooplankton distribution, abundance, community composition, and seasonal timing at interannual and longer time scales. For example, multidecadal sampling in the Gulf of Maine and Georges Bank has made it possible to detect changes in copepod community composition that have impacts on higher trophic levels (Pershing et al. 2005; Kane 2007). Indicators of these shifts in community composition have been developed and are being tracked to provide information about the state of the Gulf of Maine ecosystem (Ecosystem Assessment Program 2009).

The review of past and ongoing multiyear zooplankton sampling programs presented here provides an overview of the zooplankton monitoring data that are available from the Gulf of Maine region and waters upstream of the gulf on the western Scotian Shelf. We also recommend additional sampling to provide observations at time scales, in regions, and of properties that are not adequately resolved at present, and we discuss additional approaches to generate knowledge for management of marine resources from monitoring data.

Monitoring Programs

Surveys of zooplankton in the Gulf of Maine started a century ago with Bigelow's surveys conducted between 1912 and 1922 (Table 1; Bigelow 1926). Bigelow identified the dominant meso- and macrozooplankton species of the gulf and described the spatial distributions and abundance of a wide range of species (Bigelow 1926). He also described a seasonal cycle of low zooplankton abundance during winter and high abundance during spring. These early studies were followed by research that focused on specific fishery species and their links to phytoplankton and zooplankton in the Gulf of Maine (e.g., copepods, Fish 1936; haddock *Melanogrammus aeglefinus*,, Walford 1938).

The next stage of zooplankton and pelagic nekton studies in the Gulf of Maine was the establishment of research and monitoring programs, with the goal of describing and understanding zooplankton and pelagic dynamics in the ecosystem. Some of these programs had definite start and end dates while others continue to the present, with minor modifications. The first such large-scale program sampled with Continuous Plankton Recorders (CPRs) along several transects in the northwest Atlantic and was started in 1961 by the predecessors of the Sir Alister Hardy Foundation for Ocean Science (SAHFOS) (Jossi et al. 2003). The CPR surveys provide a relative index of abundance of mesozooplankton species, including larvae of benthic organisms and macrozooplankton. Microzooplankton and large phytoplankton species that adhere to the CPR silk are also included in counts. The CPR sampling activities continue under the auspices of SAHFOS and the Northeast Fisheries Science Center Ecosystem Monitoring (EcoMon) Program (see below).

In 1968, the International Commission for the Northwest Atlantic Fisheries (ICNAF) began a study of the early life history of Atlantic herring *Clupea harengus* in the Gulf of Maine, Georges Bank, and Nantucket Shoals regions (Lough and Bolz 1980). The ICNAF operated for 10 years and sampled more than 5,000 stations, but sampling was limited to fall and winter because of its focus on Atlantic herring. Sampling programs which focused on larval Atlantic herring were also run by Fisheries and Oceans Canada in the Bay of Fundy, starting in 1969 and then annually from 1972 to 1998, and on Georges Bank from 1987 to 1995 (Iles et al. 1985; Corey and Milne 1987; Melvin and Stephenson 2007). The Bay of Fundy larval Atlantic herring survey was repeated in 2009. These surveys focused mainly on larval Atlantic herring, while other zooplankton taxa were enumerated at varying levels of taxonomic resolution.

In the late 1970s, ICNAF sampling was replaced by the Scotian Shelf Ichthyoplankton Program (SSIP) in Canada. This program sampled about 2,300 stations on the Scotian Shelf, upstream of the Gulf of Maine (O'Boyle et al. 1984; Shackell and Frank 2000). The SSIP sample enumeration focused on ichthyoplankton, with non-ichthyoplankton such as copepods and euphausiids enumerated at varying taxonomic levels. A

TABLE 1. Gulf of Maine and Scotian Shelf zooplankton monitoring programs.

Program	Year Start	End	Temporal resolution	Number of resolution	Region	Gear	Mesh	Description
Bigelow	1912	1925	Irregular, year-round	~20–60 stations	Gulf of Maine, Georges Bank, Massachusetts and Cape Cod Bays	Various	Various	Bigelow (1926)
Fish	1931	1932	Monthly	~30 stations	Gulf of Maine, Bay of Fundy	Pump; 0.5 m net	153 μm (No. 10 silk)	Fish (1936)
Walford	1931	1932	Spring	~50 stations	Georges Bank	1 m ring net	505 μm	Walford (1938)
Continuous Plankton Recorder	1961	Ongoing	Monthly	10 nm	Gulf of Maine	Continuous plankton recorder	270 μm	Jossi et al. (2003)
ICNAF Larval Herring Survey	1968	1978	Fall, winter	50–150 stations	Georges Bank and Nantucket Shoals	0.2 m and 0.6 m ring net	333 and 505 μm	Lough and Bolz (1980)
DFO Larval Herring Program	1972	1998	Spring (1975–1983) Summer (1977–1991) Fall (1972–1998)	116–163 stations, 10 nm resolution	Bay of Fundy	0.61 m bongo net	333 and 505 μm	Iles et al. (1985); Corey and Milne (1987)
DFO Larval Herring Program	2009	2009	Fall	10 nm resolution	Bay of Fundy	0.61 m bongo net	333 and 505 μm	Protocol as in Iles et al. (1985); M. Power, Fisheries and Oceans Canada, personal communication
DFO Larval Herring Program	1987	1995	Fall	20 mi resolution	Georges Bank	0.61 m bongo net	333 and 505 μm	Melvin and Stephenson (2007)

Table 1. Continued.

Program	Year Start	End	Temporal resolution	Number of resolution	Region	Gear	Mesh	Description
SSIP	1978	1982	2–9 annually	27–199 stations per cruise	Scotian Shelf, eastern Gulf of Maine	0.61 m bongo net	333 μm	O'Boyle et al. (1984); Shackell and Frank (2000)
MARMAP	1977	1987	6–8 times annually	50–200 stations	Northeast U.S. shelf ecosystem	0.6 m bongo net	333 and 505 μm	Sibunka and Silverman (1989)
EcoMon	1988	Ongoing	6 times annually	60 stations	Northeast U.S. shelf ecosystem	0.61 m bongo net	333 μm	Sibunka and Silverman (1989)
OPEN	1990	1992	Monthly .	45 stations	Western Scotian Shelf	Rectangular midwater trawl	64 and 333 μm	McLaren and Avendaño (1995)
MWRA	1992	Ongoing	6 times 17 times annually	15 stations 2 stations	Boston Harbor, Massachusetts and Cape Cod Bays	0.5 m ring net	102 μm	Turner (1994); Turner et al. (2006)
U.S. GLOBEC	1995	1999	Monthly (Jan–Jun)	38–41 stations	Georges Bank	Pump, 1 m^2 MOCNESS	35 and 150 μm	Durbin and Casas (2006)
Maritimes AZMP	1998	Ongoing	3 times annually	6–10 stations on each of 4 transects	Scotian Shelf	0.75 m ring net	202 μm	Therriault, J.-C. et al. (1998); Mitchell et al. (2002)
Maritimes AZMP	1998	Ongoing	Semi-monthly[a]	2 stations	Inshore, central Scotian Shelf, northwest Bay of Fundy	0.75 m ring net	202 μm	Therriault, J.-C. et al. (1998); Mitchell et al. (2002)
Bedford Basin	1998	2002[b]	Weekly	1 station	Bedford Basin, Nova Scotia	0.75 m ring net	202 μm	Li and Harrison (2008)
COOA	2002	2007	Monthly	4–6 stations	Western Gulf of Maine	¼ m^2 MOCNESS 0.75 m ring net	150 μm 202 μm	Manning and Bucklin (2005); www.cooa.unh.edu
PULSE	2003	2008	2–3 times monthly	2 stations	Western Gulf of Maine	0.75 m ring net	202 μm	Runge and Jones (2012, this volume)

[a] Bay of Fundy station sampled once per month since 2003.
[b] Samples collected but not counted from 2002 to present.

selection of SSIP samples from 1980 to 1982 was later reanalyzed at higher taxonomic resolution. In the United States, the Marine Resources Monitoring, Assessment, and Prediction (MARMAP) Program followed ICNAF. The MARMAP Program ran from 1977 to 1987 and sampled more than 5,000 stations in the Gulf of Maine region in all seasons (Sibunka and Silverman 1989). Aspects of this program continue today as the Ecosystem Monitoring (EcoMon) Program, which samples 60 stations six times per year in the Gulf of Maine system. Several gear-related studies have compared the sampling methodologies of ICNAF, MARMAP and EcoMon, and relative catchability has been quantified (Colton et al. 1980; Johnson and Morse 1994; Kane and Anderson 2007), allowing combination of the data sets (e.g., Richardson et al. 2010).

Fisheries and Oceans Canada began monitoring zooplankton on the Scotian Shelf and in the Bay of Fundy in 1998 as part of the Atlantic Zone Monitoring Program (AZMP; Therriault et al. 1998; Mitchell et al. 2002). This program performs broadscale surveys primarily along four transects on the Scotian Shelf in the spring and fall each year, and additional samples are taken at stations on the Scotian Shelf and eastern Gulf of Maine in the summer in association with groundfish survey cruises. In addition, the program collects samples twice a month throughout the year at a station on the inshore central Scotian Shelf and once a month at a station in the northwestern coastal Bay of Fundy. Similar sampling programs are performed by the Quebec and Newfoundland regions of Fisheries and Oceans Canada, upstream of the Scotian Shelf in the Gulf of Saint Lawrence and on the Newfoundland Shelf (Therriault et al. 1998). The AZMP provides taxonomic resolution mostly at the species level, focusing on dominant mesozooplankton species.

Two programs sampled zooplankton at monthly or higher frequency in the western Gulf of Maine in the 2000s. The first program, University of New Hampshire's Center of Excellence for Coastal Ocean Observation and Analysis (COOA; www.cooa.unh.edu) sampled zooplankton monthly from 2002 to 2007 at stations along a transect from the coastal shelf to Wilkinson basin in the deep western gulf (Manning and Bucklin 2005). This program collected vertically stratified samples using a MOCNESS (multiple opening and closing net and environmental sensing system). Adult copepods were enumerated to the species level, and non-copepods were identified and enumerated at a coarse level of taxonomic resolution. In the later years of the program, COOA also included ring-net sampling, using AZMP sampling and enumeration protocols, and bongo-net sampling of ichthyoplankton. The second program, PULSE (Partnership for Pelagic Ecosystem Monitoring in the Gulf of Maine), sampled zooplankton two to three times per month from 2003 to 2008 at two stations in the coastal western Gulf of Maine, using AZMP sampling and enumeration protocols.

We are aware of only two ongoing nearshore zooplankton monitoring programs, the Massachusetts Water Resources Authority (MWRA) program that has sampled in Boston Harbor, Massachusetts Bay, and Cape Cod Bay since 1992 (Turner 1994) and the zooplankton time series that has been collected in Bedford basin in Halifax Harbor, Nova Scotia since 1998 (only samples collected between 1998 and 2002 have been counted; Li and Harrison 2008). In addition to these ongoing programs, many short- and longer-term zooplankton sampling efforts have been performed in nearshore waters throughout the Gulf of Maine system since the early decades of the 20th century. For example, the Passamaquoddy Bay zooplankton community and its seasonal, interannual, and spatial variability patterns have been extensively sampled (Legaré and Maclellan 1960; Legaré 1961; and references therein). Zooplankton studies in the upper Bay of Fundy were reviewed by Daborn (1984).

These long-term monitoring programs have served as the framework for numerous large-scale research programs. For example, the OPEN (Ocean Production Enhancement Network) program, which investigated the processes that control the survival, growth, reproduction, and distribution of fish and shellfish, sampled zooplankton and ichthyoplankton on Western Bank in 1991 and 1992 (O'Dor and Thompson 1996). The U.S. GLOBEC (Global Ocean Ecosystem Dynamics) program was a multiyear (1995–1999), multi-investigator effort to understand the dynamics of Atlantic cod *Gadus morhua*, haddock, and populations of zooplankton (notably the prominent planktonic copepods *Calanus finmarchicus* and *Pseudocalanus* spp.) on Georges Bank (Wiebe et al. 2002). GLOBEC was the most intensive program studying the pelagic habitats of the Gulf of Maine region.

Recommendations

Long-term plankton monitoring has yielded significant insights into the dynamics of the Gulf of Maine ecosystem. These data have allowed the basic temporal and spatial characteristics of plankton to be defined (Bigelow 1926; O'Boyle et al. 1984; Sherman et al. 1987). From these baselines, shifts in distribution and community composition have been identified, for example, a shift to greater dominance of small copepod species in the 1990s (Pershing et al. 2005; Kane 2007) that was associated with decreased cod larval survivorship and increased haddock larval survivorship (Mountain and Kane 2010). In addition, abundance estimates of resource species have been generated from collections of meroplankton or the early life stages (e.g., Smith and Morse 1993; Melvin and Stephenson 2007; Richardson et al. 2010), and these estimates have been used in stock assessments. Continuation of ongoing U.S. and Canadian monitoring programs represents the minimum of effort needed to track interannual changes in the planktonic ecosystem (Pepin et al. 2005; Ecosystem Assessment Program 2009). However, there are a number of additional activities that would complement these basic monitoring programs and contribute to a better understanding of the Gulf of Maine ecosystem.

Much of the ongoing zooplankton sampling effort in the Gulf of Maine is performed at relatively coarse temporal scales. The EcoMon program samples every 1.5–3 months, CPRs are towed monthly (Kane 2009), and the Bay of Fundy AZMP station is sampled monthly. This sampling frequency imposes a limit on phenological changes that can be detected. There are numerous studies that suggest that climate-driven changes in phenology are ecologically important (Cushing 1990; Platt et al. 2003; Ji et al. 2010) and that the timing of production cycles is variable in the Gulf of Maine (Ji et al. 2007). Currently, only the AZMP station on the central Scotian Shelf is sampled more frequently than monthly (every 2 weeks, Therriault et al. 1998), following the 2008 termination of high-frequency sampling by the PULSE program in the western Gulf of Maine (Runge and Jones 2012). Monthly zooplankton sampling along a western Gulf of Maine transect by the University of New Hampshire's Coastal Ocean Observing Center was terminated in 2007 (www.cooa.unh.edu). To monitor changes in phenology, sampling at greater than monthly frequency, using the AZMP protocol, could be extended to sites throughout the Gulf of Maine. In addition to the current AZMP sites, fixed sites that are critical to understanding larger-scale physical dynamics in the ecosystem (e.g., sentinel sites) could be selected for high frequency biological and physical sampling, taking advantage of advances in ocean observing technology (e.g., Greenfield et al. 2006).

Most ongoing zooplankton monitoring in the Gulf of Maine is focused on offshore areas, with the exception of the AZMP site in Bedford basin and the MWRA sampling in Massachusetts Bay. The EcoMon Program samples waters deeper than ~15 m; 33% of samples are in waters < 50 m. The CPR samples at coarse spatial scales (16 km), and typically the CPR is deployed and retrieved once the ship is out of the coastal zone. There are numerous aspects of nearshore zooplankton and ichthyoplankton dynamics that are important to management, including the abundance and distribution of larval bay scallops, hard clams, and soft clams, zooplankton productivity and grazing in estuarine systems, and ichthyoplankton and shellfish transport and retention in nearshore nursery areas (Turner 1994; Lazzari et al. 2003; Campbell et al. 2005). In addition, while zooplankton monitoring programs have focused on estimating regional-scale, average values of zooplankton biomass and abundance, the availability of zooplankton prey for higher trophic level organisms such as marine mammals and phalaropes can also be influenced by physical processes that concentrate prey in certain "hot spot" locations (Baumgartner and Mate 2003; Brown and Gaskin 2008; Stevick et al. 2008). Sustained plankton monitoring in nearshore areas that provide critical habitat for fish and shellfish, and additional physical and biological sampling in critical feeding areas for upper trophic level species, could yield significant insights into the dynamics of these species.

Current monitoring programs sample mesozooplankton (zooplankton between 0.2 and 20.0 mm, Sieburth et al. 1978) more effectively than other size-groups. Macrozooplankton taxa such as euphausiids (krill) and gelatinous zooplankton are also important components of the Gulf of Maine ecosystem (Bigelow 1926; Link et al. 2008), but there is very little information regarding their distribution and abundance over time. Globally, gelatinous zooplankton populations may be increasing (Richardson et al. 2009), and on the northeast U.S. Continental Shelf, Link and Ford (2006) found a dramatic increase in the occurrence of ctenophores in the guts of spiny dogfish *Squalus acanthias*. Standard net-based sampling is not adequate to quantitatively sample euphausiids, which have a strong escape response from nets, or gelatinous zooplankton, which is often destroyed through contact with the net surface. Alternative techniques, including acoustic (Warren et al. 2001) and video-based systems (Benfield et al. 2003), are needed to adequately sample these taxa.

Identification of zooplankton and meroplankton is time-consuming and difficult. Genetic techniques can allow identification when traditional techniques do not (Bucklin et al. 1998; Vandersea et al. 2008). These techniques can also be integrated into in situ remote sampling systems (Greenfield et al. 2006). In addition, genetic techniques can be used to identify populations (Ruzzante et al. 1999) and parents (Almany et al. 2007), allowing more detailed information than species identification. These approaches should be used more broadly in plankton monitoring programs to take advantage of existing sampling efforts and to yield information on population structure that can be used to address ecological and management-related questions.

Monitoring of phytoplankton and microbial production is essential for understanding the drivers of zooplankton variability and the processes controlling energy transfer through the Gulf of Maine food web. Current monitoring of phytoplankton and microbial production in the Gulf of Maine is insufficient. While satellite observations provide a tremendous amount of information regarding surface chlorophyll (Thomas et al. 2003), and methods exist for estimating overall productivity from satellite data (Behrenfeld and Falkowski 1997), similar information for subsurface phytoplankton, species composition, and microbial production throughout the system is lacking in ongoing U.S. Gulf of Maine monitoring programs. Process sampling has shown complex three-dimensional dynamics in the distribution of chlorophyll (Franks 1995; O'Reilly and Zetlin 1998), but satellites can only resolve the two-dimensional surface signature of these dynamics. There is evidence of a shift in phytoplankton species composition toward smaller taxa that is not resolved in satellite measurements (Morán et al. 2010), but the impact of this shift on ecosystem function is unclear. Finally, trophic modeling indicates an important role for microbial dynamics in the Gulf of Maine (Steele et al. 2007; Link et al. 2008), and measurements on the Scotian Shelf and in the Labrador Sea indicate multiyear trends in microbial plankton (Li et al. 2006). Phytoplankton and microbial abundance in the water column are measured on the Scotian Shelf by AZMP, but current monitoring efforts by EcoMon in the Gulf of Maine do not address primary productivity and microbial dynamics.

Generation of knowledge about Gulf of Maine ecosystem dynamics from monitoring data requires synthesis of physical, chemical, and biological data from multiple programs. Effective data management is critical to preserving data and making it available for data synthesis efforts, and it is an essential first step toward data synthesis. Efforts to systematically manage current and past Gulf of Maine monitoring data have increased in recent years, but data management effort is inconsistent among programs. In some cases, past data may already be lost (e.g., Bigelow 1926). In other cases, data are held by individual researchers and are not available to the larger scientific community. Even data that are "available" often require close interaction with the data holders for access and interpretation (EcoMon, MARMAP). There have been marked improvements in recent years. Fisheries and Oceans Canada (DFO) archives biological and chemical monitoring data from ongoing programs in the BioChem database (www.meds-sdmm.dfo-mpo.gc.ca/biochem/biochem_e.html); physical and ocean color data are archived in different databases (www.mar.dfo-mpo.gc.ca/science/ocean/database/introduction.html). There are ongoing efforts to archive data from past DFO and university-led Canadian sampling programs in the BioChem and physical databases. In the United States, the Biological and Chemical Oceanography Data Management Office (BCO-DMO, www.bco-dmo.org) is a repository for JGOFS (Joint Global Ocean Flux Sutdy) and GLOBEC data, and plans are underway to add EcoMon hydrographic and zooplankton data. We envision a publicly accessible, actively updated plankton data inventory for the Gulf of Maine, with links to extant databases. This paper creates such an inventory of multiyear zooplankton sampling programs (Table 1) that will be available on the NERACOOS (Northeastern Regional Association of Coastal Ocean Observing System) Web site (www.neracoos.org/projects/zooplankton). We invite data producers and users to contact the authors with additional data sources and corrections to the inventory.

In addition to zooplankton monitoring data, zooplankton samples from many programs are themselves archived. For example, aliquots of EcoMon zooplankton samples are archived at the Morski Instytut Rybacki in Gdynia, Poland and the Northeast Fisheries Science Center in Narragansett, Rhode Island, and ichthyoplankton is archived at the Northeast Fisheries Science Center in Narragansett, Rhode Island. Samples from the Maritimes Region AZMP program are archived at the Bedford Institute of Oceanography in Dartmouth, Nova Scotia. Many plankton samples from the SSIP are archived in the Canadian Museum of Nature in Ottawa, Ontario, and nearly all fall samples from the Canadian Larval Herring Survey cruises from 1972 to 1998 are archived at the Atlantic Reference Centre of the Huntsman Marine Laboratory, St. Andrews, New Brunswick. Most samples are stored in a formalin solution. Further analyses on archived zooplankton samples could yield useful information, for example, higher taxonomic resolution data (e.g., SSIP sample reanalysis presented in Sameoto and Herman [1992]) or genetic information about population structure (e.g., Kirby et al. 2007).

Numerous studies have shown that the flows from upstream into the Gulf of Maine system are dominant drivers of ecosystem dynamics. Studies using oxygen isotopes demonstrate that much of the water in the Gulf of Maine system has a high latitude source (e.g., Labrador Sea), although the specific source remains unknown (Fairbanks 1982; Chapman and Beardsley 1989). Variability in the transport of water out of high latitudes in the North Atlantic can be traced along the coastal current system from the Labrador Sea, along the Scotian Shelf, through the Gulf of Maine, and into the Middle Atlantic Bight (Belkin et al. 1998; Smith et al. 2001). Changes in upstream flow cause variability in salinity in the Gulf of Maine (Mountain 2004), which has been linked to changes in primary production (Ji et al. 2007), as well as zooplankton abundance and community structure (Pershing et al. 2005; Kane 2007). There is also recent evidence that the pathways by which water enters the Gulf of Maine have shifted in the past decade, with decreased volume transport through the Northeast Channel and increased inflow along the inner Scotian Shelf (Smith et al. 2012, this volume). Monitoring the physical and chemical inflow to the Gulf of Maine system, primarily volume transport, temperature, salinity, nutrients, dissolved oxygen, and aspects of the carbon system (e.g., partial pressure CO_2, dissolved inorganic carbon, total alkalinity), is a high priority for understanding the environmental drivers of zooplankton variability. Nutrients and dissolved oxygen are measured in the AZMP surveys on the Scotian Shelf, but they have only recently been added to the EcoMon program. These observations will be made in EcoMon for the next 3 years, but there is no sustained support for them after this initial period. Monitoring of chemical properties related to ocean acidification has recently been added to both the Scotian Shelf AZMP and EcoMon programs on a short-term basis.

There have been clear changes in the pelagic ecosystem of the Gulf of Maine linked to larger-scale forcing (Pershing et al. 2005; Kane 2007). These changes involve hydrographic properties, as well as phytoplankton and zooplankton. Our current monitoring provides a system to identify when a change has occurred, but there remains a need to improve the mechanistic understanding of the structure and function of the Gulf of Maine pelagic ecosystem and, ultimately, to develop the ability to predict future change. In addition to continuation of monitoring, process-oriented research and modeling approaches are needed to improve mechanistic understanding of ecosystem variability. The ECOHAB-GOM (Ecology of

Harmful Algal Blooms-Gulf of Maine; Anderson et al. 2005) and Georges Bank GLOBEC (Wiebe et al. 2001) programs are examples of intensive, multi-investigator research programs that incorporated multiyear, broadscale sampling, process studies, and modeling.

Although there have been many panels, workshops, and initiatives on the Gulf of Maine ecosystem, there is currently no synthetic community consensus on the critical oceanographic questions facing the Gulf of Maine ecosystem. Such a consensus would be useful for coordinating observing activities, the development of large-scale process and modeling studies, and smaller-scale research conducted by individual investigators. Data synthesis activities could also be coordinated by identifying the data and scientific understanding required to implement ecosystem approaches to management and then prioritizing these activities in a regional science plan. Development of predictive capability will be challenging since observational prediction makes the questionable assumption that the future response of the ecosystem to change will be the same as past responses. Clearly, there is a need to develop tools for dynamical prediction with skill at multiple time scales. On a global scale, IPCC (Intergovernmental Panel on Climate Change) class models currently simulate future climate state decades in the future, and efforts to improve decadal prediction are currently underway (Keenlyside et al. 2008). On a local scale, the work of Ji et al. (2007) and others demonstrates that dynamical models have skill in reproducing observed characteristics of plankton dynamics in the Gulf of Maine. Combining these global and local-scale models into regional earth-system models (e.g., Henson et al. 2009) will improve our ability to forecast the state of the Gulf of Maine ecosystem in the future; such forecasts would contribute greatly to ecosystem-based management.

Current and future monitoring designs should be evaluated using formal tools, including observing system simulation experiments (OSSEs). These were developed in meteorology as a method to formally assess and optimize observing strategies for use in numerical weather forecasting (Arnold and Dey 1986). A dynamic model is used to represent perfect measurement of the system or "truth." The model fields are then sampled with various designs, and these samples are used to reconstruct the "true" fields using the same analyses as if the sampling were real. Comparison of the reconstructed and modeled ("truth") fields are then compared, providing a quantitative evaluation of the sampling strategies. These tools are slowly moving into biological oceanographic applications, and in one of the first examples, McGillicuddy et al. (2001) assessed the sampling plan of the U.S. GLOBEC program on Georges Bank. They concluded that the lack of synopticity and relatively sparse sampling design contributed to the majority of error in the calculation of distribution and abundance maps for Calanus finmarchicus. A recent study on optimizing survey strategies for the abundance and distribution of the copepod species Pseudocalanus spp. on Georges Bank using OSSEs demonstrated that an optimized sampling strategy (using a variance quadtree method) yielded more accurate estimates of areal abundance than a simple random sampling strategy (Lin et al. 2010). The studies of McGillicuddy et al. (2001) and Lin et al. demonstrate the power of OSSEs to develop more effective zooplankton sampling strategies. Now these tools need to be applied to the ongoing zooplankton monitoring programs in the Gulf of Maine region.

Gulf of Maine waters fall under multiple jurisdictions, including the Canadian and U.S. federal governments and state and provincial governments of Massachusetts, New Hampshire, Maine, New Brunswick, and Nova Scotia. There are efforts to cooperate on ocean science and management issues across jurisdictions, for example, the Gulf of Maine Council (www.gulfofmaine.org) and NERACOOS (www.neracoos.org), both of which include U.S. and Canadian scientists. United States surveys (e.g., EcoMon) occur in Canadian waters with permission, and Canadian surveys occur in U.S. water with permission (DFO Larval Herring Program on Georges Bank). A number of exploited resources are jointly assessed through the Transboundary Resources Assessment Committee (TRAC), including Atlantic herring and Atlantic mackerel *Scomber scombrus*; information from plankton monitoring programs has recently been presented at TRAC meetings related to the status of these two stocks (TRAC 2009, 2010). Finally, the ICES Working Group on the Northwest Atlantic Regional Sea was formed in 2010 to help coordinate marine science at the scale of the northwest Atlantic continental shelf system. This greater degree of coordination is needed as we work to better understand the basin-scale and regional-scale influences on the Gulf of Maine ecosystem.

Acknowledgments

We thank Jack Jossi, Mary Kennedy, Bill Li, Chris Manning, Mike Power, David Richardson, Jeff Runge, Jackie Spry, and Jeff Turner for their help in locating and clarifying information on monitoring programs and Donna Johnson, Glen Harrison, and Jeff Runge for helpful comments on the manuscript.

References

Almany, G. R., M. L. Berumen, S. R. Thorrold, S. Planes, and G. P. Jones. 2007. Local replenishment of coral reef fish populations in a marine reserve. Science 316:742–744.

Anderson, D. M., D. W. Townsend, D. J. McGillicuddy, Jr., and J. T. Turner, editors. 2005. The ecology and oceanography of toxic *Alexandrium fundyense* blooms in the Gulf of Maine. Deep Sea Research Part II 52:2365–2876.

Arnold, C. P., and C. H. Dey. 1986. Observing-systems simulation experiments: past, present, and future. Bulletin of the American Meteorology Society 67:687–695.

Baumgartner, M. F., and B. R. Mate. 2003. Summertime foraging ecology of North Atlantic right whales. Marine Ecology Progress Series 264:123–135.

Behrenfeld, M. J., and P. G. Falkowski. 1997. Photosynthetic rates derived from satellite-based chlorophyll concentration. Limnology and Oceanography 42:1–20.

Belkin, I. M., S. Levitus, J. Antonov, and S.-A. Malmberg. 1998. "Great salinity anomalies" in the North Atlantic. Progress in Oceanography 41:1–68.

Benfield, M. C., A. C. Lavery, P. H. Wiebe, C. H. Greene, T. K. Stanton, and N. J. Copley. 2003. Distributions of physonect siphonulae in the Gulf of Maine and their potential as important sources of acoustic scattering. Canadian Journal of Fisheries and Aquatic Sciences 60:759–772.

Bigelow, H. B. 1926. Plankton of the offshore waters of the Gulf of Maine. U.S. Bureau of Fisheries Bulletin 40(2). U.S. Government Printing Office, Washington, D.C.

Brown, R. G. B., and D. E. Gaskin. 2008. The pelagic ecology of the grey and red-necked phalaropes *Phalaropes fulicarius* and *P. lobatus* in the Bay of Fundy, eastern Canada. Ibis 130:234–250.

Bucklin, A., A. M. Bentley, and S. P. Franzen. 1998. Distribution and relative abundance of *Pseudocalanus moultoni* and *P. newmani* (Copepoda: Calanoida) on Georges Bank using molecular identification of sibling species. Marine Biology 132:7–106.

Campbell, R. G., G. J. Teegarden, A. D. Cembella, and E. G. Durbin. 2005. Zooplankton grazing impacts on *Alexandrium* spp. in the nearshore environment of the Gulf of Maine. Deep-Sea Research Part II 52:2817–2833.

Chapman, D. C., and R. C. Beardsley. 1989. On the origin of shelf water in the Middle Atlantic Bight. Journal of Physical Oceanography 19:384–391.

Colton, J. B., J. R. Green, R. R. Byron, and J. L. Frisella. 1980. Bongo net retention rates as effects by towing speed and mesh size. Canadian Journal of Fisheries and Aquatic Sciences 37:606–623.

Corey, S., and W. R. Milne. 1987. Recurrent groups of zooplankton in the Bay of Fundy and southwest Nova Scotia regions, Canada. Canadian Journal of Zoology 65:2400–2405.

Cushing, D. H. 1990. Plankton production and year-class strength in fish populations: an update of the match/mismatch hypothesis. Advances in Marine Biology 26:249–293.

Daborn, G. R. 1984. Zooplankton studies in the upper Bay of Fundy since 1976. Canadian Technical Report of Fisheries and Aquatic Sciences 1256:135–162.

Durbin, E. G., and M. C. Casas. 2006. Abundance and spatial distribution of copepods on Georges Bank during the winter/spring period. Deep-Sea Research Part II 53:2537–2569.

Ecosystem Assessment Program. 2009. Ecosystem assessment report for the Northeast U.S. Continental Shelf large marine ecosystem. National Oceanic and Atmospheric Administration, National Marine Fisheries Service, Northeast Fisheries Science Center, Reference Document 09–11, Woods Hole, Massachusetts.

Fairbanks, R. G. 1982. The origin of continental shelf and slope water in the New York Bight and Gulf of Maine: evidence from H2 18O/H2 16O ratio measurements. Journal of Geophysical Research 87:5796–5808.

Fish, C. J. 1936. The biology of *Calanus finmarchicus* in the Gulf of Maine and Bay of Fundy. Biological Bulletin 70:118–170.

Franks, P. J. S. 1995. Thin layers of phytoplankton: a model of formation by near-inertial wave shear. Deep-Sea Research Part I 42:75–83.

Greenfield, D. J., R. Marin, S. Jensen, E. Massion, B. Roman, J. Feldman, and C. A. Scholin. 2006. Application of environmental sample processor (ESP) methodology for quantifying *Pseudonitzschia australis* using ribosomal RAN-targeted probes in sandwich and fluorescent in situ hybridization formats. Limnology and Oceanography Methods 4:426–435.

Henson, S. A., J. P. Dunne, and J. L. Sarmiento. 2009. Decadal variability in North Atlantic plankton bloom. Journal of Geophysical Research 114: C04013. DOI:10.1029/2008JC005139.

Iles T. D., M. J. Power, and R. L. Stephenson. 1985. Evaluation of the use of larval herring survey data to tune herring stock assessments in the Bay of Fundy/Gulf of Maine. Northwest Atlantic Fisheries Organization, Scientific Council Research Document 85/107, Dartmouth, Nova Scotia.

Ji, R., C. S. Davis, C. Chen, D. W. Townsend, D. G. Mountain, and R. C. Beardsley. 2007. Influence of ocean freshening on shelf phytoplankton dynamics. Geophysical Research Letters 34:L24607. DOI:10.1029/2007/GL032010.

Ji, R., M. Edwards, D. L. Mackas, J. A. Runge, and A. C. Thomas. 2010. Marine plankton phenology and life history in a changing climate: current research and future directions. Journal of Plankton Research 32:1355–1368.

Johnson, D. L., and W. W. Morse. 1994. Net extrusion of larval fish: corrections factors for 0.333 mm versus 0.505 mm mesh bongo nets. Northwest Atlantic Fisheries Organization Scientific Council Studies 20:85–92.

Jossi, J. W., A. W. G. John, and D. Sameoto. 2003. Continuous plankton recorder sampling off the east coast of North America: history and status. Progress in Oceanography 58:313–325.

Kane, J. 2007. Zooplankton abundance trends on Georges Bank, 1977–2004. ICES Journal of Marine Science 64:909–919.

Kane, J. 2009. A comparison of two zooplankton time series data collected in the Gulf of Maine. Journal of Plankton Research 31:249–259.

Kane, J., and J. L. Anderson. 2007. Effect of towing speed on retention of zooplankton in bongo nets. Fishery Bulletin (Washington DC) 105:440–444.

Keenlyside, N. S., M. Latif, J. Jungclaus, L. Kornblueh, and E. Roeckner. 2008. Advancing decadal-scale climate prediction in the North Atlantic sector. Nature (London) 453:84–88.

Kirby, R. R., J. A. Lindley, and S. D. Batten. 2007. Spatial heterogeneity and genetic variation in the copepod *Neocalanus cristatus* along two transects in the North Pacific sampled by the Continuous Plankton Recorder. Journal of Plankton Research 29:97–106.

Lazzari, M. A., S. Sherman, and J. K. Kanwit. 2003. Nursery use of shallow habitats by epibenthic fishes in Maine nearshore waters. Estuarine Coastal and Shelf Science 56:73–84.

Legaré, J. E. H. 1961. The zooplankton of the Passamaquoddy region. Fisheries Research Board of Canada Manuscript Report Series (Biological) 707.

Legaré, J. E. H., and D. C. Maclellan. 1960. A qualitative and quantitative study of the plankton of the Quoddy region in 1957 and 1958 with special reference to the food of the herring. Journal of the Fisheries Research Board of Canada 17:409–448.

Li, W. K. W., W. G. Harrison, and E. J. H. Head. 2006. Coherent sign switching in multiyear trends of microbial plankton. Science 311:1157–1160.

Li, W. K. W., and W. G. Harrison. 2008. Propagation of an atmospheric climate signal to phytoplankton in a small marine basin. Limnology and Oceanography 53:1734–1745.

Lin, P., R. Ji, C. S. Davis, and D. J. McGillicuddy. 2010. Optimizing plankton survey strategies using observing system simulation experiments. Journal of Marine Systems 82:187–194.

Link, J. S., and M. D. Ford. 2006. Widespread and persistent increase of Ctenophora in the continental shelf ecosystem off NE USA. Marine Ecology Progress Series 320:153–159.

Link, J. S., W. Overholtz, J. O'Reilly, J. Green, D. Dow, D. Palka, C. Legault, J. Vitaliano, V. Guida, M. Fogarty, J. Brodziak, L. Methratta, W. Stockhausen, L. Col, and C. Griswold. 2008. The Northeast U.S. continental shelf Energy Modeling and Analysis exercise (EMAX): ecological network model development and basic ecosystem metrics. Journal of Marine Systems 74:453–474.

Lough, R. G., and G. R. Bolz. 1980. A description of the sampling methods, and larval herring (*Clupea harengus* L.) data report for surveys conducted from 1968 to 1978 in the Georges Bank and Gulf of Maine areas. National Oceanic and Atmospheric Administration, National Marine Fisheries Service, Northeast Fisheries Science Center, Document 79–60, Woods Hole, Massachusetts.

Mackas, D. L., and G. Beaugrand. 2009. Comparisons of zooplankton time series. Journal of Marine Systems 79:286–304.

Manning, C. A., and A. Bucklin. 2005. Multivariate analysis of the copepod community of nearshore waters in the western Gulf of Maine. Marine Ecology Progress Series 292:233–249.

McGillicuddy, D. J., D. R. Lynch, P. Wiebe, J. Runge, E. G. Durbin, W. C. Gentleman, and C. S. Davis. 2001. Evaluating the synopticity of the US GLOBEC Georges Bank broad-scale sampling pattern with observational system simulation experiments. Deep-Sea Research Part II 48:483–499.

McLaren, I. A., and P. Avendaño. 1995. Prey field and diet of larval cod on Western Bank, Scotian Shelf. Canadian Journal of Fisheries and Aquatic Sciences 52:448–463.

Melvin, G. D., and R. L. Stephenson. 2007. The dynamics of a recovering fish stock: Georges Bank herring. ICES Journal of Marine Science 64:69–82.

Mitchell, M. R., G. Harrison, K. Pauley, A. Gagné, G. Maillet, and P. Strain. 2002. Atlantic Zonal Monitoring Program sampling protocol. Canadian Technical Report of Hydrography and Ocean Sciences 223.

Morán, X. A. G., Á. López-Urrutia, A. Calco-Díaz, and W. K. W. Li. 2010. Increasing importance of small phytoplankton in a warmer ocean. Global Change Biology 16:1137–1144.

Mountain, D. G. 2004. Variability of the water properties in NAFO Subareas 5 and 6 during the 1990s. Journal of Northwest Atlantic Fishery Science 34:103–112.

Mountain, D. G., and J. Kane. 2010. Major changes in the Georges Bank ecosystem, 1980s to the 1990s. Marine Ecology Progress Series 398:81–91.

O'Boyle, R. N., M. Sinclair, R. J. Conover, K. H. Mann, and A. C. Kohler. 1984. Temporal and spatial distribution of ichthyoplankton communities of the Scotian Shelf in relation to biological, hydrological, and physiological features. Rapports et Proces-Verbaux des Reunions Conseil International pour l'Exploration de la Mer 183:27–40.

O'Dor, R. K., and R. Thompson. 1996. Ocean Production Enhancement Network (OPEN): fourth annual scientific meeting contributions. Marine Biology 124:591–592.

O'Reilly, J. E., and C. Zetlin. 1998. Seasonal, horizontal, and vertical distribution of phytoplankton chlorophyll a in the northeast U.S. continental shelf ecosystem. NOAA Technical Report NMFS 139:1–287.

Pepin, P., B. Petrie, J.-C. Therriault, S. Narayanan, G. Harrison, J. Chassé, E. Colbourne, D. Gilbert, D. Gregory, M. Harvey, G. Maillet, M. Mitchell, and M. Starr. 2005. The Atlantic Zone Monitoring Program (AZMP): review of 1998–2003. Canadian Technical Report of Hydrography and Ocean Science 242.

Pershing, A. J., C. H. Greene, J. W. Jossi, L. O'Brien, J. K. T. Brodziak, and B. A. Bailey. 2005. Interdecadal variability in the Gulf of Maine zooplankton community, with potential impacts on fish recruitment. ICES Journal of Marine Science 62:1511–1523.

Platt, T., C. Fuentes-Yaco, and K. T. Frank. 2003. Spring algal bloom and larval fish survival. Nature (London) 423:398–399.

Richardson, A. J., A. Bakun, G. C. Hays, and M. J. Gibbons. 2009. The jellyfish joyride: causes, consequences and management responses to a more gelatinous future. Trends in Ecology and Evolution 24:312–322.

Richardson, D. E., J. A. Hare, W. J. Overholtz, and D. L. Johnson. 2010. Development of long-term larval indices for Atlantic herring (*Clupea harengus*) on the northeast US continental shelf. ICES Journal of Marine Science 67:617–627.

Runge, J. A., and R. J. Jones. 2012. Results of a collaborative monitoring program of coastal zooplankton and ichthyoplankton in the western Gulf of Maine: 2003–2008. Pages 345–359 *in* R. L. Stephenson, J. H. Annala, J. A. Runge, and M. Hall-Arber, editors. Advancing an ecosystem approach in the Gulf of Maine. American Fisheries Society, Symposium 79, Bethesda, Maryland.

Ruzzante, D. E., C. T. Taggart, and D. Cook. 1999. A review of the evidence for genetic structure of cod (*Gadus morhua*) populations in the NW Atlantic and population affinities of larval cod off Newfoundland and the Gulf of St. Lawrence. Fisheries Research 43:79–97.

Sameoto, D. D., and A. W. Herman. 1992. Effect of the outflow from the Gulf of St. Lawrence on Nova Scotia shelf zooplankton. Canadian Journal of Fisheries and Aquatic Sciences 49:857–869.

Shackell, N. L., and K. T. Frank. 2000. Larval fish diversity on the Scotian Shelf. Canadian Journal of Fisheries and Aquatic Sciences 57:1747–1760.

Sherman, K., W. G. Smith, J. R. Green, E. B. Cohen, M. S. Berman, K. A. Marti, and J. R. Goulet. 1987. Zooplankton production and the fisheries of the northeastern shelf. Pages 268–282 *in* R. H. Backus and D. W. Bourne, editors. Georges Bank. MIT Press, Cambridge, Massachusetts.

Sibunka, J. D., and M. J. Silverman. 1989. MARMAP surveys of the continental shelf from Cape Hatteras, North Carolina, to Cape Sable, Nova Scotia (1977–1983), Atlas no. 1, summary of operations. National Oceanic and Atmospheric Administration, National Marine Fisheries Service, Northeast Fisheries Center, NMFS-F/NEC-68, Woods Hole, Massachusetts.

Sieburth, J. M., V. Smetecek, and J. Lenz. 1978. Pelagic ecosystem structure: heterotrophic compartments of plankton and their relationship to plankton size fractions. Limnology and Oceanography 23:1256–1263.

Smith, P. C., R. W. Houghton, R. G. Fairbanks, and D. G. Mountain. 2001. Interannual variability of boundary fluxes and water mass properties in the Gulf of Maine and on Georges Bank: 1993–1997. Deep-Sea Research Part II 48:37–70.

Smith, P. C., N. R. Pettigrew, P. Yeats, D. W. Townsend, and G. Han. 2012. Regime shift in the Gulf of Maine. Pages 185–203 *in* R. L. Stephenson, J. H. Annala, J. A. Runge, and M. Hall-Arber, editors. Advancing an ecosystem approach in the Gulf of Maine. American Fisheries Society, Symposium 79, Bethesda, Maryland..

Smith, W. G., and W. W. Morse. 1993. Larval distribution patterns: early signals for the collapse/recovery of Atlantic herring *Clupea harengus* in the Georges Banks area. Fishery Bulletin (Washington, DC) 91:338–347.

Steele, J. H., J. S. Collie, J. J. Bisagni, D. J. Gifford, M. J. Fogarty, J. S. Link, B. K. Sullivan, M. E. Sieracki, A. R. Beet, D. G. Mountain, E. G. Durbin, D. Palka, and W. T. Stockhausen. 2007. Balancing end-to-end budgets of the Georges Bank ecosystem. Progress in Oceanography 74:423–448.

Stevick, P. T., L. S. Incze, S. D. Krause, S. Rosen, N. Wolff, and A. Baukus. 2008. Trophic relationships and oceanography on and around a small offshore bank. Marine Ecology Progress Series 363:15–28.

Therriault, J.-C., B. Petrie, P. Pepin, J. Gagnon, D. Gregory, J. Helbig, A. Herman, D. Lefaivre, M. Mitchell, B. Pelchat, J. Runge, and D. Sameoto. 1998. Proposal for a northwest Atlantic zonal monitoring program. Canadian Technical Report of Hydrography and Ocean Sciences 194.

Thomas, A. C., D. W. Townsend, and R. Weatherbee. 2003. Satellite-measured phytoplankton variability in the Gulf of Maine. Continental Shelf Research 23:971–989.

TRAC (Transboundary Resources Assessment Committee). 2009. Proceedings of the Transboundary Resources Assessment Committee (TRAC): Gulf of Maine/Georges Bank herring, eastern Georges Bank cod and haddock, Georges Bank yellowtail flounder. Transboundary Resources Assessment Committee Proceedings 2009/01. Available: http://www2.mar.dfo-mpo.gc.ca/science/trac/proceedings/TRAC_pro_2009_01.pdf. (February 2012)

TRAC (Transboundary Resources Assessment Committee). 2010. Proceedings of the Transboundary Resources Assessment Committee (TRAC) mackerel benchmark assessment. Transboundary Resources Assessment Committee Proceedings 2010/03. Available: http://www2.mar.dfo-mpo.gc.ca/science/trac/meetings/mtgtrac-mackbnchoct2009.pdf. (February 2012)

Turner, J. T. 1994. Planktonic copepods of Boston Harbor, Massachusetts Bay, and Cape Cod Bay, 1992. Hydrobiologica 292–293:405–413.

Turner, J. T., D. G. Borkman, and C. D. Hunt. 2006. Zooplankton of Massachusetts Bay, USA, 1992–2003: relationships between the copepod *Calanus finmarchicus* and the North Atlantic oscillation. Marine Ecology Progress Series 311:115–124.

Vandersea M. W., R. W. Litaker, K. E. Marancik, J. A. Hare, H. J. Walsh, S. Lem, M A. West, D. M. Wyanski, E. H. Laban, and P. A. Tester. 2008. Larval identification of larval sea basses *Centropristis* spp. using ribosomal DNA-specific molecular PCR assays. Fishery Bulletin 106:183–193.

Walford. L. A. 1938. Effect of currents on distribution and survival of the eggs and larvae of the haddock (*Melanogrammus aeglefinus*) on Georges Bank. Bulletin of the Bureau of Fisheries 49(20):1–73.

Warren, J. D., T. K. Stanton, M. C. Benfield, P. H. Wiebe, D. Chu, and M. Sutor. 2001. In situ measurements of acoustic target strengths of gas-bearing siphonophores. ICES Journal of Marine Science 58:740–749.

Wiebe, P. H., R. C. Beardsley, A. C. Bucklin, and D. G. Mountain, editors. 2001. Coupled biological and physical studies of plankton populations on Georges Bank and related North Atlantic regions. Deep-Sea Research Part II 48: 1–684.

Wiebe, P. H., R. Beardsley, D. Mountain, and A. Bucklin. 2002. U.S. GLOBEC Northwest Atlantic/Georges Bank Program. Oceanography 15:13–29.

American Fisheries Society Symposium 79:219–234, 2012

Quantifying Watershed-Based Influences on the Gulf of Maine Ecosystem: The Saint John River Basin

GLENN BENOY*

Environment Canada and Agriculture & Agri-Food Canada, Potato Research Centre, 850 Lincoln Road, Fredericton, New Brunswick, E3B 4Z7, Canada

ERIC LUIKER AND JOSEPH CULP

Environment Canada and Canadian Rivers Institute, Department of Biology University of New Brunswick, 10 Bailey Drive, Fredericton, New Brunswick, E3B 5A3, Canada

Abstract.—Estuarine and coastal ecosystems of the Gulf of Maine continue to be degraded by excessive loadings of sediments, nutrients, and contaminants derived from surrounding watersheds. The Saint John River basin is the largest basin in the Gulf of Maine, and within it there are a significant number of major industries along the main stem of the river and vast expanses of land-based activities of forestry and potato production along many of the river valleys and floodplains. Water quality and loading of sediments and nutrients have changed over the past few hundred years, with the most important changes coming with the expansion of agriculture and pulp and paper processing operations since the 1950s. Several studies are discussed in this chapter that outline the identification and quantification of watershed-based activities that influence the Saint John River ecosystem. Using export coefficient modeling, nonpoint sources of nitrogen and phosphorus to the Bay of Fundy are shown to be three to four times that of point sources. Few studies explicitly couple river dynamics to estimates of load to the Saint John River estuary and the Bay of Fundy. With high-quality geographic information on land coverage, land usage and human activities, and robust water quantity and quality monitoring programs, analytical models can be developed to help evaluate policy options and chart pathways towards a more integrated understanding and management of the basin and its receiving waters.

Introduction

Over the past several decades, increasing attention has been devoted to understanding and estimating the contribution of land-based activities as determinants of the condition of estuarine, coastal, and marine ecosystems (Hopkinson and Vallino 1995; Hale and Westhead 2012, this volume). Activities such as land conversion from forest to agriculture, urbanization, and industrialization change the composition of the landscape and generally increase the transfer of natural and nonnatural substances from watersheds to surface waters. Perhaps the most acknowledged and conspicuous manifestation of these changes is the excessive loading of nitrogen and phosphorus to marine environments, also known as cultural eutrophication (Nixon 1995; Smith et al. 2006). Other associated, and usually interacting, changes include excessive loading of sediments and contaminants and alteration of flow dynamics due to both land-based activities and climate change.

Like most coastal catchments around the world, the Gulf of Maine catchment has been subjected to a wide variety of anthropogenic stressors that have altered and threaten to further modify marine ecosystem structure and function and compromise the sustainability of marine-based economic endeavors. Along the eastern seaboard of the United States, several

* Corresponding author:glenn.benoy@ec.gc.ca

studies have documented relationships between landscape metrics across watershed gradients of human activity and nutrient and sediment discharges (Weller et al. 2003), sediment contamination (Paul et al. 2002), and estuarine condition (Rodriguez et al. 2007). In the Quoddy region of the Gulf of Maine over the past 200 years, these changes in estuarine condition have included a steady loss of sea grass and wetland habitat, decreased wildlife biodiversity (i.e., mammals and birds), severely depleted stocks of fishes, and replacement of rockweed-based marine communities to algal-based marine communities as a result of eutrophication (Lotze and Milewski 2004). Reconstructions of nutrient loadings in the northeastern United States, and elsewhere in North America, indicate that while many estuaries have experienced steady increases over the past century, others have experienced decreases since the introduction of wastewater treatment efforts and pollution abatement programs concurrent with recognition of human-caused eutrophication in the latter part of the 20th century (Roman et al. 2000; Lotze et al. 2006). However, the overall reductions in point source pollution, through technological and regulatory measures, are compromised by the ubiquity of nonpoint or diffuse pollution sources across landscapes.

River systems link catchments and watershed-based human activities to estuarine and coastal ecosystems. Assessment of river hydrology, water quality, and ecological condition enables evaluation of anthropogenic stressors from individual catchments before downstream mixing of fresh and marine waters. Meaningful cross-comparisons of catchment influences on coastal water bodies, such as the Gulf of Maine, are dependent on the quality of spatial data that accurately represent landscape features, such as physiography, climate, land cover, and land use, among others. This is complicated by the fact that many river basins, indeed most large river basins, cross jurisdictional and sovereign boundaries.

The Saint John River basin is one such river. It is the largest basin in the Gulf of Maine catchment and it includes land in Maine, New Brunswick, and Quebec. Sections of the main stem river serve as an international boundary between Canada and the United States. Many aspects of the Saint John River and its basin have been studied to understand and improve water quality and ecological condition of the river and its tributaries, as a function of forest harvest and pulp and paper mills (Galloway et al. 2003) and potato agriculture (Gray et al. 2005). Munkittrick et al. (2007) argued that a framework for assimilative capacity for the Saint John River was needed to protect aquatic ecosystems in the watershed and sustain water availability for human needs. However, a basin-wide assessment of watershed-based human activities that could be used to model subsidies of sediments, nutrients, and contaminants to the river, and ultimately, to the Bay of Fundy and the Gulf of Maine, has yet to be completed.

Using the Saint John River basin as an example, this chapter covers nonpoint or diffuse pollution sources through the lens of landscape ecology. The objective is to characterize and quantify sources of nutrients from the basin and estimate loadings to the river and, ultimately, the Gulf of Maine. First, a history of the watershed is presented that covers the dominant eras of human activity from European settlement through present day. It includes a review of research to date on watershed-based influences on the Saint John River. Second, watershed-based activities are characterized across subbasins at different spatial scales to derive landscape gradients of anthropogenic activity (i.e., primarily agriculture, but also forestry and urbanization) that can be used to predict watershed exports of sediments, nutrients and contaminants. A numerical technique, known as export-coefficient modeling, is used to calculate nutrient exports from watersheds. Third, the contribution of nonpoint or diffuse sources of nutrients to overall Saint John River basin exports to the Bay of Fundy are estimated. The chapter begins with a brief description of the Saint John River basin, its major natural features, the major human activities that exist in the watershed, and a review of research into the environmental quality of the river. The chapter ends with recommendations for advancing research into quantifying and modeling watershed-based influences on the Saint John River and its estuary.

The research presented here represents an initial step towards the development of predictive export coefficient and watershed simulation models that can be used to prioritize pollution abatement strategies and forecast changes to sediment, nutrient, and pollutant loading in riverine and coastal receiving waters, according to alternative management scenarios for watersheds of the Gulf of Maine (Munkittrick et al. 2007). It also serves to identify knowledge and data gaps that hamper attempts to better understand watershed-based influences and processes.

Saint John River and Basin: Background and Context

Maliseet native peoples of New England and eastern Canada called the Saint John River *Wolastoq* or "beautiful river" (Blair 2001). The river is 673 km long and has a basin area of 55,110 km^2, making it the largest basin in the Gulf of Maine watershed (Table 1; Cunjak and Newbury 2005). Mean annual discharge at the mouth is about 1,110 m^3/s (Cunjak and Newbury 2005). Mean annual runoff rates ranges from 65 cm in the northern headwaters to 90 cm in southern New Brunswick, reflecting geographical differences in precipitation regimes. A significant portion of this precipitation is stored as snowpack, especially in headwater regions away from the marine environment and at higher elevation. As a result, high runoff rates during the snowmelt period (April–May) can cause flooding in ice-jam-prone reaches and downstream low-lying areas. Most of the basin resides in northern Maine and western and southern New Brunswick, though a portion of the headwater subbasins are in southern Quebec (Figure 1). Among the major tributaries that contribute to the Saint John River are the Madawaska River in Quebec, the Allagash and Aroostook rivers in Maine, and the Kennebecasis, Nashwaak, and Tobique rivers in New Brunswick (Cunjak and Newbury 2005). There are three major hydroelectric dams at Grand Falls, Beechwood, and Mactaquac that interrupt the natural flow regime of the river between the headwaters in the Appalachian highlands and the downstream lower reaches in the southern coastal zone. After the river passes through the city of Fredericton (17 m above sea level), near the uppermost location of tidal influence (Mactaquac Dam), the river drains into the Bay of Fundy at the city of Saint John.

TABLE 1. Watershed area and river length of major watersheds of the Gulf of Maine.

Watershed	River length (km)	Watershed area (km^2)
Annapolis	120	2,000
Shubenacadie	72	2,800
Petitcodiac	129	2,831
Saint John	673	55,110
Magaguadavic	129	1,812
St. Croix	102	3,900
Penobscot	560	22,300
Kennebec	240	15,200
Androscoggin	287	9,100
Saco	216	4,410
Piscataqua	62	3,870
Merrimack	177	12,005
Others		43,779
Total		179,007

Despite the size and history of the river and its basin, relatively little attention was devoted to water quality prior to the mid-20th century (but see Dominy 1973; Watt 1973). For mainly economic and recreational reasons, most research on the Saint John River since the mid-20th century has focused on the population dynamics of migratory fish and their habitats, particularly the emblematic Atlantic salmon *Salmo salar*. However, during the 1960s, there was a realization that the Saint John River was in very poor condition (DNHWC 1961; Sprague 1964). Results from monitoring surveys triggered provincial, state, and federal jurisdictions to establish the Saint John River Basin Board, which was responsible for an intensive multiyear study, in the 1970s, of many aspects of this watershed, including water quality, hydrology, and various ecological studies from headwaters to the estuary. Since then, research on the Saint John River has been sporadic, and there was a period between the mid-1990s to early 2000s when there was no monitoring program on the river.

More recently, with the emergence of ecosystem-based approaches to research and management of natural capital, including aquatic resources, over the past quarter century, there has been a surge of field and experimental research on

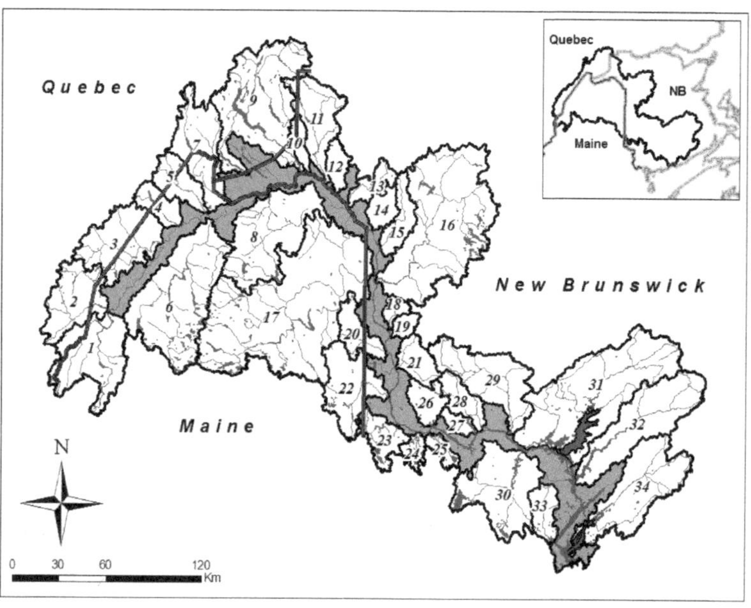

FIGURE 1. Map of Saint John River basin, including delineations of major subbasins that drain into the main stem of the river and incremental areas (in grey). Numbers that label each of the delineated subbasins correspond to names in Table 2.

factors affecting water quality and how changes in water quality affect aquatic biota. Initially, these studies focused on so-called end-of-pipe pollution point sources, such as pulp and paper mills and municipal sewage treatment plants. Such point source effluents can have major impacts on water chemistry near discharge locations (Culp et al. 2003; Galloway et al. 2003), but it is unclear how far downstream and over what time periods these impacts persist. Research from the 1970s showed that organic pollution from mills and food processing plants in the mid-reaches of the Saint John River exceeded their assimilative capacity, resulting in extreme heterotrophy (Watt 1973). However, these assessments tended to focus on concentrations of nutrients and contaminants, rather than loads or yields.

While these studies were useful for their intended purpose and have resulted in many restorative and protective policies, they only indirectly contribute to our understanding of how watershed-based human activities in the Saint John River basin affect the river estuary and the Bay of Fundy coastal ecosystem. Estimates of sediment, nutrient, and contaminant load (sometimes known as yield) are necessary to quantify the losses or exports of sediments, nutrients, and contaminants from the basin and interpret them as contributions to marine waters.

Research has since diversified to include consideration of human activities that generate nonpoint source diffuse pollution, such as agriculture and forest harvest, and their associated communities. European settlement, forestry, and agriculture have been the dominant land uses and human activities in the Saint John River basin. Though forestry remains a major component of the Maine and New Brunswick economies, agriculture, and, in particular, expansion of row-crop potato cultivation since the mid-20th century along the Saint John River valley, is frequently linked to the ecological condition of streams and rivers. With increasing availability of georeferenced spatial data, more detailed descriptions of watershed composition are possible. By comparing watersheds across gradients of land cover and human activity, predictions can be made regarding sediment, nutrient, and contaminant exports. Further, by summing across watersheds, estimates can be made of the contribution of nonpoint or diffuse sources of pollutants in the Saint John River basin to the Bay of Fundy and the Gulf of Maine.

Quantifying Watershed-Based Influences: Export Coefficient Modeling of Nonpoint or Diffuse Sources

This section looks at three studies that have attempted to construct land coverage and land use gradients within the Saint John River basin. For all three studies, a numerical technique known as export-coefficient modeling is used to estimate exports of nutrients from watersheds. Export-coefficient modeling predicts the total amount of nutrients (or sediment, contaminants, etc.), usually expressed as monthly or annual load at any location within the stream and river drainage network of a catchment, as a function of export from all sources above that location (Johnes 1996). Specific export coefficients are either derived from field studies or based on literature values. The approach has successfully modeled a wide variety of catchments, as evidenced by good agreement between modeled and empirical water quality data (Johnes et al. 1996; Hanrahan et al. 2001; Daly et al. 2002). Where available, water quality data are used to interpret these gradients and generate estimates of sediment and nutrient load to the river and the Bay of Fundy.

Basin-Wide Characterization of Land Coverage and Land Uses

The first study represents an attempt to characterize landscape properties of the entire Saint John River basin, estimate export coefficients, and relate them to water quality. A geographic information system (ArcGIS) was employed to combine physiographic, land use, and socioeconomic data sets from New Brunswick, Québec, and Maine. Using a combined digital elevation model for the region, the basin was divided into 34 subbasins according to major tributaries that drain into the main stem of the Saint John River (Figure 1; Table 2). These delineations formed a comparative analysis, and they included most of the basin (approximately 95%). River reaches and areas between confluences of tributaries and the Saint John River that were dominated by many small streams and catchments (sometimes known as incremental watersheds) were not included. Some of these incremental watersheds include urbanized areas along the main stem of the river, which can result in greater numbers of point sources. By not including these point sources, this study emphasizes nutrient exports from diffuse sources; point sources in incremental watersheds are covered in a subsequent section.

The 34 subbasins of the Saint John River basin ranged in size from 191 km^2 for the Monquart Stream subbasin to 6,327 km^2 for the Aroostook River subbasin. Most subbasins were less than 1,000 km^2 (median = 761 km^2). All of the subbasins have forest coverage greater than 50%; 29 of the 34 are greater than 70%. The composition of the forest ranges from a dominance of coniferous species in the higher elevation, more northern headwaters to a mix of coniferous and deciduous species along the main river valley and in central and southern portions of the basin. Cultivated land is the only other major type of land cover by area. It ranges from less than 1% in headwater subbasins (e.g., Allagash River, Saint John River headwaters, and Little Black River) to greater than 20% for the Big Presque Island Stream and Monquart Stream. The dominant cash crop throughout the basin is potato, in rotation with barley and clover, among other crops. Other forms of agriculture within the basin are horticultural and specialty crops (e.g., cranberry, blueberry), as well as livestock, mainly dairy.

Across the delineated watersheds, the population is estimated at approximately 242,149; 76% lives in New Brunswick and Quebec (2006 Census, Statistics Canada), with the balance living in Maine (2008 U.S. Census Bureau). The total number of people living in the basin is greater, but many of these people live in areas that are outside of the delineated watersheds and adjacent to the main stem of the river (e.g., cities of Edmundston and Fredericton that exist in incremental watersheds).

Using export-coefficient modeling, estimates of nitrogen and phosphorus exports were calculated for the major sources of nutrients for each of the subbasins. In the absence of specific validated export coefficients for the Saint John River basin, coefficients were obtained from literature sources that best matched the hydroclimatologic and physiographic conditions of the

TABLE 2. Key statistics for delineated subbasins of the Saint John River (SJR) basin. Subbasins start near the headwaters of the Saint John River (SJR) and end near the mouth. n.d. = no data.

Subbasin name	Total area (km^2)	Cultivated land (%)	Forested land (%)	Total population	Population density (#/km^2)
SJR headwaters	1,840	0.9	78.0	2,166	1.18
Northwest Branch SJR	1,308	7.3	82.8	6,685	5.11
Big Black River	1,632	5.5	79.5	6,065	3.72
Pocwock Stream	843	0.1	84.6	1,630	1.93
Little Black River	678	0.5	85.2	321	0.47
Allagash River	3,198	0.1	74.6	n.d.	n.d.
Baker Brook	1,471	6.9	80.9	7,339	4.99
Fish River	2,305	2.9	77.7	6,107	2.65
Madawaska River	3,047	10.2	78.1	16,398	5.38
Iroquois River	264	9.6	81.8	16,643	62.85
Green River	1,167	1.5	88.7	1,706	1.46
Quisibis River	330	6.0	79.7	n.d.	n.d.
Grande River	385	5.3	64.0	1,039	2.69
Little River	383	19.0	61.3	4,329	11.30
Salmon River	573	15.7	70.9	1,688	2.95
Tobique River	4,329	1.6	74.9	5,690	1.31
Aroostook River	6,326	6.0	76.5	37,914	5.99
Monquart River	190	20.1	72.0	n.d.	n.d.
Shikatehawk River	201	12.4	76.9	n.d.	n.d.
Big Presque Isle Stream	602	24.1	50.0	6,191	10.27
Becaguimec Stream	526	15.6	75.0	2,736	5.20
Meduxnekeag River	1,335	10.0	70.0	16,221	12.15
Eel River	595	10.5	70.8	155	0.26
Shogomoc River	242	1.2	76.5	555	2.29
Pokiok Stream	228	7.9	69.4	879	3.84
Nackawic Stream	478	10.5	76.1	2,881	6.02
Mactaquac Stream	220	13.1	68.2	1,215	5.52
Keswick River	521	8.9	80.9	3,159	6.05
Nashwaak River	1,707	3.5	81.0	7,591	4.44
Oromocto River	2,025	4.2	74.1	19,004	9.38
Jemseg River	3,776	2.1	81.2	14,500	3.84
Canaan River	2,167	4.8	84.2	5,528	2.55
Nerepsis River	504	2.1	73.5	n.d.	n.d.
Kennebecasis River	2,086	16.6	71.3	45,814	21.96
Average		7.8	75.6		7.2
Total	47,484			242,149	

Gulf of Maine region (Beaulac and Reckhow 1982; Frink 1991). For row-crop agriculture, an export coefficient of 9.1 kg/ha/year was used for nitrogen and 2.3 kg/ha/year for phosphorus. For forests, an export coefficient of 2.5 kg/ha/year was used for nitrogen and 0.23 kg/ha/year for phosphorus. Under the assumption that sewage receives both primary and secondary treatment, values of 2.14 kg/year/person for nitrogen and 0.38 kg/year/person for phosphorus were used to estimate human contributions to watershed exports.

Through export-coefficient modeling, exports of 12,122.9 metric tons of nitrogen and 1,554.7 metric tons of phosphorus were estimated for the delineated watersheds in the Saint John River basin on an annual basis (Table 3). Though these values are underestimates of the total export from the basin, as they do not include point sources along major tributaries and the main stem of the river, they do capture major human activities within the delineated watersheds, including agriculture and human settlements. Nitrogen exports are proportional to the percentage of forest cover in the basin. In contrast, phosphorus exports from cultivated lands are greatly underestimated by the percent of agricultural cover. When nitrogen and phosphorus export coefficients are plotted together (Figure 2), basins with relatively larger proportions of agriculture tend to have larger exports of phosphorus (i.e., above the line), and basins dominated by forest cover tend to have relatively larger exports of nitrogen (i.e., below the line).

Water quality data sets for nitrogen (mg/L) and phosphorus (mg/L) were assembled from provincial (New Brunswick Department of the Environment) and federal (Environment Canada's National Water Research Institute) sampling programs, mainly from 2003 to 2005. Because these sampling programs were not conducted to compare water quality across subbasins, not all subbasins have associated water quality data. Data set time-series were restricted to the past decade to enable investigation of water quality as a function of contemporary human activities. The frequency of water sample collections was not common across water quality data sets, although most samples were from base flow or low-flow periods (i.e., generally not during snowmelt or flood periods).

After winnowing down the subbasins to only those that have water quality data, 12 of the original 34 subbasins remained. Significant positive relationships were found between percent cultivated land and concentrations of both total nitrogen (Figure 3A; $R^2 = 0.36, p = 0.038$) and total phosphorus (Figure 3B; $R^2 = 0.32, p = 0.052$). In contrast, no significant relationships were observed between export coefficients and concentrations of nitrogen and phosphorus (Figure 3C, 3D). Measures of nutrient export and nutrient concentration are not necessarily correlated across watersheds because they describe different watershed parameters. Export describes the total mass of nutrients that is lost or yielded from a geographic area over a period of time, whereas concentration describes the quantity of nutrients measured as a function of water volume at a specified time.

Regional Scale Estimates of Nutrient Loading and Assessment of Water Quality

The second study covers an area within the Saint John River basin near Grand Falls, New Brunswick. It is commonly referred to as the potato belt, owing to the wide-spread distribution of intensive potato production. The data used for the estimation of nutrient loads were derived from an Environment Canada and Agriculture & Agri-Food Canada (AAFC) interdepartmen-

TABLE 3. Summary of export coefficient values for total nitrogen (N) and total phosphorus (P) calculated for delineated subbasins across the Saint John River basin. "Inhabitants" only includes human populations in the delineated watersheds.

Source	Total N (metric tons/year)	Total N (%)[a]	Total P (metric tons/year)	Total P (%)[a]
Cultivated land	2,457.4	20.3	621.1	40.0
Forested land	9,147.3	75.5	841.6	54.1
Inhabitants	518.2	4.3	92.0	5.9
Total	12,122.9		1,554.7	

[a] May not add up to 100% due to rounding.

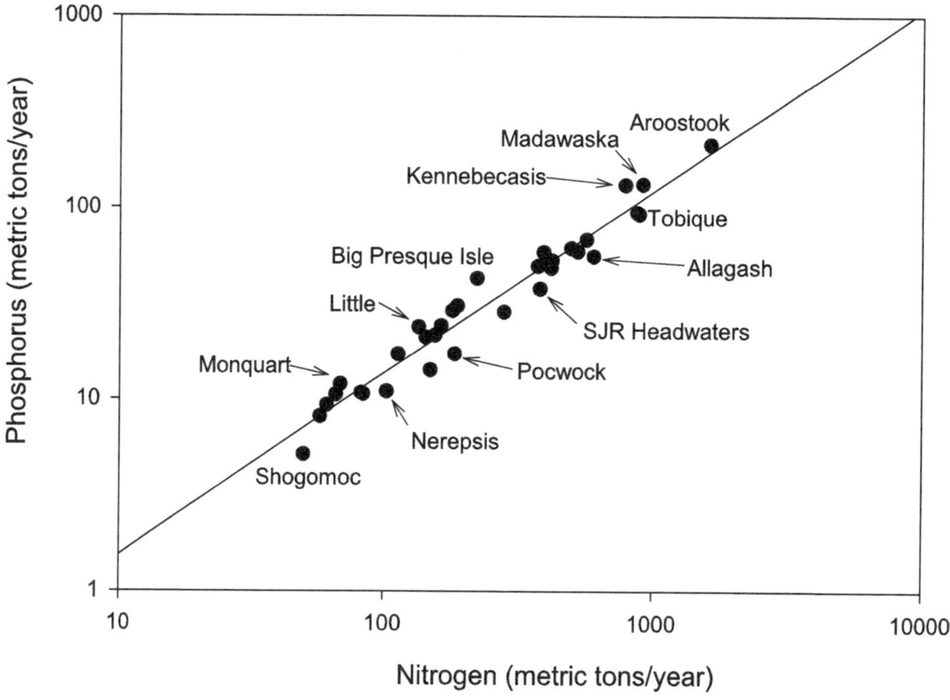

FIGURE 2. Plot of nitrogen and phosphorus export coefficients for 34 delineated watersheds in the Saint John River basin. A selection of watersheds are identified for reference.

tal research program called the National Agri-Environmental Standards Initiative (NAESI) and the same export coefficients that were used for the basin-wide analysis. The NAESI was conducted from 2004 to 2008, and it involved extensive monitoring and sampling of streams in the Grand Falls area to characterize the impacts of agriculture on water quality and aquatic communities (Chambers et al. 2009). Fourteen watersheds were delineated for this study (Table 4). They ranged in size from 5.9 to 31.1 km^2, and agricultural coverage ranged from 0.7 to 83.0%. Population levels were not available for this set of watershed delineations. By using smaller watersheds than those used for the basin-wide estimation of nutrient exports, and by targeting a specific geographic area that has limited human activity other than row-crop agriculture, predictions of nutrient loads to streams and rivers should be more accurate.

The watersheds delineated for the NAESI program totaled an area of 254.9 km^2, approximately the size of the smallest subbasins delineated for the basin-wide analysis (Table 4). Average percent cultivated land across the 14 watersheds was 41.2%, much higher than that for watersheds across the Saint John River basin. However, the range was much greater: a span of 82.3% for the NAESI watersheds, compared to a span of 24.0% for the entire basin. Using export coefficients, a total of 130.5 metric tons of nitrogen and 26.8 metric tons of phosphorus are exported from the NAESI watersheds on an annual basis. Overall, 70.1% of the nitrogen and 86.9% of the phosphorus comes from cultivated lands, the remainder primarily coming from forested land. Given the constrained range in area for the NAESI watersheds compared to the whole Saint John River basin, and the much broader range of agricultural coverage across watersheds, the contribution of nitrogen and phosphorus from agricultural lands is much higher. Total nutrient exports from the NAESI watersheds approximate those from the Little Black, Keswick, and Nackawic subbasins, except that the areas for these subbasins are 50–100% greater.

During the NAESI program, monthly water samples were collected from each of the

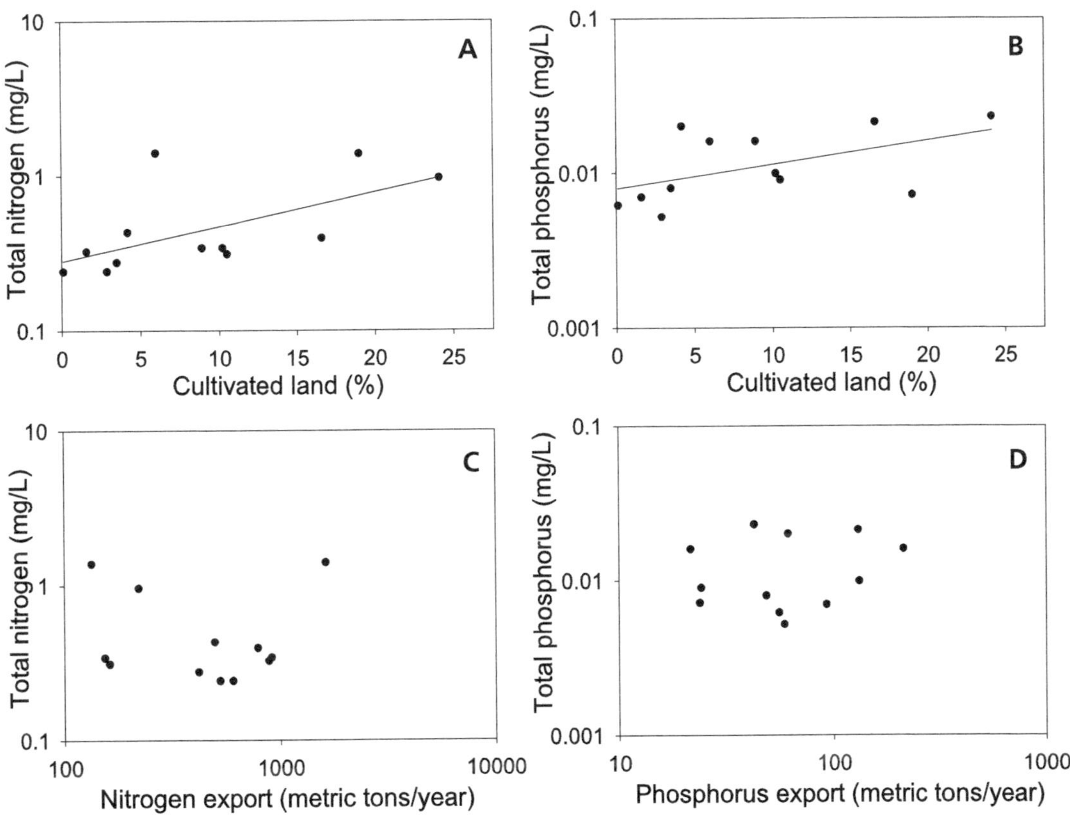

FIGURE 3. Relationships between cultivated land and nutrient concentrations (A and B) for a subset of subbasins (n = 12) in the Saint John River basin and relationships between export coefficients and concentrations of total nitrogen (C) and total phosphorus (D) for the same subbasins.

study watersheds from 2005 to 2007. Positive relationships were found between percent cultivated land across watersheds and concentrations of both total nitrogen ($R^2 = 0.82, p < 0.0001$) and total phosphorus ($R^2 = 0.75, p < 0.0001$). When nutrient concentrations were regressed against exports generated through export coefficient modeling, a significant relationship was only found for total phosphorus ($R^2 = 0.35, p = 0.025$). That stronger relationships were found with phosphorus than nitrogen is not surprising; phosphorus is a more "conservative" element that has a more straightforward pathway from upland sources to surface waters than nitrogen. Nitrogen can undergo biogeochemical transformation as it is transported across landscapes and within aquatic systems. Further, nitrogen loading is subsidized by atmospheric deposition and loss to groundwater flow paths.

Fine-Scale Evaluation of Nutrient Exports Using Field Data

The third study is from a much more detailed investigation of sediment and pollutant loads through high-frequency stream flow and water quality time-series data in relation to land use (i.e., potato farming; Chow et al. 2010). It involves the comparison of three watersheds in the same northwestern region of New Brunswick as that used for the NAESI program. The first watershed is the Black Brook Watershed (also included in the NAESI program; 14.5 km^2 and 76.9% cultivated land), a long-term beneficial management practice (BMP) development and evaluation watershed part of AAFC's Watershed Evaluation of BMPs (WEBs) program. The second watershed is the Upper Little River Watershed (196 km^2 and 3.7% cultivated land), a primarily forested

Table 4. Key statistics for watersheds examined during the National Agri-Environmental Standards Initiative watersheds near Grand Falls, New Brunswick and calculated nutrient export coefficients. Watersheds are listed according to percent area under cultivation.

Watershed	Area (km^2)	Cultivated land (%)	Forested land (%)	Total N (metric tons/year)	Total P (metric tons/year)
Sutherland Brook	28.1	0.7	99.3	7.2	0.7
Foley Brook	22.6	1.3	98.7	5.8	0.6
Bogin Brook	14.3	1.9	98.1	3.8	0.4
Little Beaver Brook	13.2	4.9	95.1	3.7	0.4
Otter Slide Brook	8.1	25.8	74.2	3.4	0.6
Ten Mile Brook	9.5	27.7	72.3	4.2	0.8
Dead Brook	24.1	45.8	54.2	13.3	2.8
Outlet Brook	37.2	47.6	52.4	21.0	4.5
Power Creek	24.8	49.0	51.0	14.2	3.1
Poitras Brook	9.7	66.8	33.2	6.7	1.6
Mooney Brook	21.5	67.6	32.4	15.0	3.5
Black Brook	14.5	76.9	23.1	11.0	2.6
Falls Brook	14.8	78.4	21.6	11.4	2.7
Godin Brook	12.5	83.0	17.0	10.0	2.4
Average		41.2	58.8		
Total	254.9		130.5	26.8	

watershed. Both watersheds are within the Little River Watershed (383.0 km^2 and 16.2% cultivated land), which is the third watershed and is also a primarily forested watershed.

Similar to the other two studies, export coefficient modeling was used to estimate exports of nutrients from the watersheds. In addition, continuous measurements from 2003 to 2007 of stream discharge and concentrations and nitrogen (nitrate-nitrogen or $NO_{3\text{-}}N$) and phosphorus (orthophosphate or ortho-P) were used to generate empirically based estimates of nutrient export from the watersheds. This represented a unique opportunity for evaluation and validation of the export coefficient modeling approach across a series of watersheds that vary greatly according to land-use composition.

Estimates of nitrogen and phosphorus exports through export coefficient modeling and field measurements were remarkably similar (Figure 4A, 4B). Exports were highest for the Little River watershed, followed by the Upper Little River watershed and the Black Brook watershed. Though the Black Brook watershed is much smaller than the other two, its contribution of nutrients to overall export is disproportionately high due to the large extent of cultivated lands. Between the two approaches for estimating nutrient exports, export coefficient estimates were consistently lower for nitrogen and consistently higher for phosphorus than those calculated through field measurements. Discrepancies between approaches for both nutrients may be due to differences in the species or fraction of the nutrient being estimated: total nitrogen versus nitrate and total phosphorus versus orthophosphate. This represents a plausible explanation for phosphorus discrepancies as field measurements were lower, but not for nitrogen. The discrepancy in export estimates for nitrogen can be explained partially through the absence of additional sources, such as atmospheric deposition. In addition, it is also possible that for both nutrients the literature-based export coefficients used either underestimated (nitrogen) or overestimated (phosphorus) actual exports. Nonetheless, the consistency between approaches suggests that export coefficient modeling is a valuable analytical tool for estimating loads of nutrients in the Saint John River basin and that

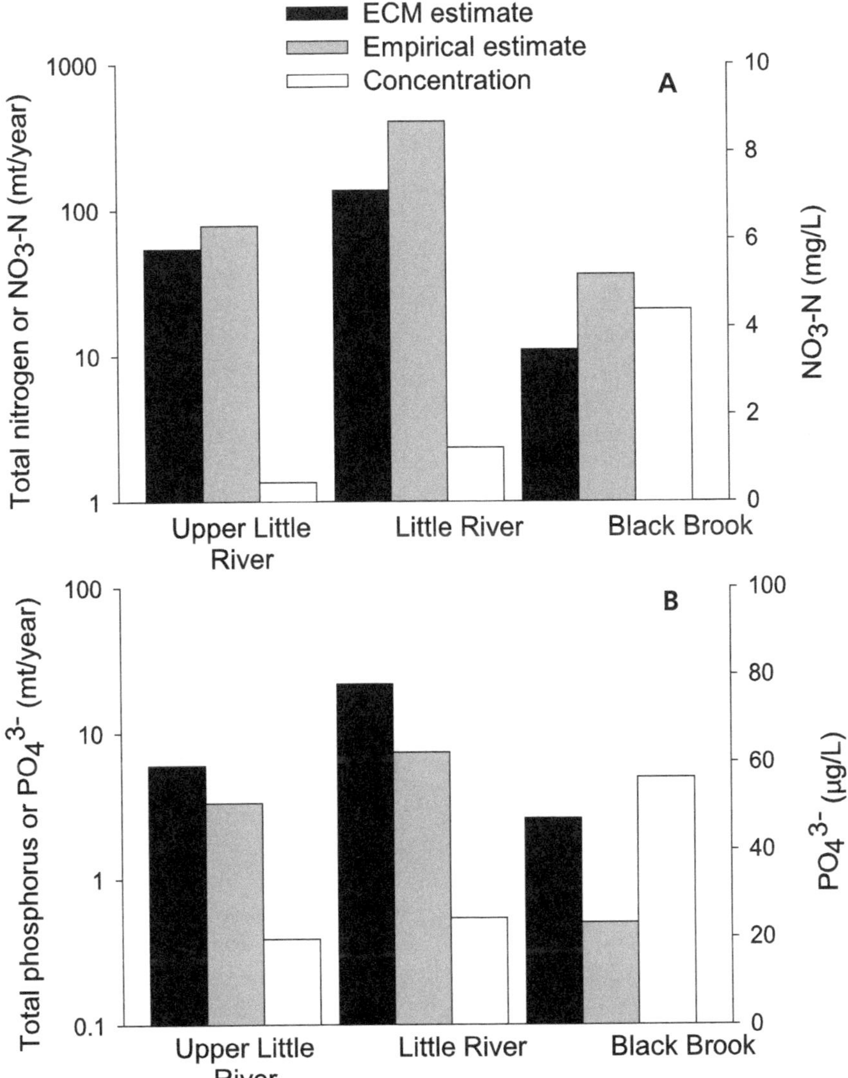

FIGURE 4. Comparison of total nitrogen (A) and total phosphorus (B) exports calculated using export coefficient modeling (ECM estimate) and field-based empirical measurements of continuous discharge and nutrient concentration (empirical estimate). The ECM estimates are expressed as total nitrogen and total phosphorus, and empirical estimates are expressed as nitrate-nitrogen ($NO_{3-}N$) and orthophosphate (PO_4^{3-}). Long-term stream water quality concentration values are also included for each watershed (unshaded bars in both panels).

field measurements can be used to evaluate export coefficients.

Long-term water quality data for the three watersheds show that larger extents of cultivated lands within watersheds results in higher concentrations of nitrate and orthophosphate, consistent with the other two studies (Figure 4A, 4B). As evidenced by all three studies, there are no a priori reasons why estimates or measurements of nutrient export should be similar to nutrient concentrations, as they reflect different characterizations of watershed features and human activities. While concentrations of nutrients tend to be used for assessments of environmental quality, and exports or loads of nutrients tend to be used for evaluating land-use change or management interventions, when both sets of measurements are considered together, a more comprehensive picture of watershed condition can be developed and interpreted at multiple scales.

Estimating Exports to the Bay of Fundy from all Watershed-Based Sources

In addition to the land-based human activities of forestry, agriculture, and urban development that exist within the Saint John River basin, there are a variety of point sources within the river network of the basin that can also degrade water quality and ecological condition and contribute sediments, effluents, and pollutants to downstream receiving waters. These point sources include pulp and paper mills and saw mills, municipal and nonmunicipal wastewater treatment plants, food processing plants, and aquaculture hatcheries. The largest of these point sources are located adjacent to or near the main stem of the river, either within New Brunswick or along the stretch of the river that separates New Brunswick from Maine. By summing both point and nonpoint sources across the Saint John River basin, an estimate of total nutrient export to the Bay of Fundy can be made.

From the basin-wide export coefficient modeling analysis, 11,605 metric tons of nitrogen and 1,463 metric tons of phosphorus are exported annually from cultivated lands and forest lands in the delineated watersheds (Table 3). Assuming that the ratio of cultivated land to forested land is the same in the incremental watersheds (5% of the total area of the Saint John River basin) as the delineated watersheds (95% of the total), this means that the nitrogen export from these watershed-based land uses is approximately 12,216 metric tons per year and the phosphorus export is 1,540 metric tons per year. From this same analysis, nutrient exports from humans are 518 metric tons of nitrogen per year and 92 metric tons of phosphorus per year. These amounts are based on a calculation of 242,149 inhabitants in the delineated watersheds. From combined census data for Maine, New Brunswick, and Quebec, it is estimated that the total population is 386,896. This total does not include the city of Saint John, as its municipal waste is discharged downstream of the Reversing Falls, directly into the estuary. By combining these estimates of the number of inhabitants in the Saint John River basin, the export for nitrogen is 827 metric tons per year and the export for phosphorus is 147 metric tons per year. Though the incremental watersheds only cover 5% of the basin area, they contain 37% of the population.

Contributions from industrial point sources to overall export rates from the Saint John River basin are estimated to be 2,978 metric tons of nitrogen per year and 555 metric tons of phosphorus per year (E. Luiker, Environment Canada, unpublished data). These point sources include pulp and paper mills and food processing plants (mainly potato). Therefore, the estimate for overall export of nitrogen to the Bay of Fundy is approximately 16,021 metric tons per year and for phosphorus, it is 2,242 metric tons per year (Table 5). Nonpoint sources are the dominant sources of nutrients to the Bay of Fundy. For nitrogen, more than 75% of the export estimate is from nonpoint sources, and for phosphorus the portion is almost 70%. Point sources contribute almost 20% of the nitrogen load and almost 25%

TABLE 5. Overall estimates of total nitrogen (N) and total phosphorus (P) exports from the Saint John River basin to the Bay of Fundy.

Source	Total N (metric tons/year)	Total N (%)[a]	Total P (metric tons/year)	Total P (%)[a]
Nonpoint	12,216	76.2	1,540	68.7
Point	2,978	18.6	555	24.8
Inhabitants	827	5.2	147	6.6
Total	16,021		2,242	

[a] May not add up to 100% due to rounding.

of the phosphorus load. The smallest fraction of nutrient loading is from inhabitants of the basin. Though point-source fractions of total nutrient loads are estimated to be much less than those of nonpoint sources, they may not directly translate into effects on estuarine and coastal ecosystems. Forms of nitrogen and phosphorus in point-source effluents tend to be more bioavailable to aquatic biota than forms from nonpoint sources (Chambers et al. 2001), meaning that more detailed characterization of aquatic chemistry is necessary to predict effects.

Future Directions for Understanding and Quantifying Watershed-Based Influences

Historical information on water quality and loading rates of sediments, nutrients, and contaminants provide reference points for what the condition of the Saint John River could be. Through the adoption of an adaptive research and management (ARM) framework, policies and practices designed to improve the ecological condition of the Saint John River, and reduce its contribution to estuarine and coastal sedimentation and eutrophication, can be tracked and evaluated. Modeling approaches, including export coefficient modeling, have great potential for integrating research and management at watershed scales and evaluating policies and practices, but their reliability and credibility are critically dependent on the availability of discharge and water quality data at useful temporal and spatial scales.

River monitoring programs play a vital role in ARM. Long-term data sets enable tracking of net changes in environmental quality over time. Within these long-term data sets, more frequent measurements of discharge and water quality during key phases of the year (e.g., snowmelt and spring freshet, postharvest bare ground in the fall) can resolve time frames when loading to rivers is greatest. Similarly, in addition to downstream monitoring stations that capture the entire Saint John River basin, targeted monitoring stations should be maintained at key locations along the main stem of the river and its tributaries to help isolate the contributions of specific human activities. Examples include the provincial water quality data sets generated by the province of New Brunswick to assess water quality and the long-term continuous monitoring water sampling program in the Little River and Black Brook Watershed maintained by AAFC. Recent advances in the emerging field of environmental fingerprinting (i.e., source-tracing of sediments and pollutants) can help to further resolve and quantify the contributions of specific human activities (Dubé et al. 2005; Krishnappan et al. 2009).

Conclusions and Recommendations

The three nonpoint or diffuse sources studies spanned a wide range of spatial scales in the Saint John River basin, and they revealed several strengths and weaknesses associated with quantifying watershed-based influences. First, export coefficient modeling is a useful tool for estimating relative differences in exports between watersheds. Second, to improve export estimates, regional export coefficients need to be developed for the Bay of Fundy and Gulf of Maine watersheds. Third, though large-scale assessments of human activities are benefitting from increased availability of high-quality, geospatial data, they are hampered by limited availability of water quality data for interpretation of estimates and model validation. Fourth, while fine-scale studies tend to have robust historical land use and water quality data sets, they are not easily "scaled up" to larger regions. When combined with estimates of nutrient and contaminant discharges from point sources, quantification of exports from watershed-based activities will be enhanced.

Few studies to date have integrated research on the Saint John River with its influence on receiving waters—estuarine and coastal ecosystems of the Bay of Fundy and the Gulf of Maine. The focus on the assessment of specific human activities, such as potato production systems and pulp and paper mill operations, on river condition has resulted in improved management practices and the adoption of pollution abatement technologies. The next step is to extract estimates of sediment, nutrient, and contaminant loading from these studies and design future studies that explicitly couple river dynamics to mass-balance models of riverine contributions to marine ecosystems.

Such studies are not without precedent along the northeast Atlantic seaboard. Roman et al. (2000) estimated total nitrogen and total phosphorus annual loads to estuaries of the United States from Maine to New York and also estimated source contributions. Higher loading rates near the southern extent of the Gulf of Maine were related to a shift from desirable sea grass communities to less desirable macroalgal communities. Similarly, Howarth et al. (1991) modeled land-use changes in the Hudson River watershed and found that inputs of sediment and organic carbon strongly affected estuarine metabolism. Application of these types of studies would contribute to a more holistic understanding of the Saint John River basin and its receiving waters; however, as noted by Munkittrick et al. (2007), the lack of baseline studies in the estuary of the Saint John River presents a challenge for the comparison of present and historical states.

The following recommendations are made to guide research directions for incorporating watershed-based influences into the understanding and assessment of the Bay of Fundy and the Gulf of Maine:

- Joint projects should be initiated between river and marine scientists to establish research objectives for understanding watershed-based influences on the Bay of Fundy and the Gulf of Maine and to explore mechanisms for supporting research;
- Monitoring and sampling stations should be established and maintained within the Saint John River basin, at multiple spatial scales, to track temporal changes in discharge and water quality;
- Geospatial data sets should be combined across jurisdictions to enable whole-basin characterizations of climatic, physiographic, and human features;
- Models should be used to evaluate changes in discharge and water quality as a function of policy introductions and changes, and the implementation of improved management practices;
- Data sets generated from river models should be coupled to estuarine and coastal ecosystem models; and
- Research products, such as reports, papers, and presentations, should be communicated to relevant public, private, and nonprofit organizations and agencies to promote this integrated approach.

Acknowledgments

We thank Sheldon Hann of the Agri-Environmental Services Branch of Agriculture & Agri-Food Canada for geographic information systems assistance and preparation of Figure 1 and Matthew Dickson of the New Brunswick Department of the Environment for information on point sources in the province of New Brunswick. The chapter benefitted from comments and suggestions provided by two reviewers. We also thank the Gulf of Maine Symposium scientific steering committee, and specifically Robert Stephenson, for showing interest in our research. This work was supported through funds from Agriculture & Agri-Food Canada, Environment Canada, Environmental Trust Fund of New Brunswick, and a National Sciences and Engineering Research Council Discovery Grant to JMC.

References

Beaulac, M. N., and K. H. Reckhow. 1982. An examination of land use-nutrient export relationships. Water Resources Bulletin 18:1013–1024.

Blair, S. 2001. The Wolastoq and its people: a study in commemoration. New Brunswick Manuscripts in Archaeology 31. A report prepared for the New Brunswick Culture and Sports Secretariat. Archaeological Services of New Brunswick, Fredericton, New Brunswick.

Chambers, P. A., M. Guy, S. S. Dixit, G. A. Benoy, R. B. Brua, J. M. Culp, D. McGoldrick, B. L. Upsdell, and C. Vis. 2009. Nitrogen and phosphorus standards to protect the ecological condition of Canadian streams, rivers and coastal waters. Environment Canada, National Agri-Environmental Standards Initiative Synthesis Report No. 11, Gatineau, Québec.

Chambers P. A., M. Guy, E. S. Roberts, M. N. Charlton, R. Kent, C. Gagnon, G. Grove, and N. Foster. 2001. Nutrients and their impact on the Canadian environment. Agriculture and Agri-Food Canada, Environment Canada, Fisheries and Oceans Canada, Health Canada, and Natural Resources Canada. Issued by Environment Canada, Gatineau, Québec.

Chow, L., Z. Xing, G. A. Benoy, H. W. Rees, F. Meng, Y. Jiang, and J.-L. Daigle. 2010. Hydrology and water quality across gradients of agricultural intensity in the Little River Watershed area, New Brunswick, Canada. Journal of Soil and Water Conservation 66:71–84.

Culp, J. M., K. J. Cash, N. E. Glozier, and R. B. Brua. 2003. Effects of pulp mill effluent on benthic assemblages in mesocosms along the Saint John River, Canada. Environmental Toxicology and Chemistry 22:2916–2925.

Cunjak, R. A., and R. W. Newbury. 2005. Atlantic rivers—Canada. Pages 938–980 *in* A. C. Benke and C. E. Cushing, editors. Rivers of North America. Academic Press, San Diego, California.

Daly, K., P. Mills, B. Coulter, and M. McGarrigle. 2002. Modeling phosphorus concentrations in Irish rivers using land use, soil type, and soil phosphorus data. Journal of Environmental Quality 31:590–599.

DNHWC (Department of National Health and Welfare Canada). 1961. Survey report: effects of pollution—St. John River, New Brunswick and Maine. DNHWC, Public Health Engineering Division, Ottawa.

Dominy, C. L. 1973. Recent changes in Atlantic salmon (*Salmo salar*) runs in the light of environmental changes in the Saint John River, New Brunswick, Canada. Biological Conservation 5:105–113.

Dubé, M., G. A. Benoy, S. Blenkinsopp, J.-M. Ferone, R. Brua, and L. I. Wassenaar. 2005. Application of multi-stable isotope (13C, 15N, 34S, 37Cl) assays to assess spatial separation of fish (longnose sucker *Catostomus catostomus*) in an area receiving complex effluents. Water Quality Research Journal of Canada 40:275–287.

Frink, C. R. 1991. Estimating nutrient exports to estuaries. Journal of Environmental Quality 20:717–724.

Galloway, B. J., K. R. Munkittrick, S. Currie, M. A. Gray, R. A. Curry, and C. S. Wood. 2003. Examination of the responses of slimy sculpin (*Cottus cognatus*) and white sucker (*Catostomus commersoni*) collected on the Saint John River (Canada) downstream of pulp mill, paper mill, and sewage discharges. Environmental Toxicology and Chemistry 22:2898–2907.

Gray, M. A., R. A. Curry, and K. R. Munkittrick. 2005. Impacts of nonpoint inputs from potato farming on populations of slimy sculpin (*Cottus cognatus*). Environmental Toxicology and Chemistry 24:2291–2298.

Hale, S. S., and M. Westhead. 2012. Ecosystem services in the Gulf of Maine. Pages 1–8 *in* R. L. Stephenson, J. H. Annala, J. A. Runge, and M. Hall-Arber, editors. Advancing an ecosystem approach in the Gulf of Maine. American Fisheries Society, Symposium 79, Bethesda, Maryland.

Hanrahan, G., M. Gledhill, W. A. House, and P. J. Worsfold. 2001. Phosphorus loading in the Frome catchment, UK: seasonal refinement of the coefficient modeling approach. Journal of Environmental Quality 30:1738–1746.

Hopkinson, C. S., Jr., and J. J. Vallino. 1995. The relationships among man's activities in watersheds and estuaries: a model of runoff effects on patterns of estuarine community metabolism. Estuaries 18:598–621.

Howarth, R. W., J. R. Fruci, and D. Sherman. 1991. Inputs of sediment and carbon to an estuarine ecosystem: influence of land use. Ecological Applications 1:27–39.

Johnes, P. 1996. Evaluation and management of the impact of land use change on the nitrogen and phosphorus load delivered to surface waters: the export coefficient modelling approach. Journal of Hydrology 183:323–349.

Johnes, P., B. Moss, and G. Phillips. 1996. The determination of total nitrogen and total phosphorus concentrations in freshwaters from land use, stock headage and population data: testing of a model for use in conservation and water quality management. Freshwater Biology 36:451–473.

Krishnappan, B. G., P. A. Chambers, G. A. Benoy, and J. Culp. 2009. Sediment source tracking: state-of-the-art. Canadian Journal of Civil Engineering 36:1622–1633.

Lotze, H. K., and I. Milewski. 2004. Two centuries of multiple human impacts and successive changes in a North Atlantic food web. Ecological Applications 14:1428–1447.

Lotze, H. K., H. S. Lenihan, B. J. Bourque, R. H. Bradbury, R. G. Cooke, M. C. Kay, S. M. Kidwell, M. X. Kirby, C. H. Peterson, and J. B. C. Jackson. 2006. Depletion, degradation, and recovery potential of estuaries and coastal seas. Science 312:1806–1809.

Munkittrick, K. R., R. A. Curry, J. M. Culp, R. A. Cunjak, D. L. MacLatchy, K. A. Kidd, S. E. Dalton, D. J. Baird, and R. Newbury. 2007. The Saint John River. Pages 129–145 *in* C. King, J. Ramkisson, M. Clüsener-Godt, and Z. Adeel, editors. Water and ecosystems: managing water in diverse ecosystems to ensure human well-being. United Nations University, Institute for Water, Environment and Health, Hamilton, Ontario.

Nixon, S. W. 1995. Coastal marine eutrophication: a definition, social causes, and future concerns. Ophelia 41:199–219.

Paul, J. F., R. L. Comeleo, and J. Copeland. 2002. Landscape metrics and estuarine sediment contamination in the mid-Atlantic and southern New England regions. Journal of Environmental Quality 31:836–845.

Rodriguez, W., P. V. August, Y. Wang, J. F. Paul, A. Gold, and N. Rubinstein. 2007. Empirical relationships between land use/cover and estuarine condition in the northeastern United States. Landscape Ecology 22:403–417.

Roman, C. T., N. Jaworski, F. T. Short, S. Findlay, and R. S. Warren. 2000. Estuaries of the northeastern United States: habitat and land use signatures. Estuaries 23:743–764.

Smith, V. H., S. B. Joye, and R. W. Howarth. 2006. Eutrophication of freshwater and marine ecosystems. Limnology and Oceanography 51:351–355.

Sprague, J. B. 1964. Chemical Surveys of the Saint John River—tributaries, impoundments and estuary in 1959 and 1960. Fisheries Research Board of Canada Manuscript Report Series 181.

Watt, W. D. 1973. Aquatic ecology of the St. John River, volume 1: background, design, methods, nutrients, bacteria, phytoplankton and primary production. St. John River Basin Board, Report 15f, Fredericton, New Brunswick.

Weller, D. E., T. E. Jordan, D. L. Correll, and Z.-J. Liu. 2003. Effects of land-use change on nutrient discharges from the Patuxent River watershed. Estuaries 26:244–266.

American Fisheries Society Symposium 79:235–241, 2012

Anthropogenic and External Influences on the Gulf of Maine: Workshop Summary

Madeleine Hall-Arber* and Judith Pederson
MIT Sea Grant College Program
77 Massachusetts Avenue E38-300, Cambridge, Massachusetts 02139, USA

Peter Wells
Dalhousie University
School for Resource and Environmental Studies and Marine Affairs Program
Faculty of Management, and International Ocean Institute
6414 Coburg Road, Halifax, Nova Scotia B3H 4R2, Canada

Abstract.—The goal of this session was to provide a synthesis of the major pressures being exerted on the Gulf of Maine (GOM), including the Bay of Fundy, that constrain the achievement of ecosystem objectives or the desired state of valued attributes (ecological, social/cultural, and economic). The GOM boasts a diverse ecosystem, much changed ecologically over the past 400 years and long the subject of intense scrutiny by a host of universities, marine labs, and research centers in both the United States and Canada. While the presentations in the session noted that this ecosystem must cope with stressors that range from climate change to pharmaceuticals and nutrient loading to overfishing, the presentations also presented an historical perspective with lessons for resiliency, including successes of finer-scale and spatial management as well as collaborations in research and communication. It was recognized that a concerted effort to raise environmental literacy amongst the gulf's residents is essential to ensure that the full value of its ecosystems and resources is recognized and protected for future generations.

Introduction

A theme session at the 2009 Gulf of Maine Symposium, "Anthropogenic and External Influences on the Gulf of Maine Ecosystem," covered a broad array of topics that were examined from two perspectives. The first set of presenters addressed issues related to climate change and impacts of other stressors on the ecosystem; the second set of presenters addressed resilience and adaptation of the ecosystem to such pressures. The poster presentations explored in more depth several different types of pressures on species and ecosystems, use of history in analyzing changes, and carbon flux in the ecosystem. In this summary, based on the papers presented in the session, we review developments largely in the context of the past decade (2000–2010), evaluate our ability to manage, and provide recommendations to shape environmental research in the future.

* Corresponding author: arber@mit.edu

Historical Perspective: How Far Have We Come in Our Understanding of Influences?

The first Gulf of Maine conference in December 1989 focused on "sustaining our common heritage," a theme returned to with several presentations in 2009 (Konrad et al. 1989). In both the earlier conference and this one, themes included historic uses of the Gulf of Maine, such as for fisheries and waste disposal, and the importance of analyzing faunal communities over time (Hale 2012; Hare and Kane 2012; both this volume). The fate and effects of nutrients and contaminants have also been a theme in all of the past four regional symposia. What is perhaps the most significant change in our perspectives on the Gulf of Maine is the recognition that ecosystem management (in this context, also referred to as integrated coastal and ocean management) is critically important but, to be effective, it must consider intertwined ecological and social factors and be operationally feasible (Holling 2001; Colding et al. 2003; Pikitch et al. 2004; DeLauer 2012, this volume; Feurt 2012, this volume).

For example, historical records identify changes in species abundance over time. Indeed, 400-year-old archeological records mark changes in fisheries and other fauna that can be resolved in near decadal chronologies as discussed by T. Willis (University of Southern Maine). Many of these changes cannot be explained without reference to human resource use. Fisheries have been a major driver of change (Lotze et al. 2004), as seen historically in other parts of the world's oceans (Roberts 2007, 2012). In addition, anthropogenic influences and external forces are closely linked to other drivers, such as climate change, chemical pollution, nutrient loading, and invasive species (Caddy 1993; Brander et al. 2010).

From pre-European settlement times to the present, the pressures of colonization, clearing of land, building sawmills and ships, industrialization, and increased runoff from agricultural lands with consequent eutrophication have affected the watersheds and coastal areas of the Gulf of Maine and the Bay of Fundy (Smith et al. 1999; Anderson et al. 2008). E. Luiker (National Water Research Institute, Burlington, Ontario) explained that increased nutrient loading, for instance, has led to a downgrading of the St. John River, with the largest watershed in the region, from low impact to high impact throughout most of its length (Munkittrick et al. 2007; Benoy et al. 2012, this volume). It was not until the 1970s, with the advent of many industrial effluent regulations and guidelines, that efforts were made to reverse the ongoing pollution of fresh, estuarine, and marine waters. The reversal of past contamination, however, is slow, particularly for persistent organic pollutants (e.g., PCBs, DDT; Wells 2005). Gulfwatch, a 19-year-old monitoring program that analyzes the levels of chemical contamination in blue mussels *Mytilus edulis* throughout the Gulf of Maine, reveals that some contaminants, but not all, are diminishing. These results, discussed by S. Jones (University of New Hampshire), show that the time required for reversing contamination and pollution may be long, even when regulations are in place and enforced (Chase et al. 2001).

In addition to understanding recovery times from pollution and other disturbance, fine-scale information and spatial patterns are noteworthy needs. Broadscale fisheries data on abundance and diversity over time are necessary for fisheries management, but finer-scale data may help evaluate why some areas are more productive or recover more quickly than others (Ames 1997). Recent work, for example, suggests that the value of forage fish (e.g., alewives *Alosa pseudoharengus*, blueback herring *A. aestivalis*, and American shad *A. sapidissima*) to groundfish, such as Atlantic cod *Gadus morhua*, in inshore habitats has been underappreciated and that fishing may have altered forage fish diets (Goode 2006). Similarly, changes in top predators have altered species abundance, if not total biomass, but the data are incomplete to fully test the hypothesis of unequal abundance and diversity in northern and southern portions of the Gulf of Maine.

A 40-year study of marine benthic communities in the Gulf of Maine demonstrated a major shift in species and their abundance. The pressures identified by speaker L. Harris (University of New Hampshire) as causing the observed changes included the development of a fishery in response to an overpopulation of green sea urchins *Strongylocentrotus droebachiensis*, a major benthic omnivore, followed by overfishing of the urchins, which led to competitive advantages of nonnative species that displaced other native populations (Harris et al. 2001; Harris and Tyrrell 2001). The observed ecological shift was dramatic and will have long-lasting impacts on other living resources (Dijkstra and Harris 2009).

Although finfish aquaculture has been a part of the Gulf of Maine and Bay of Fundy history for approximately 30 years, it was only briefly addressed in past symposia. Aquaculture has the potential to supplement wild harvests, thus helping to fulfill a growing societal appetite for seafood, but the culture of certain species (e.g., Atlantic salmon *Salmo salar*) can have negative ecological impacts in coastal waters (Hargrave 2005; Burt and Wells 2010). Eutrophication, for example, is a common concern for aquaculture facilities, but the release of antibiotics and other pharmaceuticals to the waters has drawn increased attention, as has negative effects on wild populations (Holmer et al. 2005; Ford and Myers 2008). L. Burridge (Fisheries and Oceans Canada) noted that many of the drugs used to combat parasites on fish may also impact lobster and crabs, and their sublethal effects on other organisms is largely unknown but widely suspected. Health hazards to humans from the use of antibiotics and other drugs in the management of aquaculture remain a concern, despite regulatory controls (Burridge and Haya 1995; Haya et al. 2001). These issues are likely to be topics of discussion in future meetings.

Management: How Well Are We Situated to Inform Management?

During the past decade and more, researchers have focused on biophysical and socioeconomic impacts of activities in the Gulf of Maine, the utility and value of ecosystem services, and insights gained from the aforementioned hazard and risk assessment, and monitoring (Wiggin and Mooers 1992; Wallace and Braasch 1996). The session pointed to the continuing challenges to ecosystem health of climate change, pollutants, cumulative impacts, and overuse of natural resources but also featured signs of ecosystem resilience, recovery and adaptation pertinent to management strategies.

Climate change, rarely mentioned or investigated in the past decade around the gulf, has emerged from a political gulag and is recognized by both Canada and the United States as a high-priority problem today. Keynote speaker D. van Proosdij (Saint Mary's University) suggested that the debate is no longer about whether climate change is occurring, but rather whether we can slow the change and/or adapt. Communities and governments are undertaking pilot projects to evaluate adaptation options. Increasing storm surges, sea level rise, more violent storms, and changes in precipitation are likely to have significant impacts on coastal habitats and societal infrastructure, especially in low lying areas (e.g., flood plains, estuaries; Scavia et al. 2002). Since each geographic area around the gulf is unique, and tidal amplitudes especially vary enormously north to south, a variety of approaches is needed to support and sustain coastal communities. In addition to encouraging community adaptive responses, governments are increasingly pressing for reductions in the production of carbon and greenhouse gases in order to slow change.

T. Callaghan (Massachusetts Coastal Zone Management Program) explained that Massachusetts has undertaken an ocean management planning effort to assist with balancing conflicting uses of the oceans (Massachusetts Ocean Management Plan 2009). The marine spatial planning effort examined the potential of the ocean to support alternative energy generation, as well as natural resources within coastal Massachusetts' waters. Although the plan is not yet implemented, the goal of the planning effort is to support environmentally compatible decisions and meet societal needs (The Massachusetts Executive Office of Energy and Environmental Affairs 2009). (For additional information on marine spatial planning and its potential use in management, see Douvere 2008.) Similarly, efforts to harness tidal energy through siting turbines in the Bay of Fundy reflect the desire to obtain energy with fewer negative environmental consequences than traditional sources. Speaker A. Redden (Acadian Centre for Estuarine Research) noted that one recently funded project in the Minas Channel in the upper Bay of Fundy is examining models and impacts on fisheries, benthos, and sediments, as well as the technical aspects of the effects of ice and debris on the turbines themselves. Data from these studies will inform future decisions regarding the feasibility of tidal energy extraction in macro-tidal parts of the gulf.

Research on carbon flux at the water–air interface suggests that despite the Gulf of Maine's high plankton productivity, during the winter months there is a net escape of CO_2 from vertical mixing of the waters and release of dissolved inorganic carbon (Townsend 1998; Townsend and Thomas 2001). When the data from the entire U.S. northeastern seaboard are analyzed, we will have a greater understanding of the role of oceans in carbon flux. This research has reached a new state of urgency given that atmospheric levels of CO_2 are now well above the 350 ppm threshold determined by the IPCC's (Intergovernmental Panel on Climate Change) latest assessment report.

Recommendations for Future Research: What Shall We Do Now?

According to the speakers in this session, progress has been made towards resolving some of the negative pressures and achieving desired goals for the Gulf of Maine; nevertheless, they reminded participants that additional work is needed. For example, we should develop conceptual models and an operational framework for managing the Gulf of Maine and Bay of Fundy in the context of unpredictable ecosystem change, especially given the rate of climate change now occurring. Scientific research, long-term monitoring programs, more frequent state of environment assessments, interdisciplinary and interagency collaboration, and communication are all crucial.

Communication and joint action among stakeholders is essential for progress in achieving the sustainability of the ecosystems of the Gulf of Maine. B. Thorne-Miller (Northwest Atlantic Marine Alliance) posited that sharing knowledge of the history of the region, of the booms and busts in fisheries, and of changes in habitat between fishermen and scientists may lead to innovative approaches to fisheries management (see shared fishermen and scientists observations in Dorsey and Pederson [1998]). In the past decade, steps towards successful collaboration and communication between scientists and fishermen have been taken through the cooperative research projects sponsored by the Northeast Consortium, Gulf of Maine Research Institute, NOAA Fisheries, and the Canadian Fisherman and Scientists Society (Feeney et al. 2010). In increasing numbers of cases, conversations among fishermen and scientists involved in collaborative work has been mutually beneficial, but the challenge of quantifying qualitative data for use in models remains elusive.

The critical imperative of ecosystem-based management is generally recognized by scientists and managers. On both coasts, scientists have reported on the status of the ecosystem and/or status of knowledge about the ecosystem and have urged that appropriate responses be taken (Ecosystem Assessment Program 2009; Lester et al. 2010). Action plans have been developed by, for example, the Gulf of Maine Council on the Marine Environment (www.gulfofmaine.org), but funding for individual programs is precarious and coordination among them has not yet been achieved. One of the primary challenges is implementation of such coordination, despite regional jurisdictional and institutional complexity (Wells 2010).

Public education, outreach, and communication among the other stakeholders for the Gulf of Maine, and Bay of Fundy, are also vital. In the past, keynote speaker M. Chiarappa (Western Michigan University) explained, environmental history was melded with fisheries history to create compelling stories useful for public education (Chiarappa 2012, this volume). Writers and artists, fishermen and scientists worked in concert to develop stories about, or depict images of, the fisheries contribution to the economic and ecological health of both countries, and the northeast U.S. and Maritime regions. Then, for a time, stories about the negative role of fisheries in marine ecosystem health were used to educate the public about the unintended consequences of over use of marine living resources. Today, new perspectives on fisheries and resource management are creating new opportunities for public education. More generally, a visioning and spatial planning process for the Gulf of Maine, with input from a diversity of interests, should be undertaken. However, a concerted effort to raise environmental literacy is essential at the same time, to ensure that the full value of the gulf's ecosystems and resources is recognized and protected for future generations of coastal inhabitants.

References

Ames, E. P. 1997. Cod and haddock spawning grounds of the Gulf of Maine. National Resource, Agriculture, and Engineering Service, NRAES 118, Ithaca, New York.

Anderson, D. M., J. M. Burkholder, W. P. Cochlan, P. M. Gilbert, C. J. Gobler, C. A. Heil, R. M. Kudela, M. L. Parsons, J. E. Jack Rensel, D. W. Townsend, V. L. Trainer, and G. A. Vargo. 2008. Harmful algal blooms and eutrophication: examining linkages from selected coastal regions of the United States. Harmful Algae 8:39–53.

Benoy, G., E. Luiker, and J. Culp. 2012. Quantifying watershed-based influences on the Gulf of Maine ecosystem: the Saint John River basin. Pages 219–234 *in* R. L. Stephenson, J. H. Annala, J. A. Runge, and M. Hall-Arber, editors.

Advancing an ecosystem approach in the Gulf of Maine. American Fisheries Society, Symposium 79, Bethesda, Maryland.

Brander, K., L. W. Botsford, L. Ciannelli, M. J. Fogarty, M. R. Heath, B. Planque, L. J. Shannon, and K. Wieland. 2010. Human impacts on marine ecosystems. Pages 41–71 *in* M. Barange, J. G. Field, R. P. Harris, E. E. Hofmann and R. I. Perry, editors. Marine ecosystems and global change. Oxford University Press, Oxford, UK.

Burridge, L. E., and K. Haya. 1995. A Review of Di-n-butylphthalate in the aquatic environment: concerns regarding its use in salmonid aquaculture. Journal of the World Aquaculture Society 26:1–13.

Burt, M. D. B., and P. G. Wells. 2010. Threats to the health of the Bay of Fundy: potential problems posed by pollutants. Bay of Fundy Ecosystem Partnership, BoFEP Technical Report No. 5, Wolfville, Nova Scotia.

Caddy, J. F. 1993. Toward a comparative evaluation of human impacts on fishery ecosystems of enclosed and semi-enclosed seas. Reviews in Fisheries Science 11:57–95.

Chase M. E., S. H. Jones, P. Hennigar, J. Sowles, G. C. H. Harding, K. Freeman, P. G. Wells, C. Krahforst, K. Coombs, R. Crawford, J. Pederson, and D. Taylor. 2001. Gulfwatch: monitoring spatial and temporal patterns of trace metal and organic contaminants in the Gulf of Maine (1991–1997) with the blue mussel, *Mytilus edulis* L. Marine Pollution Bulletin 42:491–505.

Chiarappa, M. J. 2012. Usable pasts of harvested waters: the working role of American fisheries history. Pages 243–259 *in* R. L. Stephenson, J. H. Annala, J. A. Runge, and M. Hall-Arber, editors. Advancing an ecosystem approach in the Gulf of Maine. American Fisheries Society, Symposium 79, Bethesda, Maryland.

Colding, J., C. Folke, and T. Elmqvist. 2003. Social institutions in ecosystem management and biodiversity conservation. Tropical Ecology 44:25–41.

DeLauer, V., S. Ryan, I. Babb, P. Taylor, and P. DiBona. 2012. Linking science to management and policy through strategic communication. Pages 89–101 *in* R. L. Stephenson, J. H. Annala, J. A. Runge, and M. Hall-Arber, editors. Advancing an ecosystem approach in the Gulf of Maine. American Fisheries Society, Symposium 79, Bethesda, Maryland.

Dijkstra, J. A., and L. G. Harris. 2009. Maintenance of diversity altered by a shift in dominant species: implications for species coexistence. Marine Ecology Progress Series 387:71–80.

Dorsey, E. M., and J. Pederson, editors. 1998. Effects of fishing gear on the sea floor of New England. Conservation Law Foundation, Boston.

Douvere, F. 2008. The importance of marine spatial planning in advancing ecosystem-based sea use management. Marine Policy 32:762–771.

Ecosystem Assessment Program. 2009. Ecosystem Assessment Report for the northeast U.S. continental shelf large marine ecosystem. National Oceanic and Atmospheric Administration, National Marine Fisheries Service, Northeast Fish Science Center, Reference Document 09–11, Woods Hole, Massachusetts.

Feeney, R. G., K. J. La Valley, and M. Hall-Arber. 2010. Assessing stakeholder perspectives on the impacts of a decade of collaborative fisheries research in the Gulf of Maine and Georges Bank. Marine and Coastal Fisheries: Dynamics, Management, and Ecosystem Science 2:205–216.

Feurt, C. B. 2012. Collaborative learning strategies to overcome barriers to ecosystem management in coastal watersheds of the Gulf of Maine. Pages 73–88 *in* R. L. Stephenson, J. H. Annala, J. A. Runge, and M. Hall-Arber, editors. Advancing an ecosystem approach in the Gulf of Maine. American Fisheries Society, Symposium 79, Bethesda, Maryland.

Ford, J. S., and R. A. Myers. 2008. A global assessment of salmon aquaculture impacts on wild salmonids. PLoS (Public Library of Science) Biology [online serial] 6(2):e33. DOI: 10.1371/journal.pbio.0060033.

Goode, A. 2006. The plight and outlook for migratory fish in the Gulf of Maine. Journal of Contemporary Water Research and Education 134:23–28.

Hale, S. S. 2012. Spatial patterns of subtidal benthic invertebrates and environmental factors in the nearshore Gulf of Maine. Pages 167–183 *in* R. L. Stephenson, J. H. Annala, J. A. Runge, and M. Hall-Arber, editors. Advancing an ecosystem approach in the Gulf of Maine. American Fisheries Society, Symposium 79, Bethesda, Maryland.

Hare, J. A., and J. Kane. 2012. Zooplankton of the Gulf of Maine—a changing perspective. Pages 115–137 *in* R. L. Stephenson, J. H. Annala, J. A. Runge, and M. Hall-Arber, editors. Advancing an ecosystem approach in the Gulf of Maine. American Fisheries Society, Symposium 79, Bethesda, Maryland.

Hargrave, B. T., editor. 2005. Environmental effects of marine finfish aquaculture. Handbook of Environmental Chemistry, volume 5. Springer, London.

Harris, L. G., M. C. Tyrrell, C. T. Williams, C. G. Sisson, S. Chavanich, and C. M. Chester. 2001. Declining sea urchin recruitment in the Gulf of Maine: is overharvesting to blame? Pages 439–444 *in* M. Barker, editor. Proceedings of the 10th International Echinoderm Conference. A.A. Balkema, Rotterdam, Netherlands.

Harris, L. G., and M. C. Tyrrell. 2001. Changing community states in the Gulf of Maine: synergism between invaders, overfishing and climate change. Biological Invasions 3:9–21.

Haya, K., L. E. Burridge, and B. D. Chang. 2001. Environmental impact of chemical wastes produced by the salmon aquaculture industry. ICES Journal of Marine Science 58:492–496.

Holling, C. S. 2001. Understanding the complexity of economic, ecological, and social systems. Ecosystems 4:390–405.

Holmer, M., D. Wildish, and B. Hargrave. 2005. Organic enrichment from marine finfish aquaculture and effects on sediment biogeochemical processes. Pages 181–206 *in* B. Hargrave, editor. The handbook of environmental chemistry. Volume 5M: the handbook of environmental chemistry. Springer, Heidelberg, Germany.

Konrad , V., S. Ballard, R. Erb, and A. Morin, editors. 1989 The Gulf of Maine: sustaining our common heritage. Maine State Planning Office, Augusta.

Lester, S. E., K. L. McLeod, H. Tallis, M. Ruckelshaus, B. S. Halpern, P. S. Levin, F. P. Chavez, C. Pomeroy, B. J. McCay, C. Costello, S. D. Gaines, A. J. Mace, J. A. Barth, D. L. Fluharty, and J. K. Parrish. 2010. Science in support of ecosystem-based management for the US West Coast and beyond. Biological Conservation 143:576–587.

Lotze, H. K., I. Milewski, and B. Worm. 2004. Two hundred years of ecosystem change in the outer Bay of Fundy. Part II—a history of contaminants: sources and potential effects. Pages 327–333 *in* P. G. Wells, G. R. Daborn, J. A. Percy, J. Harvey, and S. J. Rolston, editors. Health of the Bay of Fundy: assessing key issues. Environment Canada, Atlantic Region, Occasional Report No. 21, Dartmouth, Nova Scotia.

The Massachusetts Executive Office of Energy and Environmental Affairs. 2009. Massachusetts Ocean Management Plan. Available: http://www.mass.gov/eea/ocean-coastal-management/mass-ocean-plan (June 2010).

Munkittrick, K. R., R. A. Curry, J. M. Culp, R. A. Cunjak, D. L MacLatchy, K. A. Kidd, S. E. Dalton, D. J. Baird, and R. Newbury. 2007. The Saint John River. Pages 129–145 *in* C. King, J, Ramkissoon, M. Clüsener-Godt, and Z. Adeel, editors. Water and ecosystems: managing water in diverse ecosystems to ensure human well-being. United Nations University Institute for Water, Environment & Health, Hamilton, Ontario.

Nye, J. 2010. Climate change and its effect on ecosystems, habitats and biota. State of the Gulf of Maine report. Available: www.gulfofmaine.org/state-of-the-gulf/docs/climate-change-and-its-effects-on-ecosystems-habitats-and-biota.pdf (June 2012).

Pikitch, E. K., C. Santora, E. A. Babcock, A. Bakun, R. Bonfil, D. O. Conover, P. Dayton, P. Doukakis, D. Fluharty, B. Heneman, E. D. Houde, J. Link, P. A. Livingston, M. Mangel, M. K. McAllister, J. Pope, K. J. Sainsbury. 2004. Policy forum ecology: ecosystem-based fishery management. Science 305:346–347.

Roberts, C. M. 2007. The unnatural history of the sea. Island Press, Washington, D.C.

Roberts, C. M. 2012. The ocean of life. The fate of man and the sea. Viking, Penguin Group, New York.

Scavia, D., J. C. Field, D. F. Boesch, R. W. Buddemeier, V. Burkett, D. R. Cayan, M. Fogarty, M. A. Harwell, R. W. Howarth, C. Mason, D. J. Reed, T. C. Royer, A. H. Sallenger, and J. G. Titus. 2002. Climate change impacts on U.S. coastal and marine ecosystems. Estuaries 25:149–164.

Smith, V. H., G. D. Tilman, and J. C. Nekola. 1999. Eutrophication: impacts of excess nutrient inputs on freshwater, marine, and terrestrial ecosystems. Environmental Pollution 100:179–196.

Townsend, D. W. 1998. Sources and cycling of nitrogen in the Gulf of Maine. Journal of Marine Systems 16:283–295.

Townsend, D. W., and A. C. Thomas. 2001. Winter-spring transition of phytoplankton chlorophyll and inorganic nutrients on Georges Bank. Deep-Sea Research Part II 48:199–214.

Wallace, G. T., and E. F. Braasch, editors. 1996. Proceedings of the Gulf of Maine ecosystem dynamics: scientific symposium and workshop. Regional Association for Research of the Gulf of Maine, RARGOM Report 97–1, Portland.

Walmsley, D. 2010. Climate change and its effect

on humans. State of the Gulf of Maine report. Available: www.gulfofmaine.org/state-of-the-gulf/docs/climate-change-and-its-effects-on-humans.pdf (June 2012).

Wells, P. G. 2005. Assessing marine ecosystem health: concepts and indicators with reference to the Bay of Fundy and Gulf of Maine, northwest Atlantic. Pages 395–430 *in* S. E. Jørgensen, R. Costanza, and Fu-Liu Xu, editors. Handbook of ecological indicators for assessment of ecosystem health. CRC Press, New York.

Wells, P. G. 2010. Emerging issues—circa 2010. State of the Gulf of Maine report. Available: www.gulfofmaine.org/state-of-the-gulf/docs/emerging-issues.pdf (June 2012).

Wiggin, J., and C. N. K. Mooers, editors. 1992. Proceedings of the Gulf of Maine Scientific Workshop. Urban Harbors Institute, Boston.

American Fisheries Society Symposium 79:243–259, 2012

Usable Pasts of Harvested Waters: The Working Role of American Fisheries History

Michael J. Chiarappa*
Quinnipiac University, Department of History
275 Mount Carmel Avenue, Hamden, Connecticut 06518, USA

Abstract.—The advent of modern fisheries research during the second half of the 19th century was striking in its historical and ethnographic orientation, a precedent set by such pioneering work as George Perkins Marsh's *Man and Nature* and the collective labor of the U.S. Fish Commission and certain state fish commissions that followed its lead. This approach served to provide more than limited context or introductory remarks for scientific studies but, with compelling clarity, took seriously the historical and cultural experiences of fishing communities in an effort to structure wide public discourse on the pressing concerns confronting the use of fisheries resources. Hoping to employ knowledge of fisheries history and occupational culture in the service of publicly engaged, progressive policy and management, these investigations reached audiences not just through government reports, but also through popular periodicals and fisheries exhibitions. Today, the work of environmental and cultural history—in conjunction with their vital interdisciplinary links to oral history, anthropology, geography, field documentation, and museology—is revitalizing this tradition and establishing important patterns in how fisheries issues are communicated and deliberated in society. Similar to earlier periods, the implications of these contemporary initiatives are important for those stakeholders wishing to participate in the public culture that frames current fisheries life.

Introduction

For those who are interested in, or focused on, the anthropogenic or human factors dimensions of fisheries, there is an increasingly important convergence between these concerns and the field of environmental history. Given that fisheries research—whether practiced by scientists, social scientists, or humanities scholars—is so tethered to officially collected historical data, or the unofficially recorded and transmitted environmental work experience of fishermen themselves, such assertions of an unfolding, productive, and publicly applied relationship between environmental history, fisheries history, and the broader realms of fisheries science may come as a surprise. But movement is afoot, gradual yet compellingly steady in its force, to achieve rapprochement between the historical questions that marine ecologists pose of fisheries resources and the work of environmental historians to historicize the motivations, actions, and effects of the various stakeholders who participate in marine and freshwater fisheries. In 2006, Jeffrey Bolster cogently delineated this sentiment by suggesting that past and present deliberations on fisheries stood to gain by balancing the empirical demands of marine ecology with the task of fisheries historians to "create compelling accounts of the changing nature of marine environments in which contradictory human aspirations, values, behaviors, and institutions play central roles" (Bolster 2006).

Bolster's statement provided much needed clarification of environmental history's relationship to fisheries history and came on the heels of a variety of developments that touted the use of history in fisheries research—the "Putting Fishers

* Corresponding author: michael.chiarappa@quinnipiac.edu

Knowledge to Work" conference in 2001 (Haggan et al. 2007), the recognition of Ted Ames's use of fishermen's oral history to better manage fishing grounds (Wilkinson 2006; Ames 2007), and, of course, the international scope of the History of Marine Animal Populations project (HMAP; Starkey et al. 2008).[1] Without minimizing environmental history's capacity to interface with marine ecology and fisheries science, Bolster (2006), nonetheless, emphasized that environmental historians approach past fisheries with an awareness "that long-gone contexts and now invisible contingencies affected the past in such a way that it was qualitatively different." Perhaps, most importantly, from an empirical standpoint that steers a course through the quagmires of overly positivist and determinist interpretation, this sweeping commentary allows us to envision the prospect of synchronizing the environmental and occupational legacies of fisheries history with, in Charles Rosenberg's words, "the way in which we frame and contextualize not only our scientific knowledge, but the social uses of the authority of science" (Rosenberg 1997). If environmental history allows us to better envision the matrix that historically, culturally, and scientifically shaped modern fisheries from the latter 19th century to the present, then it also stands to shed light on how the authority of that mix projected the enterprise into the ethos and temperament of American life. Bolster (2006) states that environmental history offers us the possibility of telling "new stories" of fisheries history, as well as new "storytelling capabilities" to reach these narrative ends. The intertwined fate of environmental history and fisheries history is apparent enough, but for some researchers, particularly public historians, who want to see these stories used in public policy and public education (Figures 1–3), there is a third rail to be considered—the tradition of an actively constructed public life of fisheries history and its civic intentions (Attwood et al. 2008). Without doubt, environmental history's applications beyond traditional history pedagogy put it in a position to offer greater clarity and perspective on the tradition of a public life of fisheries history

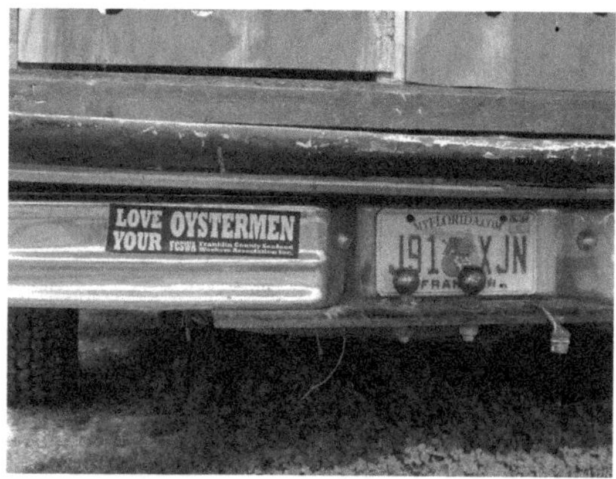

FIGURE 1. Apalachicola, Florida. Bumper stickers such as this—ubiquitous throughout the United States—speak to the unrelenting public discourse on American fisheries. (Photograph by author)

FIGURE 2. Essex, Massachusetts. This clam sculpture is evidence of how fisheries, and certain marine species, assume iconic status in communities most reliant on them. (Photograph by author)

in the United States. This should come as no surprise since the use of fisheries resources are awash in, and of consequence to, so many affairs that are unavoidably public; in the American context, they are so culturally entrenched as a liberating spectre of America's promise that their social reach can hardly be underestimated.

America's commercial fisheries took their place in the country's rapidly expanding economy in the latter 19th century, accompanied by historical and ethnographic narratives that affirmed natural resource use as an enriching environmental touchstone for citizens of all stations. Spe-

[1] Although HMAP's goals have been admirable, a number of historians noted its methodological and interpretive flaws. See Payne 2009.

FIGURE 3. "Fishtown," Leland, Michigan. Although Lake Michigan's fisheries are only a miniscule shadow of their prolific past, Leland, Michigan's desire to preserve its only surviving fishery and its supporting buildings reflects the community's broader engagement with fisheries resources and how the occupation shaped collective identity. (Poster courtesy of the Fishtown Preservation Society)

THE MAKING OF A MUSEUM.

SPENCER F. BAIRD. (FROM A PHOTOGRAPH BY T. W. SMILLIE, SMITHSONIAN INSTITUTION.)

FIGURE 4. Spencer Fullerton Baird. Baird's efforts both within and outside of government were critical in igniting broader societal interest in the fate of U.S. fisheries. (From Ingersoll 1885)

cifically, in a truly modern sense, these narratives lent authoritative power to a usable past that was constantly emerging, and authenticated—in distinctly historicist terms—a public culture whose discourse sought to progressively advance—with broad societal effect—the commercial and scientific ambitions of American fisheries. The U.S. Fish Commission, established under the leadership of Spencer F. Baird (Figure 4) in 1871, assumed the preeminent role in disseminating this information, its widely circulated annual reports seizing on the contextual poignancy of fisheries history and fisheries culture to not only frame discussion of the commission's scientific investigations, but to make the results more immediately understood by the reading public (Allard 1978; Pauly 2000). The classical training of 19th century scientists facilitated such a practical union, as did the commingling of an intellectual climate seeking to make history relevant to present problems, a pattern mutually reinforced by "a group of natural and social scientists...who believed scientific expertise in government could be a major force for the commonweal" (Ross 1991).

During the final quarter of the 19th century, personnel employed by the U.S. Fish Commission, in a manner strikingly similar to that displayed by John Wesley Powell and his staff at the U.S. Geological Survey (Pauly 2000), ably capitalized on the qualitative reach of social scientific inquiry in capturing public attention by presenting their work in such popular serial publications as *Harpers' Monthly*, *Century Magazine*, and *Scribners*. Reporting on American shad *Alosa sapidissima* and alewife *A. pseudoharengus* fisheries in *Century Magazine* in 1880, James Milner, one of the leading field investigators for the U.S. Fish Commission in the 1870s, epitomized this mode of public reporting and credited the ethnographic sweep of his account, and its revelations of the environmental basis for each fishery's iconic status, to a shared authority between himself (the scientist) and his informants (the fishermen): "Get into their confidence, so that talk will flow unrestrainedly, and there are few who would not gain attention while recounting their observations and conclusions in the mysteries of the waters and of fish life" (Milner 1880). Ironically, the promise of such shared authority among the myriad stakeholders whose interests converge in fisheries history, fisheries anthropology, and fisheries science would diminish during the first half of the 20th century, only to find resurgence in the late 20th century under the rising demands of environmentalism and the challenges of implementing ecosystem management regimes.[2]

However, in contrast, throughout the second half of the 19th century, a broad public forum saw the historic and ethnographic documentation of fisheries as an essential baseline for rigorous empirical management. Amid a context of unprecedented change and newfound national ambition, these viewpoints coalesced around the warnings posed by George Perkins Marsh regarding environmental degradation and the anxious anticipation leading up to Frederick Jackson Turner's declaration that America's relationship with nature—its environmental experience—defined its national temperament. Not surprising, this fisheries constituency—lay and professional, populist, academically eclectic—ran its course in tandem with Turner's quest "to fuse nature and history" (Ross 1991), a collective exchange whose diffuse sentiment imbued fisheries with the earmark of natural inheritance or natural patrimony, and was set on retrieving a continuously serviceable past whose social scientific and humanistic bearings might facilitate biologically sustainable fisheries compatible with America's mythic ecological expectations.

Bringing programmatic cohesion to a nationally relevant public discussion on American fisheries posed serious challenges, not the least of which was delivering source material that might provide common ground for a diverse citizenry with widely varied connections to aquatic resources, be they as a consumer, marketing agent, or a fisherman. Not confined by precedent, but mindful of American society's expectations for the use of natural resources, a group of government and nongovernment personnel showed little reticence in bringing fisheries history and fisheries anthropology into the fold of nascent environmental management, and no one was more influential in forging this direction than George Brown Goode. Hired in 1872 to assist Baird with fish and fisheries research being coordinated through the Smithsonian Institution and the U.S Fish Commission, Goode's formidable skills, as both an ichthyologist and historian, equipped him to manage scientific, economic, and cultural information and, perhaps of greater consequence, gave him the intellectual poise to cultivate presentational formats that would enhance the utility of this data for a broad spectrum of Americans (Figure 5; Goode 1888; Allard 1978; Pauly 2000).

In 1878, Goode became assistant director of the Smithsonian's National Museum and, from this post, eventually oversaw the preparation of a project entitled *The Fisheries and the Fishery Industries of the United States* (Figure 6)—a vast, multivolume inventory that scrupulously documented fish, fisheries and their harvest methods, fishery communities, and the occupational culture and history of fisheries, as well as a host of biological, ethnographic, historical, and economic details useful for evaluating the posture of American fisheries in relation to domestic and international markets (Goode 1884–1887; Pauly 2000). Environmental historians tout the value of the Goode reports and do so with reasonable cause; few, if any, government or nongovernment sponsored accounts of resource extractive industries match their exhaustive coverage, their accessible written narrative, and the communicative power of theirlavish pictorial components (Figure 7). However, the civic boldness of this collection, declared by Spencer F. Baird to be "descriptive, historical, and natural history papers...enriched by a sufficient amount of statistical detail to render them as useful as possible for the class of reader and students for whom they were intended," and Goode's ac-

[2] For more on how fisheries research might benefit from the historian's concept of "shared authority," see Frisch 1990.

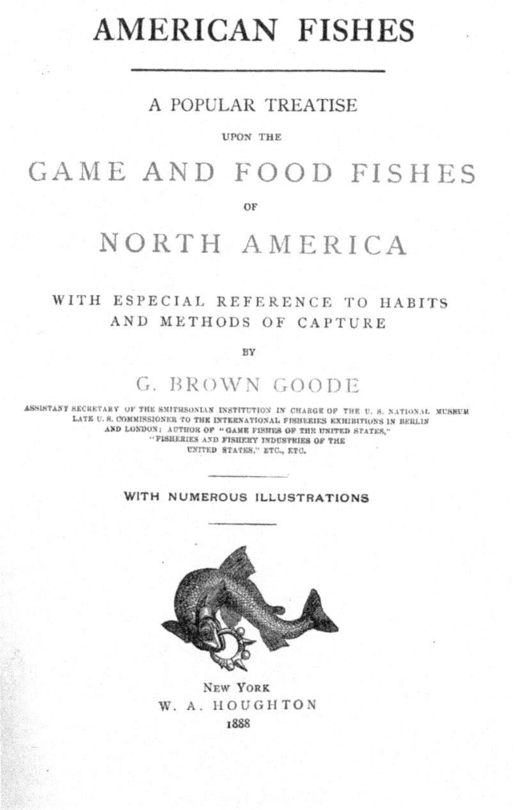

AMERICAN FISHES

A POPULAR TREATISE

UPON THE

GAME AND FOOD FISHES

OF

NORTH AMERICA

WITH ESPECIAL REFERENCE TO HABITS
AND METHODS OF CAPTURE

BY

G. BROWN GOODE

ASSISTANT SECRETARY OF THE SMITHSONIAN INSTITUTION IN CHARGE OF THE U. S. NATIONAL MUSEUM
LATE U. S. COMMISSIONER TO THE INTERNATIONAL FISHERIES EXHIBITIONS IN BERLIN
AND LONDON; AUTHOR OF "GAME FISHES OF THE UNITED STATES,"
"FISHERIES AND FISHERY INDUSTRIES OF THE
UNITED STATES," ETC., ETC.

WITH NUMEROUS ILLUSTRATIONS

NEW YORK
W. A. HOUGHTON
1888

FIGURE 5. George Brown Goode's (1888) *American Fishes*. Spencer Fullerton Baird may have inspired a public culture around U.S. fisheries, but few matched his principal assistant, George Brown Goode, in implementing this vision.

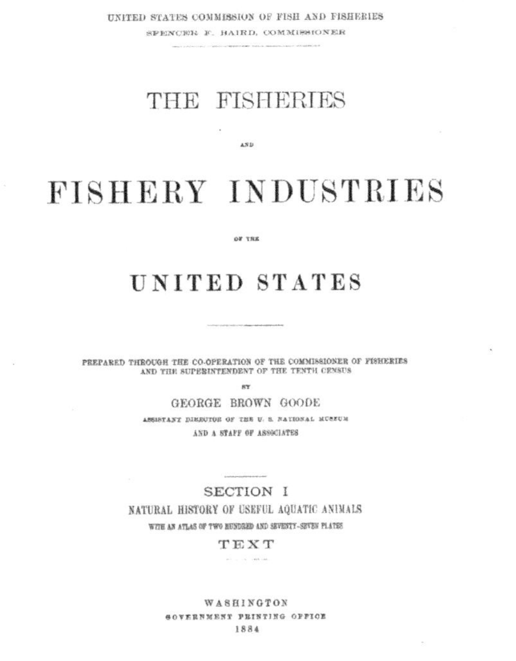

UNITED STATES COMMISSION OF FISH AND FISHERIES
SPENCER F. BAIRD, COMMISSIONER

THE FISHERIES

AND

FISHERY INDUSTRIES

OF THE

UNITED STATES

PREPARED THROUGH THE CO-OPERATION OF THE COMMISSIONER OF FISHERIES
AND THE SUPERINTENDENT OF THE TENTH CENSUS

BY

GEORGE BROWN GOODE

ASSISTANT DIRECTOR OF THE U. S. NATIONAL MUSEUM

AND A STAFF OF ASSOCIATES

SECTION I

NATURAL HISTORY OF USEFUL AQUATIC ANIMALS

WITH AN ATLAS OF TWO HUNDRED AND SEVENTY-SEVEN PLATES

TEXT

WASHINGTON
GOVERNMENT PRINTING OFFICE
1884

FIGURE 6. George Brown Goode et al.'s (1884–1887) *The Fisheries and the Fishery Industries of the United States*. These volumes set a precedent, still active today, which aims to achieve productive dialogue between scientists, the fishing industry, and the broader public.

knowledgment that it was "intended especially for the use of the reading public" (Goode 1884) should also arouse the attention of environmental historians. Quite simply, fisheries history was not just being recorded; it was also being used to frame the occupation's scientific representation in the latest round of American environmental politics. To this end, Goode's *Fisheries and Fishery Industries*—widely distributed and motivated by American society's deeply entrenched optimism in the civically unifying use of natural resources—sought to cultivate a public culture around American fisheries that would advance their efficient use in the broadest national interest, a prospect Philip Pauly (2000) characterizes as "a fund of reliable knowledge that the government and individual citizens could use…designed to provide authoritative reference points for private initiatives and public discussions." This program's historical and cultural narrative helped translate its scientific idealism, a set of aspirations ennobled by Baird's description of "a large share of the most important work…done as volunteer labor by officers of the National Museum and Fish Commission" (Goode 1884). But Goode's coordination and editorial management of *Fisheries and Fishery Industries* was also striking a fine balance and really charting a uniquely American course in pursuing these synthetic presentational goals for fisheries management. His scientific literacy required him, or one of his team, to infuse an environmental critique into the U.S. Fish Commission's message, one that would sharpen the rising stakes of the government's public trust responsibility

THE BANK TRAWL-LINE COD FISHERY.

Dory crew of cod fishermen catching birds for bait. (Sect. v, vol. i, p. 152.)

Drawing by H. W. Elliott and Capt. J. W. Collins.

PLATE 29.

FIGURE 7. Line drawing from Section V of George Brown Goode's (1884–1887) *The Fisheries and Fishery Industries of the United States*. To advance greater fisheries literacy, the Goode reports skillfully used lavish pictorial formats, giving wide audiences an unprecedented view of how fisheries were actually executed.

for managing fisheries resources (Figure 8; Collins 1897). At the same time, his inclinations as an historian—he was a founding member of the American Historical Association—enabled him to see the antecedents of fishery problems but made him sensitive to a populist historical narrative that would play to the purported virtues of the American environmental imagination and use collective memory to enshrine the connections between fishermen, their rootedness in the American experience, and their need to ensure their future survival (Allard 1978; Kammen 1991; Bolster 2006). When his longtime colleague Richard Rathbun summarized the first 20 years of the U.S. Fish Commission's work in *Century Magazine* in 1892, fisheries history was portrayed as reverberating through public life as a substantiation of the agency's scientific agenda, a connection the author emphasized in his conclusion by stating, "Science stands, therefore, between nature and the fisheries as a willing and helpful agent, powerful in its influence to promote the general good" (Rathbun 1892).

Goode's effectiveness in orchestrating the content and format of *The Fisheries and the Fishery Industries of the United States* corresponded with his desire to achieve similar, arguably greater, public outcomes through museum exhibition of the same content contained in these volumes. As one of America's pioneering museologists, Goode came to the post of overseeing the multivolume U.S. Fish Commission compendium fresh from finishing the National Museum's permanent exhibit in 1881, as well as participating in the preparation of the U.S. Fish Commission's exhibits at the Centennial Exposition in 1876 and those mounted for the International Fishery Expositions held at Berlin in 1880 and London in 1883.

BY JOSEPH WILLIAM COLLINS.

TO one familiar with the New England coast for the past forty years or so, nothing is more painfully apparent than the change that has occurred in its deep-sea fisheries. Four or five decades ago nearly every cove or harbor on mainland or isle from Connecticut to eastern Maine was a site for curing fish, or for "fitting out" vessels for the mackerel fishery, or for voyages to the ocean banks in pursuit of cod or halibut.

Harbor rivalled harbor in fleets of sturdy, trim-built, and gayly painted fishing-vessels, and the wealth and consequence of many coast towns were dependent on their piscatorial navies. While modern "sharp-shooters," with their low hulls, long raking masts, and gilded filigree-work, rather ostentatiously elbowed the older types out of the way on Long Island Sound and in the larger Massachusetts ports, many a veteran sea-toiler was still content to sail his round-bowed "jigger" or pinky, and even the "Chebacco-boat" was occasionally in evidence in some of the out-of-the-way coves "down East." Indeed, though these coves could not compete with the larger ports, many of them claimed distinction for what had been accomplished in their restricted limits. Through thrift and adventurous enterprise not a few of them had attained marked success. In unsuspected nooks, lying cozily quiet under a declining summer sun, that threw shadows of wooded heights and rocky points upon the placid water, one came upon little piers, storehouses, and flake-yards, redolent of the odors that characterize the industry to which they were devoted, and it scarcely required further evidence to convey the information that here fares of fish were received, and cured by careful and experienced hands. If the vessels were not there, one instinctively knew that they were away at sea collecting finny treasures, and erelong the eyes of watching women—mothers, wives, sweethearts, and daughters—would be gladdened by returning sails, that swept gracefully into the home port and came to rest at the pier, while the hardy fishers disappeared through many devious paths—winding among bowlders and beneath balsam-scented firs and pines—toward their cottage homes in the near vicinity.

The tragedies of the sea, that occasionally brought mourning and distress to those whose loved ones had gone forth to exact tribute from old ocean's living wealth, were the dark shades of the picture. Nevertheless, these threw into

FIGURE 8. Collins's writing is emblematic of the reach—using journalistic venues such as *Harper's*—of commentators who had practical fishing backgrounds as well as experience working for the U.S. Fish Commission (Collins 1897).

True to Goode's aim of making it an American museum, the exhibit plan was broadly inclusive in its zoological and artifact collections, yet he subjected these items to rigorous classification and analytic presentation in order to better chart humanity's evolving relationship with the natural world (Ingersoll 1885; Allard 1978). These points of emphasis distinguished Goode's interdisciplinary museological theory and its use in the representation of American fisheries. A public culture or public life of American fisheries loomed large in this scheme, one where, in Goode's words (Ingersoll 1885),

> the idea of public education has been predominant...Materials are gathered that they may serve as a basis for scientific thought. Objects that have fulfilled this purpose or have acquired historical significance are treasured up... A higher purpose calls for the administration of these objects in such a manner that masses of people instead of a few should be profited by their existence.

When Ernest Ingersoll, a contributor to *The Fisheries and Fishery Industries of the United States*, reviewed the National Museum for *Century Magazine* in 1885, he characterized this representational stance as the "Museum's high aim"—one Goode saw as having "wider application" and, unlike other museums, not wedded to "a single idea...by design or chance limited in their scope." Ingersoll remarked that the institution's integrated pedagogical presentation of fisheries benefited from the use of materials previously prepared by the U.S. Fish Commission for the Centennial Exposition and the International Fishery Expositions, Goode being involved in each to varying degrees (Ingersoll 1885). By the time the U.S. Fish Commission began preparing its fisheries exhibition for the 1893 World's Columbian Exposition in Chicago, Goode's presentational strategies were firmly entrenched among a number of staff members who claimed more than 10 years of experience in either exhibition design, fish cultural displays, or exposition layout, most notably the work of Joseph W. Collins (Figure 9). These personnel exhibited remarkable skill in collaborating with fisheries communities to achieve these objectives, such as the diorama Gloucester's fisheries prepared for the 1893 Columbian Exposition, a presentation whose precision embodied an optimistically crafted public culture seeking to convey the ecological scope of American fisheries (Bean 1896).

The public life of American fisheries history, and its service to popularly oriented ethnographic investigation, was contingent on the occupation's interplay with government, science, commerce, and the crosscurrents of more widely brokered cultural expression in American society. This was evident in the type of writing that U.S. Fish Commission personnel did for the popular press, both in magazine and book forms, and was affirmed in the commission's final report on its activities at the World's Columbian Exposition, where it acknowledged contributions of illustrations for its exhibit from some of the country's most popular, widely circulated print media—Harper and Bros., Scribner and Company, the Outing Magazine Company, the Cosmopolitan Magazine Company, the Century Magazine, and Frank Leslie's Publishing Monthly Magazine (Bean 1896). The effect of this precedent did not abate in the early 20th century, but instead accommodated the presentational shift of American fisheries into new media and publication formats—particularly those conditioned by a visual culture intoxicated by the possibilities of the photographic image—that were sympathetic to America's environmental imagination, a new stage in the use of appropriately historicized and folklorized fisheries portraits that heeded the nation's sublime regard for natural abundance yet held to an abiding belief in the noble benefits to be gained by the practical exploitation of these resources.

Outing Magazine, typically dedicated to running articles on athletics and outdoor recreation, published a considerable array of historically charged, ethnographic accounts of American commercial fisheries in the early 20th century. Accompanied by photography that portrayed a fishery's land and water settings, readers were guided through the fisherman's working life—along with the publicly beneficial working history that informed his or her place in nature—and were left with the unmistakable impression that such an environmental experience honed wisdom and endurance essential to maintaining America's

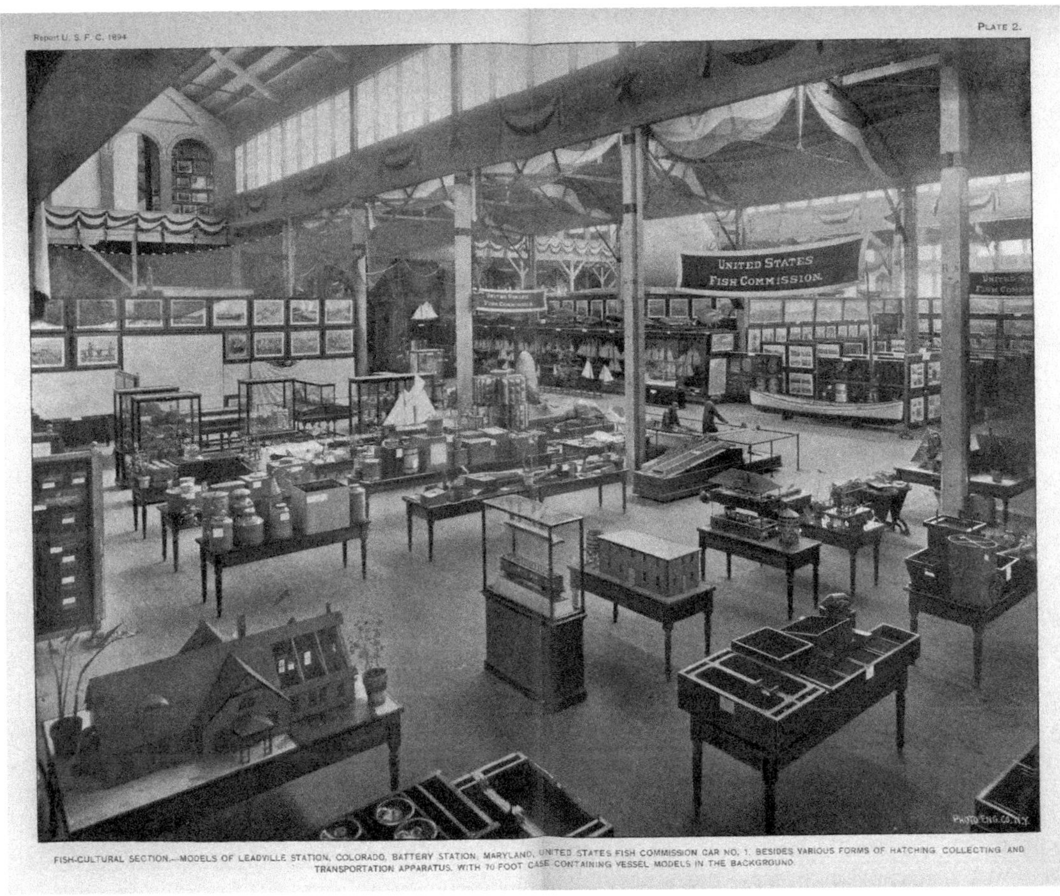

FIGURE 9. U.S. Fish Commission exhibit at the 1893 World's Columbian Exposition at Chicago. Aside from popular writing, no other expressive format was arguably more useful in advancing the public culture of fisheries than exhibitions–some academic in scope, some trade or promotionally oriented, and some a combination of both (Bean 1896).

distinct national character. Such journalism declared oystermen working in the New York metropolitan area as "the best American yeomanry" while, in another vein, castigated the recreational trend that was threatening Gloucester, Massachusetts, into being "ashamed of the smell of gurry, grown into a false estheticism, and generated into cheap politics like many another city" (Figure 10; Slocum 1903; Stansbury 1903). Although clearly containing an anti-modern polemic that justified their inclusion among articles touting the strenuous sporting life and back-to-nature activities, their tone sought to remind readers that America's commercial fishermen, along with a select number of other occupations, were imbued with the historic standing to vicariously engage audiences in calibrating society's environmental benchmarks—articles outlining the rudimentary details of a fishery's ecological dynamics, yet given an ethnographic tint in picture and prose that spoke longingly of the nation's need to reconnect to the natural resources that sustained it.

In a similar vein, but with much greater emphasis on an elaborately stylized photographic format, *National Geographic* aimed to make the occupational, environmental, and scientific contexts of fisheries understandable to the general public in a world where information was becoming increasingly sequestered, and rendered inaccessible by professionalization. This arbiting role, enhanced by the National Geographic Society's close ties to the United States Federal Govern-

"The oystermen of the Kills represent the best American yeomanry."

FIGURE 10. Images such as this (from Stansbury 1903), declaring oystermen working in the New York metropolitan area as "the best American yeomanry," valorized commercial fishing–a useful trope in garnering attention for fisheries and the deliberations that engulfed them.

ment, led it to photographically project American fisheries with a decidedly nationalistic hue and an approving nod to their modern, progressive commercial development (Lutz and Collins 1993). While on assignment for *National Geographic*, well-known fisheries journalist and photographer Frederick William Wallace captured this sentiment when he described "Every Banks fishing schooner is a sort of seafaring democracy" and, being Canadian, made the magazine attentive to the transnational sympathies that linked fishermen across territorial borders (Wallace 1924). As a carefully crafted exercise in image construction, the magazine's black and white and its color photography estheticized these applied goals in depicting any one, or all, of the stakeholders in America's fishing communities and represented a new manifestation of the visual tools Goode saw as so important for the public's appreciation of the interrelated workings of fishermen, scientists, and consumers.

Widely circulated photography allowed a viewing public to virtually insert itself into the daily round of the fisheries environment and advanced the progressive notion that such historicized imagery might promote the awareness needed to ensure the occupation's economic and ecological health. While these progressively informed uses of history portrayed American fisheries as having admirably evolved, particularly in contrast to those in other parts of the world, such a working knowledge of past experience was equally valued for the public lessons to be learned from the occupation's most palpable living history—its demanding, always rugged engagement with nature. That history could be used to plan and build better fisheries, or at least make memory more operable in those public spheres where fisheries deliberations were germane, appealed to journalists and fiction writers with ever-greater urgency over the course of the first half of the 20th century. The work of some authors, such as James Connolly's (Figure 11) fiction and nonfiction accounts of Gloucester's fishermen, canonized the occupation and its marine environment, securing it a niche in America's geographic and environmental imagination that continues to exert a pull on how the community uses its fishery resources today (Connolly 1914, 1927; Schulten 2001). Other authors, hardly immune from indulging in nostalgia or environmental triumphalism, worked with, or were inspired by, progressive efforts by federal and state government to improve America's fisheries. Highly prescriptive, these writings were principally intended to socialize young men into being modern, scientifically informed fishermen who valued progressive fishery policy and government efforts to implement, manage, and enforce it. Such works typically feature a young man in a heroic or noble role, such as Colin Dare in Francis Rolt-Wheeler's (1912) *The Boy with the U.S. Fisheries*, who accompanies the U.S. Bureau of Fisheries in a variety of fishery and fishery science situations with the hope of someday joining the agency (Figure 12). Similarly, Lewis Theiss's (1922) *The Young Wireless Operator with the Oyster Fleet* (Fig-

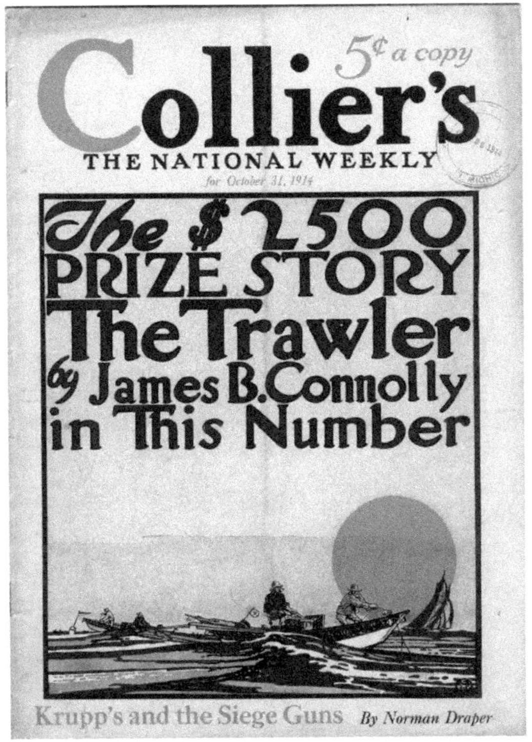

FIGURE 11. Cover of *Collier's Magazine*, volume 54 (1913). Such a cover reveals just how influential James Connolly was in stoking America's environmental imagination, so much so that he was asked to testify before Congressional committees investigating fisheries and formulating policy, even though he was not a fisherman or a scientist.

FIGURE 12. Covers of Francis Rolt-Wheeler's *The Boy with the U.S. Fisheries* (1912) and Lewis Theiss's *The Young Wireless Operator with the Oyster Fleet* (1922). The writings of both Rolt-Wheeler and Theiss were intended to socialize a new cohort of fisheries professionals who would bring a modern, progressive vision to the industry.

ure 12), inspired by Julius and Thurlow Nelson's pioneering work on oyster cultivation, traces Alec Cunningham's odyssey in becoming an exemplar of modern oyster farming methods on New Jersey's Delaware Bay (Nelson 1889, 1921; Woodward and Waller 1932). This writing tradition, particularly in nonfiction forms, remained vibrant throughout the second half of the 20th century and gained new momentum as American society approached, and entered, the new millennium and grappled with an intensifying fisheries crisis (Matthiessen 1986; Junger 1997; Kurlansky 1997; Carey 1999; Woodward 2004; Greenberg 2010).

Along with other resource extractive industries, commercial fishing captured America's fascination with the indelible connections between work and environment, a set of circumstances that drove society to consume the history and culture of the occupation with much the same vigor it expended on consuming the fish themselves. Specifically, American society's appropriation of the fisheries environment into tourism and artistic expression has exerted a lasting effect on the occupation's posture in the very public arena of environmental and cultural politics (Figure 13). For many tourists who visit fishery communities only once in their lives or return to these locales on a yearly basis, these venues often assume the status of pilgrimage sites where, through participation in an experience economy, they develop a sense of cultural ownership in the occupation and the conviction that they understand the ecological issues that consume it—even when their arrival in these communities often threatens or compromises the fisheries on both of these fronts. This does not imply that tourism was a consistently negative influence; in fact, fisheries were adept at collaborating in, or at least peripherally benefiting from, the occupational and environmental exposure provided by the tourism industry.

Tourist activity oriented toward occupational settings had its origins with the dawn of industrialization, and fisheries offered America a ripe context for indulging in their paradoxical love affair with that which was naturally endowed yet rendered economically manageable through ennobled labor and technological innovation. At the same time, tourism provided the apparatus for greater numbers of artists to pursue subject matter that was symbolic of America's defining relationship with the natural environment, particularly in a variety of rich regional settings. Starting with visual artists such as Winslow Homer and Thomas Eakins to writers such as Rudyard Kipling, John Steinbeck, and Ernest Hemingway, there emerged a coexistence between fishermen, artists, and art colonies that endures to the present and, perhaps more than some might want to admit, creates an artistic portal through which the history and culture of fisheries reaches a broad segment of American life in ways that are sometimes compelling and, at other times, are flawed or highly problematic (Curtis 2008; Denenberg et al. 2009). Artistic expression further inserted itself into the historical narrative of the fisheries environment in the form of public art (Figure 14) and in the form of photography's social documentary tradition, two

FIGURE 13. Early 20th century Cape Ann, Massachusetts tourist brochure. Tourist locations, along with the art colonies that frequently accompanied them, played no small part in bringing a variety of fisheries and fishing places to the attention of a national audience. (Author's collection)

representational formats that continue to color and exert considerable force in how public audiences interpret or view fisheries in the context of natural resource policy formation and its effect on

FIGURE 14. Gloucester Fisherman's Memorial, Gloucester, Massachusetts; Waterman's Memorial, Rock Hall, Maryland; Sponge Diver's Memorial, Tarpon Springs, Florida. Few occupations match commercial fishing's reliance on "working" memory—whether as a tool in pursuing particular marine species or in fortifying the daily cultural endurance demanded by the occupation. These monuments and commemorative sites embody this wide ranging sentiment and often reinforce the occupation's strident posture in American environmental politics. (Photographs by author)

the occupation (Taylor 2006).[3]

The environmental history of American fisheries as a form of public discourse or civic engagement builds on these precedents today. The narrative is now increasingly nuanced, couching the environmental history of commercial fisheries amid the complex competing claims and value sets that have been unleashed by the stress that capitalism exerts on fish stocks. Furthermore, environmental history has been an instrumental voice in tempering monolithic approaches to fisheries history and, instead, has brought greater public awareness of the shared and geographically distinct cultural, economic, and ecological factors that drive the local/regional fisheries context and the sometimes contested relations that engulf them (Newell and Ommer 1999; Pauly 2009). Writers, journalists, and publicly oriented environmental historians are reminding us of the fruitful relationship that once existed among scientists, fishermen, and policymakers and how we might build on this historical awareness to create new management schemes that play to stakeholder strengths rather than divide them. Delivery of these perspectives is not only coming from the academic community, but also from more popularly oriented work such as David Dobbs' *The Great Gulf* (2000) and Michael Crocker's *Sharing the Ocean: Stories of Science, Politics, and Ownership from America's Oldest Industry* (2008). Environmental history's openness to interdisciplinary driven research and interpretation has empowered the handling of the technology and material culture used by fisheries and has, over the past 12 years, resulted in museum exhibitions in small venues, such as the 1996 "Making Waves" exhibit at the Cape Ann Museum in Gloucester and the "Lobsta" exhibit at the Penobscot Marine Museum, to much more ambitious projects, such as "When Cod Was King" at New York City's South Street Seaport Museum and the Canadian Museum of Civilization's "Life Lines" exhibit that compellingly displayed and interpreted the range of fisheries in Canada's maritime provinces. If the use of technological and material culture source material has sharpened interpretation of the environmental history of fisheries, the same can be said with equal emphasis of the rigorous methodologies and interpretive frameworks that inform the collection and use of oral history (Bartch et al. 2009). Mystic Seaport Museum's oral history project on Stonington, Connecticut's fishing fleet (Calabretta et al. 2004) is emblematic of such work, as was Western Michigan University's Fish for All oral history project (Figure 15) that actively collaborated with scientists, commercial fishermen, recreational fishermen, and Native American fishers to examine the contested legacy of Great Lakes fishery policy (Chiarappa and Szylvian 2003). More recently, much stands to be gained from the unprecedented collection of oral history assembled under the National Oceanic and Atmospheric Administration's Voices from the Fisheries Project.[4] Each of these projects should cause us to take pause when we think about the vast body of oral testimony that was collected by the U.S.–Canadian International Joint Commission in the early 1890s and, from the American side, largely sits unused on the shelves at the National Archives in Washington, D.C. But researchers can also take heart that oral history research does not just collect data, but also has the added dividend of being a process that fosters discourse and exchange among those vested in the fisheries. Upon reflection, it is the public life or public culture that coalesces around the convergence of fisheries history and environmental history that offers all fisheries stakeholders the shared perspective to frame research and management questions with the broadest, most ecologically constructive outcomes in mind. This applied use of the past creates a "working" history, a tool with daily and long-term implications that has been playing out for centuries in the lives of those who claim a connection to fisheries. The latest Gulf of Maine symposium is testament to this process, an event granting historical authority to diverse fishery perspectives that hope to advance ecosystem management in ambitious new directions. But if the 2009 Gulf of Maine Symposium represents yet another evolution of a "working" history to guide fisheries management and policy, it will do so by being cognizant of its collective place among the vast historical perspectives that ripple through

For the latest historical treatment of the intersection of fisheries, environment, and tourism, see Chiang 2008.

For more on NOAA's Voices from the Fisheries Project, see www.st.nmfs.noaa.gov/voicesfromthefisheries.

FIGURE 18. Pamphlet used to announce the Fish for All exhibit. (Author's collection)

the seemingly limitless network of fishermen, scientists, and consumers. Indeed, ecosystem-based management—epitomized in its very name—represents a synthetic approach to handling marine resources that desperately needs not only a revised relationship between government, scientists, and fishermen, but, just as important, broader engagement with society at large. Ecosystem-based management is aiming, with good reason, to reimagine a new ethos for fisheries, one where environmental accountability is recognized and shared among an unprecedented number of stakeholders. But are some of the elements of this vision that new? Fisheries history shows us that the occupation has a tradition of fostering a public culture—an inclusive civic rubric. While its goals were sometimes misplaced and overly romantic—empowering unsustainable practices and overly aggressive market strategies—at its best, it did create source materials, forums, and a measure of consciousness necessary for debating the occupation's pressing issues. Heeding and scrutinizing this cultural and scientific inheritance, ecosystem-based management can use it to help frame fisheries in a new public culture that lives within nature's limitations.

In sum, the advent of modern American fisheries research during the second half of the 19th century was striking in its historical and ethnographic orientation, a precedent set by such pioneering work as George Perkins Marsh's *Man and Nature* and the collective labor of the U.S. Fish Commission and certain state fish commissions that followed its lead. This approach served to provide more than limited context or introductory remarks for scientific studies but, through various presentational formats, took seriously the historical and cultural experience of fishing communities in an effort to structure wide public discourse on the pressing concerns confronting the use of fisheries resources. Hoping to employ knowledge of fisheries history and occupational culture in the service of publicly engaged progressive policy and management, these investigations reached audiences not just through government reports, but also through popular periodicals and fisheries exhibitions. This precedent held sway throughout the 20th century, and while its force abated at times, its consistent presence has made fisheries one of American society's most historically charged, and fiercely debated, natural resource-using occupations. Today, the work of environmental and cultural history—in conjunction with their vital interdisciplinary links to oral history, anthropology, geography, field documentation, and museology—is assessing the strengths and weaknesses of this tradition and building on these precedents, establishing new patterns in how fisheries issues are communicated and deliberated in society. Similar to earlier periods, the implications of these contemporary initiatives are important for those stakeholders wishing to more ably use historical insights when participating in the public culture that frames current fisheries life.

References

Allard, D. C. 1978. Spencer Fullerton Baird and the U.S. Fish Commission. Arno Press, New York.

Ames, T. 2007. Putting fishers' knowledge to work: reconstructing the Gulf of Maine cod spawning grounds on the basis of local ecological knowledge. Pages 353–363 *in* Fishers' knowledge in fisheries science and management. UNESCO Publishing, Paris.

Attwood, B., D. Chakrabarty, and C. Lomnitz. 2008. The public life of history. Public Culture 20:1–4.

Bartch, J., S. Abbott-Jamieson, and J. Whitmore. 2009. Voices from the fisheries handbook: preserving local fisheries knowledge, linking generations, and improving environmental literacy. U.S. Department of Commerce, National Oceanic and Atmospheric Administration, National Marine Fisheries Service, Silver Spring, Maryland.

Bean, T. H. 1896. Report of the representative of the United States Fish Commission at the World's Columbian Exposition. Pages 177–196 *in* Report of the commissioner for the year ending June 30, 1894: United States Commission of Fish and Fisheries. Government Printing Office, Washington, D.C.

Bolster, W. J. 2006. Opportunities in marine environmental history. Environmental History 11:567–597.

Calabretta, F., G. S. Gordinier, and J. O. Jensen. 2004. Fishing out of Stonington: voices of the fishing families of Stonington, Connecticut. Mystic Seaport Museum, Mystic, Connecticut.

Carey, R. A. 1999. Against the tide: the fate of the New England fisherman. Houghton Mifflin Company, New York.

Chiang, C. Y. 2008. Shaping the shoreline: fisheries and tourism on the Monterey coast. University of Washington Press, Seattle.

Chiarappa, M. J., and K. M. Szylvian. 2003. Fish for all: an oral history of multiple claims and divided sentiment on Lake Michigan. Michigan State University Press, East Lansing.

Collins, J. W. 1897. Decadence of the New England deep sea fisheries. Harper's New Monthly Magazine 94:608–625.

Connolly, J. B. 1914. The trawler. Collier's 54:5–8 31:33–34.

Connolly, J. B. 1927. The book of the Gloucester fishermen. The John Day Company, New York City.

Curtis, J. A. 2008. Rocky Neck Art Colony, 1850–1950, Gloucester, Massachusetts. Rocky Neck Art Colony, Gloucester, Massachusetts.

Denenberg, T., A.K. Lansing, and S. Danly. 2009. Call of the coast: art colonies of New England. Yale University Press, New Haven, Connecticut.

Frisch, M. 1990. A shared authority: essays on the

craft and meaning of oral and public history. State University of New York Press, Albany.

Goode, G. B., editor. 1884–1887. The fisheries and fishery industries of the United States, 7 volumes. Government Printing Office, Washington, D.C.

Goode, G. B. 1888. American fishes: a popular treatise upon the game and food fishes of North America. W. A. Houghton, New York.

Greenberg, P. 2010. Four fish: the future of the last wild food. The Penguin Press, New York.

Haggan, N., B. Neis, and I. G. Baird, editors. 2007. Fishers' knowledge in fisheries science and management. UNESCO Publishing, Paris.

Ingersoll, E. 1885. The making of a museum. The Century 34:354–369.

Junger, S. 1997. The perfect storm: a true story of men against the sea. W.W. Norton and Company, New York.

Kammen, M. 1991. Mystic chords of memory: the transformation of tradition in American Culture. Vintage Books, New York.

Kurlansky, M. 1997. Cod: a biography of the fish that changed the world. Penguin, New York.

Lutz, C. A., and J. L. Collins. 1993. Reading *National Geographic*. The University of Chicago Press, Chicago.

Marsh, G. P. 1965. Man and nature: or, physical geography as modified by human action. Harvard University Press, Cambridge, Massachusetts.

Matthiessen, P. 1986. Men's lives: the surfmen and baymen of the South Fork. Random House, New York.

Milner, J. W. 1880. The shad and the alewife. Harper's New Monthly Magazine 60:845–856.

Nelson, J. 1889. Report of the biologist: oyster interests of New Jersey. Pages 163–201 *in* Ninth Annual Report of the New Jersey State Agricultural Experiment Station and the First Annual Report of the New Jersey Agricultural College Experiment Station for the year 1888. The John L. Murphy Publishing Company, Trenton, New Jersey.

Nelson, T. C. 1921. Aids to successful oyster culture. New Jersey Agricultural Experiment Stations, Bulletin 351, New Brunswick.

Newell, D., and R. E. Ommer, editors. 1999. Fishing places, fishing people: traditions and issues in Canadian small-scale fisheries. University of Toronto Press, Toronto.

Pauly, D. 2009. Aquacalypse now: the end of fish. The New Republic 240:24–27.

Pauly, P. J. 2000. Biologists and the promise of American life: from Meriwether Lewis to Alfred Kinsey. Princeton University Press, Princeton, New Jersey.

Payne, B. J. 2009. Review of Oceans past: management insights from the history of marine animal populations. International Journal of Maritime History 20:403–405.

Rathbun, R. 1892. The United States Fish Commission: some of its work. The Century Illustrated Monthly Magazine 43:679–697.

Rolt-Wheeler, F. 1912. The boy with the U.S. fisheries. Lothrop, Lee, and Shepard, Boston.

Ross, D. 1991. The origins of American social science. Cambridge University Press, Cambridge, UK.

Rosenberg, C. E. 1997. No other gods: on science and American social thought. The Johns Hopkins University Press, Baltimore, Maryland.

Schulten, S. 2001. The geographical imagination in America, 1880–1950. The University of Chicago Press, Chicago.

Slocum, V. J. 1903. The fishermen of Gloucester. Outing Magazine 42:587–597.

Stansbury, C. F. 1903. An oyster of the great kills. Outing Magazine 42:19–31.

Starkey, D. J., P. Holm, and M. Barnard, editors. 2008. Oceans past: management insights from the history of marine animal populations. Earthscan, London and Sterling, Virginia.

Taylor, M. B. 2006. A camera on the banks: Frederick William Wallace and the fishermen of Nova Scotia. Goose Lane Editions, Fredericton, New Brunswick.

Theiss, L. E. 1922. The young wireless operator with the oyster fleet. W.A. Wilde Company, Chicago and Boston.

Wallace, F. W. 1924. Life on the Grand Banks. Pages 225–238 *in* The book of fishes: game fishes, food fishes, shellfish and curious citizens of American ocean shores, lakes, and rivers. The National Geographic Society, Washington, D.C.

Wilkinson, A. 2006. The lobsterman: how Ted Ames turned oral history into science. The New Yorker (July 31):56–65.

Woodward, C. 2004. The lobster coast: rebels, rusticators, and the struggle for a forgotten frontier. Viking, New York.

Woodward, C. R., and I. N. Waller. 1932. New Jersey's Agricultural Experiment Station, 1880–1930. New Jersey Agricultural Experiment Station, New Brunswick, New Jersey.

American Fisheries Society Symposium 79:261–272, 2012

Advances in Understanding Ecosystem Structure and Function in the Gulf of Maine

Michael J. Fogarty*
Northeast Fisheries Science Center, National Marine Fisheries Service
Woods Hole, Massachusetts 02543, USA

David W. Townsend
School of Marine Sciences, University of Maine
Orono, Maine 04469, USA

Emily Klein
Institute for the Study of Earth, Oceans, and Space
University of New Hampshire, Durham, New Hampshire 03824, USA

Abstract.—Here we summarize presentations given at the theme session "Structure and Function of the Gulf of Maine System" of the 2009 Gulf of Maine Symposium—Advancing Ecosystem Research for the Future of the Gulf, covering a broad spectrum of multidisciplinary research underway in one of the world's most intensively studied marine systems. Our objective was to attempt a synthesis of the current ecological and oceanographic understanding of the Gulf of Maine and, in particular, to document progress in these areas since the 1996 Gulf of Maine Ecosystem Dynamics Symposium more than a decade earlier. Presentations at the session covered issues ranging from habitat structure and function, biodiversity, population structure, trophic ecology, the intersection of the biological, chemical and physical oceanography of the region, and the dynamics of economically important species. Important strides in characterizing the broader dimensions of biodiversity in the region, the establishment of new sampling programs and the availability of new sensor arrays, and the renewed emphasis synthesis and integration to meet the emerging needs for ecosystem-based management in the gulf have all contributed to a deepened appreciation of its dynamical structure. The critical importance of the ecosystem goods and services provided by the gulf, and the factors affecting the sustainable delivery of these services, was clearly demonstrated in the course of the session. The papers presented at the session made it clear how far we have come and how far we need to go to ensure the sustainable delivery of these services into the future.

Introduction

An understanding of ecosystem structure and function requires identification and enumeration of the species comprising the system, as well as information on the interrelationships among these species. We further require documentation of patterns of energy flow and utilization, spatial characteristics and interchange among system components, and the role of external stressors, including climate and fishing in system dynamics. The theme session "Structure and Function of the Gulf of Maine System" at the 2009 Gulf of Maine Symposium—Advancing Ecosystem Research for the Future of the Gulf provided new and important insights into these issues. Here, ecosystem structure refers to fundamental organi-

* Corresponding author: michael.fogarty@noaa.gov

zational elements of the system, including the species present and their attributes (e.g., abundance, distribution, and ecological roles), and aspects of the physical environment (nutrients and their distribution/availability, environmental characteristics, etc.) that affect the biological components. Factors affecting ecosystem function include the interrelationships among species, related considerations of the transfer of energy among system components, nutrient cycling, and stabilizing mechanisms. The interwoven issues of biological diversity, interspecific interactions, and functional redundancy of species or species groups can hold important implications for the overall stability and functioning of the system and its likely response to natural and anthropogenic forcing factors. A recurrent theme of dramatic temporal and spatial change on a range of scales within the gulf emerged in presentations throughout the session. This theme was neatly foreshadowed in the keynote addresses by K. Frank (Fisheries and Oceans Canada, Bedford Institute of Oceanography) and J. Hare (National Marine Fisheries Service, Northeast Fisheries Science Center) that opened our session. As shown by Hare, the Gulf of Maine (GoM) has undergone dramatic alteration in fundamental structural aspects that can be tracked on multidecadal to centennial time scales (see also Hare and Kane 2012, this volume). Change on seasonal to interannual time scales is no less evident, and emerging issues, such as the possibility of important change in phenology, deserve urgent attention. The prospect of future climate change in the gulf highlights the need to understand its current status and past changes as related to anthropogenic and natural forcing factors. In complex systems, surprise is to be expected, and the possibility of rapid change to alternate stable states must be anticipated, as beautifully illustrated by Frank at the symposium.

This theme session covered a broad spectrum of topics, including advances in understanding of (1) patterns of biodiversity (largely focused on species richness); (2) nutrient dynamics, biogeochemical cycling, and biophysical coupling; (3) plankton community dynamics; and (4) upper trophic level and fishery resource dynamics. Our objective was to document progress in these areas since the 1996 Gulf of Maine Ecosystem Dynamics Symposium. Synthesizing recent research on the structure and function of the ecosystem has assumed greater importance, as the impetus to move toward ecosystem-based management (EBM) has gained momentum at both the national and international levels.

An important dimension to the progress since the last symposium, and very much evident in this theme session, is the development of enhanced sampling and analytical tools and the implementation of new research and monitoring programs in the GoM. Advances in genomics have made possible the identification of previously underrepresented microbial components. The now routine use of satellite observations for estimation of chlorophyll concentration on fine spatial and temporal scales has revolutionized our ability to document critical ecosystem processes related to bloom dynamics and overall levels of productivity. Advanced in situ sampling tools, ranging from gliders to coastal observatories, have provided important adjuncts to traditional sampling devices in the gulf. Isotopic signatures have been examined in investigations ranging from identification of GoM water mass characteristics to diet composition and trophodynamics. The establishment of new monitoring programs, such as the Gulf of Maine North Atlantic Time Series (1998–present), since the last symposium, has allowed fine-scale resolution of nutrients, hydrography, chlorophyll concentration, and primary production along a transect in the gulf. Finally, advances in high-end computing resources have opened important avenues for the development of coupled physical-biological models with data assimilation capabilities.

In the following, we summarize findings related to each of the four major theme areas identified above and place them in the context of recently published sources of information for this system. We then turn our attention to the implications for EBM in the region.

Patterns of Biodiversity

Since the last GoM symposium, our understanding of patterns of species richness, and its distribution in space and time, has increased substantially with the implementation of a major research program established under the auspices of the Census of Marine Life (CoML; Incze et al., in press). The Gulf of Maine is an extensively studied system with a long history of ecological research conducted by an impressive concentration of research facilities distributed along its coast. As a result, the species composition of mid- and upper-trophic levels, in particular, is well known, and the ecological roles of many of these species have been intensively studied. What is strikingly evident in the years since the last GoM symposium is the increase in knowledge of other components of the system. As noted by L. Incze (University of Southern Maine) and colleagues at the symposium, as a result of the CoML initiative, more than 50,000 new viral and bacterial operational taxonomic units have now been recognized (W. Li, Fisheries and Oceans Canada, Bedford Institute of Oceanography, unpublished data), which represents 20% of the maximum global estimate of bacterioplankton diversity. In addition, a Gulf of Maine Register of Marine Species (GoMRMS) has now been created by CoML, with a provisional total of more than 4,000 species represented. Important new insights have also been gleaned from studies conducted on a broad spectrum of spatial and temporal scales. These include a new CoML Discovery Corridor initiative, encompassing a broad swath from the intertidal through deep basins in the gulf to the edge of the Continental Shelf and beyond (Figure 1), as described at the symposium by P. Lawton (Fisheries and Oceans Canada) and colleagues.

Cobscook Bay, Maine (Figure 1) has long been recognized as a macroinvertebrate biodiversity hotspot (e.g., Larsen 2004). In our session, T. Trott (Suffolk University) confirmed that macroinvertebrate species richness at this site is indeed higher than any other sampled site in the GoM and is comparable to that of estimates for the highest biodiversity sites globally available for comparison from the Arctic to subtropical systems.

Increased sampling continues to augment the list of macroinvertebrates known to inhabit the GoM. A. Holmes and G. Pohle (Huntsman Marine Science Center) noted 38 species, previously unreported in the gulf, as part of the Discovery Corridor initiative, in samples collected in Jordan basin (Figure 1). S. Hale (Environmental Protection Agency) provided new insights into nearshore biogeographical patterns of macroinvertebrates in the gulf at the meeting (see Hale 2012, this volume). He reported faunal breaks within the Acadian province in relation to environmental factors, dominated by temperature patterns.

In an earlier study to examine changes in benthic biodiversity in the gulf, Link (2004) used fish stomachs as samplers of benthic communities for the northeast shelf to look at relative abundance and distribution of major benthic macrofauna, compared to surveys from earlier in the 1900s. Link found relatively stable levels of biodiversity of benthic fauna found in predator stomachs, but with declines evident in caprellids, cumaceans, and pagurids.

For zooplankton communities, sustained monitoring in the GoM, based on the Continuous Plankton Recorder (CPR) program by the National Marine Fisheries Service, has shown variation in decadal-scale species composition, as reported by Hare in his keynote address at the session. The CPR transect in the GoM runs from Boston, Massachusetts to Yarmouth, Nova Scotia (Figure 1). N. Record and A. Pershing (University of Maine) further described east–west gradients in plankton species diversity (with highest diversity in the west along the CPR Boston–Yarmouth transect: see Figure 1) and an increase in biodiversity, peaking in the 1990s.

Changes in the diversity of fish species on decadal to centennial scales have been documented in the Gulf of Maine. It has been possible to examine evidence of temporal changes in biodiversity of fish through a careful comparison of samples collected in two research programs separated by a century, as described by J. Cournane and S. Claesson (University of New Hampshire) at the session. Higher levels of fish biodiversity were found in the most recent sampling period (using the National Marine Fisheries Service trawl survey database), relative to an earlier survey conducted by the U.S. Fish Commission (principally using beam trawls). However, the biodiversity in combined fish and invertebrate catches was higher in the earlier period (for further information, see Cournane 2010). The sampling efficiencies of the sampling gears for benthic and fish species differed markedly for the two periods, explaining a substantial part of the observed differences, although the effects of fishing and environmental change also appear to be important. Further work is required to disentangle these important sources of variation.

The History of Marine Animal Populations (HMAP) project of CoML has also provided cru-

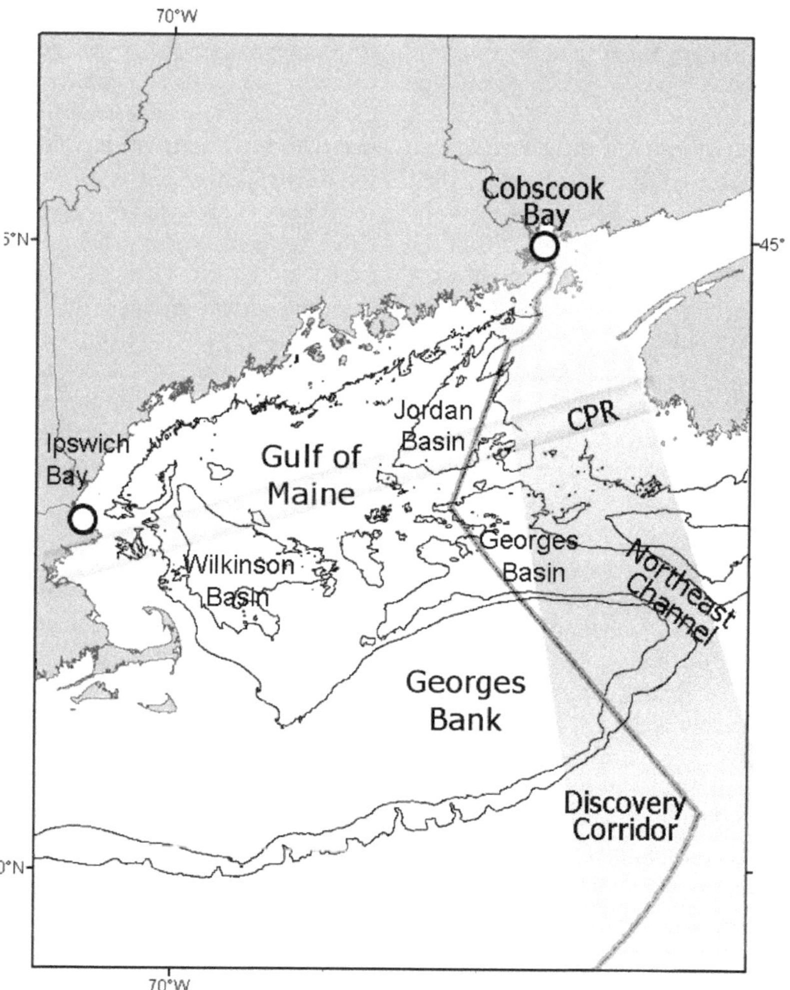

FIGURE 1. Map of the Gulf of Maine with major topographic features identified. The locations for the Continuous Plankton Recorder (CPR) transect (Boston, Massachusetts to Yarmouth, Nova Scotia) and the Census of Marine Life Discovery Corridor are indicated. Location of coastal areas (Ispwich Bay and Cobscook Bay) mentioned in the text are also shown.

cial insights into changes in system structure in the GoM on centennial time scales (e.g., Jackson et al. 2001; Bourque et al. 2008). This program has demonstrated an overall loss in complexity, biodiversity, and abundance of larger fish. In particular, coastal GoM systems, such as kelp forests (Jackson et al. 2001; Steneck et al. 2004; Bourque et al. 2008) and estuaries (Leavenworth 2008), have seen dramatic losses in species abundance and ecosystem complexity on centennial time scales. More recently (over the past four decades), however, a general increase in nekton species richness has been found for the GoM and Georges Bank (Link et al. 2002), reflecting, in part, a recent redistribution of more southerly species into the region as water temperatures have increased, particularly during the past decade

Nutrient Dynamics, Biogeochemical Cycling, and Biophysical Coupling

Our understanding of the general oceanography of the GoM and Georges Bank, as well as the coastal and estuarine systems in the region, has without doubt greatly advanced since the last GoM workshop in 1996, particularly with respect to biophysical processes and sources, and cycling of carbon and nutrients. The general features of the system, including specific discussions of these issues, have been reviewed by Townsend et al. (2006). That review covers many of the advances made since 1996 as part of several major research programs, both on Georges Bank (the U.S. GLOBEC program) and in the greater GoM, in relation to red tides (National Oceanic and Atmospheric Organization's ECOHAB [Ecology and Oceanography of Harmful Algal Blooms] program). There is, however, an additional wealth of information available not included in that review. For example, the review does not include the Gulf of Maine Ocean Observing System (GoMOOS), which only became operational in the spring of 2001, as well as results of numerous other smaller research projects that have been completed and published. Special issues of the journals *Deep-Sea Research Part II* and *Journal of Geophysical Research* have been produced that detail much—but not all—of the research results from the U.S. GLOBEC Program (Wiebe et al. 2001, 2006; Beardsley et al. 2003), and the GoM ECOHAB research program (Anderson et al. 2005). Indeed, those volumes alone contain more than 1,700 pages of articles that have significantly advanced our level of understanding since 1996. Furthermore, the online availability of ocean mooring data, satellite imagery, and physical circulation models (GoMOOS and University of Maine) makes any synopsis of where we currently stand, with respect to our scientific advancement, a rapidly moving target.

While constituting a large and complex oceanographic system, the gulf operates almost in oceanographic isolation from the open northwest Atlantic Ocean, in that it is a semi-enclosed shelf sea. It has been known since Bigelow (1927) that at depths greater than 100 m, the exchange of waters between the gulf and the North Atlantic is limited, with much of exchange confined to the deep (>220 m) Northeast Channel (Figure 1) that separates Georges Bank from Browns Bank and the Scotian Shelf. However, more recent studies reported at this workshop are beginning to show that this exchange may be changing in overall character, driven by the indirect effects of global warming and increased melting in the Arctic (Townsend et al. 2010). It now appears that shelf waters from Nova Scotia, with origins in the Labrador Sea, are becoming more important to the GoM's water properties, including its dissolved inorganic nutrient loads, than the deep influxes through the Northeast Channel.

The gulf's physical characteristics, with its deep basins and limited deepwater exchanges with the open Atlantic, are coupled with other important features and processes that act together to control the general oceanography of the gulf, including nutrient fluxes and biological productivity. These features and processes include vertical mixing by tides; the seasonal cycle of heating and cooling, which leads to winter convection and vertical stratification in summer; pressure gradients from density contrasts set up by deepwater inflows and lower salinity waters; and influxes of the cold, but fresher waters associated with Scotian Shelf water. The tides in the GoM are among the highest in the world and consequently generate swift tidal currents. Tidal ranges decrease from northeast to southwest, and the resulting differences in intensity of tidal mixing exert a strong influence on the spatial pattern of hydrographic structure in the gulf, nutrient delivery to the euphotic zone, benthic–pelagic coupling, and, ultimately, the total biological productivity.

The mean circulation in the GoM–Georges Bank region is generally cyclonic, driven by density contrasts between slope waters residing in the three offshore basins and fresher waters along the coast that are fed principally by discharges from the St. John, Penobscot, Kennebec/Androscoggin, and Merrimac rivers. However, river discharges account for only about half of the freshwater budget for the GoM; the remaining half enters the gulf as a surface flow of relatively cold, low salinity Scotian Shelf waters. The resulting surface circulation in the gulf is thus dominated by a buoyancy-driven coastal current system that flows counterclockwise around its edges. This system has been argued to be important to the overall nutrient budget and biological oceanography of the GoM. J. Churchill (Woods Hole Oceanographic Institution) and colleagues demonstrated the importance of wind-driven processes on recruitment of cod originating in Ipswich Bay in the western gulf. Downwelling, in particular, was associated with high recruitment success.

Levels of primary production in the GoM's offshore waters, the least productive areas in the GoM, are reported to average about 270 gC/m^2/year, based on ^{14}C bottle incubations (O'Reilly and Busch 1984; O'Reilly et al. 1987). More recent estimates are available based on satellite-derived sources. Estimates in phytoplankton species composition, based on CPR samples (Hare, personal communication), document a shift in dominance from diatoms to dinoflagellates, notably in the past decade. In light of apparent changes in the nutrient dynamics reported by D. Townsend (University of Maine), and the fundamental dependence on bloom productivity of particle fluxes to deep waters and the benthos, as reported at this symposium by C. Pilskaln (University of Massachusetts-Dartmouth), primary production processes deserve careful reconsideration for the gulf. Satellite-measured variability in the timing and intensity of the spring phytoplankton bloom, as reported at the symposium by A. Thomas (University of Maine) and colleagues, would also benefit from more detailed studies of phytoplankton productivity in relation to remotely sensed chlorophyll biomass estimates. Total phytoplankton productivity is currently being measured along commercial ferry transects across the offshore gulf, as reported by B. Balch (Bigelow Institute for Ocean Sciences) in this symposium. Those data, as well as yet to be proposed or initiated inshore productivity measurements, will be especially important to modeling efforts to assess growing concerns of artificial nutrient enrichment and eutrophication, as well as dynamics of red tides in the gulf (Anderson et al. 2008) and on Georges Bank, as reported at this meeting by D. Anderson (Woods Hole Oceanographic Institution) and colleagues.

Biological-physical modeling efforts have advanced rapidly in the years following the 1996 workshop, stimulated by the large research programs just mentioned and facilitated by dramatic increases in computing capabilities. Online circulation models are available at the University of Maine (http://rocky.umeoce.maine.edu/Gulf-of-Maine/GoM.htm), Dartmouth College (www-nml.dartmouth.edu/circmods/GoM.html), and the University of Massachusetts, Dartmouth (http://fvcom.smast.umassd.edu/research_projects/GB/index.html). These models have been coupled with biological models of plankton dynamics at the University of Maine (http://rocky.umeoce.maine.edu/Gulf-of-Maine/GoM.htm) and Woods Hole Oceanographic Institution (www.whoi.edu/hpb/viewPage.do?id = 1200). Models have also been applied to many specific research questions, such as the actions of internal waves on phytoplankton productivity, as presented at this symposium by Z. Lai (University of Massachusetts-Dartmouth) and colleagues, and the variability in timing and intensity of the spring phytoplankton bloom by R. Ji (Woods Hole Oceanographic Institution).

Plankton Community Dynamics

Research on plankton communities in the GoM has a long and storied history, beginning with the pioneering work of Henry Bigelow in the early decades of the past century. Bigelow's (1926) seminal study of the GoM provided an important impetus to the establishment of the long-term monitoring programs noted above. His identification of the copepod *Calanus finmarchicus* as a keystone species at the nexus of the food web in the gulf remains a vital research area. At the symposium, R. Jones and J. Runge (University of Maine) described a 5-year monitoring program, with high temporal resolution, that documented seasonal patterns in zooplankton and ichthyoplankton communities in the western GoM. This program highlighted the dominant role of *C. finmarchicus* in the system and changes in the abundance of this species. F. Maps (University of Maine) and colleagues described an individual-based model of *C. finmarchicus* in the GoM that has been used to explore the interplay of environmental factors and food reserves in determining the timing of diapause. This species is at the southern extent of its range in the GoM, and if temperatures increase under climate change, particularly in the deepwater layer where diapause occurs, the overwintering success of this species could be adversely affected. The loss or decline of a keystone species can have important direct and indirect effects on ecosystem structure by affecting energy pathways.

Our understanding of the role and relative importance of gelatinous zooplankton in the GoM is still in an early stage of development. Large-scale increases in these taxa have been noted in many parts of the world oceans. These increases have been associated with anthropogenic effects related to overfishing and climate change. Link and Ford (2006) found a major increase in gelatinous ctenophores in spiny dogfish *Squalus acanthias* stomachs across the GoM from 1981 to 2000. They suggested possible consequences, including increased competition with and predation on other species, decreased fish recruitment, and increases in ctenophore predators. Altered food webs, as a result of changes in abundance of gelatinous zooplankton, have been noted in other areas of the world ocean (Mills 2001).

Connections between plankton communities and climate and physical processes have now been established. The GoM, and the greater northeast shelf ecosystem, is heavily affected by remote environmental forcing (Greene and Pershing 2007). In terms of plankton dynamics, increases in the North Atlantic oscillation and northward movement of the Gulf Stream and decreases in Labrador Slope water and salinity may result in decadal scale alterations in zooplankton community composition (Ecosystem Assessment Program 2009). Greene and Pershing (2007) indicated that decreases in salinity in the 1990s enhanced water column stratification, which led to increases in productivity and phytoplankton biomass. In addition, a reorganization of zooplankton communities, with an increase in smaller species, were also noted during this time (Frank et al. 2005; Greene and Pershing 2007).

Further alterations in lower trophic levels may have been due to top-down effects. For example, although increased abundance of the juvenile stages of *C. finmarchicus* have been observed in CPR samples over the past decade, the adult stages decreased. As reported by J. Stockwell (Gulf of Maine Research Institute) and colleagues and W. Golet (University of New Hampshire) and colleagues at the meeting, this may be due to size-selective predation by herring populations, which were also on the rise in the 1990s (see, also, Greene and Pershing 2007). Frank et al. (2005) noted possible trophic cascades as a result of changes in groundfish, benthic crustacean, and plankton populations on the nearly Scotian Shelf. These authors attributed these changes to predator release resulting from overfishing of cod and allowing an increase in planktivorous fish (Frank et al. 2005). However, the impacts of this cascade on lower trophic level dynamics are uncertain. Greene and Pershing (2007) propose that this shift may also have been attributable to changes in oceanographic conditions related to the influence of the Labrador Current.

Upper Trophic Level and Fishery Resource Species

Presentations at the workshop, dealing with upper trophic level species, covered a range of species and taxonomic groups. A common theme throughout was the critical role of trophodynamics in the distribution and abundance of these species. No overview of the GoM ecosystem, however, could be complete without special consideration of two iconic species, Atlantic cod *Gadus morhua* and American lobster *Homarus americanus*, and both species received considerable attention at the session. Fisheries for these species have strongly shaped the history of the region and determined the character of local fishing communities. Evidence presented by K. Wilson and T. Willis (University of Southern Maine) at the session suggests that once numerous, local, nearshore populations of cod have been decimated through overfishing in the gulf by the turn of the last century (see also Ames 2004). This decline was possibly exacerbated by declines in river herring populations, a principal prey resource. There is some evidence of a recent resurgence of cod over the past decade in these areas, and studies have been undertaken to examine the possible role of recovery of river herring in some Maine rivers. An examination of cod feeding inside and outside the western Gulf of Maine fishery closed area has revealed higher levels in prey availability and growth rates within the closed area, but also in modes of feeding (higher levels of benthic feeding within the closure), as shown by G. Sherwood and J. Grabowski (Gulf of Maine Research Institute) at the session. Closed areas as a tactical management tool, therefore, seem to have implications for trophodynamics, as well as abundance and demography.

As cod populations in the gulf underwent long-term declines over the past century or more, lobster fishing gradually replaced groundfisheries in these coastal environments as a dominant resource species. Lobster populations and catches have increased markedly over the past several decades, as shown by D. Cowan (The Lobster Conservancy) at the meeting. A reduction in predation on lobsters by fish predators is among the hypotheses under consideration for the increase. Since 1993, the establishment of a long-term intertidal monitoring program has both documented the increase in juvenile lobsters and explored mechanisms for increased utilization of hard substrate habitats through shelter-sharing. These observations have enhanced our understanding of patterns of shelter use and overall carrying capacity of the system. The observed increase of lobster abundance may reflect broader patterns of habitat use overall under reduced risk of predation. It has been proposed that energy subsidies provided through herring bait in lobster traps may also have played a role in the increase (Grabowski et al. 2010).

Information on other fish species, including forage species such as Atlantic herring *Clupea harengus* and apex predators such as bluefin tuna *Thunnus thynnus*, was also highlighted during the symposium. Detailed studies of ontogentic shifts in diet composition for herring by Stockwell and colleagues at the session demonstrated a propensity of larger herring to feed selectively on copepods. These high-energy-content prey have potentially strong effects on herring bioenergetics. In related observations, Golet and colleagues documented declines in condition factors in tuna, an extremely valuable commercial and recreational species, apparently as a result of lower energy content of herring, an important prey item for tuna. In turn, this change was linked to a decline in availability of adult *C. finmarchicus*, with a resulting decline in herring condition factors.

For another historically important species, Atlantic salmon *Salmo salar*, K. Friedland et al. (2012, this volume) described the interplay of environmental conditions and predator distribution patterns as a predation gauntlet for salmon smolts as they exit river systems and enter the marine environment.

Historical records document sharp declines of many GoM stocks, including Atlantic halibut *Hippoglossus hippoglossus* (Grasso 2008) and Atlantic cod *Gadus morhua* (Rosenberg et al. 2005), as well as whales, seabirds, sturgeons *Acipenser* spp., and anadromous fish such as alewives *Alosa pseudoharengus* and Atlantic salmon *Salmo salar* (Bolster 2008; Leavenworth 2008). Depleted local stocks resulted in fishing restrictions as early as 1688 (Bolster 2008; Leavenworth 2008). The HMAP Gulf of Maine Cod Project has shown dramatic declines in mean trophic level, species richness, abundance, and habitat quality in the area now covered by the Stellwagen Bank National Marine Sanctuary and the GoM proper (Claesson and Rosenberg 2009).

Embedded within each of the issues described above is the overarching theme of trophodynamics and energy flow. This topic has been extensively studied in the gulf, and several generations of network models have been developed over the past three decades. These models depend heavily on information on feeding interactions at upper trophic levels. It was demonstrated by A. Bundy (Fisheries and Oceans Canada, Bedford Institute of Oceanography) at the session that long-term diet composition studies conducted by both Fisheries and Oceans Canada and the National Marine Fisheries Service can be effectively combined to provide broad spatial coverage and long-term trajectories of change in consumption patterns (see Link and Bundy 2012, this volume).

Implications for Ecosystem-Based Management

It is now widely appreciated that a more holistic approach to management is needed that accounts for the full spectrum of human impacts in the marine environment and the implications for ecosystem services that these systems provide (USCOP 2004). The EBM embodies several key attributes: (1) it is place-based and entails the development of integrated management plans for defined ecological regions, (2) it considers humans as integral components of the ecosystem, and (3) it requires an understanding of the interrelationships among the components of the system and the environment. Adoption of EBM strategies for the GoM will require the implementation of new regulatory and legislative frameworks. These include confronting trade-offs among and within different ocean-use sectors. It will further require the development of appropriate governance structures and close cooperation between the United States and Canada.

An ecosystem overview report for the GoM has been developed by Fisheries and Oceans Canada and the National Marine Fisheries Service to provide an overview of our current understanding of the GoM (broadly defined to include the Bay of Fundy and Georges Bank; East Coast Aquatics 2011). This document is intended to provide an overall context for critical issues associated with EBM for the region.

To move forward with EBM in this region, it will be necessary not only to define spatial management units, but to determine the fishery production potential for these units, to specify sustainable exploitation rates for the units as a whole, and to grapple to allocation strategies among different stakeholder groups. Research strategies to lay the groundwork for this approach have been developed at the Northeast Fisheries Science Center and presented to the New England Fishery Management Council for their consideration (see www.nefsc.noaa.gov/ecosys/EBFMbrochure.pdf). The council is responsible for fishery management within the U.S. exclusive economic zone of the GoM.

In all of this, it must be remembered that the GoM was subject to important alteration long before detailed scientific studies were undertaken. These include the dramatic reduction of whale populations, the decimation of anadromous fish stocks due to obstruction of rivers, habitat loss and overfishing, and overfishing of once dominant species such as halibut. We must be aware of the implications of these factors for overall system productivity in any consideration of what might be possible.

With this backdrop, it is clear that the rich history of research in the GoM provides a strong foundation for moving toward EBM in this region. The gulf is a semi-enclosed continental shelf sea with distinctive physical characteristics relative to adjacent regions such as Georges Bank and the Scotian Shelf. Reasonable arguments can accordingly be made for the GoM proper as a spatial unit for EBM. Indeed, an analysis of spatial patterns of physiography, hydrography/oceanography, and lower trophic level dynamics (chlorophyll and primary production) point to features in the western and central GoM that distinguish this area relative to Georges Bank and the eastern GoM/Scotian Shelf (Fogarty et al. 2012, this volume). The nearshore GoM emerges as a distinct subregion of the gulf. Primary production is relatively low in the central gulf, particularly over the deep basin areas relative to adjacent regions, coloring our expectations for potential yield in the deeper regions, which stand in sharp contrast to the more highly productive coastal regions.

Results from this session provide further insight into questions related to the appropriate spatial scales for management. Based on these investigations, it is clear that distinctive variations in physical and ecological characteristics exist in the eastern and western GoM. It is further evident that, for a number of reasons, the nearshore GoM may require special attention that goes beyond the physical and lower trophic level considerations noted above in defining subregions. The diversity of human activities on the coast, and in the immediate coastal zone, requires consideration of cumulative impacts of fishing, pollution, and habitat alteration/destruction. The nearshore region is also substantially affected by watershed influences that affect productivity patterns and other characteristics. Collectively, these and other concerns suggest that it is possible that nested spatial structures for management could be recognized within the gulf to account for these differences.

With respect to understanding interrelationships among parts of the system and with the environment, again it is clear that we have much to build on. There is important evidence of bottom-up control in the system with effects throughout the food web. A critical issue that remains is understanding potential changes in nutrient regimes and possible changes in the role of the microbial food web in energy transfer in this system. The impressive gains in identifying microbial components of the system during the symposium must be continued and the role of this component in system productivity fully explored.

Evidence of changing stratification patterns in the gulf, reviewed at the session, holds important implications for the relative balance of productivity in the pelagic and benthic components of the system. Our understanding of benthic communities in the gulf has also greatly benefited from renewed attention, as amply demonstrated at the symposium. This effort must be expanded to permit a fuller understanding of issues such the role of benthic–pelagic coupling in overall system structure under changing environmental conditions—increased stratification can impede energy flow among benthic and pelagic subsystems. The immensely complex physiographic structure of the gulf substantially increases the sampling difficulties involved, but the task is essential.

Results presented at the session show that the system has undergone regime shifts related to climate and physical forcing affecting nutrient dynamics, with attendant consequences for primary production, zooplankton community composition, and fish community structure. The effects of far-field forcing on oceanographic properties and productivity of the GoM and, in particular, changes attributable to the influence of the Labrador Current under changing climate conditions (Greene and Pershing 2007), will require a dynamic view of system productivity patterns and sustainable exploitation rates. These considerations will necessarily play an important role in devising management strategies in an ecosystem context. Basic biology of ecologically and economically important species and aspects of community structure of nektonic, planktonic, and benthic assemblages were all highlighted. In particular, for EBM, understanding shifts in productivity states will be essential in adjusting sustainable exploitation strategies at the ecosystem level to account for variation in environmental states.

Continued emphasis on synthesis of the rich body of research in the gulf, and the development of models for integration of this information, and predicting the effects of changing environmental conditions and the implications of alternative management actions are essential. Important strides have been made, particularly in development of coupled biophysical models. Linkage of numerical hydrodynamic models to a broader array of ecological models will be necessary to place the modeling efforts in service to management. For example, we must further develop approaches that link our hydrodynamic models to general circulation models to evaluate the potential impacts of climate change on the gulf.

Our session highlighted the rich information base available for the ecology of secondary producers and of higher trophic levels in the gulf. This information can be integrated into coupled physical-biological models and in other multispecies or ecosystem models. Ecosystem models have, in fact, been developed for the GoM to explore the implications of alternative fishery management options. These include static energy flow models (Link et al. 2007, 2008), dynamic ecosystem models (Overholtz and Link 2009), and multispecies production models (Fogarty et al., in press). In addition, single-species models for key resource species of the gulf, such as herring, have been evaluated in an ecosystem context (Overholtz et al. 2008). These, and other models now in development, provide the foundation to evaluate approaches to EBM for this system.

Finally, it must be recognized that we are still in the very early stages of understanding how best to integrate the human dimension into EBM in the gulf. We need to understand the role of humans as part of the ecosystem and how changes in the GoM affect human communities dependent on the gulf for work, recreation, transport, and other activities. Important work is now underway to understand human use patterns in space and time in the GoM, and ways to define fishing communities (Olson 2005; St. Martin 2006; St. Martin and Hall-Arber 2008). Climate-related changes in the gulf are likely to result in changes in the distribution patterns of resource species and their abundance. The factors will have direct implications for fishing patterns and will certainly contribute to the long history of change in the gulf. We need both to be able to anticipate shifting patterns of ecosystem structure and function and to implement adaptive management strategies to deal with both expected and unexpected change in system dynamics.

References

Ames, E. 2004. Atlantic cod stock structure in the Gulf of Maine. Fisheries 29:10–28.

Anderson, D. M., J. M. Burkholder, W. P. Cochlan, P. M. Glibert, C. J. Gobler, C. A. Heilf, R. Kudela, M. L. Parsons, J. E. Rensel, D. W. Townsend, V. L. Trainer, and G. A. Vargo. 2008. Harmful algal blooms and eutrophication: examining linkages from selected coastal regions of the United States. Harmful Algae 8:39–53.

Anderson, D. M., D. W. Townsend, D. J. McGillicuddy, and J. T. Turner, editors. 2005. The ecology and oceanography of toxic *Alexandrium fundyense* blooms in the Gulf of Maine. Deep-Sea Research Part II 52:2365–2876.

Beardsley, R. C., P. C. Smith, and C. M. Lee, editors. 2003. Physical processes on Georges Bank

(GLOBEC). Journal Geophysical Research 108(C11):8000.

Bigelow, H. B. 1926. Plankton of the offshore waters of the Gulf of Maine. Bulletin of the United States Bureau of Fisheries 40:1–509.

Bigelow, H.B. 1927. The physical oceanography of the Gulf of Maine. Bulletin of the United States Bureau of Fisheries 40:511–1027.

Bolster, W. J. 2008. Putting the ocean in Atlantic history: maritime communities and marine ecology in the northwest Atlantic, 1500–1800. American Historical Review 113:19–47.

Bourque, B. J., B. J. Johnson, and R.S. Steneck. 2008. Possible prehistoric fishing effects on coastal marine food webs in the Gulf of Maine. Pages 165–185 *in* T. C. Rick and J. Erlandson, editors. Human impacts on ancient marine ecosystems. University of California Press, Berkeley.

Claesson, S. H., and A. A Rosenberg 2009. Stellwagen Bank marine historical ecology. University of New Hampshire, Office of National Marine Sanctuaries, Final Report, Durham.

Cournane, J. M. 2010. Spatial management of groundfish resources in the Gulf of Maine and Georges Bank. Doctoral dissertation. University of New Hampshire, Durham.

East Coast Aquatics. 2011. Gulf of Maine Ecosystem Overview Report. Canadian Technical Report of Fisheries and Aquatic Sciences 2946.

Ecosystem Assessment Program. 2009. Ecosystem assessment report for the northeast US continental shelf large marine ecosystem. U.S. Department of Commerce, National Marine Fisheries Service, Northeast Fisheries Science Center, Ref Doc 09–11, Woods Hole, Massachusetts.

Fogarty, M. J., K. D. Friedland, L. Col, R. Gamble, J. Hare, K. Hyde, J. S. Link, S. Lucey, H. Liu, J. Nye, W. J. Overholtz, D. Richardson, B. Rountree, and M. Taylor. 2012. Status of the northeast U.S. Continental Shelf large marine ecosystem: an indicator-based approach. Pages 139–165 *in* R. L. Stephenson, J. H. Annala, J. A. Runge, and M. Hall-Arber, editors. Advancing an ecosystem approach in the Gulf of Maine. American Fisheries Society, Symposium 79, Bethesda, Maryland.

Fogarty, M. J., W. J. Overholtz, and J. S. Link. In press. Aggregate-species production models for the demersal fish complex of the Gulf of Maine. Marine Ecology Progress Series.

Frank, K. T., B. Petrie, J. S. Choi, and W. C. Leggett. 2005. Trophic cascades in a formerly cod-dominated ecosystem. Science 308:1621–1623.

Friedland, K. D., J. P. Manning, and J. S. Link. 2012. Thermal phenological factors affecting the survival of Atlantic salmon in the Gulf of Maine. Pages 393–407 *in* R. L. Stephenson, J. H. Annala, J. A. Runge, and M. Hall-Arber, editors. Advancing an ecosystem approach in the Gulf of Maine. American Fisheries Society, Symposium 79, Bethesda, Maryland.

Grabowski, J. H., E. J. Ciesceri, A. J. Baukus, J. Gaudette, M. Weber, and P. O. Yund. 2010. Use of herring bait to farm lobsters in the Gulf of Maine. PLoS (Public Library of Science) Biology [online serial] 5:e10188. DOI: 10.1371/journal.pone.0010188.

Grasso, G. M. 2008. What appeared limitless plenty: the rise and fall of the nineteenth-century Atlantic halibut fishery. Environmental History 13:66–91.

Greene, C. H., and A. J. Pershing. 2007. Climate drives sea change. Science 315:1084–1085.

Hale, S. S. 2012. Spatial patterns of subtidal benthic invertebrates and environmental factors in the nearshore Gulf of Maine. Pages 167–183 *in* R. L. Stephenson, J. H. Annala, J. A. Runge, and M. Hall-Arber, editors. Advancing an ecosystem approach in the Gulf of Maine. American Fisheries Society, Symposium 79, Bethesda, Maryland.

Hare, J. A., and J. Kane. 2012. Zooplankton of the Gulf of Maine—a changing perspective. Pages 115–137 *in* R. L. Stephenson, J. H. Annala, J. A. Runge, and M. Hall-Arber, editors. Advancing an ecosystem approach in the Gulf of Maine. American Fisheries Society, Symposium 79, Bethesda, Maryland.

Incze, L. S., P. Lawton, S. L. Ellis, and N. H. Wolff. 2010. Biodiversity knowledge and its application in the Gulf of Maine area. Pages 43–64 *in* A. McIntyre, editor. Life in the world's oceans: diversity, distribution and abundance. Wiley Blackwell Scientific Publications, London.

Jackson, J. B. C., M. X. Kirby, W. H. Berger, K. A. Bjorndal, L. W. Botsford, B. J. Bourque, R. H. Bradbury, R. Cooke, J. Erlandson, J. A. Estes, T. P. Hughes, S. Kidwell, C. B. Lange, H. S. Lenihan, J. M. Pandolfi, C. H. Peterson, R. S. Steneck, M. J. Tegner, and R. R. Warner. 2001. Historical overfishing and the recent collapse of coastal ecosystems. Science 293:629–638.

Larsen, P. F., editor. 2004. Ecosystem modeling in Cobscook Bay, Maine: a boreal, macrotidal estuary. Northeastern Naturalist 11(Special Issue 2):1–12.

Leavenworth, W. B. 2008. The changing landscape of maritime resources in seventeenth-century New England. International Journal of Maritime History 20:33–62.

Link, J. S. 2004. Using fish stomachs as samplers of the benthos: integrating long-term and broad scales. Marine Ecology Progress Series 269:265–275.

Link, J. S., J. K. T. Brodziak, S. F. Edwards, W. J. Overholtz, D. Mountain, J. W. Jossi, T. D. Smith, and M. J. Fogarty. 2002. Marine ecosystem assessment in a fisheries management context. Canadian Journal of Fisheries and Aquatic Sciences 59:1429–1440.

Link, J. S., and A. Bundy. 2012. Ecosystem modeling in the Gulf of Maine region: towards an ecosystem approach to fisheries. Pages 281–310 *in* R. L. Stephenson, J. H. Annala, J. A. Runge, and M. Hall-Arber, editors. Advancing an ecosystem approach in the Gulf of Maine. American Fisheries Society, Symposium 79, Bethesda, Maryland.

Link, J. S., and M. D. Ford. 2006. Widespread and persistent increase of Ctenophora in the continental shelf ecosystem off NE USA. Marine Ecology Progress Series 320:153–159.

Link, J., J. O'Reilly, M. J. Fogarty, D. Dow, J. Vitaliano, C. Legault, W. J. Overholtz, J. R. Green, D. Palka, V. Guida, and J. Brodziak. 2007. Energy flow on Georges Bank revisited: the energy modeling and analysis exercise in historical context. Journal of Northwest Atlantic Fishery Science 39:83–101.

Link, J., W. Overholtz, J. O'Reilly, J. Green, D. Dow, D. Palka, C., Legault, J. Vitaliano, V. Guida, M. Fogarty, J. Brodziak, E. Methratta, W. Stockhausen, L. Col, G. Waring, and C. Griswold. 2008. An overview of EMAX: the northeast U.S. continental shelf ecological network. Journal of Marine Systems 74:453–474.

Mills, C. E. 2001. Jellyfish blooms: are populations increasing globally in response to changing ocean conditions? Hydrobiolgia 451:55–68.

Olson, J. 2005. Re-placing the space of community: a story of cultural politics, policies, and fisheries management. Anthropological Quarterly 78:247–68.

O'Reilly, J. E., and D. A. Busch. 1984. Phytoplankton primary production on the northwestern Atlantic Shelf. Rapports et Proces-Verbaux des Reunions Conseil International pour l'Exploration de la Mer 183:255–268.

O'Reilly, J. E., C. Evans-Zetlin, and D. A. Busch. 1987. Primary production. Pages 220–233 *in* R. H. Backus, editor. 1987 Georges Bank. MIT Press, Cambridge, Massachusetts.

Overholtz, W. J., L. D. Jacobson, and J. S. Link. 2008. An ecosystem approach for assessment advice and biological reference points for the Gulf of Maine–Georges Bank Atlantic herring complex. North American Journal of Fisheries Management 28:247–257.

Overholtz, W. J., and J. S. Link. 2009. A simulation model to explore the response of the Gulf of Maine food web to large-scale environmental and ecological changes. Ecological Modelling 220:2491–2502.

Rosenberg, A. A., W. J. Bolster, K. E. Alexander, W. B. Leavenworth, A. B. Cooper, and M. G. McKenzie. 2005. The history of ocean resources: modeling cod biomass using historical records. Frontiers in Ecology and the Environment 3:84–90.

St. Martin, K. 2006. The impact of "community" on fisheries management in the U.S. Northeast. Geoforum 37:169–184.

St. Martin, K., and M. Hall-Arber. 2008. The missing layer: geo-technologies, communities, and implications for marine spatial planning. Marine Policy 32:779–786.

Steneck, R. S., J. Vavrinec, and A. V. Leland. 2004. Accelerating trophic-level dysfunction in kelp forest ecosystems of the western North Atlantic. Ecosystems 7:323–332.

Townsend, D. W., A. C. Thomas, L. M. Mayer, M. Thomas, and J. Quinlan. 2006. Oceanography of the northwest Atlantic Continental Shelf. Pages 119–168 *in* A. R. Robinson and K. H. Brink, editors. The sea, volume 14. Harvard University Press, Cambridge, Massachusetts.

Townsend, D. W., N. D. Rebuck, M. A. Thomas, L. Karp-Boss, and R. M. Gettings. 2010. A changing nutrient regime in the Gulf of Maine. Continental Shelf Research 30:820–832.

USCOP (U.S. Commision on Ocean Policy). 2004. An ocean blueprint for the 21st century. USCOP, Final Report, Washington, D.C.

Wiebe, P. H., R. C. Beardsley, A. C. Bucklin, and D. G. Mountain, editors. 2001. Coupled biological and physical studies of plankton populations: Georges Bank and related North Atlantic regions. Deep-Sea Research Part II 48:1–684.

Wiebe, P. H., R. C. Beardsley, D. G. Mountain, and R. G. Lough, editors. 2006. Dynamics of plankton and larval fish populations on Georges Bank, the North Atlantic U.S. GLOBEC study site. Deep-Sea Research Part II 53:2455–2832.

American Fisheries Society Symposium 79:273–280, 2012

Seafloor Mapping for Ecosystem Management in the Gulf of Maine

Jonathan H. Grabowski*
Northeastern University, Marine Science Center
430 Nahant Road, Nahant, Massachusetts 01908, USA

Tracy Hart
Gulf of Maine Mapping Initiative, Gulf of Maine Council on the Marine Enrivonment
84 Marginal Way, Suite 600, Portland, Maine 04101, USA

Abstract.—The Gulf of Maine's seafloor provides a wide array of valuable ecosystem services, including provision of habitat for commercially and ecologically important mammals, seabirds, fish, and invertebrates. Implementing ecosystem-based management will require improved information about the habitats of economically and ecologically important species and the impacts of different human activities, such as fish harvesting, offshore energy development, and shipping, to balance these competing needs. Currently, there is limited high resolution seabed substrate information in the Gulf of Maine, especially in the U.S. portion, because of the high cost of multi-beam echo sounder surveys. Moreover, this lack of coverage limits the ability of managers to use seafloor substrate information in ecosystem management activities, such as fisheries management, that require more holistic coverage of the bioregion. Therefore, the potential need for seafloor mapping in this region is enormous given the value of accurate seafloor information to managers in charge of minimizing impacts to and sustaining the ecosystem services provided by benthic habitat in the Gulf of Maine.

How Far We Have Come: The Current State of Mapping to Support Ecosystem Management

The Gulf of Maine ecosystem contains a diverse array of geological and biological substrates that serve as habitat for commercially and ecologically important mammals, seabirds, fish, and invertebrates. Managing this system effectively will require balancing the habitat needs of these species with ongoing and emerging human activities such as fish harvesting, offshore energy development, and shipping. The potential need for seafloor mapping in this region is enormous when one considers the importance of accurate seafloor information to managers in charge of minimizing impacts to, and sustaining the resources of, the ocean.

* Corresponding author: j.grabowski@neu.edu

Seabed Mapping in the Gulf of Maine

Recent advances in seabed mapping have afforded scientists the ability to examine very small seafloor features over wide swathes of the ocean (Todd and Greene 2008). While video and grab surveys have proven useful methods to examine seafloor characteristics at discrete locations, acoustic imaging techniques have resulted in the collection of continuous data over much coarser spatial areas. Several acoustic techniques have been developed over the past couple of decades, ranging from single-beam and multi-beam echo sounders to side-scan sonar and also electro-optical approaches such as LIDAR (light detection and ranging). Multi-beam echo sounders, in particular, have been used extensively to provide bathymetry and seafloor morphology at many scales. Each of the acoustic techniques can also produce information on signal strength (backscatter) and other acoustic characteristics that can be used to characterize seafloor roughness and substrate type (Mayer and Fonseca 2007). Specifically, several analytical approaches (e.g., Geocoder, Multiview) have been developed to analyze backscatter in order to evaluate seafloor habitat properties. These technological advances were quickly recognized as especially valuable for use in coastal and offshore development projects and fisheries management (Todd and Greene 2008).

The amount of effort devoted, to date, to collecting geophysical seafloor data in the Gulf of Maine by Canada and the United States has differed substantially (Gulf of Maine Mapping Initiative coverage map, www.gulfofmaine.org/gommi/coverage-map.php). The vast majority of the Canadian portion of the Gulf of Maine has been surveyed using multi-beam echo sounders, and the remaining portion, that has yet to be surveyed, consists largely of deep muddy basins. In contrast, far less of the U.S. portion of the Gulf of Maine has been surveyed, with efforts focused primarily on offshore ledge features (i.e., Cashes Ledge, Jeffreys Ledge, and Phippennies Ledge) and coastal waters. Of further concern is that backscatter information was not processed and presented for several of these data sets. Thus, the utility of these data are limited to interpreting bathymetry and visualizing seafloor morphology.

Canadian scientists have been more successful at surveying seafloor characteristics of their portion of the Gulf of Maine, in part because of the partnerships that were forged with the scallop fishing industry. The scallop industry paid for a significant portion of the cost of multi-beam surveys off of southwestern Nova Scotia because the maps that were generated helped them target scallop beds with greater precision. While these maps may have heightened pressure on the fishery by helping fishers identify scallop beds, they have reduced the spatial extent of bottom contact, which consequently has reduced the extent of bottom habitat impacted by scallop dredging in Canada. The Canadian government historically has prioritized the collection of seafloor geophysical data because of its utility to management efforts. Unfortunately, U.S. scientists and managers have been much less successful at garnering federal support for mapping efforts in spite of several concerted efforts over the past two decades. Furthermore, U.S. scientists have been less successful at developing partnerships with industry to motivate high-resolution acoustic mapping surveys. However, the U.S. scallop fishery has collaborated with scientists at the University of Massachusetts Dartmouth to contribute extensively to efforts to map bottom habitat on Georges Bank via video surveys (B. Harris, Alaska Pacific University, personal communication). These surveys are currently used by fisheries managers and scientists to assess habitat vulnerability from fishing gear impacts.

The Use of Seabed Mapping Efforts in Ecosystem-Based Management

Although seafloor mapping is a relatively new technology, it has already been used as an effective tool in assessing and minimizing ecosystem impacts of human activities. For example, seafloor substrate information from high-resolution acoustic data are essential for assessing the impacts of, and siting activities such as, bottom cable installation, mining, dredging, and offshore energy development projects. Current tidal energy projects in the Bay of Fundy have utilized existing high-resolution seafloor characteristics to select sites that are stable enough to support development activities with minimal impact to the seafloor substrates (Todd and Shaw 2009). Efforts to develop offshore energy projects will also benefit heavily from proactive mapping efforts that guide these projects to choose suitable bottom and minimize impacts. For example, seafloor mapping can be utilized to delineate coastal features, such as drowned lakes, that likely contain preserved cultural features of historical value (Kelley et al. 2010). Features such as these should be considered by managers when siting future offshore energy development projects in the Gulf of Maine. Currently, the state of Maine is considering several offshore wind energy pilot projects, which could result in commercial-scale offshore wind development activities by the end of this decade. This growing push to harvest wind off the coast of Maine highlights the need for accurate seafloor information to guide these development activities.

Another human activity that can pose a major threat to seafloor habitat, and consequently could benefit from advances in seafloor mapping activities, is fishing. Existing seafloor habitat information has recently been used by the Habitat Plan Development Team (PDT) of the New England Fishery Management Council to assess the vulnerability of seafloor habitat to fishing gear impacts in the Gulf of Maine (New England Fishery Management Council Habitat Plan Development Team 2010). This assessment integrates an understanding of how different gears contact the bottom with the susceptibility and recovery of the geological substrate features and biological communities to each gear type. Assessments such as these will be used increasingly by fisheries managers in an attempt to minimize the impacts of fishing activities to the Gulf of Maine ecosystem and consequently will be integral to attempts to shift to an ecosystem-based management (EBM) approach regionally, but their results can only be as good as their input data.

The accuracy of the Habitat PDTs vulnerability assessment hinges, in part, on the quality of substrate information available. Existing video survey and sediment grain size sample data were used to map substrate types as a part of the assessment, which was sufficient for areas such as Georges Bank and the Middle Atlantic Bight, where extensive video surveys have been conducted over the past two decades (Stokesbury et al. 2004; Adams et al. 2008). However, in the Gulf of Maine, video and substrate information are far less common. High-resolution geophysical acoustic seabed data that have been collected in the Gulf of Maine were not used because the data have not been made readily available and do not exist region-wide at a scale that is applicable to fisheries management. Thus, while the vulnerability assessment represents a significant step forward in fisheries impact assessment, the quality of the predictions generated by this model for the Gulf of Maine region could be vastly improved by better seafloor information.

In addition to benefiting the assessment of bottom impacts from fishing activities, seafloor substrate information is being used to mitigate conflicts among fisheries that overlap spatially. For example, seafloor maps can be combined with fishery data to help understand species-bottom type associations (Smith et al. 2009). This type of information is currently being utilized in southwestern Nova Scotia to minimize negative interactions among scallop and lobster fisheries. Integration of the management of fisheries that are spatially linked, either by natural processes or anthropogenic activities, is a pivotal step in shifting to an EBM approach, and mapping can provide the framework for this integration.

Seafloor information is also of value for fisheries management because it can be used to examine how species are distributed by depth, season, and habitat. Moreover, it can be used to examine seafloor attributes associated with key life history stages, such as spawning and recruitment, and to ultimately designate essential fish habitat (EFH; Kostylev et al. 2005). Therefore, geophysical techniques that permit characterization of features common on the seafloor, such as geological and biological substrate components, are extremely valuable in defining and delineating EFH (Kvitek et al. 1999). Fisheries ecologists have used an understanding of seafloor components to study EFH in the Gulf of Maine. For example, scientists have effectively described groundfish habitat associations in the Gulf of Maine and Georges Bank for several important species such as Atlantic cod *Gadus morhua* and haddock *Melanogrammus aeglefinus* (Lough et al. 1989; Tupper and Boutilier 1995; Auster and Lindholm 2005; Lindholm et al. 2007). Collectively, these studies have contributed greatly to our understanding of fine-scale habitat requirements; however, acoustic mapping is needed to extend the results of these studies to the entire Gulf of Maine/Georges Bank region. Fisheries ecologists have incorporated this type of information into models to examine the parameters that dictate EFH on regional spatial scales that are more appropriately matched with fisheries management (Pickrill and Kostylev 2007). Yet, similar to efforts described above to model fisheries gear impacts, these modeling efforts rely heavily on the availability of high-resolution seafloor information.

Efforts to integrate management of all of the above activities into marine spatial planning rely on seafloor information as a necessary foundation for these efforts. Seafloor mapping has been conducted throughout the state of Massachusetts coastal waters in order to facilitate marine spatial planning efforts that are ongoing locally (D. Sampson, Coastal Zone Management). Mapping of seabed characteristics (substrate, depth, roughness, etc.) is often a necessary initial step in managing marine ecosystems and the goods and services that they provide. Marine spatial planning has been used effectively internationally to manage coastal resources among neighboring countries (e.g., SeaZone BLAST North Sea mapping project) and promises to be a valuable tool that will assist coordination of EBM efforts among the various private, state, provincial, and national interests throughout the Gulf of Maine region.

How Well Positioned Are We to Implement Ecosystem-Based Management? Seafloor Mapping as a Catalyst for Ecosystem-Based Management and Marine Spatial Planning

Implementing EBM will require improved information about the habitats of economically and ecologically important species and the impacts of different human activities. Yet, currently, there is limited seabed substrate information in the Gulf of Maine, especially in the United States. This lack of coverage limits the ability of managers to use substrate information in ecosystem management activities, such as fisheries management, that require more holistic coverage of this bioregion. The high cost of multi-beam acoustic surveys have rendered fisheries industries, research, and managers either incapable of, or unwilling to, encumber the costs associated with completing the remaining seafloor mapping efforts in the Gulf of Maine. Therefore, other partnerships will need to be developed. Offshore energy efforts offer a potential funding source that could grow as pressure to develop wind energy in the Gulf of Maine mounts. For example, the oil industry funded mapping of a large portion of the northern Gulf of Mexico using multi-beam acoustic imaging, and the data are publically available. The Gulf of Maine region would benefit from marine spatial planning efforts that balance this development pressure with other resource management needs, in order to assure that future multi-beam surveys for development purposes serve EBM efforts more broadly.

More holistic seafloor data on biological and geological substrate features would facilitate shifting to an ecosystem approach to fisheries management for a variety of reasons. The lack of available substrate information in this region limits fisheries managers' ability to (1) define essential fish habitat and design closures that protect EFH, (2) assess habitat vulnerability from fishing and other human activities that impact the seafloor; and (3) evaluate how habitat affects fish productivity and integrate seafloor parameters into the stock assessment models used to manage regional fisheries (Hart and Grabowski 2009). How can EFH be protected, or gear impacts reduced, without maps that clearly delineate seafloor habitat characteristics? Most current stock assessment models are not spatially oriented, in part because of the paucity of high-resolution seafloor data. Furthermore, stock assessment models will likely remain as is until detailed spatial information is available. Consequently, the absence of such basic information on appropriate spatial scales severely hinders the use of habitat in fisheries management more extensively.

While absolutely necessary for fisheries management, better geophysical seabed information only provides the basis for further investigations. Achieving an understanding of how seafloor substrates affect fish productivity (i.e., level 4 EFH) requires examining the linkages between seabed substrates and life history parameters such as survival, growth, and ultimately productivity. Efforts to model habitat suitability have identified the link between habitats in terms of their "scope for growth" for important biota (Kostylev and Hannah 2007). The habitat types modeled from interactions of physical variables also reflect sensitivity of seabed habitats and communities to human impacts such as fishing. This approach is currently being used in the Canadian portion of the Gulf of Maine (Kostylev and Hannah 2007) and is a potential mechanism to begin incorporating habitat features and impacts to habitat into efforts to manage the productivity of fisheries.

In addition to supporting economically important fisheries, seafloor habitats in the Gulf of Maine provide a wide variety of ecosystem goods and services such as maintaining biodiversity, processing and cycling of nutrients, and providing mineral and chemical resources. Shifting to an EBM approach will require managers to balance human uses in the marine environment with sustaining these ecosystem services. Therefore, efforts to map the ecosystem goods and services associated with seafloor habitat throughout the Gulf of Maine will facilitate the shift to an EBM approach. A better understanding of how these goods and services are spatially distributed, and their associations with specific seafloor substrates, is currently needed.

There are several regions of the Gulf of Maine and Georges Bank that have been closed to different types of commercial fishing gear over the past two decades. Originally designed to reduce mortality to groundfish such as Atlantic cod and haddock, many of these closures have been morphed into habitat closures in recognition of their putative importance. Meanwhile, the New England Fishery Management Council has selected several habitat areas of particular concern (HPAC) over the past 5 years. These areas were selected based on several criteria, and the management intent of HPACs is currently unclear. Management of existing area closures and HPACs would profit greatly from greater seafloor information. In particular, more holistic seabed information from throughout the Gulf of Maine would benefit ongoing and future attempts to assess if these closures are providing important habitat for groundfish species as intended. Moreover, more holistic seabed coverage would permit scientists and managers to evaluate whether the closures and HPACs need to be modified in order to enhance management of the Gulf of Maine ecosystem and its associated resources.

One of the challenges facing the integration of seafloor geophysical information into management is choosing the appropriate spatial scale at which to collect seabed data in order to meet both scientific and management objectives. Scientists often must weigh the trade-off between using existing fine-scale substrate information from sediment samples versus the costs of collecting continuous acoustic information on seafloor substrates. The geophysical data provide continuous coverage and superior resolution of seafloor geology but are not available in most areas. Furthermore, the cost of acquiring this information currently limits more extensive seafloor mapping. Meanwhile, using only existing point data, which are available at no cost, can result in misinterpreting much of the seafloor, which, invariably, will limit our ability to manage it effectively (Barnhardt and Sampson 1998). This problem will become even more apparent as marine spatial planning expands from state waters to federal waters of the Gulf of Maine in the United States where mapping information is sparse. Seafloor information is often the foundation upon which marine spatial planning is based, and attempts to manage and sustain marine resources and ecosystems will require accurate seafloor information.

Prioritizing Future Seafloor Mapping Efforts to Support Ecosystem-Based Management in the Gulf of Maine

The greatest limitation to integrating seafloor information into EBM efforts in the Gulf of Maine is the lack of high-resolution, continuous seafloor information throughout the region. While the benefits of such maps are most apparent for fisheries management, marine spatial planning efforts aimed at balancing the different types of human activities in the Gulf of Maine also are limited by the paucity of existing seafloor information. Of further concern is that current data have been collected under widely varying standards, which is completely understandable given the variety of funding sources and rationales for why these mapping efforts were conducted. The upshot of this situation is that some of the existing data sets may not be broadly applicable to the needs regionally. Moreover, this problem highlights the need to catalyze funding to support mapping throughout the Gulf of Maine using standardized methods that maximize the utility of these data and to make them publicly available.

Achieving appropriate data coverage to enable more effective EBM in the Gulf of Maine will require a coalition of scientists and managers within the region to work collectively towards generating funding from either private industry undertaking development activities (i.e., offshore wind energy), federal agency-administered sources, or federally appropriated support targeted for mapping efforts. The cost of a regional effort would be extensive and is difficult to pinpoint because of the highly variable costs associated with collecting multi-beam echo sounder data. Given the fact that EBM would be best supported by seafloor classification data, the cost of analyzing multi-beam data would likely be higher than previous efforts that were focused primarily on collecting bathymetric information. Therefore, the timeline for achieving extensive enough coverage in the Gulf of Maine is likely at least a decade off unless pressure to develop offshore sites increases quickly, significant support is achieved either regionally or nationally for mapping the oceans, and/or substantial cost reductions are achieved in mapping technology.

Further development and application of models that help develop the linkages between habitat features and ecosystem services such as EFH are a top priority. For instance, seafloor substrate information could be used to identify essential fish habitat for many commercially and ecologically important species by mapping the spatial extent of seabed features that provide EFH for a particular species (P. C. Valentine, U.S. Geological Survey). Valentine used the Stellwagen Bank National Marine Sanctuary as a model system and examined the spatial extent of several groundfish species using existing information about fish habitat associations, such as rugosity and grain size. Kostylev and Hannah (2007) have developed models that integrate multiple environmental fields into a single map to delineate the distribution of habitats where organisms with particular life history traits are likely to persist. They have also used this approach to develop habitat sensitivity indices in order to evaluate how different human uses will impact species of concern. The model developed by the New England Fishery Management Council Habitat Plan Development Team (2010) permits analysis of how future area-based management efforts will influence fishing impacts to habitats throughout the Gulf of Maine. Given their utility for understanding the relationship between seafloor characteristics, their associated ecosystem functions, and their vulnerability, future funding should be directed towards these types of modeling efforts.

Interdisciplinary studies that test the assumptions developed above in the field are needed to validate these models and to redirect hypotheses that are not supported by empirical work. Mapping efforts have been coupled with assessments of ecosystem services associated with key offshore features such as Cashes Ledge in the Gulf of Maine and Stanton Bank off the coast of Ireland (Brown et al. 2010; McGonigle et al. 2011). For example, seafloor habitat information is being used in these studies to develop a better understanding of how habitats influence key life history characteristics of cod. These studies are intended to inform managers about whether marine closures, such as the Cashes Ledge Closure Area, are recovering and performing important ecosystem functions (i.e., provide nursery habitat for fish). Moreover, this empirical information could be used to test habitat suitability models developed for groundfish species in the Gulf of Maine.

The field of acoustic seafloor mapping is relatively young, and data acquisition technologies and analytical approaches are evolving rapidly. For example, effective automatic methods are still being developed for segmentation of hydroacoustic, remote sensing data acquired by multi-beam echo sounders in order to generate quantitative estimates of the spatial distribution of seafloor relief, bottom type, and composition (Fonseca and Mayer 2007). Furthermore, echo sounder data acquisition capacity has matched increases in computer and data storage technology so that multi-beam technologies are currently capable of collecting at much higher resolutions than 5 years ago. Future technological and analytical advances will likely facilitate the shift to an EBM approach in the Gulf of Maine by making it possible to increase the quantity and quality of seafloor information available to managers, but only if decisions are made soon to advance the cause of seafloor mapping to improve its management.

Acknowledgments

Brian Todd, Craig Brown, and Page Valentine provided valuable feedback and insights on this manuscript. The ideas presented in this manuscript benefited from many discussions and presentations at two regional workshops: Integrating Seafloor Mapping and Benthic Ecology into Fisheries Management in the Gulf of Maine, April 15–16, 2009, Portland, Maine, USA, and

Technical Workshop on Seafloor Mapping for Ecosystem Management in the Gulf of Maine, Gulf of Maine Science Symposium, October 4–9, 2009, St. Andrews, New Brunswick, Canada.

References

Adams, C. F., B. P. Harris, and K. D. E. Stokesbury. 2008. Geostatistical comparison of two independent video surveys of sea scallop abundance in the Elephant Trunk closed area USA. ICES Journal of Marine Science 65:995–1003.

Auster, P. J., and J. Lindholm. 2005. The ecology of fishes on deep boulder reefs in the western Gulf of Maine. Pages 91–109 *in* J. Godfrey and S. Schumway, editors. Diving for science 2005. Proceedings of the American Academy of Underwater Sciences. Connecticut Sea Grant, Groton.

Barnhardt, W. A., J. T. Kelley, S. M. Dickson and D. F. Belknap.1998. Mapping the Gulf of Maine with side-scan sonar: a new bottom-type classification for complex seafloors. Journal of Coastal Research 14:646–659.

Brown, C. J., S. J. Smith, P. Lawton, and J. T. Anderson. 2010. Benthic habitat mapping: a review of progress towards improved understanding of the spatial ecology of the seafloor using acoustic techniques. Estuarine Coastal and Shelf Science 92:502–520.

Fonseca, L., and L. Mayer. 2007. Remote estimation of surficial seafloor properties through the application Angular Range Analysis to multibeam sonar data. Marine Geophysical Research 28:119–126.

Hart, T. E., and J. H. Grabowski. 2009. Integrating seafloor mapping and benthic ecology into fisheries management in the Gulf of Maine. Available: www.gulfofmaine.org/gommi. (November 2011).

Kelley, J. T., D. F. Belknap, and S. Claesson. 2010. Drowned coastal deposits with associated archaeological remains from a sea-level "slowstand": northwestern Gulf of Maine, USA. Geology 38:695–698.

Kostylev V. E., and C. G. Hannah 2007. Process-driven characterization and mapping of seabed habitats. Pages 171–184 *in* B. J. Todd and H. G. Greene, editors. Mapping the seafloor for habitat characterization. Geological Association of Canada, Special Paper 47, St. John's, Newfoundland.

Kostylev, V. E., B. J. Todd, O. Longva, and P. C. Valentine. 2005. Characterization of benthic habitat on northeastern Georges Bank, Canada. Pages 141–152 *in* P. W. Barnes and J. P. Thomas, editors. Benthic habitats and the effects of fishing. American Fisheries Society, Symposium 41, Bethesda, Maryland.

Kvitek, R., P. Iampietro, E. Sandoval, M. Castleton, C. Bretz, T. Manouki, and A. Green. 1999. Final report: early implementation of nearshore ecosystem database project. California State University, Monterey Bay.

Lindholm, J., P. J. Auster, and A. Knight. 2007. Site fidelity and movement of adult Atlantic cod *Gadus morhua* at deep boulder reefs in the western Gulf of Maine, USA. Marine Ecology Progress Series 342:239–247.

Lough, R. G., P. C. Valentine, D. C. Potter, P. J. Auditore, G. R. Bolz, J. N. Neilson, and R. I. Perry. 1989. Ecology and distribution of juvenile cod and haddock in relation to sediment type and bottom currents on eastern Georges Bank. Marine Ecology Progress Series 56:1–12.

McGonigle, C., J. H. Grabowski, C. J. Brown, T. Weber, and R. Quinn. 2011. Detection of deep water benthic macroalgae using image-based classification techniques on multibeam backscatter at Cashes Ledge, Gulf of Maine, USA. Estuarine Coastal and Shelf Science 91:87–101.

New England Fishery Management Council Habitat Plan Development Team. 2010. Essential Fish Habitat (EFH) Omnibus Amendment. The Swept Area Seabed Impact (SASI) model: a tool for analyzing the effects of fishing on essential fish habitat. Available: www.nefmc.org/habitat/index.html (January 2012).

Pickrill, R. A., and V. E. Kostylev. 2007, Habitat mapping and national seafloor mapping strategies in Canada. Pages 483–495 *in* B. J. Todd and H. G. Greene, editors. Mapping the seafloor for habitat characterization. Geological Association of Canada, Special Paper 47, St. John's, Newfoundland.

Smith, S. J., J. Black, B. J. Todd, V. E. Kostylev, and M. J. Lundy. 2009. The impact of commercial fishing on the determination of habitat associations for sea scallops (*Placopecten magellanicus*). ICES Journal of Marine Sciences 66:2043–2051.

Stokesbury, K. D. E., B. P. Harris, M. C. Marino, I. I., and J. I. Nogueira. 2004. Estimation of sea scallop abundance using a video survey in offshore USA waters. Journal of Shellfish Research 23:33–44.

Todd B. J., and H. G. Greene, editors. 2008. Mapping the seafloor for habitat characterization. Geological Association of Canada, Special Paper 47, St. John's, Newfoundland.

Todd, B. J, and J. Shaw. 2009. International year of planet earth 5. Applications of seafloor mapping on the Canadian Atlantic continental shelf. Geoscience Canada 36(2):81–94.

Tupper, M., and R. G. Boutilier. 1995. Effects of habitat on settlement, growth, and postsettlement survival of Atlantic cod (*Gadus morhua*). Canadian Journal of Fisheries and Aquatic Sciences 52:1834–1841.

American Fisheries Society Symposium 79:281–310, 2012

Ecosystem Modeling in the Gulf of Maine Region: Towards an Ecosystem Approach to Fisheries

Jason S. Link*
National Marine Fisheries Service, Northeast Fisheries Science Center
166 Water Street, Woods Hole, Massachusetts 02543, USA

Alida Bundy
Fisheries and Oceans Canada, Bedford Institute of Oceanography
Post Office Box 1006, Dartmouth, Nova Scotia B2Y 4A2, Canada

Abstract.—As we move towards an ecosystem approach to fisheries (EAF) in the Gulf of Maine (GoM), it is valuable to collectively gauge where we have been, where we are now, and where we anticipate we might be headed with respect to ecosystem modeling. We do so by providing a brief history of ecosystem modeling in the GoM region, focused on a set of network models at various points in time over the past 70 years. We then describe current and ongoing ecosystem modeling efforts in the GoM region, with a particular emphasis on how they are being used in a living marine resource (LMR) management context. We then discuss how such models could be used to advance an EAF in the near term with a focus on the appropriate application of classes of models for addressing various types of research and management questions. Finally, we highlight major lessons learned from our modeling endeavors in an LMR context in the GoM region, so that we and other regions around the world can continue to move towards an EAF.

Introduction

There have been numerous calls for an ecosystem approach to fisheries (EAF; e.g., Larkin 1996; Link 2002b, 2002c; Garcia et al. 2003; Browman and Stergiou 2004, 2005; Pikitch et al. 2004). There are many methods, tools and approaches that can be used to implement an EAF, including a wide range of analytical, indicator, framework, and governance considerations. One of the more important tools among these approaches is the use of models.

The Gulf of Maine (GoM) region is one of the better studied regions of the world in terms of oceanography, fisheries, and marine biology. Here, we define the GoM region more broadly than the sensu strictu GoM as the entire surrounding region, inclusive of these contiguous shelf and bay ecosystems found in both Canada and the United States (Figure 1). This focus on the GoM region has led to substantial data collections, analytical efforts, and synthesis studies (e.g., Backus 1987; Parsons 1993; Boreman et al. 1997; Fogarty and Murawski 1998; Breeze 2002; Breeze et al. 2002; Link and Brodziak 2002; Zwanenburg et al. 2002, 2006). From this solid foundation, one can readily glean generic principles that could be of utility in other marine ecosystems and broadly assist in the implementation of an EAF.

The use of models has recognized value in marine science (Fennel and Neumann 2004; Megrey and Moksness 2009a). Modeling approaches provide a means to collate and integrate a broad array of data, provide a way to synthesize a suite of information, allow one to evaluate the relative importance of several concurrent processes, allow one to test hypotheses concerning ocean system structure and functioning, force one to formalize hypotheses, and provide the basis for predictions

* Corresponding author: jason.link@noaa.gov

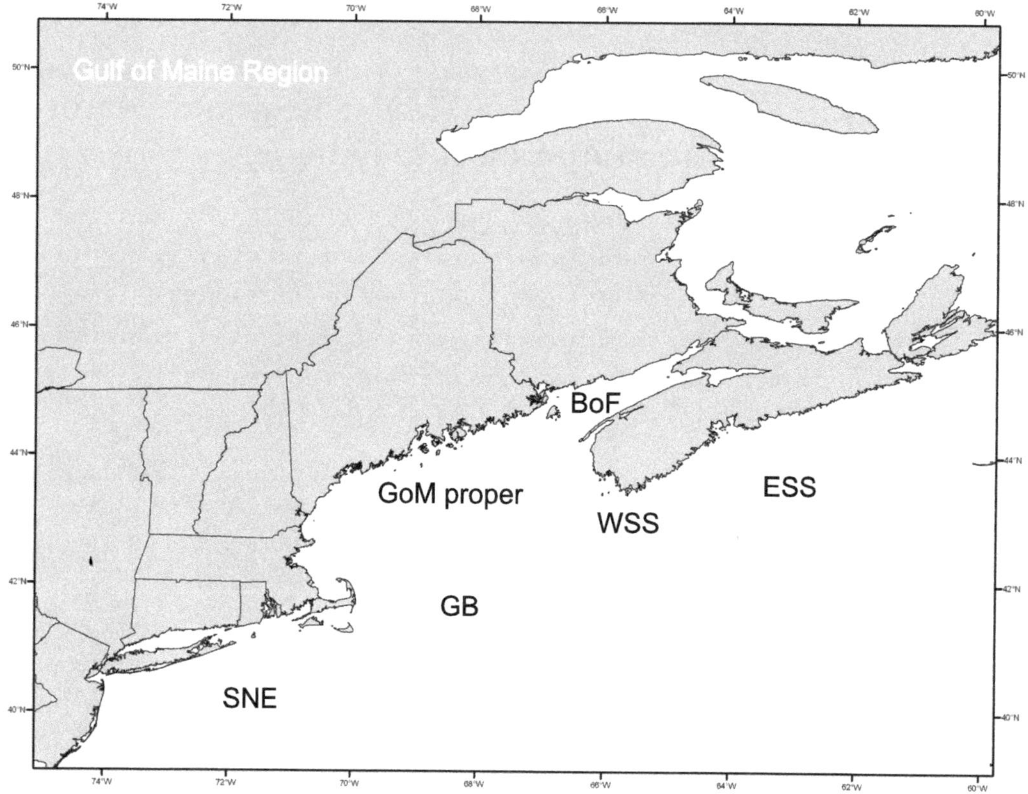

Figure 1. Map of the broader Gulf of Maine region. GoM = Gulf of Maine, GB = Georges Bank, SNE = southern New England, BoF = Bay of Fundy, WSS = western Scotian Shelf, ESS = eastern Scotian Shelf.

of both scientific and resource management interest. The GoM region has had copious modeling efforts. These range from physiochemical features (e.g., Chen et al. 2001; Franks and Chen 2001; Townsend et al. 2006; Hu et al. 2008) to socioeconomic uses (e.g., Holland and Sutinen 1999; Edwards et al. 2004) of the ecosystem. Further, we recognize that many of our academic colleagues in the GoM region have produced, or are continuing to develop, some excellent multispecies and ecosystem models (e.g., Collie and DeLong 1999; Tsou and Collie 2001; Hall et al. 2006). Although cognizant of this wide range of efforts, our focus here is on the suite of modeling approaches, collectively termed ecosystem models, that relate to living marine resources (LMRs), particularly relevant to an EAF and particularly as they are beginning to be developed and used operationally by our institutes in a formal LMR evaluation context.

The intention of this work is to provide a brief history of ecosystem modeling in the GoM region, describe current and ongoing ecosystem modeling efforts, discuss how such models could be used to advance an EAF in the near term, and highlight major lessons learned for an EAF (sensu Pitcher et al. 2009).

Points in Time for Ecosystem Modeling in the Gulf of Maine Region

Science is an adaptive process that builds upon previous knowledge, incorporates news ideas and technology, and is influenced by the society and the culture in which it is embedded (David 2005). Thus, how we view and thus model ecosystems has changed in response to the changing world around us (Simmons 2000) and, like our worldview, is a product of the time and place in which we live (Thagard 1994; Castelo-Lawless 1995;

Simmons 2000). Understanding the context within which we execute science, including major legal mandates, technology, worldview and culture, may help to understand some of the influences as to how we conduct, interpret, and use science; doing so is useful since it could enable the recognition of potential biases, perspectives, models, or processes that have been, or can be, ignored (Smith 1994). Additionally, examining these changing contexts and worldviews, and the science they produce, can provide insights into understanding (scientific or otherwise) that would otherwise be obfuscated by a narrower focus.

Perhaps the most obvious change that has occurred in science, and the broader community, over the past century is the developments in technology (Beck 1999; Megrey and Moksness 2009a, 2009b). The computing tools and power available to a contemporary scientist are orders of magnitude greater than those that were available to scientists at the dawn of the computer age in the middle of the 20th century (Megrey and Moksness 2009b). Advances have been made in processing capability, hard disk storage, and RAM, as well graphics resolution. As such, computers today offer far greater technological capabilities than were available to earlier scientists. As such, this change in technology may have limited the range and type of models that fisheries scientists in the 1930s–1950s were able to construct and execute.

Here, we highlight how ecosystem modeling has changed over time in the GoM region by examining different views and applications of energy budget or network modeling efforts for Georges Bank at various points in time. These include views in the 1940s, 1980s, and the 2000s, which we then cursorily link to changes in concurrent technology, management schemes (and their implied legislative mandates), and culture. There are numerous other modeling approaches that we could have considered, but we submit that the food webs clearly show the changes in approaches to ecosystem modeling, which are reviewed at the end of this section.

The 1940s View

Context.—The network analysis constructed in the 1940s (Clarke 1946) is represented by a mechanistic image of interacting gears (Figure 2). The fisheries on Georges Bank and the western Scotian Shelf largely focused on large gadids (primarily Atlantic cod *Gadus morhua* and haddock *Melanogrammus aeglefinus*) and Acadian redfish *Sebastes fasciatus*. Many measurement and data collection capabilities were either being developed or enhanced during this era, particularly estimates of primary and secondary production. Some of the earliest measurements of marine primary production were made on Georges Bank by Riley (1941).

The representation of the network (note that neither network nor energy budget were terms in use during this era) is highly simplified and represented a limited set of taxa. This translates into a system representation of very low complexity, with fewer nodes and flows than subsequent networks (Table 1). The food web depicted represents the classical grazing food chain, with diatoms as the major primary producer. There was no mass balance constraint in the model, although using energy units does not require conservation of mass and, in many ways, mitigates this concern. The relatively small number of trophic levels in the Clarke (1946) model may also reflect the choice of energetic units as compared to other measurements requiring further constraints and information (Table 1).

The major theoretical context at the time was Lindeman's trophodynamic concept (Lindeman 1942), which noted that the energy flows among various biotic components of an ecosystem are interdependent and usually hierarchical. The concept of productivity of aquatic and marine ecosystems was novel at the time and reflected the farming perspective prevalent in that era (i.e., how to maximize production for human benefit). During this era, the simplified trophic diagram represented the conventional wisdom of both the biological oceanographic and fisheries science communities.

This simple representation of the ecosystem likely reflects the tools available to scientists at the time. The 1940s predates the development of computers and is a time when fisheries and ecosystem modeling was in its infancy; the seminal works by Ricker (1954) and Beverton and Holt (1957) were not published until the following decade. Furthermore, fisheries management was also in its infancy; the first fisheries management measures

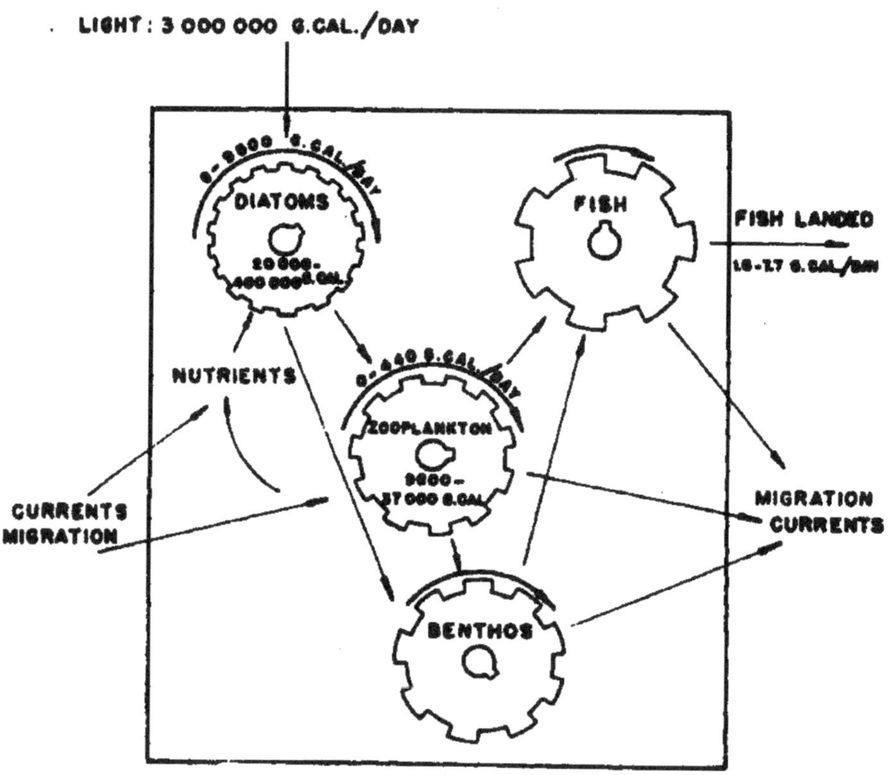

FIGURE 2. General network diagram used for Clarke (1946).

of any consequence were imposed in this region by the International Commission for the Northwest Atlantic Fisheries in the 1950s (Halliday and Pinhorn 1996). Finally, the prevailing world view of the surrounding culture was still very much an agricultural-machine perspective (Thagard 1994; Curry 2000; Simmons 2000). Collectively, these factors resulted in a limited ability to address complexity.

1940s conclusions.—One of the key observations made by Clarke (1946) was that marine ecosystems are less productive than agricultural ones. Not surprising, in hindsight, this observation reflects the agricultural view imposed upon the ocean rather than the obvious distinction between terrestrial and aquatic ecosystems or the fact that farming is effectively a highly controlled monoculture designed to maximize production.

The 1940s network primarily focused on terminology and description of the system. That Clarke (1946) was able to assemble disparate pieces of information and integrate them into a composite whole was quite remarkable for this pre-computer era. Expanding such concepts and descriptions into a quantitative framework was not trivial and, in some ways, mirrored the much more cited work by Lindeman (1942) in freshwater ecosystems. The ability to even measure production, particularly primary production, in the ocean at these scales, was fairly revolutionary at the time. In many ways, the conceptual basis for much of biological oceanography and marine ecology was established via the 1940s works on the Georges Bank food web.

The 1980s View

Context.—The network analyses constructed in the 1980s (Mills and Fournier 1979; Mills 1980; Cohen et al. 1982; Mills et al. 1984; Sissenwine et al. 1984) is represented by an image of an analog (predigital) energy flow diagram, similar to an electrical diagram (Figure 3). The fisheries on Georges Bank and the western Scotian Shelf were still primarily focused on large gadids and increasingly on flatfish, but fisheries for Acadian redfish were in decline. The emphasis of much of the Canadian and U.S. fisheries research focused on the nationalization of fisheries (particularly expelling foreign fleets from national waters and expanding the domestic fleet) and attempted to stabilize the stocks, particularly the small pelagic stocks (Atlantic herring *Clupea harengus* and Atlantic mackerel *Scomber scombrus*). This occurred via transitioning from an international to a nationally administered, regional management system. The Georges Bank modeling exercise was conducted in 1982 and repeated in 1984, with improvements in information and conceptual basis incorporated into the second edition. Similar network analyses were developed for the Scotian Shelf in Canadian waters (Mills and Fournier 1979; Mills 1980; Mills et al. 1984).

The representation of the network (or energy budget in the terminology of that time) is moderate in simplicity. The taxa shown are aggregated and represent a broader range than the 1940s scheme by Clarke (1946), yet the investigators were selective when choosing species for inclusion in this energy budget. The system representation is still of fairly low complexity, with about double the number of nodes and flows than the prior network but an order of magnitude or two less than subsequent networks (Table 1). The food web depicted still predominantly represented the classical grazer food chain but was quite progressive at the time by including bacteria and some energy flow through a microbial food web. As with the 1940s model, there was no mass balance constraint incorporated into the food web and the units were energy-based.

The major theoretical context at the time was centered on food web theory, building upon

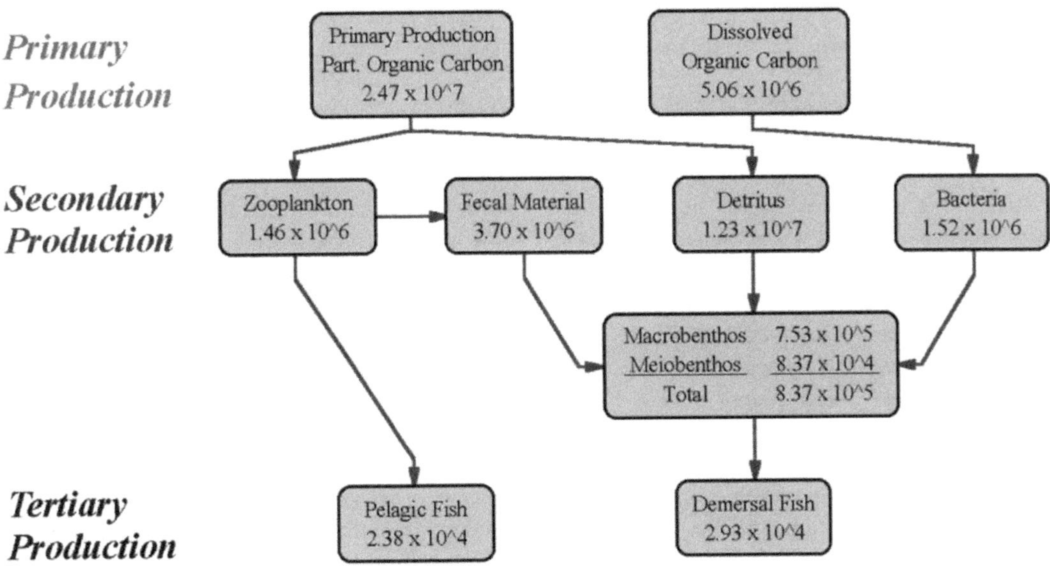

FIGURE 3. General network diagram used for Cohen et al. (1982; Sissenwine et al. 1984 not shown, but generally similar).

TABLE 1. Contrasts of the number of nodes, species interactions, and trophic levels among the different Georges Bank network studies.

	# nodes	# flows	Trophic levels
Clarke 1946	4	6	3
Cohen et al. 1982	9	9	3
Sissenwine et al. 1984	10	19	4
EMAX[a]	31	290	5

[a] See Link et al. 2006, 2008a, 2008b.

works by May (1972, 1973), the Odum brothers (e.g., Odum 1957, 1964) and Steele (1974). These theories were not only providing a template for energy flows, balances, and budgeting, but also spurred a generation of debate about the relationship between food web complexity and stability. The stability aspect was particularly germane in a fisheries context for planning of long-term yields. Also prominent in this era was the development of multispecies fisheries models (e.g., Andersen and Ursin 1977; ICES 1991) that were beginning to incorporate aspects of food web ecology into fisheries decision criteria.

Advances were also being made on the role of microbes. In the 1970s, Pomeroy (1974) noted the importance of microbes in the flow of energy and cycling of nutrients at the lower trophic levels in the oceanic food web. This paradigm shift away from the classical grazing food chain (diatoms-mesozooplankton-zooplanktivores-carnivirous fish) was later termed the microbial loop by Azam et al. (1983). Biological oceanographers focused on the role of the microbial loop in pelagic food webs and whether this was a sink for nonliving organic carbon (dissolved and particulate), which was ultimately respired, or a link to the grazing food chain through microzooplankton, mesozooplankton, and macrozooplankton. Furthermore, the Cohen et al. (1982) energy budget reflects some of this thinking, together with the role of bacteria attached to detritus (Mann 1972), on the relationship between dissolved organic carbon, particulate organic carbon, and bacteria in supporting the benthic food web based on the synthesis by Parsons and Seki (1970).

The 1980s were an important time in the histories of Canada and the United States following the extension of jurisdiction in 1977, which gave complete control of the fisheries within its exclusive economic zone to each nation. This shift led to a focus on both management of fish stocks and expansion of domestic fleets. Fisheries science was expanding, taking advantage of new computing tools, although central processing unit speed and processing speeds were low compared to today (Megrey and Moksness 2009b). Fisheries management had imposed a two-tier quota management system in 1974 (Murawski et al. 1997), which provided the nucleus for recovery of depleted stocks. This approach included explicit recognition and allowance for bycatch, discarding practices, and interspecific interactions (Brown et al. 1976). Thus, there was a growing need to understand the structure and functioning of ecosystems and the impact of fishing. Enhanced computing ability enabled more complex ecosystem models to be developed to explore these questions. Indeed, the two-tier system was a strong precursor to an ecosystem approach, but unfortunately the two-tiered system was never fully implemented (Murawski et al. 1997; Fogarty et al. 2008; NEFSC 2008). The prevailing world view at the time was effectively moving from a mechanical to an electronic basis of executing work (McCormmach 1970; Walsh 1972; Patten and Odum 1981; O'Neill 2001), with recognition that stochasticity in nature was a more realistic perspective—or at least as equally valuable—than a deterministic view.

1980s conclusions.—Three major conclusions were drawn from the studies conducted on Georges Bank in this period (Cohen et al. 1982; Sissenwine et al. 1984). First was that primary production was high for Georges Bank relative to other, comparable marine ecosystems, including the Scotian Shelf. Given this, secondary production of zooplankton was much lower than would be expected. One possible explanation for the low zooplankton production was the use of a *P/B* (secondary production to biomass) ratio of 7–8, a number now known as relatively low for this taxa group (Link et al. 2006, 2008b). It is also possible that a significant portion of the primary production is advected off the bank or exported out of the euphotic zone to support the benthic food chain, leaving the zooplankton with a lower food ration of phytoplankton and detritus. These hypotheses pointed out the need to account for advection and nutrient recycling, thus more broadly understanding the biophysical processes that determined productivity of the biota and fisheries in this ecosystem.

Second, given that primary production was high, despite the caveat noted above, secondary and tertiary production were also relatively high. Overall, Georges Bank was regarded as a very productive system relative to other marine ecosystems. A comparison with a similar energy budget constructed for the North Sea and the Scotian Shelf noted how much more productive (per unit area; by several factors, usually at least four to five) Georges Bank was (Mills 1980).

The third major conclusion, analytically speaking, was that fish recruitment and prerecruit processes were not well captured but that this information is critical to understanding overall system dynamics and energy flows. The second version of this network (Sissenwine et al. 1984) more explicitly considered prerecruit fish, recognizing their importance as food for larger fish, as competitors with pelagic fish and zooplankton, and as a critical intermediate trophic level link in the energy budget.

The Early 2000s View

Context.—The EMAX (Energy Modeling and Analysis eXercise) network for the early 2000s (Link et al. 2006, 2008a, 2008b) is represented by a circuitry diagram; in fact, the connections are so complicated that only a few flows can be usefully presented at any one time (Figure 4; just shows the network nodes sans any particular flows). The fisheries on Georges Bank and western Scotian Shelf were still largely focused on large gadids and flatfish, even though those stocks were only a fraction of their former size. Major community changes had occurred during the 1980s and 1990s. This included the mentioned declines in groundfish and significant increases in elasmobranchs. The small pelagic stocks and invertebrate stocks were the dominant fisheries in this era. The emphasis of much of the fisheries research on Georges Bank at the time addressed the trade-offs among different domestic stakeholder groups, biomass configurations, and other nontargeted species in the ecosystem. The data for the network analysis of Georges Bank were based on information compiled and updated from the late 1990s, primarily from 1996 to 2000. This network analysis utilized two software packages (Ecopath and Econetwork), reflective of different underlying modeling philosophies (Walters et al. 1997; Heymans and Baird 2000; Allesina and Bondavalli 2003; Kavanagh et al. 2004; Ulanowicz 2004; Dames and Christian 2006). We also note that these approaches were being developed for adjacent ecosystems on the Scotian Shelf (e.g., Bundy 2004, 2005b) and that there was a contemporary energy budget to the EMAX work for Georges Bank (Steele et al. 2007). The latter work emphasized different factors and nodes than EMAX but generally confirmed some of the same findings, particularly for primary and secondary production.

These models had a mass balance constraint, which represents the use of mass units rather than energy units and which explains why they resulted in a higher number of trophic levels (i.e., the constraints force biomass to be accounted for, and thus more explicitly retained, in the system; Table 1). Note that the prior eras' network models did not directly consider this constraint. The representation of this network, although a distinct simplification from other versions of the food web (e.g., Link 1999, 2002a), is much more complex than prior network configurations. The taxa included are still aggregated but are much more explicitly inclusive of a broad range of organisms, and the numbers of nodes and flows are orders of magnitude higher than in previous network analyses. The food web depicted has a more nuanced view of primary production, including both the classical grazing chain and the microbial loop, again inclusive of more taxa than solely diatoms or bacteria (e.g., microzooplankton, flagellates, a wide range of phytoplankton, and dissolved organic carbon). Network analysis examined both the

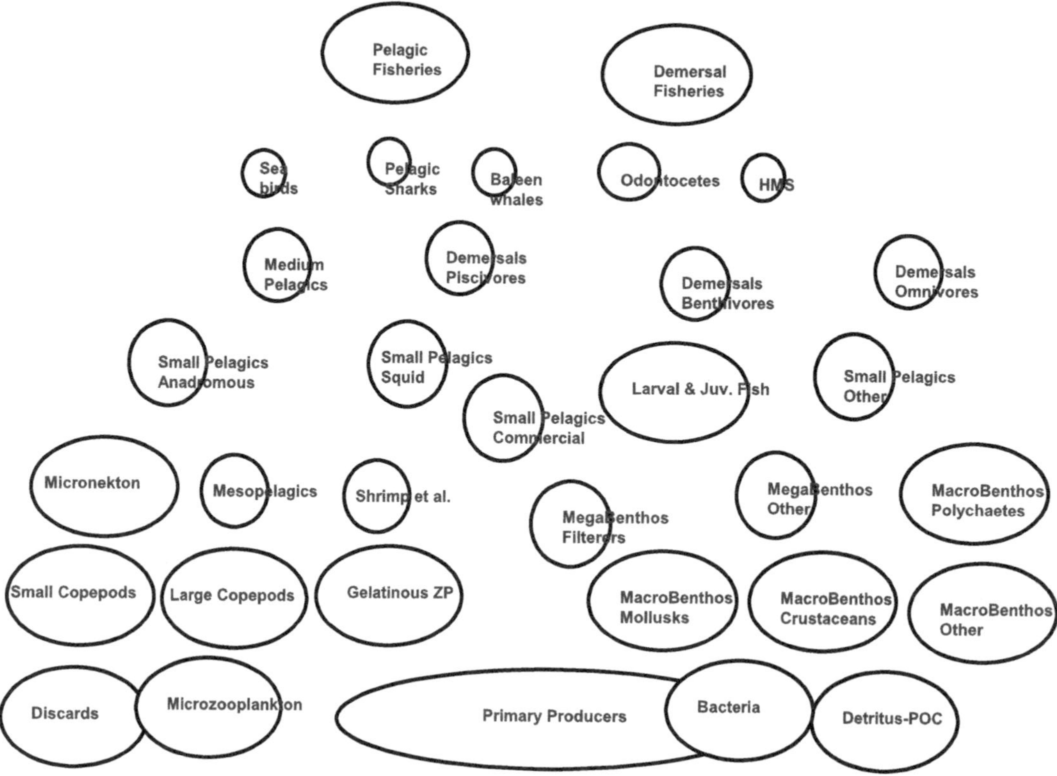

FIGURE 4. General network diagram used for EMAX (Link et al. 2006).

direct (phytoplankton-zooplankton) and indirect (detritus-benthos-demersal fish) flow pathways that result in greater system throughput and resilience in the modeled system than the classical linear energy budgets (where flows go from phytoplankton sources to various heterotrophic sinks). The effective food chain length (i.e., trophic levels) in network models is longer, in part, because of the incorporation of direct and indirect flows.

The major theoretical context at the time was centered on biocomplexity (Nicholis and Prigogine 1989; Lewin 1992; Pickett et al. 2005). This body of work pointed out the need to catalog diversity at various levels, explore the ramifications and structures of biocomplexity, and attempt to model complicated system dynamics.

By the advent of the 21st century, and as a consequence of global and local overexploitation of fish stocks and ecosystem change, an ecosystem approach to fisheries was being adopted as a more holistic approach to fisheries management (Larkin 1996; Link 2002b, 2002c; Garcia et al. 2003). Significant legislative and political emphasis has been placed on ecosystem-based management in the United States and Canada (NMFS 1999; Pitcher et al. 2009; Link et al. 2010), resulting in the need for a process-oriented approach to explore ecosystem structure and functioning and the impacts of fishing. Computing power, measured as millions of instructions per second, increased by 10,000 times since 1979 (Megrey and Moksness 2009b), and ecosystem modeling software programs have been developed to take advantage of this power and ease of access, thus enabling more complex ecosystem models to be developed and in this era. The prevailing world view at the time was circuitry- and computer-based (Murphy and Pardeck 1988; Changizi et al. 2002; Takahara et al. 2005), effectively an extension in complexity of the prior analog perspective of systems thinking to a digital perspective familiar to the modern reader. The mode of understanding things was cognizant of chaotic dynamics, effectively requiring a need for a more holistic and integrative view.

2000s conclusions.—The work from the early 2000s emphasized structure of the food web (Link et al. 2006, 2008a, 2008b; Steele et al. 2007). First, the 2000s' version of the network noted that bottom-up processes drive much of the system dynamics. The EMAX energy flow models contained a higher level of detail for the biomass and production at lower trophic levels than most network models that focus on the grazing food chain. In fact, many of the upper trophic level nodes of great commercial and conservation importance were quite minor in the overall energy flow of the system; furthermore, fisheries catch was a low proportion of overall energy flow.

Second, although estimates have improved over time for several of the network nodes, additional data gaps were identified. Many of these were at lower trophic levels (e.g., gelatinous zooplankton) that have the potential to drastically alter overall system dynamics.

Third, this network demonstrated that small pelagic fishes are important for overall energy flow in the system. Similar to the findings of the 1980s network for prerecruits, these mid trophic level forage fish serve as a critical set of links between upper trophic level biota and lower level producers.

Fourth, given the connectedness of all the species in this ecosystem, it appears that if one energy pathway is altered, another pathway compensates, such that overall changes in standing stock biomass at a given trophic level are minimized. These are all symptomatic of a highly productive and highly resilient system, confirming observations from prior Georges Bank network studies.

Finally, this work noted that many network and cybernetic system metrics can greatly elucidate overall system functioning and dynamics. These are a set of indicators, such as estimates of mixed trophic impacts, throughput, and biomass ratios across nodes that are integrative, holistic, and systemic in perspective and provide a distinct interpretation of food web dynamics than more classical taxonomic-based metrics.

Changing Food Web Views in the Gulf of Maine Region

Ecosystem modeling has moved from a food-web depiction based on farm machinery in the 1940s when the prevailing world view was largely an agricultural-machine perspective (Thagard 1994; Curry 2000; Simmons 2000) to a circuit diagram in contemporary times when the prevailing world view has become one involving widespread digital and computer applications. Our ability to conceptually and computationally address connectivity of ecosystems is much greater now than 70 years ago (Beck 1999; Megrey and Moksness 2009a, 2009b). As little as several decades ago, we did not have the theory or the computing power to examine food webs in the holistic, integrated fashion that we do today. Thus, we see a significant shift in scientific thinking from reductionist, mechanistic determinism to more connected, complex, and stochastic phenomena as we understand them today. Contemporary models, largely developed to support an EAF, have begun reconnecting the conceptual approaches of oceanographers and fishery scientists to better appreciate how the ocean food web operates. The future challenge is to continue to integrate these perspectives.

The history outlined above focused on food-web models. In fact, the continued fostering of EAF (Larkin 1996; NMFS 1999; Jennings et al. 2001; Link 2002b, 2002c; Garcia et al. 2003) requires various ecosystem modeling approaches to adequately address the various dimensions of fishing, ecology, and the environment. In a fisheries context, this has highlighted the need to consider a broader range of factors beyond fishing that could affect fish stocks and conversely a broader range of impacts from fishing beyond direct effects on targeted species. In the next section, we present ecosystem modeling approaches to that end.

Models to Provide Tactical and Strategic Living Marine Resource Management Advice

Examining the use of models to support implementation of an EAF in the United States and Canada must first consider how single-species management approaches have been adapted for this purpose and then consider multispecies methods. Integrated aggregative and ecosystem-level frameworks that may play a significant role in the future of EAF are subsequently considered. Although extensive, we do not describe the solely single species approaches that have had a long history in the GoM region. We note that ecosystem models vary in complexity from extended single-species models (i.e., single species models with add-ons such as an environmental factor or predation-caused mortality) to complex models that encompass selected aspects of the entire ecosystem.

Different types of models are used to explore different questions (Table 2), and they contribute to the scientific information required for supporting management decisions in different ways. They are often used to further conceptual understanding, such as addressing various process-driven questions. They also have the capacity to provide information required for strategic directions for management, which are long-range, broad-based, and linked to policy goals. Occasionally, ecosystem models can be used for tactical decision making, which is an evolving field of research (Plagányi 2007; Townsend et al. 2008). Management advice in Canada and the United States is often founded upon outputs from population models that assess the status of living marine resources and their associated mortalities. Multispecies models, coupled with extended stock assessment models (ESAMs), are collectively known as minimum realistic models (MRMs); these MRMs represent a subset of the ecosystem, where a limited number of species that have specific interactions with a focal species of interest are included. A suite of MRMs has been developed that include a series of ESAMs and multispecies models to explore the relative magnitude of factors that can influence mortality and growth, as compared to fishing mortality. Yet management for sustainable ecosystems needs to consider the impact of fisheries on ecosystems and vice versa. This means consideration of the impacts on target, bycatch, nontarget species, and physical damage to habitat and food webs; these questions cannot be fully explored with single-species stock assessment models. Nor can these questions all be addressed within one modeling approach. Thus, several more aggregated and system level models have also been developed, and more lately, habitat models have also begun to be developed for the GoM area (e.g., Cogan et al. 2009; Smith et al. 2009). These models do not produce tactical advice per se; rather, they are suited for strategic advice that helps to determine feasibility and probable response of the ecosystem to a range of simulated, possible management scenarios. Here, we review the modeling efforts for the GoM region, starting with the simpler extended single-species models, then moving to more complex models. Although regularly reported upon in broader contexts (ICES 2007, 2009; Townsend et al. 2008), here we note the range of these models as specifically applied to the LMR issues in the GoM region. Again, we recognize and appreciate that the many salient works on biophysical, bioeconomic, stock assessment, habitat, and other (academic) multispecies models are extant in the GoM, but our focus here is on how these ecosystem models have begun to become operational in an LMR management context.

Extended Single-Species Assessment Models

A number of ESAMs have been developed in the GoM region (Plagányi 2007; Townsend et al. 2008), through add-ons to single species formulations to account for predation, consumptive demands, or the environment in a single species assessment model. These have been both age- or stage-structured and bulk biomass or production models. The purpose of these ESAMs has ranged from providing context for stock biomass estimates, providing tuning indices, serving as sources of other mortality, informing modifications to key parameters, serving as reality checks for estimates of magnitude of population estimates, and even providing explicitly modeled estimates of predation mortality.

Single-species add-ons: predation.—These models describe the impact of predation and its effects on a stock in a single species assessment model. These models have been developed predominately for forage stocks, including Atlantic herring, Atlantic mackerel, longfin inshore squid *Loligo pealeii*, butterfish *Peprilus triacanthus*, and northern shrimp *Pandalus borealis* (NEFSC 2007a, 2007b; Overholtz and Link 2007; Overholtz et al. 2008; Link and Idoine 2009; Moustahfid et al. 2009a, 2009b). Others are planned, both in Canada and the United States. Several of these models have been used as part of formal U.S. stock assessment reviews, usually to provide context and estimates of predation mortality (M2). For the most part, predation in

TABLE 2. Major model classes as typically applied to common objectives of model use. Adapted from Townsend et al. (2008). ESAMs = extended stock assessment models; MSVPA = multispecies virtual population analysis; EMAX = Energy Modeling and Analysis eXercise; CDEENA = Comparative Dynamics of Exploited Ecosystems in the Northwest Atlantic.

		Generic model classes Specific Gulf of Maine model examples noted for some classes				
		Single species	Single species with add-ons	Multi-species	Aggregate biomass	Food web
		ESAMs: footprints	ESAMs	MSVPA, MSPROD, Allometric	AggProd, GOMAGG	EMAX, CDEENA
Major topics	Common issues and objectives					
Technological interactions	Technological interactions		x	x	x	x
	Bycatch	x	x	x	x	x
Trophic/ecological interactions	Protected species and species of interest	x	x	x	x	x
	Commercial fishing on forage species		x	x	x	x
	Trade-offs among predators being targeted		x	x	x	x
	Predation of targeted species		x	x	x	x
Physical/climate drivers	Effects of fishing on habitats					
	Habitat effects on stocks		x	x	x	x
	Climate		x	x	x	x
	Cumulative effects					x
	Toxins/bioaccumulation					x
Spatial features	Marine protected area efficacy, structure, placement					
	Range shifts		x			
	Habitat restoration strategies					
System considerations	Invasive species					
	"Ecosystem health"—sustainability, resilience				x	x

TABLE 2. Continued.

		Generic model classes Specific Gulf of Maine model examples noted for some classes				
		Single species	Single species with add-ons	Multi-species	Aggregate biomass	Food web
Major topics	Common issues and objectives	ESAMs: footprints	ESAMs	MSVPA, MSPROD, Allometric	AggProd, GOMAGG	EMAX, CDEENA
System considerations (continued)	Ecosystem status				x	x
	Biodiversity				x	x
	Underlying system carrying capacity				x	x
	Regime shifts					
Socioeconomic drivers and management	Economic issues	x			x	x
	Trade-offs among fleets			x		
	Determining reference points—systemic				x	x
	Determining reference points—single species assessments	x	x	x		
	Cumulative management effects				x	x

Table 2. Continued.

Major topics	Common issues and objectives	Generic model classes Specific Gulf of Maine model examples noted for some classes				
		Habitat	Bio-physical	Biogeo-chemical	Bio-economic	Full system
						ATLANTIS
Technological interactions	Technological interactions				x	x
	Bycatch				x	x
Trophic/ecological interactions	Protected species and species of interest		x		x	x
	Commercial fishing on forage species				x	x
	Trade-offs among predators being targeted					x
	Predation of targeted species		x			x
Physical/climate drivers	Effects of fishing on habitats	x		x		x
	Habitat effects on stocks	x	x			x
	Climate	x	x			x
	Cumulative effects	x	x			x
	Toxins/bioaccumulation			x		x
Spatial features	MPA efficacy, structure, placement	x	x		x	x
	Range shifts	x	x			x
	Habitat restoration strategies	x		x		x
System considerations	Invasive species					
	"Ecosystem health"—sustainability, resilience					x
	Ecosystem status		x			x
	Biodiversity					x
	Underlying system carrying capacity	x	x			x
	Regime shifts		x			
Socioeconomic drivers and management	Economic issues				x	x
	Trade-offs among fleets				x	x
	Determining reference points—systemic		x	x		x

Table 2. Continued.

		Generic model classes Specific Gulf of Maine model examples noted for some classes				
		Habitat	Bio-physical	Biogeo-chemical	Bio-economic	Full system
Major topics	Common issues and objectives					ATLANTIS
Socioeconomic drivers and management (continued)	Determining reference points—single species assessments					
	Cumulative management effects				x	x

these models is considered as an additional "fleet." That is, predation by other species (than the target stock) is treated collectively, but explicitly, as another source of removals. The data required, in addition to the usual survey and fisheries catch data, are abundance of predators of the stock of interest, predator stomach contents, estimated consumption rates, and diet composition estimates.

The positive aspects of this approach are that such models are relatively simple conceptually and operationally, use extant data, are implemented in a familiar assessment and management context, provide familiar (albeit modified) model outputs amenable to calculating biological reference points (BRPs), improve the biological realism of assessment models, and help to inform and improve stock assessments for species that may have been difficult to assess in the past. The negative aspects are that, like all minimal realistic models, they may be missing a suite of complex interactions and nonlinear responses, caused by not including the full suite of interactions involved in a real-world ecosystem. They also have the potential to be controversial, by producing more conservative BRPs and emphasizing the potential for competition between predators and fleets that target these stocks, without having a fuller modeling capability to fully address these trade-off issues.

Single species add-ons: ecological footprints.—The models in this category attempt to account for the amount of food eaten by a fish stock. These estimates of energetic requirements (i.e., consumptive demands), at a given abundance level, are then contrasted to estimates of the amount of food known to be available in the ecosystem from surveys and mass-balance system models. In many ways, this is the same calculation as noted above for predatory removals; the difference here is that, instead of summing across all predators feeding on a stock of fish, here it is summed across all species serving as prey for the fish.

These models have been developed for a wide range of U.S. groundfish, elasmobranch, and pelagic fish species (Link and Garrison 2002; NEFSC 2007b; Tyrrell et al. 2007; Link and Sosebee 2008), with similar work being developed for some Canadian stocks. Estimates for a few sets of stocks (e.g., the skate complex, NEFSC 2007a; Link and Sosebee 2008; spiny dogfish *Squalus acanthias*, pollock *Pollachius virens*, goosefish *Lophius americanus*, Northeast Fisheries Science Center, unpublished data) have gone through a formal stock assessment model review[1]; others are in various stages of development. In addition to survey and fisheries catch data, the data required are abundance of the focal stock, stomach content and diet composition estimates, and consumption estimates. The positive and negative aspects of this approach are similar to those outlined above for predatory applications of this approach.

Single species add-ons: environmental considerations.—In the United States, scientists have begun to incorporate environmental considerations into population models, but not yet in a fully operational mode. These include changes in carrying capacity (*K*), population growth rates (*r*), stock–recruitment relationships, or stock distribution relative to environmental conditions (Keyl and Wolff 2008). These have been done, or are being done, for a wide range of fish, mammal, and invertebrate species. With environmental terms in population models, it is possible to forecast the response of a population to climate change, thereby providing a long-term forecast that can inform EAF (Fogarty et al. 2008; Hare et al. 2010). Brander and Mohn (2004) incorporated the North Atlantic oscillation (NAO) into stock–recruitment models of 13 Atlantic cod stocks in the North Atlantic, recommending that medium- and long-term stock assessments consider likely future states of the NAO in areas where the NAO had a strong effect. Currently, none of these models have been through formal model review nor explicitly incorporated into a review process that directly informs management. Such modeling remains an active area of research and development.

In addition to the needs of a standard stock assessment, these approaches require appropriately (spatiotemporal) scaled environmental data, such as temperature, depth, and salinity, and the associated monitoring data products.

[1] Here and throughout, when we note "formal model review," we mean that although the model may be in the peer-reviewed literature, the model has not been evaluated by a review panel as to its behavior, dynamics, diagnostics, and implementation all to determine even if it is suitable for use in a LMR management context in the first instance and its results for an actual implementation applied to a particular situation thereafter.

The advantages of this approach are that the environmental data are usually available, and relating them to stock dynamics typically takes advantage of commonly established statistical methods. These approaches also improve the biological realism of assessment models and allow for consideration of dynamics driven by factors typically outside of usual assessment considerations. The chief drawback of this approach is that the data are often auto-correlated without definitive causal mechanisms; similarly, environmental correlates have a noted history of decoupling with additional data, and the data may also often be collinear and, short of exhaustive multivariate analysis, are difficult to untangle for useful stock projection.

Multispecies Virtual Population Analysis

Multispecies virtual population analysis (MSVPA) is one of a suite of multispecies models that focuses on age-structured populations of commercial importance. The MSVPA approach was developed within the International Council for the Exploration of the Sea context in Europe and is, in effect, a series of single-species VPAs linked together via a feeding model. The modeling approach has the ability to provide short-term forecasts. Most typically, the model examines the stock dynamics of multiple species that are both predators and prey, particularly exploring the role of predatory removals of stocks relative to fishery removals.

An extended version of MSVPA (MSVPA-X) has been developed in the GoM region, which, among other improvements, includes predators without age-structured assessment data and has multiple forms of VPAs for each species, thus enhancing the flexibility of the approach. The MSVPA-X models have been applied to two subsystems in the northwest Atlantic and are being developed for a third (Garrison and Link 2004; NEFSC 2006; Tyrrell et al. 2008; Garrison et al. 2010). An MSVPA-X model for the mid-Atlantic region emphasizes Atlantic menhaden *Brevoortia tyrannus* as prey with three main predators and has gone through extensive and formal model review (NEFSC 2006). Outputs from that model have informed single species assessments, particularly by providing time series of predation mortalities for the assessment of menhaden. A second MSVPA-X model applies to the southern New England–Georges Bank–Gulf of Maine ecosystem (Tyrrell et al. 2008). It involves 19 species and emphasizes Atlantic herring and Atlantic mackerel as the major prey. The results have contextually informed single species assessments for herring and mackerel. The third application applies to the southwestern Scotian Shelf/Bay of Fundy area, with a focus on Atlantic herring as prey (Guénette and Stephenson, in press).

The data required for this approach include abundance estimates for predators that eat the stock of interest, stomach contents, consumption estimates, and diet composition estimates (in addition to survey and fisheries catch data).

The positive aspects of this approach mirror those of the single species add-on with predation; namely, it uses extant data, is implemented in a familiar assessment and management context, improves the biological realism of assessment models, and helps to inform and improve stock assessment outputs. The key negative facets of this approach are that it is quite data-intensive and there is no feedback loop between predator and prey.

Bioenergetic-Allometric Models

Bioenergetic-allometric models are based on the modeling framework originally developed by Yodzis and Innes (1992). This approach describes the dynamics of predator–prey systems in terms of biomass and relies on allometric relationships between vital rates (e.g., respiration, maximum consumption) and individual body mass to reduce and/or constrain the number of parameters to be estimated. The theory behind this approach has been expanded in two critical aspects. On the grounds provided by the metabolic theory of ecology (Gillooly et al. 2001; Brown et al. 2004), Vasseur and McCann (2005) incorporated the effect of temperature on vital rates for ectotherm species. More recently, De Roos et al. (2008) develop a way of incorporating juvenile and adult stages, which captures most of the behavior of more complex physiologically structured models (e.g., food-dependent growth and maturation).

Data requirements for these models include time series estimates of biomass and catch, feeding relationships, and individual body mass. This approach provides a high degree of flexibility for handling model complexity. Based on earlier work by Yodzis (1998) and Koen-Alonso and Yodzis (2005), bioenergetic-allometric models are currently being developed for Atlantic Canadian systems. Koen-Alonso and Bundy (2009) and Koen-Alonso et al. (2008) are developing a five-species model to explore the dynamics of core species of the eastern Scotian Shelf marine community. This work allows for comparisons with the results of previous modeling exercises for this system (e.g., Bundy 2005a, 2005b).

The pros of this approach are that the model structure can extend from a minimal realistic format to entire food webs; it uses allometric relationships, which reduce its parameter needs; it can explore different dynamics (static, local, and global); and it includes the possibility of incorporating temperature, and other physical variables, into the basic allometric equations using first principles. The cons of this approach are the inability of the approach to address age structure, its potential to produce complex and potentially unstable dynamics, and its inability to handle management objectives in terms of goals or thresholds for the biomass of the species. Furthermore, it is data-intensive and not entirely user-friendly.

Multispecies Production Models: MS-PROD

A multispecies extension of the Schaefer production model has been developed in the United States to include both predation and competition. The model is a simulation tool and incorporates a wide range of what are primarily ecological processes (Link 2003; Gamble and Link 2009). The chief aim of this model is to simulate the relative importance of predation, intra-guild competition, inter-guild competition, and fisheries removals.

An MS-PROD model has been parameterized for 25 species from the Georges Bank region, with empirically based values that can be used to explore sensitivities and scenarios for different considerations. It was not designed to be directly used for management advice. Nonetheless, it has proved useful in providing contextual information regarding the ecosystems influenced by fisheries and for simulating options for LMR management. The data required are initial biomass estimates, carrying capacities, predation and competition interaction terms, growth rates, and fishery removals. Again, some simulation results have been used to provide context to management of LMR (NEFSC 2008).

The desirable aspects of this approach include explicitly accounting for ecological processes, in addition to the effects of fisheries, and inclusion of lower trophic level processes that can be directly linked to estimates of carrying capacity. Limitations include the fact that some of the parameters, although empirically derived, are difficult to estimate. Another negative is that, like most multispecies models, it is parameter intensive but less so than many other multispecies models, given the simplicity of the model equation structure.

Multispecies Production Models: Agg-PROD

This type of model is effectively the same as the MS-PROD model noted above but initialized for aggregate groups of species (i.e., many species are not represented individually). The interactions with other ecosystem components in these groups have been parameterized, both as functional guilds and taxonomically related species. The one distinction from MS-PROD is that this model simulates BRPs and production at a more systemic or group level, rather than species level. This model could be useful for considering two-tier quotas by which there are both limits per stock individually and for a full group of stocks collectively. The data needs, pros, and cons are the same as MS-PROD, with the caveat that amalgamation of parameters across groups remains a challenge. Again, simulation results have been used as contextual information for management of the U.S. fisheries influence on LMRs in the GoM region (NEFSC 2008).

Food Web Models

The Ecopath with Ecosim (EwE) model has been widely used to describe aquatic systems and to explore the impacts of fishing ecosystems (Christensen and Pauly 1993; Christensen et al. 2005; Arkema and Sambouri 2012, this volume). It is composed of a mass balance model (Ecopath; Polovina 1984; Pauly et al. 2000; Christensen et al. 2005) from which temporal (Ecosim) and spatial (Ecospace) dynamic simulations can be developed (Walters et al. 1997). Mass balance (Ecopath) models have been developed for many regions across the northwest Atlantic: the Newfoundland-Labrador Shelf (Bundy et al. 2000; Heymans 2003), the northern and southern Gulf of St. Lawrence (Morissette et al. 2009), the Bay of Fundy, western Scotian Shelf (Araújo and Bundy 2011), the eastern Scotian Shelf (Bundy 2004, 2005b); and, for the Gulf of Maine, Georges Bank, southern New England, and middle Atlantic Bight ecosystems (Link et al. 2006, 2008a, 2008b). The U.S. ecosystems were similarly modeled using the Econetwork software (Kavanagh et al. 2004; Ulanowicz 2004; Dames and Christian. 2006). Ecosim models have been developed for the Newfoundland-Labrador Shelf (Bundy 2001), the Bay of Fundy, western Scotian Shelf (Araújo and Bundy 2011), and the eastern Scotian Shelf (Bundy 2005a).

Data requirements for these models include estimates of biomass, production and consumption rates, catch, and diets. These food web models in the GoM region have been developed under specific projects—CDEENA (Comparative Dynamics of Exploited Ecosystems in the Northwest Atlantic) and the Maritimes Region Ecosystem Research Initiative in Canada and EMAX in the United States.

These models have been used to further understanding of ecosystem structure and functioning, explore hypotheses concerning ecosystem change, used as a basis for comparative studies (spatial and temporal), used to provide ecosystem indicators, and used in various simulated perturbation experiments. Performance measures and metrics such as throughput, total flow, biomass ratios (e.g., pelagic fishes to zooplankton), and trophic reference points (i.e., marine mammal biomass to pelagic fish biomass) can be tracked and compared with empirical information over the simulated time horizon. The use of these models remains an active area of research. Some results have been used as contextual information in a LMR management context (Bundy et al. 2008; NEFSC 2008), and planning is underway to use these models in a management strategy evaluation (MSE) context in the GOM region.

The big advantages of this approach are that it encompasses the whole ecosystem and is conceptually simple, versatile, accessible, and adaptable. The cons of this approach are that because of the widespread availability EwE, it can be misused. It also requires a plethora of data and parameters to initialize the model that are not routinely collected, either in terms of process or taxa group, in a fisheries context.

GOMAGG

A dynamic simulation model of the GoM ecosystem has been constructed, with the system partitioned into 16 aggregated biomass nodes spanning the entire trophic scale from primary production to seabirds and marine mammals (Overholtz and Link 2009). Parameters from the Ecopath model of the GoM ecosystem were used to construct a simulation model using recipient-controlled equations to model the flow of biomass, and the biomass update equation used in Ecosim, to model the annual biomass transition. As with EwE, GOMAGG produces performance measures that can compared with empirical information over the simulated time period. The model has been used to evaluate how the GoM ecosystem might respond to large- and small-scale changes to the trophic components and system drivers, specifically events such as climate change, fishing scenarios, and system response to changes in the biomass of lower and upper trophic levels.

GOMAGG has not been through a formal model review. This remains a research tool and has not yet been used in informing management, but GOMAGG simulation results have informed other modeling efforts. Other production, production capacity, and similarly aggregated dynamic models are also under exploration (NEFSC 2008). The pros of this approach are that it examines the food web dynamically and utilizes extant model structures and data. It has the ability to simulate a wide range of scenarios. The chief negatives of this approach are that it is not entirely user-friendly and is can be difficult to validate some scenarios and inputs.

ATLANTIS

ATLANTIS (Fulton et al. 2004) is by far the largest, most complicated model in use in the GoM region. Generically, ATLANTIS integrates physical, chemical, ecological, and fisheries dynamics in a three-dimensional, spatially explicit domain. In addition to ecological interactions, it contains environmental components, including a simulated ocean with all its complex dynamics, a simulated monitoring and assessment process, a simulated set of ocean uses (namely fishing), and a simulated management process. The dynamics represented in the model range from solar radiation to hydrodynamics, and it includes nutrient processes, growth (with age structure), feeding, settling, sinking, migration, fishery captures, fleet dynamics, market valuation, regulation, and feedback among the various components of the model, as appropriate.

The ATLANTIS application of the northeast U.S. Continental Shelf ecosystem (Link et al. 2011) is composed of 30 regional boxes, five depth layers per box, 12-h time steps for 50 years, 45 biological groups, and 16 fisheries. Model parameterization and initialization required more than 60,000 parameters and 140,000 initial values to estimate. A first level of calibration has been completed to ensure that basic biophysical processes match observed dynamics. Second- and third-level calibrations have also been completed, thus ensuring that fishing processes (catch and effort, respectively) are reasonable. Future scenarios of different management strategies will follow the third level of calibration.

Although parameterized, initialized, and loosely tuned to empirical values, the ATLANTIS application of the U.S. northeast is too complex and was not designed to provide specific tactical management advice for a particular stock (e.g., a quota or effort limit). Rather, ATLANTIS is not only a research tool, but a simulator to guide strategic management decisions and broader concerns. For instance, it has been used in other contexts to provide multispecies fisheries and multisector ocean uses in support of an EAM (Smith et al. 2007). The northeast U.S ATLANTIS application has not been through a formal model review, although documentation of key parameters and calibration is extant (Link et al. 2011). It will likely serve as a key operating model in future MSE applications.

The advantage of ATLANTIS is that it can incorporate multiple forms of myriad processes, can emphasize those considerations and processes most appropriate for a given system, and can be used to evaluate management decisions to provide insight into what might happen in a real system (i.e., as an operating model in an MSE context). Another advantage is that it covers a wide range of biota and is flexible or adaptive to a range of key factors. The chief negative aspect of ATLANTIS is that it is unwieldy in its complexity and takes an inordinate amount of time to parameterize, initialize, calibrate, and run any particular application. Additionally, the validation routines and capabilities of ATLANTIS are minimal at best, requiring much further improvement.

Gulf of Maine Ecosystem Modeling Summaries

A range of ecosystem models are extant in the GoM region, from MRMs to EwE to ATLANTIS. It is also noted that there are other models in this region not specifically discussed here that are being constructed by some partners and colleagues in the region (e.g., Collie and DeLong 1999; Tsou and Collie 2001; Hall et al. 2006; Steele et al. 2007). Although not described in any detail in this work, habitat and biophysical models address yet other sets of issues (Chen et al. 2001; Franks and Chen 2001; Townsend et al. 2006; Hu et al. 2008; Cogan et al. 2009; Smith et al. 2009). The use of these models is increasing with time, as the data to support (i.e., initialize, parameterize, calibrate, and validate) them is quite extensive for the GoM region, the computing power to execute models that can handle increasing ranges of complexity is readily available (Megrey and Moksness 2009a), and the need to consider more than one isolated species is readily apparent.

What has been interesting to note is that in an LMR context, many of these models have advanced well beyond the research tool stage and are being applied to inform management advice (Link et al. 2010). Many of the proof of concepts, feasibilities, identification of robust functional forms, and basic model sensitivities have been undertaken. What remains is to take the gamut of ecosystem models and characterize their uncertainties and utilities (Link et al. 2010) for further inclusion in LMR management and to better support an EAF. Exactly how such models shall be used is the next important consideration.

Appropriate Uses of Ecosystem Models in the Gulf of Maine Region

Ecosystem models in the GoM region have been used to further understanding of ecosystem structure and functioning, as central pieces of broader comparative studies (spatial and temporal), to develop ecosystem indicators, and in perturbation "virtual" experiments. While these models continue as an active area of applied and basic research, some have been used in a LMR management context (such as for the groundfish carrying capacity issue [NEFSC 2008]). Others (e.g., MRMs) are now explicitly used in a stock assessment context.

Given the wide array of ecosystem models developed and in use in the GoM region, the question begs, "What is the most appropriate use of these models for any given situation?"

The management of LMRs requires both tactical (i.e., what is the value and level of a BRP to determine stock status?) and strategic advice (i.e., what management strategies are feasible, viable, and likely to best achieve management objectives?). Certain models are better suited to address some questions better than others; prior efforts (Plagányi 2007; Townsend et al. 2008) have attempted to map the type of model to its best use for research and management applications. We note this mapping as applied to example models from the GoM region (Table 2). We assert that a generic "ecosystem modeling" activity, if applied to the wrong type of question for which a particular ecosystem model class was not designed, could actually dampen further efforts to implement these models and thus hamper their ability to better support EAF and ultimately lead to potentially spurious or negative management implications.

Models like the ESAMs and multispecies approaches noted above better address issues such as specific physical/climate drivers, trophic interactions, or technical interactions. Some of the aggregated, allometric, or GOMAGG approaches can also explore these types of questions but may be better suited to the examination of system carrying capacity and related systemic issues. Food web models, such as EwE or Econetwork, can explore trophic issues in particular and can be used to explore management trade-offs among species. By their nature, full system models such as ATLANTIS can explore a wide range of questions but are best suited to explore a range of trade-offs among multiple system components beyond solely the biota, particularly elucidating the relative prominence among a myriad of processes; all as applied across a range of management strategy evaluation exercises.

Lessons Learned and Possibilities for the Future

We trust that this presentation of ecosystem models in the GoM region has demonstrated how such models have changed over time. We also hope that the use and development of such models continues. We additionally trust that the full range of models presented demonstrates a wide range of existing tools available for application for an EAF. We discuss some of the experiences with these models we have observed in the GoM region to perhaps shed insight into how they could be used and be further developed.

It is relatively easy to see the value of MRMs as tools to assist application of an EAF. Yet, somewhat surprisingly, the information from MRMs has only rarely been utilized in a fisheries-management context specifically directed to stock assessments, despite the large amount of effort applied to that end (e.g., NEFSC 2006, 2007a, 2007b). Essentially, the information is there, the underlying mechanisms are mostly understood, and the data are mostly no less certain than other data used in the assessment and management process. Certainly, there are aspects of estimation and precision uncertainty that can increase by including additional data on predator–prey interactions, but these are largely outweighed by the decreases in process, magnitude, and accuracy uncertainty that are associated with including this extra information. We suspect that a lack of familiarity and "comfort" with these approaches, along with ill-informed expectations over precision, has mainly precluded their inclusion in the stock assessment process. To be fair, it could also simply be a healthy respect for the limits of modeling, single species and multispecies alike. We also suspect that, particularly for models that include environmental factors, the challenge of predicting future states has limited their use. However, the skill of environmental models is improving and the ability to couple climate, environmental, and population models is developing rapidly (Hollowed et al. 2009; Fulton 2010; Hare et al. 2010).

All of that said, we are encouraged that such "ancillary" information has been evaluated in the stock assessment process to provide "contextual" assessments that are reviewed along with the primary assessment. Certainly more research is required, but what is encouraging is that much of this work is now at the stage of focusing on sensitivity analyses or model diagnostics, having already accomplished proof of concept and understanding of basic, underlying mechanisms.

From these and related observations, it has become apparent to the authors that distinct venues for evaluating ecosystem models beyond MRMs are required (Link et al. 2010). Perhaps a bigger factor is management institutions structured along single species lines that cannot easily accommodate multispecies advice. This highlights a dimension of EAF that has not been raised here—the need for management institutions to adapt to accommodate an EAF. Certainly, some of the MRMs could and have been incorporated into existing stock assessment review frameworks. Yet many of the more aggregative, food web, and full system modeling approaches need to be evaluated by a subtly but importantly different set of expertise. These models are quite distinct from those for solely single species protected species, or targeted species approaches currently used to support LMR management. Additionally, the sources and types of uncertainty are distinct and require review panels more familiar with the nuances of this broader array of considerations. Further, the outcomes of these models, being largely strategic, are addressing different terms of reference than what is currently done in typical LMR assessment review venues. Review panels more familiar with the nuances of this broader array of considerations are needed. We assert that general familiarity with these models is nascent but growing, and we support efforts to develop modeling capacity, as well as standardized and codified use of ecosystem models in this context (e.g., Plagányi 2007; Townsend et al. 2008; Link 2010; Link et al. 2010).

Importantly, these models have brought a better understanding of what we do not know. In many respects, these modeling efforts have served as a veritable catalog of disparate data sets from which we can identify major data gaps, a valuable outcome of these efforts. These observations particularly center on species groups such as those associated with the microbial loop, krill, most benthos (on a synoptic, broadscale, real-time fashion), mesopelagics, gelatinous zooplankton, and seabirds. Clearly, further work to understand and monitor those species will be invaluable for further insights into ecosystem functioning. It is sobering that a taxa group such as gelatinous zooplankton, which has typically not received much attention in LMR contexts, can potentially influence total system dynamics, including LMRs of interest. Further, the vital rates of many species are known at only a cursory level, but a better resolution of such rates will lead to enhanced parameterization and development of the ecosystem models. In addition, further exploration of the socioeconomic elements of EAF, particularly related to trade-offs, also merits further attention.

We reiterate a critical point of EAF and the use of these models: confronting trade-offs. We have begun to do so but need to further expand use of these models to explore the range of feasible ecosystem configurations relative to national policies, laws, and objectives in both countries of the GoM region. What has emerged from the modeling thus far is that EAF is not apt to be an optimization exercise, but rather an approach to avoid undesired ecosystem states and to identify those management approaches that are most robust for LMR management. Ignoring trade-offs among objectives is not prudent; doing so could result in unanticipated consequences as a result of unexamined trade-offs among the biota of, or sectors exploiting, an ecosystem. We assert that ecosystem models provide a tool to explicitly state and explore the range of viable options among potential trade-offs in species, harvest, and management tools. As we have noted, this will require establishment of new institutional and formal processes to express and discuss these trade-offs, which is as important as the actual modeling.

Exploring the use of model outputs at the aggregate or systemic levels is something that is sorely needed and requires much greater attention in the near future. This is an excellent example of needing to address biological trade-offs. For instance, a recent assessment meeting in the United States noted that the sum of all species, B_{MSY}, was greater than that as modeled in aggregate, for the system (NEFSC 2008). This confirms several prior studies (Garrod 1973; Pope 1975; Brown et al. 1976; May 1976; Pope 1979). Exploring the use of system level or aggregate group level BRPs remains an important feature of many of these models.

Similarly, there are also plans to apply management strategy evaluation (MSE) to the fisheries in some of these ecosystems. MSE (Smith et al. 1999, 2007; Sainsbury et al. 2000) takes what is known now, places that information in an adaptive framework, simulates a range of management options or scenarios from a wide range of operating models, and then reports the outcomes of these virtual experiments. The goal of doing this is to identify management options that are robust to uncertainty and will meet as many of the legislative mandates as possible, while affording managers the flexibility to adapt to changing conditions. In eastern Canada, MSE has been used for Greenland halibut *Reinhardtius hippoglossoides* (Miller et al. 2008; Shelton and Miller 2009), and there are plans to extend this to additional species and species groups in both countries. In the northeast United States, several preliminary discussions have occurred with the regional fisheries management councils (both in the Mid-Atlantic and New England) and their supporting scientific and statistical committees (SSC). The councils' SSCs have a keen interest in ecosystem approaches, as doing so affords the opportunity for enhanced coordination across all managed species, as well as holding the prospect for actually simplifying the (assessment) process, particularly if a more aggregated production approach is considered. Similarly, in Canada, the Fisheries Resource Conservation Council, which advises the Fisheries Minister on research and assessment priorities, advocates an EAF. Many of the models noted here could serve as the operating model in an MSE context. Using an MSE allows those management institutions to test drive various options before their actual implementation.

Both countries are also moving towards some form of integrated management. Integrated management (IM) has been operating at a pilot level in Canada and integrated ecosystem assessments (IEAs) are planned for U.S. ecosystems. At their core, IMs and IEAs seek to assess the status of an ecosystem, cognizant of the major drivers or pressures influencing that system and its status relative to pre-established thresholds (Levin et al. 2009). Ecosystem modeling is an integral part in the support of IMs and IEAs. The IMs and IEAs are meant to be inclusive of the wider range of factors and processes that influence large marine ecosystems and their component LMRs, but a major distinction from EAF is that these approaches have a much broader inclusion of other ocean-use sectors beyond fisheries. It is clear that much further work is required to support these multi-sectoral efforts, but some of the preliminary full-system models described above could be adapted to address these more inclusive considerations.

Finally, we assert the need to work in the realm of the possible, cognizant of what remains to be done but not letting perceived limitations hamper progress. One does not need perfect knowledge of every process to model and manage LMRs from an ecosystem perspective. We reiterate that the knowledge base and modeling infrastructure to do an EAF in the GoM region exists in Canada and the United States; doing an EAF is feasible, now, with information, tools, and approaches that are available and tractable. A recent evaluation of progress in implementing ecosystem-based management of fisheries in 33 countries placed the United States and Canada in the top ranks across a number of different criteria (Pitcher et al. 2009), indicating that both countries are doing relatively well in steps towards implementing EAF. We readily admit that several steps to that end remain, and as we continue to move towards ecosystem approaches to fisheries management, ecosystem models will be a central feature of such an approach.

We conclude by quoting Francis of Assisi, a thought highly germane for ecosystem modeling in the Gulf of Maine region: "Start by doing what's necessary; then do what's possible; and suddenly you are doing the impossible."

Acknowledgments

We thank R. Stephenson who invited us to contribute this synopsis of ecosystem modeling for the Gulf of Maine symposium proceedings. We thank personnel at both NMFS NEFSC and DFO Maritimes laboratories who have collected much of the requisite data and have maintained some of the world's premier fisheries databases upon which these models were largely built. J. Link thanks the EMAX team, particularly J. O'Reilly, D. Dow, and W. Overholtz whose discussion provided much of the background for the historical considerations noted in this work. J. Link also acknowledges the contributions of R. Gamble, M.

Fogarty, W. Overholtz, J. Hare, K. Friedland, and J. Manderson for their prior discussions and sharing of concepts, data, and other works that have helped to shape many of themes noted herein. A. Bundy thanks the CDEENA team for their ecosystem modeling support. A. Bundy also acknowledges the contributions of N. Shackell, A. Cook, M. Koen-Alsonso, K. Trzcinski, P. Fanning, R. Mohn, and D. Duplisea for their prior discussions that have also helped to elucidate many of themes noted herein. We also acknowledge the several initiatives, specifically ERI in Canada and CAMEO in the United States, that have partially led to and maintained the CANUSE I and II interactions of which this work was a result. Finally, we thank two reviewers, R. O'Boyle and E. Fulton, who provided useful comments on prior drafts of this work.

References

Allesina, S. and C. Bondavalli. 2003. Steady state of ecosystem flow networks: a comparison between balancing procedures. Ecological Modelling 165:221–229.

Andersen, K. P., and E. Ursin. 1977. A multispecies extension to the Beverton and Holt theory, with accounts of phosphorus circulation and primary production. Meddelelser fra Danmarks Fiskeri-og Havundersogelser 7:319–435.

Araújo, J. N., and A. Bundy. 2011. Description of three ecopath with ecosim ecosystem models developed for the Bay of Fundy, western Scotian Shelf and NAFO Division 4X. Canadian Technical Report of Fisheries and Aquatic Sciences 2952.

Arkema, K. K., and J. F. Samhouri. 2012. Linking ecosystem health and services to inform marine ecosystem-based management. Pages 9–25 *in* R. L. Stephenson, J. H. Annala, J. A. Runge, and M. Hall-Arber, editors. Advancing an ecosystem approach in the Gulf of Maine. American Fisheries Society, Symposium 79, Bethesda, Maryland.

Azam, F., T. Fenchel, J. G. Field, J. S. Gray, L. A. Meyer-Reil, and F. Thingstad. 1983. The ecological role of water-column microbes in the sea. Marine Ecology Progress Series 10:257–263.

Backus, G. H., editor. 1987. Georges Bank. MIT Press, Cambridge, Massachusetts.

Beck, M. B. 1999. Coping with ever larger problems, models, and databases. Water Science Technology 39:1–11.

Beverton, R. J. H., and S. J. Holt. 1957. On the dynamics of exploited fish populations. Fisheries Investment Series 2, volume 19. U.K. Ministry of Agriculture and Fisheries, London.

Boreman, J., B. S. Nakashima, J. A. Wilson, and R. L. Kendall, editors. 1997. Northwest Atlantic groundfish: perspectives on a fishery collapse. American Fisheries Society, Bethesda, Maryland.

Brander, K., and R. Mohn, 2004. Effect of the North Atlantic oscillation on recruitment of Atlantic cod (*Gadus morhua*). Canadian Journal of Fisheries and Aquatic Sciences 61:1558–1564.

Breeze, H. 2002. Commercial fisheries of the Sable Gully and surrounding region: historical and present activities. Canadian Manuscript Report of Fisheries and Aquatic Sciences 2612.

Breeze, H., D. G. Fenton, R. J. Rutherford, and M. A. Silva. 2002. The Scotian Shelf: an ecological overview for ocean planning. Canadian Technical Report of Fisheries and Aquatic Sciences 2393.

Browman, H. I., and K. I. Stergiou. 2004. Perpectives on ecosystem-based approaches to management of marine resources Marine Ecology Progress Series 274:269–303.

Browman, H. I., and K. I. Stergiou. 2005. Politics and soci-economics of ecosystem-based approaches to management of marine resources Marine Ecology Progress Series 300:214–296.

Brown, B. E., J. A. Brennan, M. D. Grosslein, E. G. Heyerdahl, and R. C. Hennemuth. 1976. The effect of fishing on the marine finfish biomass in the northwest Atlantic from the Gulf of Maine to Cape Hatteras. International Commission for the Northwest Atlantic Fisheries Research Bulletin 12:49–68.

Brown, J. H., A. P. Allen, V. M. Savage, and G. B. West. 2004. Toward a metabolic theory of ecology. Ecology 85:1771–1789.

Bundy, A. 2001. Fishing on ecosystems: the interplay of fishing and predation in Newfoundland-Labrador. Canadian Journal of Fisheries and Aquatic Sciences 58:1153–1167.

Bundy, A. 2004. Mass balance models of the eastern Scotian Shelf before and after the cod collapse and other ecosystem changes. Canadian Technical Report of Fisheries and Aquatic Sciences 2520.

Bundy, A. 2005a. On fishing and forcing: gauging the role of environmental and anthropogenic forces on the ecosystem dynamics of the eastern

Scotian Shelf, Canada. International Council for the Exploration of the Sea, CM 2005/M:11, Copenhagen, Denmark.

Bundy, A. 2005b. Structure and functioning of the eastern Scotian Shelf ecosystem before and after the collapse of groundfish stocks in early 1990s. Canadian Journal of Fisheries and Aquatic Sciences 62:1453–1473.

Bundy, A., S. Heymans, L. Morissette, and C. Savenkoff. 2008. Seals, cod, and forage fish: a comparative exploration of variations in the theme of stock collapse and ecosystem change in northwest Atlantic ecosystems. Pages 133–135 *in* W. D. Bowen, M. O. Hammill, M. Koen-Alonso, G. Stenson, D. P. Swain, and K. Trzcinski, editors. Proceedings of the national workshop on the impacts of seals on fish populations in eastern Canada (part 1); 12–16 November 2007. Fisheries and Oceans Canada, Canadian Science Advisory Secretariat, Proceedings Series 2008/021, Ottawa.

Bundy, A., G. Lilly, and P. Shelton, 2000. A mass balance model of the Newfoundland-Labrador shelf. Canadian Technical Report of Fisheries and Aquatic Sciences 2310.

Castelo-Lawless, T. 1995. Phenomenotechnique in historical perspective: its origins and implications for the Philosophy of Science. Philosophy of Science 62:44–59.

Changizi, M. A., M. A. MacDannald, and D. Widders. 2002. Scaling of differentiation in networks: nervous systems, organisms, ant colonies, ecosystems, businesses, universities, cities, electronic circuits, and legos. Journal of Theoretical Biology 218:215–237.

Chen, C., R. C. Beardsley, and P. J. S. Franks. 2001. A 3-D prognostic model study of the ecosystem over Georges Bank and adjacent coastal regions. Part I: physical model. Deep-Sea Research Part II 48:419–456.

Christensen, V., and D. Pauly. 1993. On steady-state modeling of ecosystems. Pages 14–19 *in* V. Christensen and D. Pauly, editors. Trophic models of aquatic ecosystems. International Center for Living Aquatic Resources Management Conference Proceedings 26, Manila, Philippines.

Christensen, V., and D. Pauly. 1993. Trophic models of aquatic ecosystems. International Center for Living Resources Management Conference Proceedings 26, Manila, Philippines.

Christensen, V., and C. Walters. 2004. Ecopath with Ecosim: methods, capabilities and limitations. Ecological Modelling 172:109–139.

Christensen, V., C. J. Walters, and D. Pauly. 2005. Ecopath with Ecosim: a user's guide. Fisheries Centre of University of British Columbia, Vancouver.

Clarke, G. L. 1946. Dynamics of production in a marine area. Ecological Monographs 16:321–337.

Cogan, C. B., B. J. Todd, P. Lawton, and T. T. Noji. 2009. The role of marine habitat mapping in ecosystem-based management. ICES Journal of Marine Science 66:2033–2042.

Cohen, E. B., M. D. Grosslein, and M. P. Sissenwine. 1982. Energy budget of Georges Bank. Canadian Special Publication of Fisheries and Aquatic Sciences 59:95–107.

Collie, J. S., and A. K. DeLong. 1999. Multispecies interactions in the Georges Bank fish community. Pages 187–210 *in* Ecosystem approaches for fisheries management. Alaska Sea Grant College Program, University of Alaska, AK-SG-99-01, Fairbanks.

Curry, J. M. 2000. Community worldview and rural systems: a study of five communities in Iowa. Annals of the Association of American Geographers 90:693–712.

Dames, J. K., and R. R. Christian. 2006. Uncertainty and the use of network analysis for ecosystem-based fishery management. Fisheries 31:331–341.

David, M. 2005. Science and society. Palgrave Macmillan, New York.

De Roos, A. M., T. Schellekens, T. van Kooten, K. van de Wolfshaar, D. Claessen, and L. Persson. 2008. Simplifying a physiologically structured population model to a stage-structured biomass model. Theoretical Population Biology 73:47–62.

Edwards, S. F., J. S. Link, and B. P. Rountree. 2004. Portfolio management of wild fish stocks. Ecological Economics 49:317–329.

Fennel, W., and T. Neumann. 2004. Introduction to the modelling of marine ecosystems. Elsevier, Amsterdam.

Fogarty, M. J., and S. A. Murawski. 1998. Large-scale disturbance and the structure of marine systems: fishery impacts on Georges Bank. Ecological Applications 8(S1):S6–S22.

Fogarty, M. J., W. J. Overholtz, and J. Link. 2008. Fishery production potential of the northeast continental shelf of the United States. National Oceanic and Atmospheric Administration, National

Marine Fisheries Service, Northeast Fisheries Science Center, Woods Hole, Massachusetts.

Franks, P. J. S., and C. Chen. 2001. A 3-D prognostic numerical model study of the Georges Bank ecosystem. Part II: biological-physical model. Deep-Sea Research Part II 48:457–482.

Fulton, E. A. 2010. Approaches to end-to-end ecosystem models. Journal of Marine Systems 81:171–183.

Fulton, E. A., A. D. M. Smith, and C. R. Johnson. 2004. Biogeochemical marine ecosystem models I: IGBEM–a model of marine bay ecosystems. Ecological Modelling 174:267–307.

Gamble, R. J., and J. S. Link. 2009. Analyzing the tradeoffs among ecological and fishing effects on an example fish community: a multispecies (fisheries) production model. Ecological Modelling 220:2570–2582.

Garcia, S. M., A. Zerbi, C. Aliaume, T. Do Chi, and G. Lasserre. 2003. The ecosystem approach to fisheries: issues, terminology, principles, institutional foundations, implementation and outlook. FAO Fisheries Technical Paper 443.

Garrison, L., and J. Link. 2004. An expanded multispecies virtual population analysis approach (MSVPA-X) to evaluate predator–prey interactions in exploited fish ecosystems. Version 1.1. Users manual and model description. Atlantic States Marine Fisheries Commission, Arlington, Virginia.

Garrison, L. P., J. S. Link, M. Cieri, P. Kilduff, A. Sharov, D. Vaughan, B. Muffley, B. Mahmoudi, and R. Latour. 2010. An expansion of the MSVPA approach for quantifying predator–prey interactions in exploited fish communities. ICES Journal of Marine Science 67:856–870.

Garrod, D. J. 1973. Memorandum on the mixed fishery problem in Subarea 5 and Statistical Area 6. International Commission for the Northwest Atlantic Fisheries Research Document 73/6.

Gillooly, J. F., J. H. Brown., G. B. West, V. M. Savage, and E. L. Charnov. 2001. Effects of size and temperature on metabolic rate. Science 293:2248–2251.

Guénette, S., and R. L. Stephenson. In press. Accounting for predators in ecosystem-based management of herring fisheries of the western Scotian shelf, Canada. In Ecosystems 2010: global progress on ecosystem-based fisheries management. Proceedings of the 26th Lowell Wakefield Fisheries Symposium, Anchorage. Alaska Sea Grant, Fairbanks.

Halliday, R. G., and A. T. Pinhorn. 1996. North Atlantic fishery management systems: a comparison of management methods and resource trends. Journal of Northwest Atlantic Fishery Science 20.

Hall, S. J., J. S. Collie, D. E. Duplisea, S. Jennings, M. Bravington, and J. Link. 2006. A length-based multi-species model for evaluating community responses to fishing. Canadian Journal of Fisheries and Aquatic Sciences 63:1344–1359.

Hare, J. A., M. Alexander, M. Fogarty, E. Williams, and J. Scott. 2010. Forecasting the dynamics of a coastal fishery species using a coupled climate-population model. Ecological Applications 20:452–464.

Heymans, J. J. 2003. Comparing the Newfoundland marine ecosystem models using information theory. Pages 62–71 *in* Ecosystem models of Newfoundland and southeastern Labrador: additional information and analysis for 'Back to the Future.' University of British Columbia, Fisheries Center Research Report 11(5), Vancouver.

Heymans, J. J., and D. Baird. 2000. Network analysis of the northern Benguela ecosystem by means of NETWRK and ECOPATH. Ecological Modelling 131:97–119.

Holland, D. S., and J. G. Sutinen. 1999. An empirical model of fleet dynamics in New England trawl fisheries. Canadian Journal of Fisheries and Aquatic Sciences 56:253–264.

Hollowed, A. B., N. A. Bond, T. K. Wilderbuer, W. T. Stockhausen, Z. T. A'Mar, R. J. Beamish, J. E. Overland, and M. J. Schirripa. 2009. A framework for modelling fish and shellfish responses to future climate change. ICES Journal of Marine Science 66:1584–1594.

Hu, S., D. W. Townsend, C. Chen, G. Cowles, R. C. Beardsley, R. Ji, and R. W. Houghton. 2008. Tidal pumping and nutrient fluxes on Georges Bank: a process-oriented modeling study. Journal of Marine Systems 74:528–544.

ICES (International Council for the Exploration of the Sea). 1991. Multispecies models relevant to management of living resources. ICES Marine Science Symposium 193.

ICES (International Council for the Exploration of the Sea). 2007. Report of the working group on multispecies assessment methods (WGSAM), 15–19 October 2007, San Sebastian, Spain. ICES, CM 2007/RMC:08, Copenhagen, Denmark.

ICES (International Council for the Exploration of the Sea). 2009. Report of the working group on multispecies assessment methods (WGSAM), 5–9 October 2009, ICES Headquarters, Copenhagen. ICES, CM 2009/RMC:10, Copenhagen, Denmark.

Jennings, S., M. J. Kaiser, and J. D. Reynolds. 2001. Marine fisheries ecology. Blackwell Scientific Publications, Oxford, UK.

Kavanagh, P., N. Newlands, V. Christensen, and D. Pauly. 2004. Automated parameter optimization for Ecopath ecosystem models. Ecological Modelling 172:141–149.

Keyl, F., and M. Wolff. 2008. Environmental variability and fisheries: what can models do? Reviews in Fish Biology and Fisheries 18:273–299.

Koen-Alonso, M., and A. Bundy. 2009. Trophodynamic modelling of core species of the eastern Scotian Shelf marine community. Pages 54–61 *in* W. D. Bowen, M. O. Hammill, M. Koen-Alonso, G. Stenson, D. P. Swain, and K. Trzcinski, editors. Proceedings of the national workshop on the impacts of seals on fish populations in eastern Canada, Fisheries and Oceans Canada, Halifax, November 24–28, 2008. (Part 2). Fisheries and Oceans Canada, Canadian Science Advisory Secretariat, Proceedings Series 2009/020, Ottawa.

Koen-Alonso, M., A. Buren, A. Bundy, and G. Stenson. 2008. Bionenergetic-allometric multispecies models:approach description and ongoing work in the Newfoundland-Labrador and eastern Scotian Shelf systems. Pages 135–142 *in* W. D. Bowen, M. O. Hammill, M. Koen-Alonso, G. Stenson, D. P. Swain, and K. Trzcinski, editors. Proceedings of the national workshop on the impacts of seals on fish populations in eastern Canada (Part 1); 12–16 November 2007. Fisheries and Oceans Canada, Canadian Science Advisory Secretariat, Proceedings Series 2008/021, Ottawa.

Koen-Alonso, M., and P. Yodzis. 2005. Multispecies modelling of some components of the northern and central Patagonia marine community, Argentina. Canadian Journal of Fisheries and Aquatic Sciences 62:1490–1512.

Larkin, P. A. 1996. Concepts and issues in marine ecosystem management. Reviews Fish Biology and Fisheries 6:139–164.

Levin P. S., M. J. Fogarty, S. A. Murawski, and D. Fluharty. 2009. Integrated ecosystem assessments: developing the scientific basis for ecosystem-based management of the ocean. PLoS (Public Library of Science) Biology [online serial] 7(1):e1000014. DOI: 10.1371/journal.pbio.1000014.

Lewin, R. 1992. Complexity: life at the edge of chaos. MacMillan Publishing Company, New York.

Lindeman, R. L. 1942. The trophic-dynamic aspect of ecology. Ecology 23:399–418.

Link, J. 1999. (Re)constructing food webs and managing fisheries. Pages 571–588 *in* Ecosystem considerations in fisheries management: proceedings of the 16th Lowell Wakefield fisheries symposium. Alaska Sea Grant, AK-SG-99-01, Fairbanks.

Link, J., W. Overholtz, J. O'Reilly, J. Green, D. Dow, D. Palka, C. Legault, J. Vitaliano, V. Guida, M. Fogarty, J. Brodziak, E. Methratta, W. Stockhausen, L. Col, G. Waring, and C. Griswold. 2008a. An overview of EMAX: the northeast U.S. Continental Shelf ecological network. Journal of Marine Systems 74:453–474.

Link, J. S. 2002a. Does food web theory work for marine ecosystems? Marine Ecology Progress Series 230:1–9.

Link, J. S. 2002b. Ecological considerations in fisheries management: when does it matter? Fisheries 27(4):10–17.

Link, J. S. 2002c. What does ecosystem-based fisheries management mean? Fisheries 27(4):18–21.

Link, J. S. 2003. A model of aggregate biomass tradeoffs. International Council for the Exploration of the Sea, CM 2003/Y08, Copenhagen, Denmark.

Link, J. S. 2010. Adding rigor to ecological network models by evaluating a set of pre-balance diagnostics: a plea for PREBAL. Ecological Modelling 221:1582–1593.

Link, J., and J. Brodziak, editors. 2002. Report on the status of the NE US Continental Shelf ecosystem. National Oceanic and Atmospheric Association, National Marine Fisheries Service, Northeast Fisheries Science Center, Reference Document 02-11, Woods Hole, Massachusetts.

Link, J. S., R. J. Gamble, and E. A. Fulton. 2011. NEUS – ATLANTIS: construction, calibration and application of an ecosystem model with ecological interactions, physiographic conditions, and fleet behavior. National Oceanic and Atmospheric Administration, National Marine Fisheries Service, Northeast Fisheries Science Center, Technical Memorandum NE-218, Woods Hole, Massachusetts.

Link, J. S., and L. P. Garrison. 2002. Changes in piscivory associated with fishing induced changes to the finfish community on Georges Bank. Fisheries Research 55:71–86.

Link, J. S., C. A. Griswold, E. T. Methratta, and J. Gunnard, editors. 2006. Documentation for the Energy Modeling and Analysis eXercise (EMAX). National Oceanic and Atmospheric Administration, National Marine Fisheries Service, Northeast Fisheries Science Center, Reference Document 06–15, Woods Hole, Massachusetts.

Link, J. S., and J. S. Idoine. 2009. Predator consumption estimates of the northern shrimp *Pandalus borealis*, with implications for estimates of population biomass in the Gulf of Maine. North American Journal of Fisheries Management 29:1567–1583.

Link, J. S., T. F. Ihde, H. Townsend, K. Osgood, M. Schirripa, D. Kobayashi, S. Gaichas, J. Field, P. Levin, K. Aydin, G. Watters, and C. Harvey, editors. 2010. Report of the 2nd national ecosystem modeling workshop (NEMoW II): bridging the credibility gap—dealing with uncertainty in ecosystem models. National Oceanic and Atmospheric Administration, National Marine Fisheries Service, Technical Memorandum, NMFS-F/SPO-102, Silver Spring, Maryland.

Link, J. S., J. O'Reilly, D. Dow, M. Fogarty, J. Vitaliano, C. Legault, W. Overholtz, J. Green, D. Palka, V. Guida, and J. Brodziak. 2008b. Comparisons of the Georges Bank ecological network: EMAX in historical context. Journal of Northwest Atlantic Fisheries Science 39:83–101.

Link, J. S., and K. Sosebee. 2008. Estimates and implicatons of skate consumption in the northeastern U.S. Continental Shelf ecosystem. North American Journal of Fisheries Management 28:649–662.

Mann, K. H. 1972. Macrophyte production and detritus food chains in coastal waters. Memoirie Istitutuo Italiano di Idrobiologia 29(Supplement):353–383.

May, A. W. 1976. Report of standing committee on research and statistics. International Commission for the Northwest Atlantic Fisheries seventh special commission meeting - September 1975. International Commission for the Northwest Atlantic Fisheries Redbook 1976:5–19.

May, R. M. 1972. Will a large complex system be stable? Nature 238:413–414.

May, R. M. 1973. Stability and complexity in model ecosystems. Princeton University Press, Princeton, New Jersey.

McCormmach, R. 1970. H. A. Lorentz and the electromagnetic view of nature. Isis 61:459–497.

Megrey, B. A., and E. Moksness, editors. 2009a. Computers in fisheries research. Springer, New York.

Megrey, B. A., and E. Moksness. 2009b. Past, present and future trends in the use of computers in fisheries research. Pages 1–30 *in* B. A. Megrey and E. Moksness, editors. Computers in fisheries research. Springer, New York.

Miller, D. C. M., P. A. Shelton, B. P. Healey, W. B. Brodie, M. J. Morgan, D. S. Butterworth, R. Alpoim, D. González, F. González, C. Fernandez, J. Ianelli, J. C. Mahé, I. Mosqueira, R. Scott, and A. Vazquez. 2008 Management strategy evaluation for Greenland halibut (*Reinhardtius hippoglossoides*) in NAFO Subarea2 and Divisions 3LK-MNO. North Atlantic Fisheries Organization Scientific Council Research Document 08/25.

Mills, E. L. 1980. The structure and dynamics of shelf and slope ecosystems off the north east coast of North America. Pages 25–48 *in* K. R. Tenore and B. C. Coull, editors. Marine benthic dynamics. University of South Carolina Press, Columbia.

Mills, E. L., and R. O. Fournier. 1979. Fish production and the marine ecosystems of the Scotian Shelf, eastern Canada. Marine Biology 54:101–108.

Mills, E. L., K. Pittman, and F. C. Tan. 1984. Food-web structure on the Scotian Shelf, eastern Canada: a study using 13C as a food- chain tracer. Rapports et Proces-Verbaux des Reunions Conseil Internationale pour l'Exploration de la Mer 183:111–118.

Morissette, L., M. Castonguay, C. Savenkoff, D. P. Swain, D. Chabot, H. Bourdages, M. M. Hammill, and J. M. Hanson. 2009. Contrasting changes between the northern and southern Gulf of St. Lawrence ecosystems associated with the collapse of ground fish stocks. Deep-Sea Research Part II 56:2117–2131.

Moustahfid, H., J. S. Link, W. J. Overholtz, and M. C. Tyrell. 2009a. The advantage of explicitly incorporating predation mortality into age-structured stock assessment models: an application for northwest Atlantic mackerel. ICES Journal of Marine Science 66:445–454.

Moustahfid, H., M. C. Tyrrell, and J. S. Link. 2009b. Accounting explicitly for predation mortality in surplus production models: an application to longfin inshore squid. North American Journal of Fisheries Management 29:1555–1566.

Murawski, S. A., J. J. Maguire, R. K. Mayo, and F. M. Serchuk. 1997. Groundfish stocks and the fishing industry. Pages 27–70 *in* J. Boreman, B. S. Nakashima, J. A. Wilson, and R. L. Kendall, editors. Northwest Atlantic groundfish: perspectives on a fishery collapse. American Fisheries Society, Bethesda, Maryland.

Murphy, J. W., and J. T. Pardeck. 1988. The computer micro-world, knowledge and social planning. Computers in Human Services 3:127–141.

NEFSC (Northeast Fisheries Science Center). 2006. 42nd northeast regional stock assessment workshop (42nd SAW) stock assessment report. National Oceanic and Atmospheric Administration, National Marine Fisheries Service, Northeast Fisheries Science Center, Reference Document 06-09b, Woods Hole, Massachusetts.

NEFSC (Northeast Fisheries Science Center). 2007b. 44th Northeast Regional Stock Assessment Workshop (44th SAW): 44th SAW assessment report. National Oceanic and Atmospheric Administration, National Marine Fisheries Service, Northeast Fisheries Science Center, Reference Document 07-10, Woods Hole, Massachusetts.

NEFSC (Northeast Fisheries Science Center). 2007a. 45th Northeast Regional Stock Assessment Workshop (45th SAW): 45th SAW assessment report. National Oceanic and Atmospheric Administration, National Marine Fisheries Service, Northeast Fisheries Science Center. Reference Document 07-16, Woods Hole, Massachusetts.

NEFSC (Northeast Fisheries Science Center). 2008. Assessment of 19 northeast groundfish stocks through 2007: report of the 3rd Groundfish Assessment Review Meeting (GARM III), Northeast Fisheries Science Center, Woods Hole, Massachusetts, August 4–8, 2008. National Oceanic and Atmospheric Administration, National Marine Fisheries Service, Northeast Fisheries Science Center, Reference Document 08-15, Woods Hole, Massachusetts.

Nicholis, G., and I. Prigogine. 1989. Exploring complexity: an introduction. W. H. Freeman, New York.

NMFS (National Marine Fisheries Service). 1999. Ecosystem-based fishery management. A report to Congress by the Ecosystems Principles Advisory Panel. National Oceanic and Atmospheric Administration, National Marine Fisheries Service, Silver Spring, Maryland.

Odum, E. P. 1964. The strategy of ecosystem development. Science 164:262–270.

Odum, H. T. 1957. Trophic structure and productivity of Silver Springs, Florida. Ecological Monographs 27:55–112.

O'Neill, R. V. 2001. Is it time to bury the ecosystem concept? (with full military honors of course!). Ecology 82:3275–3284.

Overholtz, W. J., L. D. Jacobson, and J. S. Link. 2008. Developing an ecosystem approach for assessment advice and biological reference points for the Gulf of Maine-Georges Bank herring complex: adding the impact of predation mortality. North American Journal of Fisheries Management 28:247–257.

Overholtz, W. J., and J. S. Link. 2007. Consumption impacts by marine mammals, fish, and seabirds on the Gulf of Maine-Georges Bank Atlantic herring (Clupea harengus) complex during 1977–2002. ICES Journal of Marine Science 64:83–96.

Overholtz, W. J., and J. S. Link. 2009. A simulation model to explore the response of the Gulf of Maine food web to large scale environmental and ecological changes. Ecological Modelling 220:2491–2502.

Parsons, L. S. 1993. Management of marine fisheries in Canada. Canadian Bulletin of Fisheries and Aquatic Sciences 225.

Parsons, T. R., and H. Seki. 1970. Importance and general implications of organic matter in aquatic environments. Pages 1–27 *in* D. W. Hood, editor. Organic matter in natural waters. University of Alaska, Institute of Marine Science, Occasional Publication No. 1, College.

Patten, B. C., and E. P. Odum. 1981. The cybernetic nature of ecosystems. American Naturalist 118:886–895.

Pauly, D., V, Christensen, and C. Walters. 2000. Ecopath, Ecosim and Ecospace as tools for evaluating ecosystem impact of fisheries. ICES Journal of Marine Science 57:697–706.

Pickett, S. T. A., M. L. Cadenasso, and J. M. Grove. 2005. Biocomplexity in coupled natural-human systems: a multidimensional framework. Ecosystems 8:225–232.

Pikitch, E. K., C. Santora, E. A. Babcock, A. Bakun, R. Bonfil, D. O. Conover, P. Dayton, P. Doukakis, D. Fluharty, B. Heneman, E. D. Houde, J. Link, P. Livingston, M. Mangel, M. McAllister, J. Pope, and K. J. Sainsbury. 2004. Ecosystem-based fishery management. Science 305:346–347.

Pitcher, T. J., D. Kalikoski, K. Short, D. Varkey, and G. Pramoda. 2009. An evaluation of progress in implementing ecosystem-based management of fisheries in 33 countries. Marine Policy 33:223–232.

Plagányi, É. E. 2007. Models for an ecosystem approach to fisheries. FAO Fisheries Technical Paper 477.

Polovina, J. J. 1984. Model of a coral reef ecosystem. The ECOPATH model and its application to French Frigate Shoals. Coral Reefs 3:1–11.

Pomeroy, L. R. 1974. The ocean's food web: a changing paradigm. Bioscience 24:49–54.

Pope, J. G. 1975. The application of mixed fisheries theory to the cod and redfish stocks of Subarea 2 and Division 3K. International Commission for the Northwest Atlantic Fisheries Research Document 75/IX.

Pope, J. 1979. Stock assessment in multistock fisheries, with special reference to the trawl in the Gulf of Thailand. South China Sea Fisheries Development Programme, SCS/DEV/79/19 FAO, Manila, Philippines.

Ricker, W. E. 1954. Stock and recruitment. Journal of the Fisheries Research Board of Canada 11:559–623.

Riley, G. A. 1941. Plankton studies. IV. Georges Bank. Bulletin of the Bingham Oceanographic Collection 7:1–73.

Sainsbury, K. J., A. E. Punt, and A. D. M. Smith. 2000. Design of operational management strategies for achieving fishery ecosystem objectives. ICES Journal of Marine Science 57:731–741.

Shelton, P. A., and D. C. M. Miller. 2009. Robust management strategies for rebuilding and sustaining the NAFO Subarea 2 and Divs. 3KLMNO Greenland halibut fishery. Northwest Atlantic Fisheries Organization Scientific Council Research Document 09/37, Dartmouth, Nova Scotia.

Simmons, I. 2000. Making a mark: two thousand years of ecology, economy and worldview. Journal of Biogeography 27:3–5.

Sissenwine, M. P., E. B. Cohen, and M. D. Grosslein. 1984. Structure of the Georges Bank ecosystem. Rapports et Proces-Verbaux des Reunions du Conseil International pour l'Exploration de la Mer 183:243–254.

Smith, A. D. M., E. J. Fulton, A. J. Hobday, D. C. Smith, and P. Shoulder. 2007. Scientific tools to support the practical implementation of ecosystem-based fisheries management. ICES Journal of Marine Science 64:633–639.

Smith, A. D. M., K. J. Sainsbury, and R. A. Stevens. 1999. Implementing effective fisheries-management systems: management strategy evaluation and the Australian partnership approach. ICES Journal of Marine Science 56:967–979.

Smith, S. J., J. Black, B. J. Todd, V. E. Kostylev, and M. J. Lundy. 2009. The impact of commercial fishing on the determination of habitat associations for sea scallops (*Placopecten magellanicus*, Gmelin). ICES Journal of Marine Science 66:2043–2051.

Smith, T. D. 1994. Scaling fisheries: the science of measuring the effects of fishing, 1855–1955. Cambridge University Press, New York.

Steele, J. H. 1974. The structure of marine ecosystems. Harvard University Press, Cambridge, Massachusetts.

Steele, J. H., J. S. Collie, J. J. Bisagni, D. J. Gifford, M. J. Fogarty, J. S. Link, B. K. Sullivan, M. K. Sieracki, A. R. Beet, D. G. Mountain, E. G. Durbin, D. Palka, and W. T. Stockhausen. 2007. Balancing end-to-end budgets of the Georges Bank ecosystem. Progress in Oceanography 74:423–448.

Takahara, Y., Y. M. Liu, X. H. Chen, and Y. Yano. 2005. Model theory approach to transaction processing system development. International Journal of General Systems 34:537–557.

Thagard, P. 1994. Mind, society and the growth of knowledge. Philosophy of Science 61:629–625.

Townsend, D. W., A. C. Thomas, L. M. Mayer, and M. A. Thomas. 2006. Oceanography of the northwest Atlantic Continental Shelf (1,W). Pages 119–168 *in* A. R. Robinson and K. H. Brink, editors. The sea: the global coastal ocean: interdisciplinary regional studies and syntheses. Volume 14. Harvard University Press, Cambridge, Massachusetts.

Townsend, H. M., J. S. Link, K. E. Osgood, T. Gedamke, G. M. Watters, J. J. Polovina, P. S. Levin, N. Cyr, and K. Y. Aydin, editors. 2008. Report of the National Ecosystem Modeling Workshop (NEMoW). National Oceanic and Atmospheric Administration, National Marine Fisheries Service, Technical Memorandum NMFS-F/SPO-87, Silver Spring, Maryland.

Tsou, T.-S., and J. S. Collie. 2001. Estimating predation mortality in the Georges Bank fish community. Canadian Journal of Fisheries and Aquatic Sciences 58:908–922.

Tyrrell, M. C., J. S. Link, H. Moustahfid, and B. E.

Smith. 2007. The dynamic role of pollock (*Pollachius virens*) as a predator in the northeast US Atlantic ecosystem: a multi-decadal perspective. Journal of Northwest Atlantic Fisheries Science 38:53–65.

Tyrrell, M. C., J. S. Link, H. Moustahfid, and W. J. Overholtz. 2008. Evaluating the effect of predation mortality on forage species population dynamics in the northwest Atlantic Continental Shelf ecosystem: an application using multispecies virtual population analysis. ICES Journal of Marine Science 65:1689–1700.

Ulanowicz, R. E. 2004. Quantitative methods for ecological network analysis. Computational Biology and Chemistry. 28/5- 6:321–339.

Vasseur, D. A., and K. S. McCann. 2005. A mechanistic approach for modeling temperature-dependent consumer-resource dynamics. American Naturalist 166:184–198.

Walsh, J. J. 1972. Implications of a systems approach to oceanography. Science 176:969–975.

Walters, C. J., V. Christensen, and D. Pauly. 1997. Structuring dynamic models of exploited ecosystems from trophic mass-balance assessments. Reviews in Fish Biology and Fisheries 7:139–172.

Yodzis, P. 1998. Local trophodinamics and the interaction of marine mammals and fisheries in the Benguela ecosystem. Ecology 67:635–658.

Yodzis, P., and S. Innes. 1992. Body size and consumer-resource dynamics. American Naturalist 139:1151–1175.

Zwanenburg, K. C. T., D. Bowen, A. Bundy, K. Drinkwater, K. Frank, R. N. O'Boyle, D. Sameoto, and M. Sinclair. 2002. Decadal changes in the Scotian Shelf large marine ecosystem. Pages 105–150 in K. Sherman and H. R. Skjoldal, editors. Changing states of the large marine ecosystems of the North Atlantic. Blackwell Scientific Publications, Malden, Massachusetts.

Zwanenburg, K. C. T., A. Bundy, P. Strain, W. D. Bowen, H. Breeze, S. E. Campana, C. Hannah, E. Head, and D. Gordon. 2006. Implications of ecosystem dynamics for the integrated management of the eastern Scotian Shelf. Canadian Technical Report of Fisheries and Aquatic Sciences 2652.

American Fisheries Society Symposium 79:311–320, 2012

Managing for Migrants: The Gulf of Maine as a Global "Hotspot" for Long-Distance Migrants

ANTONY W. DIAMOND*
University of New Brunswick
Post Office Box 4400, Fredericton, New Brunswick E3B 5A3, Canada

Abstract.—An ecosystem-based framework for managing the Gulf of Maine ecosystem needs to include the long-distance migrant species that use the system for only a part of their annual cycle. The numbers of cetaceans and seabirds in this category greatly outnumber those that breed within the gulf and risk being neglected in any framework that does not consider them explicitly. I review the role of the gulf in the life cycles of five species of cetacean, 18 seabirds or shorebirds, and five species of fish; the familiar iconic species, such as North Atlantic right whale *Eubalaena glacialis*, bluefin tuna *Thunnus thynnus*, phalaropes (red-necked phalarope *Phalaropus lobatus* and red phalarope *P. fulicaria*), and sandpipers *Calidris* spp., are flagships for many more species for which the Gulf of Maine plays an irreplaceable part in the annual cycle.

Introduction

Ocean management in the Gulf of Maine is currently focused on harvested species, most of which are year-round residents. Yet, the number of individuals of many species that use the gulf for part of the year, without breeding there, greatly exceeds the number that are resident, so the majority of the biomass of the system is not considered explicitly in current management frameworks. Clearly, a truly ecosystem-based management framework needs to include such migrants and their needs.

The Gulf of Maine region (see Geographic Scope, below) is of critical importance to the life cycles of many species that undertake long-distance migrations to feed there in the local summer and fall, accumulating reserves essential to either successful breeding or overwintering elsewhere. For a few other species, the system is important as a wintering area and, for some, for spring or fall migration. In this paper, I review the range of vertebrates following these patterns (i.e., for which the region is an essential part of the life cycle but is not a breeding area) and attempt to assess the likely contribution of the ecosystem to the sustainability of their populations. The intention is to make the case that the ecological importance of the gulf extends so far beyond its borders that sustainable management of the ecosystem is essential for many more species than those harvested within its borders.

* Corresponding author:diamond@unb.ca

Geographic Scope

Except where stated otherwise, "Gulf of Maine" here includes the Bay of Fundy, Northeast Channel, Massachusetts Bay, and Georges Bank (Gulf of Maine Council on the Marine Environment 2010). Where relevant, I refer specifically to the Bay of Fundy, defined as bounded to the south by a "line running northwesterly from Cape St. Mary (44°05'N) Nova Scotia, through Machias Seal Island (67°06'W), and on to Little River Head (44°39'N) in the State of Maine" (International Hydrographic Organization 1953).

Taxonomic Scope

Here, I refer to species known to spend only part of their annual cycle in the Gulf of Maine region, visiting from breeding grounds elsewhere. I address only vertebrates—marine mammals (cetaceans), marine birds, and large predatory fish. Species that visit specifically to breed, as for example several species of migratory bird and fish, are not treated, since their habitat needs for breeding are generally well known and, in most cases, are taken into account by current management programs. The list is intended to be exemplary rather than exhaustive.

Predominance of Migrants in the Annual Cycle of the Region

Species that are not resident in the region but spend all or part of their nonbreeding season there can be divided into summer residents, winter residents, spring migrants, and fall migrants. Throughout, seasons referred to are local, recognizing that some of the species cross the equator to reach the gulf and so occupy it in the austral winter. Combined, numbers and biomass of these seasonally visiting ("temporary resident") populations greatly exceed those of breeding species (whether resident or migratory); this is especially clear in cetaceans (Kenney et al. 1997) and also clearly true of all seabirds with the possible exception of herring gulls *Larus argentatus* and great black-backed gulls *L. marinus*.

Summer Residents

The most common pattern is for all, or part, of a population to spend the local summer in the gulf, traveling there from more southerly breeding areas. This is the case in four species of seabird, five species of cetacean, and several fishes, including tuna and sharks.

Marine birds.—Three species of seabird migrate enormous distances from South Atlantic or subantarctic breeding grounds to spend the summer (austral winter) here, dominating the avian biomass for several months. Great shearwaters *Puffinus gravis* breed mainly on the South Atlantic islands of Tristan da Cunha and Gough Island, with small numbers in the Falkland Islands, migrating after breeding to the offshore northwest Atlantic (including the gulf) as far as Newfoundland and the northeast Atlantic, especially off the British Isles. Recent telemetry work by Ronconi (2007a, 2007b) has identified local hot spots of feeding activity in the outer Bay of Fundy and tracked the birds' transatlantic migration to West Africa, Argentina, and the breeding grounds; current work (R. Ronconi, Dalhousie University, unpublished data) is tracking the postbreeding migration to northeastern North America. Recent population estimates suggest a world breeding population of more than 16 million individuals (Fishpool and Evans 2001); Ronconi (personal communication) estimates up to 10,000 birds using the outer Bay of Fundy in summer and fall, with an unknown multiple of this number across the Gulf of Maine as a whole.

The Atlantic population of sooty shearwater *P. griseus* breeds on islands off Tierra del Fuego and in the Falkland Islands, following a similar annual movement pattern to great shearwaters. They are usually scarcer than great shearwaters in the Bay of Fundy, but numbers have not been estimated rigorously. The world population is estimated at more than 20 million birds; the species is listed as "near threatened" globally (IUCN 2010), chiefly because of declines in the Pacific population breeding in New Zealand. The size and status of the Atlantic population are not clear. The proportion of the global population summering in the Atlantic is also unknown, as is the proportion of the Atlantic population in the western side of the Atlantic, but the relatively low numbers in the Gulf of Maine in summer likely represent a small proportion of the global population.

A third species of shearwater—Cory's shearwater *Calonectris diomedea*—is a summer visitor to the southern parts of the gulf, penetrating the Bay of Fundy extremely rarely. Its breeding grounds are in the eastern Atlantic, from the Mediterranean west to the Azores; it is thus a transatlantic visitor to the gulf, whereas the other two shearwater species are trans-equatorial migrants. Unlike great and sooty shearwaters, Cory's shearwaters visit this region during their breeding season, so must be nonbreeding individuals.

Wilson's storm petrel *Oceanites oceanicus* breeds in the Falkland Islands, islands off Tierra del Fuego, South Georgia, the South Shetland and South Sandwich islands, and along the Antarctic coastline. It has an extremely wide distribution and an apparently stable population of between 12 million and 30 million birds (BirdLife International 2010). The numbers summering in the region have not been estimated, but casual observation suggests that they are in the same order of magnitude as great shearwaters.

These four species illustrate several shortcomings in our knowledge. Not only are the numbers using the gulf uncertain (numbers of seabirds at sea are not monitored routinely), but so is the broader context of the population distribution. It is striking that great and sooty shearwaters are also abundant off the northwestern coast of Europe in the summer; it used to be thought that the same birds moved through the offshore of North American waters, crossed the North Atlantic, and spent late summer off Europe (Wynne-Edwards 1935; Murphy 1936). However, both shearwaters are abundant on both sides of the North Atlantic at the same time; furthermore, recent satellite tracking (Ronconi 2007a, 2007b) of great shearwaters marked in the outer Bay of Fundy does not show those birds crossing the North Atlantic until their fall migration, suggesting that the birds summering on opposite sides of the North Atlantic are different subsets of the Atlantic population. Furthermore, 97 of 100 great shearwater victims of fishery bycatch in the Gulf of Maine were juvenile (M. C. Martin, College of Staten Island/City University of New York, and G. Shields, National Oceanic and Atmospheric Administration, personal communication), providing additional evidence that the birds using the gulf in summer are predominantly prebreeders. Equivalent information on sooty shearwaters and Wilson's storm petrels does not yet exist. Preliminary results of tracking sooty shearwaters from breeding grounds on the Falkland Islands suggest that they are likely not contributing to the population summering in the Gulf of Maine (A. Hedd, Memorial University, personal communication).

Marine mammals.—Several species of cetaceans (whales, porpoises, and dolphins) also spend most of the northern summer feeding in the Gulf of Maine, to the extent that these enormous animals dominate the biomass of the ecosystem to a greater extent in summer than at other times of year (Kenney et al. 1997). The best known is the North Atlantic right whale *Eubalaena glacialis*, whose population of 300–350 individuals is classified as a strategic stock under the U.S. Marine Mammal Protection Act (MMPA) and classified globally as endangered (Reilly et al. 2008). While a small proportion may be resident in the Gulf of Maine, most move from calving grounds off the coasts of Florida and Georgia into the Gulf of Maine in late winter and spring. They spend the summer feeding chiefly on aggregations of older stages of the copepod *Calanus finmarchicus*, especially in the Grand Manan basin between Grand Manan Island and Nova Scotia and in Roseway basin on the southwestern Scotian Shelf (Baumgartner et al. 2003). The entire world population likely summers in the Gulf of Maine and Roseway basin.

Four other species of baleen whales follow similar patterns but with larger populations and summer distribution less restricted to the Gulf of Maine. Three are strategic stocks under the MMPA and listed as "endangered" both globally (International Union for Conservation of Nature and Natural Resources) and under the U.S. Endangered Species Act. Humpback whales *Megaptera novaeangliae* mate and calve off the West Indies, moving into the Gulf of Maine in spring and summer to feed chiefly on Atlantic herring *Clupea harengus* and sand lance *Ammodytes* sp. and, to a lesser extent, euphausiid shrimp. The total northwest Atlantic population is about 10,400 individuals, of which the Gulf of Maine stock makes up about 850 (Waring et al. 2009) or 8%. However the northwest Atlantic population comprises six distinct stocks, including that of the Gulf of Maine, which exchanges about 24% of individuals with the nearest stock on the Scotian Shelf (Waring et al. 2009). The western North Atlantic stock is classified as "not at risk" in Canada (COSEWIC Web site, www.cosewic. gc.ca).

The extent and timing of migratory behavior of fin whales *Balaenoptera physalus* and the possibility of a distinct Gulf of Maine stock, are unclear, but this species dominates cetacean biomass at all times of year (Kenney et al. 1997; DFO 2009). The most recent (2006) population estimate is under 2,300 for the western North Atlantic stock (Waring et al. 2009). Calving occurs from October to January off the mid-Atlantic U.S. coast; the Gulf of Maine and Bay of Fundy are recognized as important feeding areas at other times. Fin whales are designated as a "species of special concern" in Canada (COSEWIC 2010). Kenney et al. (1997) estimated that the fin whales using the gulf in summer represent more than two-thirds of the northeast shelf stock.

The sei whale *Balaenoptera borealis* is primarily a continental slope species, but a small proportion (up to ~380 individuals) of the Nova Scotia stock penetrates the Gulf of Maine sporadically (Waring et al. 2009). Breeding grounds appear to be unknown; the species is most abundant in the gulf in spring, representing the entire northeast shelf population according to Kenney et al. (1997).

The eastern Canadian stock of minke whales *Balaenoptera acutorostrata* is estimated at about 3,300 individuals, of which around 25% are found in the Gulf of Maine (Waring et al. 2009). The species is not considered at risk by any authority, and sources agree that numbers are highest in the Gulf of Maine in spring through fall. Very few are found in the gulf in winter, when breeding takes place, so the stock is evidently migratory.

Several species of smaller toothed cetaceans are year-round residents of the gulf; the most common are harbor porpoise *Phocoena phocoena*, white-sided dolphin *Lagenorhynchus acutus*, pilot whales *Globicephala melas*, and common dolphin *Delphinus delphinus* (DFO 2009). As year-round residents, they will not be discussed further; for gulf-wide seasonal movements of harbor porpoise tagged in the Bay of Fundy, see Read and Westgate (1997).

Teleost fish.—"Giant" Atlantic bluefin tuna *Thunnus thynnus* use the gulf to fatten up in summer (late May–October), prior to migrating to spawning grounds in either the Gulf of Mexico or the straits of Florida (Lutcavage et al. 2000); some may even cross the Atlantic to the Mediterranean (Mather et al. 1995, cited by Lutcavage et al. 1999). This is an extremely lucrative and "highly overfished" population (Lutcavage et al. 2000), currently managed by the International Commission for the Conservation of Atlantic Tunas (Wilson et al. 2005). Recent declines in body condition of tuna summering in the Gulf of Maine have emphasized the ecological importance of this region to this commercially important migratory species (Golet et al. 2007). Aerial surveys of the Gulf of Maine population suggested 45,000–51,000 (Newlands et al. 2006). The proportion of the west Atlantic stock using the Gulf of Maine is not clear.

Cartilaginous fish.—At least four species of pelagic shark are known from the Gulf of Maine in summer. Great white sharks *Carcharadon carcharias* are known chiefly from the Bay of Fundy, where most of the 32 records from Atlantic Canada come from; there is a known mating area just south of Cape Cod. The North Atlantic population has declined sharply in the past 20 years and is designated as endangered in Canada (COSEWIC 2010). Basking sharks *Cetorhinus maximus* are also summer visitors throughout the gulf, often in association with right whales (DFO 2009) and are listed as "of special concern" in Canada (COSWEIC 2010). A single aerial survey in the Bay of Fundy in September 2009 estimated 732 (242–2,208 95% confidence interval; A. J. Westgate and coworkers, University of North Carolina at Wilmington, unpublished data). Blue sharks *Prionace glauca* also occur in summer, usually south of the Bay of Fundy, and have declined over the past 20 years (DFO 2009). Shortfin mako *Isurus oxyrinchus* are late-summer and fall migrants, usually associated with warmer waters, designated as "threatened" in Canada (COSEWIC 2010). In none of these cases has the proportion of the population using the Gulf of Maine been estimated.

Fall Migrants

One of the best known examples of the importance of the region as a migratory staging area is that of a small arctic-breeding shorebird, the semipalmated sandpiper *Calidris pusilla*. This species has been declining from an estimated 3.5 million at more than 7%, annually (Donaldson et al. 2000). Up to 74% of the world population migrates through the upper Bay of Fundy in fall, where they feed on intertidal amphipods and other invertebrates in mudflats at low tide. This is a critical staging area for the birds to fatten up for a 3,200-km transoceanic flight to wintering areas in northern South America (Gratto-Trevor 1992).

The outer Bay of Fundy is also a critical fall-migration staging area for two other species of small shorebird that breed in the Arctic and winter in the tropics. The red-necked phalarope *Phalaropus lobatus* and red phalarope *P. fulicaria* are unique among shorebirds in feeding, on migration, at sea, on nekton and plankton concentrated at tidal rips and upwellings; thus, unlike (intertidal) sandpipers, they are part of the pelagic marine ecosystem. Both species breed in arctic tundra and winter at sea in equatorial regions; both are declining. Red phalaropes are also seen in large numbers on Georges Bank in spring, but the outer Bay of Fundy has long been a known hot spot for both species in fall (July–September; Gratto-Trevor 1992; Rubega et al. 2000; Tracy et al. 2002). The southwestern Bay of Fundy used to host a fall passage of up to 3 million migrants, mainly red-necked phalaropes, but these flocks declined in the mid- to late 1980s to virtually none by 1990 (Duncan 1995); considerable numbers are still found southeast of Gand Manan and off southwestern Nova Scotia.

Bonaparte's gull *Larus philadelphia* breeds in boreal North America. Flocks of tens of thousands occur among terns and phalaropes on migration in fall (July through September) or early winter (Huettmann et al. 2000) in the Bay of Fundy, en route to winter quarters along Atlantic and Gulf of Mexico coasts, south through the Caribbean to Mexico. The proportion of the global (=North American) population migrating in fall through the Bay of Fundy was estimated by Braune (1989) as up to 21%.

Arctic terns *Sterna paradisaea* and common terns *S. hirundo* breed in small numbers (ca. 4,300 and 21,000 pairs in 2008; GOMSWG 2009) on managed islands in the Gulf of Maine, but many more pass through (especially in the Bay of Fundy) on fall migration. Northern gannets *Morus bassanus* also pass through the bay and gulf in considerable numbers in fall (as well as in spring) and some throughout the winter.

Winter Residents

At least half the eastern population of harlequin duck *Histrionicus histrionicus* (there is another, much larger population on the western coast) winters in coastal waters in the Gulf of Maine, including several hundred in the Bay of Fundy (Robertson and Goudie 1999). This population is generally recognized as "at risk"; it is a "species of special concern" in Canada (COSEWIC 2010), and "threatened" in Maine (Maine Department of Inland Fisheries and Wildlife 2010).

Razorbills *Alca torda* are seabirds that breed in small numbers (about 1,000 pairs) in the outer Bay of Fundy and northern Gulf of Maine, but most of the North American population of 38,000 pairs breeds in Quebec, Newfoundland, and Labrador (Chapdelaine et al. 2001). Recent work shows that most, if not all, of the North American breeding population spends at least part of the winter (late November to early March) in the outer Bay of Fundy around Grand Manan, especially Old Proprietor and Bulkhead shoals and off Long Eddy (Huettmann et al. 2005; Clarke et al. 2010); large numbers have also been observed in winter on Georges Bank.

Two other arctic-breeding seabirds, black-legged kittiwake *Rissa tridactyla* and northern fulmar *Fulmarus glacialis*, winter in considerable numbers in the Bay of Fundy and Gulf of Maine; kittiwakes also have a very small breeding population in the bay (Corrigan and Diamond 2001).

Spring Migrants

All three species of scoters *Melanitta* spp. (a group of benthic-feeding sea duck, breeding on freshwater lakes in boreal North America) migrate in large numbers through the Bay of Fundy en route to the Gulf of St. Lawrence. In the case of black scoter *M. nigra*, numbers passing through the bay in spring represent 56–75% of the North America population. Equivalent figures for surf scoter *M. perspicillata* and white-winged scoter *M. fusca* are 12–35% and about 1%, respectively (Bond et al. 2007).

Importance of Gulf of Maine to Migratory Populations

It is clear from the above discussion that migratory, nonbreeding species make up a large proportion of the top-predator level of the food web at most times of year. Clearly, then, these populations are important to the Gulf of Maine marine ecosystem. As such, they merit serious attention in truly integrated ecosystem management. The converse of this relationship is the importance of the Gulf of Maine to these migratory populations; what proportion of the regional or global population is found in the gulf during at least part of its life cycle? This is an important component of the case for habitat conservation of birds, for example; the criteria for designation of "important bird areas" (see below) rely heavily on significant proportions of bird populations using a potential site. The spatial scale of the proportions considered is usually regional (here, for example, the northeast shelf, or northwest Atlantic), or even global, as appropriate. Table 1 shows those species, for which there are estimates of the proportion of the species, population, or stock, that use the gulf at some stage of the annual cycle. For most species, there are simply not the data to assess the proportion of a population or stock using the Gulf of Maine. The lack of such information for most other species is a serious knowledge gap but need not hinder development of protected areas in those cases where the importance of the species to the gulf ecosystem is clear.

Current Management Systems

The predominant mechanisms for ocean management consist, at present, of the usual suite of measures designed to sustain particular fisheries. Some of these may make some incidental contribution to conservation of some of the long-distance migrants described above; we might think of this as "reverse incidental take" or "collateral conservation." For example, maintenance of sustainable stocks of Atlantic herring, widely recognized as a keystone species in the Gulf of Maine food web, would be expected to contribute to conservation of all species for which herring is an important prey item; this includes most of the species listed, with the exception of harlequin ducks *Histrionicus histrionicus* and scoters *Melanitta* spp. (benthic feeders), red-necked and red phalaropes, sandpipers *Calidris* spp., North Atlantic right whales, and basking sharks (planktivores). Recent restrictions on midwater trawling for herring in the Gulf of Maine (Litteral 2010) have been based partly on minimizing interactions between the fishing fleet and whales (chiefly humpback and fin), partly to reduce bycatch of small cetaceans and also to reduce the potential loss of biomass of Atlantic herring and sand lance, which are important prey species for many long-distance migrants. Ad hoc measures for conserving particular species, such as the 2003 change to shipping lanes in the Bay of Fundy to reduce collisions with right whales, can be effective for those and other nontarget species (in this case, basking sharks and other large whales; Vanderlaan et al. 2008) but again contribute to broader goals of ecosystem conservation only incidentally. For some bird species, such as harlequin ducks, management takes the form of restrictions on hunting, but for the majority of the long-distance migrants discussed here, the threats come not directly from human persecution but indirectly through depletion of the food web or degradation of the habitat. It can reasonably be concluded that traditional species- or stock-oriented management regimes are very unlikely to achieve the broader goals of integrated ecosystem management being called for by the public and by both U.S. and Canadian federal governments.

One drawback to existing management measures is that they focus on particular stocks or populations (often) without addressing the larger ecological context or other populations sharing similar ecological needs. An approach widely used in bird conservation, which tries to assess the importance of sites in a regional or global population context, is establishing networks of important bird areas (IBAs). These are sites that meet one or

Table 1. Long-distance migratory species for which a significant proportion uses the Gulf of Maine marine ecosystem at some stage of the annual cycle. See text for details.

Species	Proportion using Gulf of Maine	Season of use	Source
Razorbill	>75% (North American population)	Winter	Clarke et al. 2010
Semipalmated sandpiper	74% (world population)	Fall	Gratto-Trevor 1992
North Atlantic right whale	>50% (world population)	Summer	Reilly et al. 2008
Harlequin duck	~50% (eastern population)	Winter	Robertson and Goudie 1999
Bonaparte's gull	~21% (eastern world population)	Fall	Braune 1989

more of three broad criteria: they "(a) hold significant numbers of one or more globally threatened species; (b) are one of a set of sites that together hold a suite of restricted-range species or biome-restricted species; or (c) have exceptionally large numbers of migratory or congregatory species" (BirdLife International 2010). In the Bay of Fundy, five existing IBAs include significant marine areas (Grand Manan Archipelago, Machias Seal Island, The Wolves Archipelago, Quoddy Region, and Brier Island and Offshore Waters; Bird Studies Canada 2010). The Machias Seal Island IBA is also recognized as globally important in the U.S. IBA system, reflecting overlapping jurisdictional interests there. Note that recognizing sites as IBAs gives no legislated protection, merely public profile, which, it is hoped, can be used to enhance protection of the site from further ecological damage. The IBAs have, however, achieved global recognition by their incorporation into United Nations millennium management goals as indicators of biodiversity conservation (United Nations 2010).

The IBA system identifies sites important for one taxonomic group (birds) but does result in recognition of sites showing at least an avian reflection of ecological richness. A comparable approach to marine conservation has been the more recent development of criteria, and candidate sites, for ecologically and biologically significant areas (EBSAs); this program represents a step forward from the IBA approach in that it considers biodiversity properties of multiple taxa and the environmental features promoting biodiversity. The EBSA program was developed specifically to meet international obligations towards the conservation of biodiversity and establishing a system of protected areas (Canadian Biodiversity Strategy 1995) and as a key component of ecosystem-based management through meeting defined ecological objectives (DFO 2004). The major criteria against which candidate sites are evaluated (Buzeta and Singh 2008) are uniqueness (few or no alternatives), aggregation (of individuals, species, or physical or oceanographic features), and fitness consequences (the degree to which an area is essential to a species' or population's life cycle). Several sites recommended in the Bay of Fundy overlap with the IBA sites listed above while also considering benthic diversity, habitats for several commercially harvested marine species, and unique or special physical characteristics (Buzeta and Singh 2008). Just as with IBAs, EBSAs offer recognition but no legislated protection or management prescriptions.

The lack of marine areas protected by legislation is in sharp contrast to the situation on land. Terrestrial ecosystem conservation frequently takes the form of setting aside protected areas where harvesting is not allowed or is strictly limited. At best, such protected areas are chosen by a rigorous and scientifically defensible analysis of the spatial distribution of natural biogeographic and ecological entities, often incorporating "enduring features" of the landscape, with the goals of representing major habitat types, including biologically distinctive areas, and recognizing biogeographic areas. The marine equivalents—marine protected areas (MPAs)—are a relatively recent phenomenon, and many in the Gulf of Maine allow multiple uses, including harvesting, or were established to protect cultural resources, such as shipwrecks; "of the protected areas in our region, almost none are fully protected, and the level of protection in the remainder is inadequate for biodiversity conservation" (Conservation Law Foundation and World Wildlife Fund Canada 2006). In the United States, MPAs have been established under several different jurisdictional frameworks: Stellwagen Bank as a marine sanctuary, right whale critical habitats under the Endangered Species Act, and essential fish habitats under fishery management regulations. The MPAs, in general, appear to have been established piecemeal, for many different purposes, without an overarching plan for a network with clearly defined goals.

Truly integrated marine ecosystem conservation requires the setting of scientifically defensible goals for sustaining the marine ecosystem in its broadest sense, including all physical and biological components of the ecosystems and the human communities that depend on the continued health of those ecosystems. It also requires a top-down approach where MPAs are planned within the appropriate spatial context. The Conservation Law Foundation and World Wildlife Fund Canada (2006) describe such an approach for a network of MPAs in the Gulf of Maine and proposed three major goals: representing major habitat types, highlighting biologically distinctive regions, and recognizing biogeographic areas. The inclusion of "biologically distinctive areas" offers the opportunity to recognize the importance of the region to species that do not breed here—the long-distance migrants that are the focus of this paper. In proposing candidate EBSAs in the Bay of Fundy, Buzeta and Singh (2008) used some information on site requirements of long-distance migrants—in particular, whales and some birds—but ultimately selected sites that are mainly coastal and so did not fully address the needs of most long-distance migrants, which are predominantly pelagic species.

The Way Forward?

The next steps would seem to be twofold: first, develop and apply a clear scientific framework for selecting candidate MPAs in the Gulf of Maine (including the Bay of Fundy), as outlined by the Conservation Law Foundation and World Wildlife Fund Canada (2006), paying at least equal attention to the ecological needs of long-distance migrants, as to sectors more traditionally considered, such as benthic communities and fish stocks; second, develop and apply legislative tools to ensure effective protection of those areas from further degradation and overharvesting. Selection of candidate sites for protection, if they are to contribute to conservation of long-distance migrants, requires new research on the spatial distribution of these species using up-to-date approaches combining tracking data with geospatial mapping. Pelagic MPAs will need to recognize the dynamic and ephemeral nature of some pelagic habitats, in some cases probably needing flexible boundaries and possibly buffer zones (e.g., Hyrenbach et al. 2000). Long-distance migrants considered here remind us that our collective responsibility for integrated ecosystem management does not end in our own region, but extends far beyond, throughout the ranges of those species for which the Gulf of Maine plays a critical role in their life cycle. Many of the long-distance migrants here are top predators, whose continued diversity is essential for the health of the food webs in which they participate (Baum and Worm 2009). As oceanic ecosystems continue to be degraded by pollution, overharvesting, acidification and warming, the importance of maintaining the full suite of biodiversity (Worm et al. 2006), and of maintaining critical habitat outside the breeding season, becomes more acute; establishing and implementing effective MPAs for conserving long-distance migrants, as well as resident populations, is becoming increasingly urgent. Long-distance migrants remind us that the consequences of failing that responsibility extend far beyond the borders of the Gulf of Maine.

Acknowledgments

I thank Rob Stephenson for the invitation to make this contribution and R. A. Ronconi and G. T. Waring for helpful comments on an earlier draft.

References

Baum, J. K., and B. Worm. 2009. Cascading top-down effects of changing predator abundances. Journal of Animal Ecology 78:699–714.

Baumgartner, M. F., T. V. N. Cole, P. J. Clapham, and B. R. Mate. 2003. North Atlantic right whale habitat in the lower Bay of Fundy and on the SW Scotian shelf during 1999–2001. Marine Ecology Progress Series 264:137–154.

BirdLife International. 2010. Important bird areas (IBAs). Available: www.birdlife.org/action/science/sites/index.html. (August 2010)

Bird Studies Canada. 2010. Canadian important bird areas. Available: www.ibacanada.com/mapviewer.jsp?lang=EN. (August 2010)

Bond, A. L., P. W. Hicklin, and M. R. Evans. 2007. Daytime spring migration of scoters (*Melanitta* spp.) in the Bay of Fundy. Waterbirds 30:566–572.

Braune, B. M. 1989. Autumn migration and comments on the breeding range of Bonaparte's gull, *Larus philadelphia*, in eastern North America. Canadian Field-Naturalist 103:524–530.

Buzeta, M-I., and R. Singh. 2008. Identification of ecologically and biologically significant areas in the Bay of Fundy, Gulf of Maine. Volume 1: Areas identified for review, and assessment of the Quoddy region. Canadian Technical Report of Fisheries and Aquatic Sciences 2788.

Canadian Biodiversity Strategy. 1995. Canada's response to the Convention on Biological Diversity. Minister of Supply and Services Canada, Ottawa.

Chapdelaine, G., A. W. Diamond, R. D. Elliot, and G. J. Robertson. 2001. Status and population trends of the razorbill in eastern North America. Canadian Wildlife Service Occasional Paper 105.

Clarke, T. C., A. W. Diamond, and J. W. Chardine. 2010. Origin of Canadian razorbills (*Alca torda*) wintering in the outer Bay of Fundy confirmed by radio-tracking. Waterbirds 33:541–545.

Conservation Law Foundation and World Wildlife Fund Canada. 2006. Marine ecosystem conservation for New England and maritime Canada: a science-based approach to identifying priority areas for conservation. Available: http://awsassets.wwf.ca/downloads/wwf_northwestatlantic_marineecosystemconservation2006.pdf. (January 2011)

Corrigan, S., and A. W. Diamond. 2001. Northern gannet, *Morus bassanus*, nesting on Whitehorse Island, New Brunswick. Canadian Field-Naturalist 115:176–177.

DFO (Department of Fisheries and Oceans). 2004. Identification of ecologically and biologically significant areas. DFO, Canadian Scientific Advisory Secretariat, Ecosystem Status Report 2004/006, Ottawa.

DFO (Department of Fisheries and Oceans). 2009. Gulf of Maine ecosystem overview report. DFO, Dartmouth, Nova Scotia.

Donaldson, G., C. Hyslop, G. Morrison, L. Dickson, and I. Davidson. 2000. Canadian Shorebird Conservation Plan. Canadian Wildlife Service, Ottawa.

Duncan, C. D. 1995. The migration of red-necked phalaropes: ecological mysteries and conservation concerns. Birding 28:482–488.

Fishpool, L. and M. Evans, editors. 2001. Important bird areas in Africa and associated islands. Pisces Publications and BirdLife International, BirdLife International Conservation Series No. 11, Newbury and Cambridge, UK.

Golet, W. J., A. B. Cooper, R. Campbell, and M. Lutcavage. 2007. Decline in condition of northern bluefin tuna (*Thunnus thynnus*) in the Gulf of Maine. Fishery Bulletin 105:390–395.

Gratto-Trevor, C. L. 1992. Semipalmated sandpiper (*Calidris pusilla*). In A. Poole and F. Gill, editors. The birds of North America, No. 6. The Birds of North America, Inc., Philadelphia.

Gulf of Maine Council on the Marine Environment. 2010. The Gulf of Maine in context. State of the Gulf of Maine report. Available: www.gulfofmaine.org/state-of-the-gulf/docs/the-gulf-of-maine-in-context.pdf. (January 2012)

GOMSWG (Gulf of Maine Seabird Working Group). 2009. Gulf of Maine Seabird Working Group annual summer meeting report. GOMSWG, Bremen, Maine.

Huettmann, F., A. W. Diamond, B. Dalzell, and K. Macintosh. 2005. Winter distribution, ecology, and movements of razorbills *Alca torda* and other auks in the outer Bay of Fundy, Atlantic Canada. Marine Ornithology 33:161–171.

Huettmann, F., K. MacIntosh, C. Stevens, T. Dean, and A. W. Diamond. 2000. A mid-winter observation of a large population of Bonaparte's gulls in the Head Harbour Passage, Passamaquoddy Bay. Canadian Field-Naturalist 114:327–330.

Hyrenbach, K. D., K. A. Forney, and P. K. Dayton. 2000. Marine protected areas and ocean basin management. Aquatic Conservation: Marine and Freshwater Ecosystems 10:437–458.

International Hydrographic Organization. 1953. Limits of oceans and seas, 3rd edition. Available: www.iho-ohi.net/iho_pubs/IHO_Download.htm. (July 2010)

IUCN (International Union for Conservation of Nature and Natural Resources). 2010. Red list of threatened species, version 2010.2. Available: www.iucnredlist.org. (July 2010)

Kenney, R., G. Scott, T. Thompson, and H. Winn. 1997. Estimates of prey consumption and trophic impacts of cetaceans in the USA northeast continental shelf ecosystem. Journal of the Northwest Atlantic Fisheries Science 22:155–171.

Litteral, J. 2010. A question of scale: midwater trawlers are transforming the herring industry. The Working Waterfront (August 25). Island Institute & Working Waterfront, Rockland, Maine.

Lutcavage, M. E., R. W. Brill, G. B. Skomal, B. C. Chase, J. L. Goldstein, and J. Tutein. 2000. Tracking adult North Atlantic bluefin tuna (*Thunnus thynnus*) in the northwestern Atlantic using ultrasonic telemetry. Marine Biology 137:347–358.

Lutcavage, M. E., R. W. Brill, G. B. Skomal, B. C. Chase, and P. W. Howey. 1999. Results of pop-up satellite tagging of spawning size class fish in the Gulf of Maine: do North Atlantic bluefin tuna spawn in the mid-Atlantic? Canadian Journal of Fisheries and Aquatic Sciences 56:173–177.

Maine Department of Inland Fisheries and Wildlife. 2010. Harlequin duck (*Histrionicus histrionicus*). Available: www.maine.gov/ifw/wildlife/species/endangered_species/harlequin_duck/index.htm. (January 2012)

Mather, F. J., III, J. M. Mason, Jr., and C. A. Jones. 1995. Historical document: life history and fisheries of Atlantic bluefin tuna. National Oceanic and Atmospheric Administration, Southeast Fisheries Science Center, Technical Memorandum NMFS-SEFSC-370, Miami.

Murphy, R. C. 1936. Oceanic birds of South America. MacMillan and American Museum of Natural History, New York.

Newlands, N. K., M. E. Lutcavage, and T. J. Pitcher. 2006. Atlantic bluefin tuna in the Gulf of Maine, 1: estimation of seasonal abundance accounting for movement, school and school-aggregation behavior. Environmental Biology of Fishes 77:177–195.

Read, A. J., and A. J. Westgate. 1997. Monitoring the movements of harbor porpoises (*Phoecena phoecena*) with satellite telemetry. Marine Biology 130:315–322.

Reilly, S. B., J. L. Bannister, P. B. Best, M. Brown, R. L. Brownell, Jr., D. S. Butterworth, P. J. Clapham, J. Cooke, G. P. Donovan, J. Urbán, and A. N. Zerbini. 2008. *Eubalaena glacialis*. In IUCN 2010. IUCN red list of threatened species. Version 2010.2. Available: www.iucnredlist.org. (July 2010)

Robertson, G. R., and R. I. Goudie. 1999. Harlequin duck (*Histrionicus histrionicus*). In A. Poole and F. Gill, editors. The birds of North America, No. 466. The Birds of North America, Inc., Philadelphia.

Ronconi, R. A. 2007a. Identifying critical marine habitat for seabirds in the Bay of Fundy: a pilot study using satellite telemetry to monitor the movements of greater shearwaters. Final Report to the New Brunswick Environmental Trust Fund. Grand Manan Whale and Seabird Research Station, Project No. 060263, Grand Manan, New Brunswick.

Ronconi, R. A. 2007b. The spectacular migration of greater shearwaters. BirdWatch Canada 39:4–7.

Rubega, M. A., D. Schamel, and D. M. Tracy. 2000. Red-necked phalarope (*Phalaropus lobatus*) In A. Poole and F. Gill, editors. The birds of North America, No. 538. The Birds of North America, Inc., Philadelphia.

Tracy, D. M., D. Schamel, and J. Dale. 2002. Red phalarope (*Phalaropus fulicarius*). In A. Poole and F. Gill, editors. The birds of North America, No. 698. The Birds of North America, Inc., Philadelphia.

United Nations. 2010. The millennium development goals report. Goal 7: ensure environmental sustainability. Available: www.un.org/millenniumgoals/environ.shtml. (August 2010)

Vanderlaan, A. S. M., C. T. Taggart, A. R. Serdynska, R. D. Kenney, and M. W. Brown. 2008. Reducing the risk of lethal encounters: vessels and right whales in the Bay of Fundy and on the Scotian Shelf. Endangered Species Research 4:283–297.

Waring, G. I., E. Josephson, K. Maze-Foley, and P. E. Rose. 2009. U.S. Atlantic and Gulf of Mexico marine mammal stock assessments—2009. National Oceanic and Atmospheric Administration, Northeast Fisheries Science Center, Technical Memorandum NMFS NE 213, Gloucester, Massachusetts.

Wilson, S. G., M. E. Lutcavage, R. W. Brill, M. P. Genovese, A. B. Cooper, and A. W. Everly. 2005. Movements of bluefin tuna (*Thunnus thynnus*) in the northwestern Atlantic Ocean recorded by pop-up satellite archival tags. Marine Biology 146:409–423.

Worm, B., E. B. Barbier, N. Beaumont, J. E. Duffy, C. Folke, B. S. Halpern, J. B. C. Jackson, H. K. Lotze, F. Micheli, S. R. Palumbi, E. Sala, K. A. Selkoe, J. J. Stachowicz, and R. Watson. 2006. Impacts of biodiversity loss on ocean ecosystem services. Science 314:787–790.

Wynne-Edwards, V. C. 1935. On the habits and distribution of birds on the North Atlantic. Proceedings of the Boston Society of Natural History 40:233–346.

American Fisheries Society Symposium 79:321–343, 2012

Stalk-Eyed Views of the Gulf of Maine—Through a Nepheloid Layer Dimly

PETER A. JUMARS*
School of Marine Sciences, University of Maine
5706 Aubert Hall, Orono, Maine 04469, USA

Abstract.—Because of partial recirculation and steep bottom slopes, the Gulf of Maine (GoM) contains steep environmental gradients in both space and time. I focus, in particular, on optical properties associated with both resources and risks. The GoM estuary-shelf systems differ from those whose fine sediments are trapped behind barrier bars; in the GoM, nepheloid layers prevail over a wide range of depths, and onshore-offshore turbidity gradients at a given water depth are also steep. Turbidity reduces predation risk. Three crustacean species that are major fish forages respond to the strong environmental gradients in resources and risks by migrating seasonally both horizontally and vertically. Northern shrimp (also known as pink shrimp) *Pandalus borealis*, sevenspine bay shrimp *Crangon septemspinosa*, and the most common mysid shrimp in the GoM, *Neomysis americana*, share both stalked eyes that appear capable of detecting polarized light and statocysts. This pair of features likely confers sun-compass navigational ability, facilitating use of multiple habitats. All three species converge on a shallow-water bloom at depths <100 m of the western GoM shelf in December–March, well before the basin-wide, climatological spring bloom in April. In addition to reaching abundant food resources, I propose that they are also using optical protection, quantified as the integral of the beam attenuation coefficient from the surface to the depth that they occupy during daylight. Spring immigration into, and fall emigration from, estuaries appear to be common in GoM sevenspine bay shrimp and *N. americana*, out of phase with their populations south of New England and with turbidity differences a likely cause. Migration studies that include measurements of turbidity are needed, however, to test the strength of the effect of optical protection on habitat use by all three species. Simultaneous sampling of estuaries and the adjacent shelf, together with trace-element tracer studies, would be very useful to resolve timing and extent of mass migrations, which likely are sensitive to turbidity change resulting from climate change. These migrations present special challenges to ecosystem-based management by using so many different habitats.

Introduction

Interplay of geologic and oceanographic processes and structures supports unusually steep environmental gradients in the Gulf of Maine (GoM), not only in space, but also over time. Asynchrony in seasonal changes across the GoM expands habitat choices substantially in terms of resources and risks that select, respectively, for and against specific patterns and timings of movement among habitats along and across the gradients. The GoM is far less enclosed than the Strait of Georgia and Puget Sound on North America's opposite and also glacially sculpted coast but is enclosed, large and topographically diverse enough to produce cyclonic, partial recirculation around its three principal basins (Jordan, Wilkinson, and Georges) and anticyclonic circulation over George's Bank at its outer boundary. Through both active migra-

* Corresponding author: jumars@maine.edu

tions and passive drift, this combination of features gives access to multiple habitats over ontogenetically and ecologically relevant distances and times (Figure 1). Time for a full circuit of any one of these gyres at a characteristic coastal current speed of 0.1 m/s (Pettigrew et al. 2005) would be on the order of one to a few months. Interaction of circulation and ontogeny is well appreciated in the GoM, in part through the U.S. GLOBEC Program (e.g., McGillicuddy et al. 2002).

Relatively steep bottom slope is perhaps GoM's most obvious gradient, with the 100-m isobath roughly 20 km from shore. Temperature also shows strong gradients in both space and time. Partial recirculation sets up particularly steep temperature gradients between the Jordan and Wilkinson basins in summer (GoM eastern and western gyres, respectively; cf. Pettigrew et al. 1998), and Wilkinson basin experiences some of the largest seasonal changes in surface temperatures in any coastal sea, as well as steep vertical gradients in summer. The existence of this thermal structure is well known (e.g., Bisagni et al. 2001), and temperature as an important driver of physiological rates and covariate with other important variables such as nutrient concentrations (e.g., Thomas et al. 2003) is indisputable. Partially for that reason, I focus here instead on other variables whose importance should have been more obvious from the striking, stalked, moveable eyes borne by the subjects of this paper.

Comparatively subtle, often overlooked, and much less often measured (with the exception of ocean color) than the aforementioned gradients are strong gradients in optical properties over both space and time. Turbidity is caused by both phytoplankton and mineral particles. Weathering and coastal deposition are often contrasted between active and passive continental margins. Steep topographies and short distances from mountains to seas in active margins deliver fine sediments at high rates that account for mid-shelf deposition of silt along the northwest coast of North America. Coastal plains estuaries of passive margins are, on average, more turbid than those that flow through younger, rockier, steeper terrains (e.g., Roman et al. 2000; their Table 2 that compares turbidities of estuaries from GoM southward and into the Gulf of Mexico). Less steep topographies and longer rivers on passive margins also lead to barrier islands and bars that trap a greater fraction of the fine sediments inside their estuaries, leaving broad

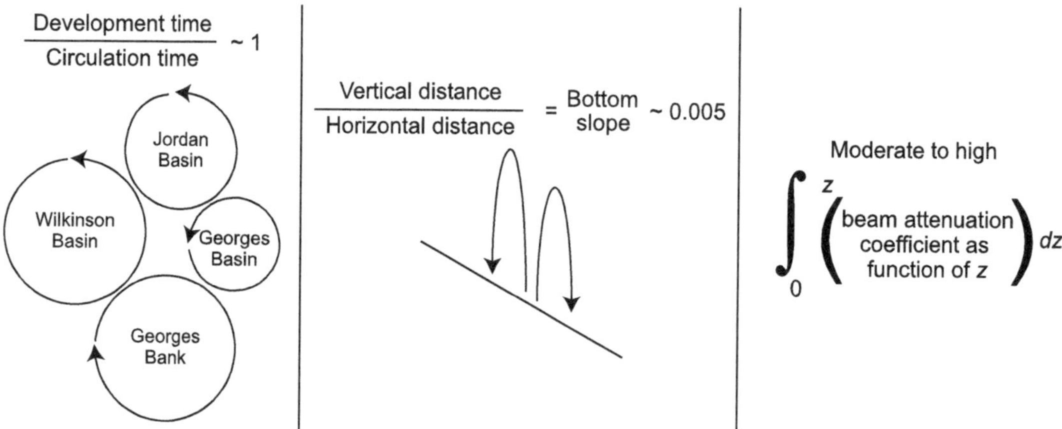

FIGURE 1. Nondimensional numbers whose combination of values makes the Gulf of Maine unique and structures its habitats. Recirculation times about the major gyres of the gulf are relevant both ecologically and developmentally. The relatively steep bottom slope juxtaposes deep and shallow habitats with high value in terms of resources and protection from predators. Optical protection from predators is high, particularly near shore, because the depth integral of the beam attenuation coefficient is moderate to high relative to waters of the same depth elsewhere, being elevated by seasonally high water column average turbidity but in particular by near-bottom nepheloid layers that favor emergent species.

expanses of relict sands on continental shelves and clearer shelf waters that support substantial benthic primary production (Jahnke et al. 2008; Grippo et al. 2009), but also bring significant risks from visual predation. The GoM falls at neither of the classic extremes. The results are clearer estuarine waters than found farther south (Roman et al. 2000), but significant export of fines into GoM, producing considerable turbidity much of the time in nepheloid layers near the seabed of the GoM over a broad range of water depths (Townsend et al. 1992) and steep onshore-offshore gradients in turbidity near the coast (Spinrad 1986).

One ecological context of turbidity is optical protection from predators. Turbidity affords protection both locally and integrally. In many ways, local turbidity acts analogously with terrestrial plant cover as a refuge. In turbid water, a small planktivore may be able to see far enough to catch enough food to grow but, at the same time, be difficult to find for its own visual predators, whose prey are more widely spaced. The situation is not unlike that of a small bird feeding within the visual protection provided by a bush from raptors and cats, except that clouds of phytoplankton are more mobile and less discrete than bushes. Local irradiance that enables vision is also affected by turbidity in all the water layers above it, and this reduced irradiance can also afford optical protection by effectively causing darkness during daytime.

Estuaries and nearshore habitats with abundant physical structure (macroalgae and sea grasses) are often considered primary nursery areas (e.g., Beck et al. 2001). I suggest in this perspective that this view carries a mid-Atlantic bias; a suite of optical characteristics, coupled with other strong gradients, makes the broader nearshore zone in the GoM take on a greater nursery function than it does elsewhere. Highly productive shelf waters provide both food and "cover." The optical gradients result in strong gradients in resources and risks and also provide navigational cues enabling efficient travel among resulting habitats.

My attempt is to view the GoM quite literally through the (stalked) eyes of three abundant, broadly distributed, crustacean species that respond to resource and risk gradients that change with developmental stage by migrating seasonally and both vertically and horizontally. The mysid shrimp *Neomysis americana* and the decapod sevenspine bay shrimp both move between the GoM shelf and its estuaries, and the decapod northern shrimp (also known as pink shrimp) *Pandalus borealis* moves between shallow and deep waters of the GoM. These emergent omnivores (i.e., species that typically spend the day on the seabed and the night in the water column) are generally underappreciated with respect to their contribution and role in the Gulf of Maine ecosystem (Johnson and Hare 2012, this volume). They interact in this obvious way (diel migration) but also in more subtle ways and on multiple time scales, with temporal and spatial variations in light.

Few data have been collected to test explicit hypotheses about turbidity versus risks and benefits to these three species in the GoM. Circumstantial evidence of important roles of turbidity in population dynamics of all three, however, invites more systematic analysis of habitat use as a function of turbidity for these three species and for others. Summarizing that circumstantial evidence gleaned from a wide-ranging literature is the primary purpose of this paper. My principal conclusion is that optical protection is an important habitat characteristic that merits attention in ecosystem-based management. Circulation and temperature changes are obvious concerns in the climate change arena. Optical properties as drivers of trophic interactions should be considered as well because they are very directly affected by changes in climate (i.e., in intensity and phenology of rainfall and runoff, in frequency of sediment transport by currents and waves, and in timing and intensity of phytoplankton blooms and crashes).

These three focal species also exemplify considerable management challenges. Because they occupy multiple habitats both vertically and horizontally as a function of time, with patterns at least as complex as those of diadromous fishes, their ecosystem-based management must encompass both benthic and pelagic environments and broad ranges of water depths. Only for the commercially significant northern shrimp are there published data on GoM abundance over periods longer than a few years. Historical abundances and interannual variability of GoM populations of the other two are poorly known, yet all three are major forage species for both demersal and pelagic fishes.

Have their populations exploded as fish populations have dwindled? A related question is whether mysids are ultimate beneficiaries of fishing down the food web, favored in particular by their extreme omnivory. Emergent omnivores have benefitted in other overfished ecosystems (e.g., Utne-Palm et al. 2010). Raising these sorts of trophic questions is a secondary purpose of this paper and adds further motivation for future measurement programs that include optical properties from the upper mixed layer to the bottom boundary layer.

Background—The Three Principal Actors

The three central characters all converge on the shallow shelf of the western GoM in December through March, well before the climatological bloom month of April for the broader GoM (Thomas et al. 2003). For a typical poikilotherm, development takes longer in colder water, yet growing beyond the gapes of a panoply of predators is no less important here than in tropical seas. In colder water, an earlier growth start carries a greater premium. The three central characters share a remarkable number of other traits (Table 1). All are omnivorous, widening doors to multiple habitats. During nocturnal excursions into the water column, all indulge in substantial carnivory on pelagic prey (including each other where the species co-occur in appropriate size ratios), particularly during gonadal maturation, suggesting that omnivory might be an excellent strategy for maintenance and some growth, but that lipid-intensive planktivory may be a common reproductive solution for emergent omnivores. All can be voracious, with sevenspine bay shrimp known for turning stereotypical predation tables on juvenile flatfishes (e.g., Taylor 2005). All three are favored foods of both benthic and pelagic fishes, but entering the pelagic realm must nonetheless increase fitness or it would cease. All engage in diel migration, at least in some seasons, and do so in a pattern consistent with visual predation as driver for the timing. All three share release from optically protected epibenthic existence at night but show no single, preferred, water column depth of activity or feeding. Their vertical distributions broaden at night, but their populations nevertheless remain more concentrated near the seabed than the sea surface. For even those individuals that remain near the seabed, however, activity levels increase when risk of visual predation declines—and increase even in otherwise similar species that do not migrate vertically (Fosså 1986). The behaviors of these species contrast sharply with diel migrators that show distinctly preferred nighttime strata, frequently the uppermost decimeter (e.g., *Anchialina agilis* cf. Macquart-Moulin and Maycas 1995), their concentration giving obvious advantage in intraspecific encounter rates for mating.

Despite these striking similarities and the three species' seasonal convergence in the coastal GoM, they show equally striking divergence in habitat use and in other respects. All are malacostracans, but the mysid is in a separate order, and the two decapod shrimps are in different superfamilies. Mysids show direct development, whereas both decapods have extended larval development in the plankton. *Neomysis americana* (biology reviewed by Wigley and Burns 1971) ranges from about 0.5 to 2.0 cm long, whereas northern shrimp (biology reviewed by Shumway et al. 1985) can reach nearly 20 cm, with sevenspine bay shrimp (biology and distribution summarized in Williams [1984] and Theroux and Wigley [1998]) reaching about 8 cm. *Neomysis americana* and sevenspine bay shrimp in the GoM typically have two generations per year (one overwintering). Northern shrimp takes 2 years to reach protandrous adulthood and 4 years to reach first female reproduction; all its stages are stenohaline. Northern shrimp is discontinuously circumboreal in both northern oceans, most commonly in 50–500 m of water, and prefers mud bottoms. The GoM is its southerly limit in the northwest Atlantic. *Neomysis americana* is common in the GoM from only a few to about 90 m deep over both mud and sand. Sevenspine bay shrimp is common from the intertidal to the outer shelf and has been collected considerably deeper but most frequently over sands. *Neomysis americana* and sevenspine bay shrimp share northern range limits in the southern portion of the Gulf of St. Lawrence, with rare specimens reported farther north, and are seasonally abundant in estuaries (typically in summer). Both extend southward to eastern Florida but show decided changes in habitat use with latitude. Why, when, and how do their distributions converge in GoM; why, when, and how do they diverge?

TABLE 1. References documenting shared characteristics of the three focal species.

Characteristic	Species		
	Pandalus borelis	*Crangon septemspinosa*	*Neomysis americana*
Winter convergence on the Gulf of Maine shelf	Clark et al. 2000	Haefner 1972; Embich 1973; Grabe 2003	Grabe 1996
Omnivory	Weinberg 1981	Wilcox and Jeffries 1974	Zagursky and Feller 1985
Pelagic carnivory	Horsted and Smidt 1956	Price 1962	Fulton et al. 1982; Winkler et al. 2003
Diel migration	Barr 1970	Locke and Corey 1988; Sato and Jumars 2008	Herman 1963; Hopkins 1965
Spreading of distribution at night	Haynes and Wigley 1969	Brown et al. 2005[a]	Hulbert 1957; Herman 1963; Sato and Jumars 2008
Eaten by nondemersal animals	Bergström 2000	Gilmurray and Daborn 1981; Blackwell et al. 1995	Friedland et al. 1998; Walter and Olney 2003
Eaten by demersal fishes	Worm and Myers 2003; Parsons 2005	Pihl et al. 1992	Lankford and Targett 1997; Link and Garrison 2002

[a] Harmon Brown (Louisiana State University, personal communication) confirms that *C. septemspinosa* specimens in this study were taken primarily in nighttime plankton tows.

A brief digression on terminology is needed as a warning to readers. Both of the decapod species utilize internal insemination and fertilization. Females extrude fertilized eggs onto their pleopods and carry them until hatching. The strictest definition of "to spawn" is to release gametes, whether eggs or sperm, directly into the water. Under this strict definition, neither decapod species spawns (nor do mysids, which brood young in a marsupium). A slightly looser definition is to extrude gametes outside the body. A looser definition still is to produce offspring in large numbers. Unfortunately, usage is uniform neither within nor between the two decapod species. For sevenspine bay shrimp, field ecological studies that report spawning season typically equate the season when eggs are carried with the season of spawning and with the season of hatching. For example, Corey (1981:26) parenthetically equates spawning with release of larvae. As a practical matter, most uses of the word "spawning" (including Corey's) for this species refer to the finding of berried females, and with the short life cycle of sevenspine bay shrimp, the imprecision in definition generally leads to imprecision in timing of no more than a month. Most authors of field ecological studies of northern shrimp, however, make a finer distinction and refer to spawning as the time of extrusion of eggs onto the pleopods. The reason is clear; time between extrusion and hatching in this species is 6 months or more (Shumway et al. 1985, Figure 5; Koeller et al. 2009, Figure 2). A notable exception in usage is Grabe's (2003). He appears to equate spawning with hatching rather than extrusion. I attempt to avoid compounding confusion by substituting less ambiguous terms (e.g., carrying eggs or hatching) that reflect the data used by the original authors and avoiding "spawning" altogether.

Interactions with Light

Light and Gravity as a Compass

The first part of the title of my perspective is borrowed from a 19th century treatise on the crustacean fauna north of Cape Cod (Smith 1879). The three focal species share moveable, stalked eyes. Lamarck and many others who followed him believed that this feature to imply taxonomic affinity, but its monophyly is contested (Bowman 1984). Although they may not imply a common ancestor, stalked eyes do appear likely in all three species to share functions in orientation. Efficient exploitation of multiple habitats over a broad range of time scales from developmental to diel requires efficient navigation. Potential use of polarization in sun-compass mode has been appreciated for some time in both mysid and decapod shrimp (Bainbridge and Waterman 1957). Turbidity generally improves an individual's ability to detect linear polarization direction (Bainbridge and Waterman 1958), and the recent discovery that some crustaceans can detect circularly polarized light (Chiou et al. 2008) has yet to be explored as a navigational tool in decapods or mysids. Both linearly and circularly polarized light can also help to detect otherwise camouflaged or transparent prey and predators and facilitate sexual signaling within species (Chiou et al. 2008). To my knowledge, use of polarization in navigation has not been established definitively in any of the three focal species, but such capabilities in other species with similarly stalked eyes have been (Goddard and Forward 1991).

All three species also carry statocysts. In both mysids (Neil 1975) and decapods (Schöne 1954, 1957), statocyst ablation reveals movement of the eye stalks to be coordinated with signals from the statocysts. This coordination suggests horizontal azimuthal direction detection via polarization (cf. Land 1980): Scanning with moveable eyes, when coupled with information on the gravity vector, potentially allows energetically efficient detection of the azimuthal polarization direction: It circumvents inertially expensive changes in swimming direction that would otherwise be needed to sample polarization directions. Even diminutive mysids show impressive homing abilities to individual coral heads and caves (Twining et al. 2000; Coma et al. 1997), and detection of onshore-offshore direction should be even easier (Goddard and Forward 1991), with the highest signal-to-noise ratio in horizontal direction at the low sun angles found near sunrise and sunset (Waterman 2005) when the three species make their vertical excursions.

Light and Resource Gradients

All students of biological oceanography—witting, willing, or not—are students of the GoM. Gran and Braarud (1935), in the context of the tidal mixing of the turbid waters of the Bay of Fundy, were early to articulate the negative effects on phytoplankton production of intense tidal mixing in waters kept unusually turbid by that same mixing. They clearly described a depth of vertical mixing beyond which positive net phytoplankton production could not occur, and they suggested it to be a few times the compensation depth. Their observations resonated with Gordon Riley (Riley 1942; Riley et al. 1949), who extended the concepts to less turbid waters in the rest of the GoM and with Harald Sverdrup (Sverdrup 1953), who generalized the concept of too much mixing through a simple mathematical formula as being deeper than the critical depth that yields a positive value for depth-integrated, net production. Whereas these concepts are widely known, and despite discussion of notable blooms where mixing is limited by bottom depth by Gran and Braarud (1935), as well as the clear focus on this very issue by Riley (1942), it is less widely appreciated that a shallow bottom can lead to an early spring bloom simply by restraining mixing depth to be shallower than the local critical depth. Quoting Townsend et al. (2006), "The spring bloom throughout the northwest Atlantic shelf region begins first in shallow inshore areas (Hitchcock and Smayda 1977; Townsend 1984; Townsend and Spinrad 1986), when the critical depth (i.e., Sverdrup 1953) exceeds the bottom depth, which can happen in winter." Indeed, both far inshore and on shallow banks, net primary production occurs even in the darkest months, evidenced in multiyear satellite imagery (Thomas et al. 2003). A necessary caveat is that both suspended sediments and proximity to land (or any very sharp change in upwelling light with distance) complicate conversion of color to chlorophyll concentrations, but the logic of blooms being earlier in shallower waters remains sound (Riley 1942).

Still more inconspicuous is the GoM's "secret garden" (Thiel 1997). On sands and muds, benthic diatoms reach peak abundances of about 100 μg chlorophyll per gram dry sediment a few meters below MLLW (mean lower low water) as what might be considered the golden brown bathtub ring of the GoM—with substantial quantities eroding, adding to turbidity and food availability to suspension feeders (Thiel 1997). In samples taken in March, May, August, and November, benthic diatom abundances were highest in March and lowest in May (Thiel 1997). Growing evidence suggests high benthic diatom production on subtidal, boreal sediments early in the year before grazer populations build and before the intervening water column is filled with light-absorbing phytoplankton. Through effective diffuse scattering by sediments, benthic microalgae then get the irradiance benefit of a scalar irradiance roughly double the magnitude of the downwelling irradiance that approaches the sediments from above (Kühl and Jørgensen 1994). Oxygen optode data, for example, reveal net production in February (Wenzhöfer and Glud 2004) but net consumption of oxygen by the seabed later in the year in a pattern that may be typical for shallow, boreal waters. The existence of a coevolved fauna that exploits this secret garden (Thiel 1997) argues for its reliability and predictability from year to year. Large benthic diatoms are available directly to benthic individuals of *N. americana*, and mobile animals that feed on this secret garden are potentially available to all three of the shrimp species treated here.

Two other sources of primary production besides phytoplankton are important in the Bay of Fundy because of its extreme range of tides. They comprise substantial intertidal areas with microphytobenthic production during low tide (e.g., Trites et al. 2005) and expansive macroalgal beds and salt marshes that export phytodetritus (e.g., Gordon et al. 1985). Because they all originate near the land–sea interface and are mixed or transported offshore as they decay and are otherwise consumed, all of these inputs from intertidal primary production contribute to multiple, strong resource gradients that potentially influence habitat choice.

Light and Risk Gradients

Prevalence of diel vertical migration attests to risk gradients from visual predators. A photon coming from a potential predator or prey can arrive at a detector or be absorbed or scattered along the way. If it is scattered, it can further degrade the remaining visual signal by brightening the background optical field (i.e., arriving at the detector from a different location and direction). Species that can selectively detect polarized light, however, can in principle remove some of this "veiling" light and recover better contrast (Schechner and Karpel 2005).

The influence of turbidity on visual feeders can be quantified as the extent to which their clear-water detection distances for prey are reduced by that turbidity. The fraction of light arriving at point B from point A is reduced in proportion to the scattering and absorption coefficients of the water (per meter) and the number of meters between point A and B. Visually feeding species that depend on a greater reactive distance to detect larger prey therefore are expected to be more sensitive to both absorption and scattering. Experiments generally confirm this prediction, with large piscivores being more sensitive than small planktivores (e.g., De Robertis et al. 2003). Planktivores often show maximal feeding rates at intermediate turbidities (e.g., Boehlert and Morgan 1985), however, perhaps because near-field prey have greater contrast against a brightened background that is free of distant visual noise. Indeed, it has become relatively common to use "green water" to enhance growth and survival of fish larvae in aquaculture settings, even when predators are intentionally excluded (e.g., Faulk and Holt 2005).

Turbidity thus can be an advantage to small individuals at the same time that it disadvantages large individuals; turbidity produced by small, scattering particles can replace one nursery function normally attributed to macrophytes that in clear water provide analogous cover from visual predation. Fiksen et al. (2002) developed this argument more fully in the context of enhancement of larval fish survival by turbidity. In a terrestrial analogy, fog replaces trees. It should be pointed out that all three of the focal shrimp species fall in the category of planktivores when they are in the water column, and even the smallest of them uses vision in capture of its animal prey (Fulton 1982). Predator sensitivity to scattering versus absorption can be quite species specific, however (e.g., Macia et al. 2003), and sensitivity to scattering can depend strongly on light intensity (e.g., Benfield and Minello 1996). Effects of turbidity on visual predation by fishes feeding on shrimp has been studied, but the sensitivity of zooplanktivores bearing nonimage-forming, compound eyes to turbidity, to my knowledge, has not.

Nevertheless, strong, whole-community patterns can be explained through optical properties. One of the most striking examples is the explanation of roughly 80% of the long-term variance in combined anchovy and sprat abundance in the Black Sea as a function of Secchi depth alone (Aksnes 2007). In terms of zooplankton abundance and size, Aksnes et al. (2004) used the absorption coefficient, a, as a measure of protection from visual predators across a dozen Norwegian fjords. Fish abundance measured over large scale by acoustic backscatter declined exponentially with increasing a ($r^2 = 0.70$). They argued that the protection provided for zooplankton from planktivorous fishes was proportional to a (m^{-1}) times water depth, D (m), producing a nondimensional measure of the extent of habitat protected from visual predators. They found that zooplankton biomass per square meter, as well as mean individual weight of zooplankters, increased linearly with this measure ($r^2 = 0.91$ and 0.82, respectively).

A generalization of their argument suggests a measure of the protective value of a daytime habitat for a species that emerges at night. Turbidity is often much higher in near-bottom nepheloid layers than in surface waters, which argues for a measure of the form

$$\int_0^D c(z)\,dz, \tag{1}$$

where $c(z)$ is the beam attenuation coefficient as a function of depth (where $c = b + a$; b is the scattering coefficient; and depth, z, is measured as positive downward from the air–sea interface). If this generalized equation is an accurate measure of optical protection, a thin but turbid layer can enhance protection substantially and potentially add even more value by reducing the distance to which an animal needs to swim for protection at dawn and the transit distance to reach prey at night. A measure of the protective value of waters at an intermediate depth, z_i, is simply the same integral, truncated at this depth of integration in place of D.

The case for optical protection is perhaps best made for *N. americana*. Hulbert (1957) showed that it reached high abundance in daytime plankton tows from the Delaware River estuary only where light intensities fell below a critical irradiance, as calculated from measured Secchi depths (roughly 5.5 times Secchi depth; his Figure 2). In the classic pattern for a coastal plains estuary, turbidity increased up river to a maximum at the low salinities where electrochemical flocculation of particles first occurred. In a 26-h sample series at the estuary mouth, where the clearest water was found, *N. americana* was caught in surface waters only after dark (Hulbert 1957). Perhaps the most striking evidence of optical protection, however, comes from the estuary of Río la Plata (Schiariti et al. 2006), where *N. americana* is an introduced species. Its abundance is predictable from the estuarine salinity gradient (Figure 2), that is, stratification and presumed turbidity, but not from salinity per se. In optically protected habitats of the Río la Plata, *N. americana* reached abundances of 2,500 m^{-3}. Similar number densities are seen in summer in mid-Atlantic estuaries (up to 3,300 m^{-3} reported by Hopkins (1965) at Indian River Inlet, Delaware). The turbidity maximum in an estuary may be a particularly high-value habitat: The local and integral measure of optical protection (beam attenuation) is large, and large copepod populations to feed upon are often nearby the estuarine turbidity maximum (e.g., Winkler et al. 2003).

Patterns of Habitat Use

Winter Convergence

Among the three central characters, evidence of migration to shallow shelf waters in midwinter is clearest for northern shrimp. Eggs are hatched in anticipation of local peak phytoplankton abundances and earlier in GoM than in any of the species' more northerly habitats (Koeller et al. 2009). No direct studies of the species' distributional dependence on optical protection have been undertaken, but some circumstantial evidence exists. Spinrad (1986) produced contour plots of tur-

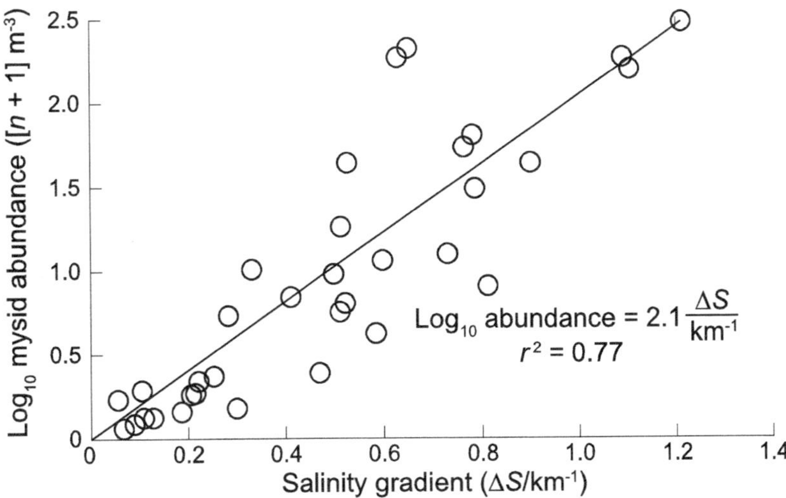

FIGURE 2. Log_{10} (number per cubic meter + 1) of *Neomysis americana* versus the local horizontal (along-channel) salinity gradient for Río la Plata between Argentina and Uruguay (redrawn from Schiariti et al. 2006; data values obtained by DataThief, www.datathief.org). The regression line explains 77% of the variance in abundance. Addition of the number one to the observed number density prevents problems with zeros in the log transformation, so actual number density is predicted by replacing the leading factor of 2.1 with 2.0.

bidity in the GoM at several water depths, and high values in the deeper western gyre correspond broadly with the fall distribution of the species (Haynes and Wigley 1969; Clark et al. 2000). Its spring distribution (Clark et al. 2000, Figure 8A) appears to reflect the shortest path to shallow water from the sites of its fall concentrations (Clark et al. 2000, Figure 8B) while still being in relatively turbid waters (Spinrad 1986). If turbidity played no role, one would expect to see hatching also on Georges Bank, but northern shrimp is rarely collected there (Clark et al. 2000). Temperature is often cited as the likely reason for its absence farther south, but northern shrimp's highest abundance in the GoM's warmest, western gyre is not consonant with that explanation. Lack of optically protective, deep basins near shore and possibly a less predictable winter bloom may prevent southward range extension. Also supporting the importance of sequential habitat use during development are strong differences in size frequencies and growth curves for populations that follow different migration paths in the Gulf of St. Lawrence (Simard and Savard 1990).

If sevenspine bay shrimp and *N. americana* migrate seasonally, they do so between waters of different depths on the shelf, within estuaries and (or) between estuaries and the shelf. Year-round marine sampling programs are much rarer than year-round estuarine sampling programs, and year-round programs that synoptically sample both estuaries and the adjacent shelf for either species do not appear to have been done in the GoM. Nevertheless, the few winter sampling programs that have been conducted near shore indicate convergence of these two species on shallow coastal waters very early in the year. Just south of the GoM, Fish (1925) reported *N. americana* juveniles and adults at Woods Hole on the southwest coast of Cape Cod from December through April. In a 3-year study of nighttime peracarid zooplankton in 20-m-deep coastal waters of New Hampshire, Grabe (1996) found *N. americana* to be the winter dominant and to be reproductive most of the year. Near this same site in weekly scuba samples through 1975 and 1976, Burreson and Allen (1978) reported leeches on *N. americana* only between December and March. A possible inference is density-dependent infestation (i.e., higher *N. americana* concentrations in shallow coastal waters in winter than summer).

Early larval stages of sevenspine bay shrimp appeared in waters over the same New Hampshire coastal depths in late February or early March (depending on the year) and disappeared near the end of November (Grabe 2003). Grabe (2003) reported *Pandalus* larvae throughout the year but did not identify them to species. According to Haynes and Wigley (1969), average catch depth of northern shrimp that have exhausted their yolk is about 90 m (Haynes and Wigley 1969), but the average depth of ovigerous shrimp shoaled to 23 m in March. Grabe (2003) expressed doubts that their larvae would be carried to his sampling sites at 15–20 m depth, even though most commercial fishers for the large, gravid females capture them in waters within the 60-m isobath (Haynes and Wigley 1969). In detailed, individual-based models of dispersal, retention, and settlement success in GoM northern shrimp, Bates (2007) found consistently greatest settlement success when modeled larvae were released at 15 or 20 m water depth, noting that periods of downwelling north winds in winter should aid in bringing larvae shoreward. Without a doubt, all three species utilize waters of the western Maine Coastal Current as larval nursery areas well before the climatological bloom times for the deep basins (Thomas et al. 2003, Figure 2). It is equally clear that both of the euryhaline species (*N. americana* and sevenspine bay shrimp) are abundant and reproductive in GoM estuaries in summer. What is much less clear in the GoM is the extent of seasonal dependence of their estuarine population dynamics on shelf populations, and vice versa, via active migrations of adults.

Comparisons and Contrasts among Species and Latitudes

Northern shrimp is stenohaline and in adulthood is a deepwater species (typically >100 m) outside its egg-carrying season, so the remaining distributional overlaps and contrasts mainly concern sevenspine bay shrimp and *N. americana* within the GoM and its estuaries, as well as between the GoM and populations elsewhere in their ranges. The idea that sevenspine bay shrimp and *N. americana* migrate offshore from GoM estuaries prevails in the literature but is based in more cases on disappearance of high abundances from estuaries or, more typically, away from particular stations in estuaries (adhering to the Maine connotation of "away"), than on any direct evidence of migration, and what is meant by offshore (where and how deep) is rarely indicated because lack of sampling makes it unknown. Not all individuals migrate between the ocean and estuaries; both species can be caught year-round, for example, on Georges Bank (Whiteley 1948). Both are smaller than northern shrimp and so gain more optical protection there than would northern shrimp (cf. De Robertis et al. 2000). Waters of northern shrimp's winter hatching grounds are more turbid than waters of Georges Bank (Spinrad 1986).

As reviewed by Jumars (2007), maximal densities of *N. americana* shoal southward even within the Gulf of Maine. U.S. National Marine Fisheries Service surveys showed maximal density in daytime benthic collections within the 30–60-m-depth band farther north (Wigley and Burns 1971), whereas in Cape Cod Bay, Maurer and Wigley (1982) reported maximal abundance in the 10–29-m-depth band. Despite being abundant in Long Island Sound (Richards and Riley 1967), *N. americana* is notably scarce in shelf waters immediately south of Long Island (Bigelow and Sears 1939; Wigley and Burns 1971), but it achieves substantial abundance inshore in the mid-Atlantic (Wigley and Burns 1971). South of Cape Hatteras it is known almost exclusively as an estuarine species, although it can be found in shallow waters of the open coast at least as far south as South Carolina as well (D. M. Allen, Baruch Marine Field Laboratory, personal communication). Shallow water may provide additional optical protection from visual predators in the form of scatter from bubbles (due to wave breaking) and high-frequency wave focusing of bright light (wave speckle) in addition to high-frequency wave resuspension, and turbid waters entering the oceans from estuaries tend to hug the shore as their turbidity declines through turbulent dispersion and settling. Given the low light intensities to which many estuarine populations of *N. americana* are adapted (e.g., Abello et al. 2005), the restriction of *N. americana* to shallow shelf waters and estuaries in the mid-Atlantic and the Georgia Bight suggests that some, and perhaps several, means of visual protection are involved.

The most spatially expansive time series for *N. americana* that bears on the issue of seasonal migration was collected in a multiyear survey of herring larvae in the Bay of Fundy. Corey (1988) analyzed *N. americana* abundances in oblique tows taken from March 1973 to November 1981, primarily in the months of March, August and November. In March, the core area containing *N. americana* was mostly upstream of a line between the Saint John River and Digby Gut and downstream of the branching of the Bay of Fundy between Chignecto Bay and Minas Channel (Figure 3). Catches were sparse. Individuals were nearly all large juveniles—many even larger than adults at other times of year. Population densities reached their observed maxima, with abundant juveniles present, in August and were displaced toward shallower, more turbid water (generally <60 m) from their March core area, mirroring at a larger, basin scale, a common pattern seen within (shallower) estuaries up and down the East Coast. It is notable that American shad *Alosa sapidissima*, which are not normally surface feeders but feed avidly as both adults and juveniles on *N. americana* throughout their overlapping ranges (Walter and Olney 2003; Hoffman et al. 2008), are apparently forced to become surface feeders in the high turbidity of upper Bay of Fundy (Dadswell et al. 1983). Also in August, locally dense patches of *N. americana* were observed at stations near the entrances to Digby Gut and St. Mary's Bay, suggestive of dispersal from seasonally growing populations farther up those estuaries. By far the broadest distribution was observed in November. Abundances and locations were most parsimoniously explained by emigration from estuaries to constitute at least part of the overwintering generation. In November, no adults were found shallower than 40 m, suggesting that migration of adults to deeper water may be a general phenomenon for individuals that leave estuaries, but giving little insight into what fractions of adults do leave estuaries. Population densities in August versus November (Figure 3) suggested emigration from St. Mary's Bay and Digby Gut, but whether animals emigrate from the Saint John River estuary and Cobscook Bay is less clear, although the November 1978 samples (Corey 1988; her Figure 4) are highly suggestive. There are year-round resident populations of *N. americana* in both of those estuaries (Pezzack and Corey 1979; Carter and Dadswell 1983), but that observation does not preclude exchange with Bay of Fundy populations either seasonally or continuously. If there is net immigration into those estuaries from the Bay of Fundy at a particular time of year, it would likely be timed so that young were released during the spring bloom, which Pezzack and Corey (1979) suggest happens in May to June in Passamaquoddy Bay.

Seasonal timing of reproduction in the Bay of Fundy contrasts sharply with that observed on the shelf at New Hampshire, where reproductive individuals are already abundant in February. As Gran and Braarud (1935:404) noted, "in the Bay of Fundy winter conditions were still prevailing during March." Within the GoM, reproduction of *N. americana* occurs earliest on the shelf in the western gyre of the GoM (Grabe 1996), followed by Georges Bank (Wigley and Burns 1971). Reproductive timing on the shelf between the Bay of Fundy and the New Hampshire coast is not as clear.

Carter and Dadswell (1983) sampled the Saint John River estuary and the tributary estuaries near its mouth bimonthly from May 1973 through May 1977, and sampled biweekly from June 1978 through June 1979, with some deviation when ice conditions precluded sampling. When ice was sound, they sampled through it. A sill at the river mouth 5 m below high tide level for the Bay of Fundy limits exchange and allows even the lower estuary to have higher temperatures in summer and lower temperatures in winter than does the nearby Bay of Fundy. *Neomysis americana* was never absent from all estuarine stations in any season. Local maxima in abundance varied from August to October–November, depending on estuarine location (Carter and Dadswell 1983, Figure 4). Notably, they sampled with a plankton net during daylight. The waters of the lower Saint John River are sufficiently turbid that at least part of the population of *N. americana* is always planktonic.

Pezzack and Corey (1979) sampled from one shallow station (1–10 m) and six deeper stations (10–50 m) in Passamaquoddy Bay twice monthly from May 1972 to August 1974. They found the highest abundances at a shallow station on the opposite side of the bay from its inlets and reported visual observations of high *N. americana* concentrations near water's edge at low tide.

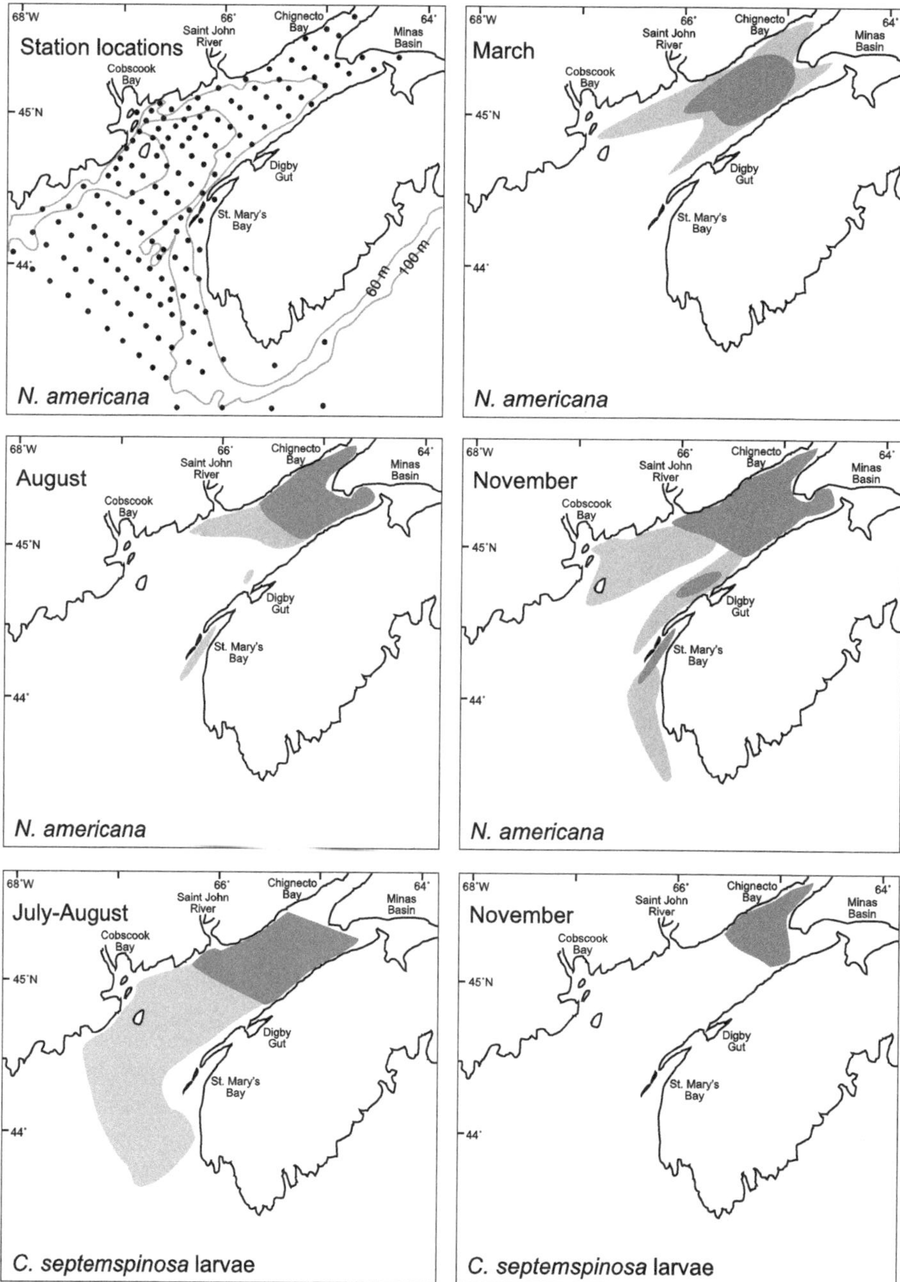

FIGURE 3. Station locations for samples from oblique plankton tows taken as part of a (larval) herring survey over several years, primarily in the months shown (upper three panels, with shading indicating relative fidelity and abundance in the mapped region). Darkly shaded regions indicate where high densities of *Neomysis americana* were observed reliably from year to year, whereas lightly shaded areas denote extensions observed in some years (redrawn based on data in Pezzack and Corey 1979). Not all extensions shown were seen in any one year. Fall emigration from estuaries is strongly suggested by the patterns. The bottom two panels show similar plots but for sevenspine bay shrimp larvae collected in neuston tows in a similar station configuration but over a smaller number of years (redrawn from Corey 1988). Larvae were absent in March, but berried adult females were captured then all along the west coast of Nova Scotia.

Abundant mysids were captured at this shallow station year round. Specimens were scarce in the deeper water samples and limited primarily to late summer and fall. Although very shallow water can sometimes be a refuge from fish predation (Ruiz et al. 1993), bird predation becomes an issue (Schneider and Harrington 1981; Blackwell et al. 1995). This counter example to generally tight association with turbidity and *N. americana* would benefit from the scrutiny of a broader spatial and seasonal coverage.

Herman (1963) sited a year-long study of vertical migration at the West Passage entrance to Narragansett Bay in 19 m of water, sampled at least monthly. He sampled a complementary station less frequently at the entrance to East Passage in 36 m of water. The limited spatial coverage does not allow direct comparison with the seasonal-spatial interactions noted in the other two estuaries, but site selection itself is revealing. The primary station was chosen "as the region of greatest abundance for *N. americana* throughout the year" (Herman 1963:229). Moreover, bottom dredge samples consistently collected more *N. americana* on flood than on ebb tides, suggesting both a source farther offshore and an abundance gradient increasing offshore. Nocturnal entry into the plankton was seen year round, but the fraction of the population staying near the bottom varied substantially from month to month.

Farther south, data from a 2-year study of leech infestation of *N. americana* in an estuary-marsh system of southern New Jersey are most parsimoniously explained by arrival of large *N. americana* individuals outside the Hereford Inlet mouth when water temperatures drop to ~10°C in autumn, with subsequent mysid migration up estuary into tidal creeks over a period of 2 weeks (Allen and Allen 1981). The leech is apparently otherwise excluded from the estuary by high temperatures. Furthermore, tidal creek population densities increased abruptly at the time of the inferred immigration, and the size-frequency distribution of *N. americana* inside the estuary shifted too quickly to be explained by in situ growth of individuals (Allen and Allen 1981). This apparent immigration occurred in November, the same month that emigration from estuaries was inferred for the Bay of Fundy by Corey (1988). The site from which leech-infested individuals outside the inlet originated is unknown but is presumed to be deeper water. Some mysids, however, were observed year-round in the estuary.

Hulburt (1957) studied *N. americana* very nearby Hereford Inlet, in the much larger Delaware River estuary, sampling primarily along its main channel. He observed the largest individuals in winter and during the cold months saw consistent up-river increase in size of the largest individuals, but this pattern held only in the main channel and not at shallower stations, perhaps because so few mysids were collected outside the (optically protective) channel. The seasonal timing of arrival of these large individuals was similar to that observed by Allen and Allen (1981), but a resident population was also always present. Hulbert (1957) suggested that the fall and winter up-estuary increase in size was consistent with import from the sea, with growth and upstream transport along the bottom in a two-layered estuarine circulation. *Neomysis americana* was absent or rare at the mouth of Delaware Bay in spring and summer when water there was clearest; the dense up-river population was essentially isolated from shelf populations at that time. The likely cause is lack of optical protection at the mouth. The spatial maximum in population density was observed near the bay mouth only in one fall cruise (October 10–November 10, 1951) that may have coincided with that year's immigration period. It must be stressed, however, that no direct evidence of migration in or out of the estuary was reported.

Williams' (1972) results for an estuary complex in North Carolina were broadly similar. He studied the Neuse and Pamlico rivers, Pamlico Sound into which they both drain, and three inlets of the barrier bar system that separates Pamlico Sound from the shallow shelf. Inlet stations were heavily populated in the cooler months, from November to May or June; presence of *N. americana* in other months was erratic. Again, it would be useful to have optical data to plot against abundance. Both rivers maintained high populations year round. Large individuals were conspicuous in the estuary from late fall to mid-May. Again, no explicit migration data or shallow shelf samples were collected.

How does sevenspine bay shrimp compare? Locke and Corey (1988) sampled nearly the same Bay of Fundy station pattern shown in Figure 3 and described by Corey (1988), but with a neuston net that sampled only the upper 18 cm during eight cruises from July 1979 through March 1983. Not all stations were occupied on each cruise. Most samples again came from the same three seasons shown for *N. americana* in Figure 3. They caught sevenspine bay shrimp larvae only in summer and fall, with a much narrower areal distribution in the fall (Figure 3). Conversely, they captured adult sevenspine bay shrimp only in March, and they were females carrying eggs. These adults were broadly distributed along the western shore of Nova Scotia except for one sample just east of Jonesport, Maine. Adults were caught only at night, and larvae were significantly more abundant then. It is quite clear that egg hatching in the Bay of Fundy lags substantially behind that seen in the western gyre of the GoM (Grabe 2003).

Because the observations he made of disappearance of sevenspine bay shrimp from shallow water (Haefner 1972) were consistent with his laboratory experiments on salinity and temperature tolerance (Haefner 1969), Haefner was particularly influential in affecting thinking about migration to deeper water in winter. The field data at Lamoine (Haefner 1972), however, were very limited spatially. The site was near the head of Frenchman Bay, north of Mount Desert Island, and observations comprised 17 beach seine samples from June 1966 through June 1967. Shrimp were scarce in January and April (no samples taken in February) and absent in March. Much more extensive sampling in the Penobscot River estuary, however, supported the idea of emigration from shallow, low-salinity waters in winter (Embich 1973). Adult shrimp were absent from all but stations at the mouth of the Penobscot River estuary in winter.

Even in GoM estuaries where sevenspine bay shrimp remains abundant in juvenile stages during winter, such as Passamaquoddy Bay, there is evidence of immigration of large, overwintering females from offshore that carry eggs in the bay from March to early June. Smaller females bear eggs in mid-July through August (Corey 1981, 1987). Given this timing, the sevenspine bay shrimp larvae observed in coastal New Hampshire (Grabe 2003) are unlikely to be emigrants from an estuarine hatch, and the presence of small juveniles in the estuaries of mid-coast Maine before the local March–June hatch is further evidence of winter release offshore (Haefner 1972; Bowdoin 2008).

In Connecticut's Mystic River estuary, however, sevenspine bay shrimp followed a quite different pattern of migration. Modlin (1980) sampled benthic populations 23 times at five stations spread along the estuary and just outside from September 1973 through September 1974, and sampled larvae from April to October 1974 and from April through June 1975. Berried females entered the estuary to release their hatch (when 2 and 3 years old) in spring (April–May) and (when 1–1.5 years old) in the fall (October). Juveniles were present in the deeper parts of the estuary year-round. Adults larger than 3 cm were transient; they were absent both in winter and after estuary water exceeded 20°C in June. Most larvae were carried out of the estuary. Postlarvae from the spring hatch appeared rapidly in the shelf area around the mouth of the estuary; postlarvae from the fall hatch did not, and presumably stayed in deeper water to overwinter.

Farther south, a single but longer egg-bearing season is reported (Corey 1987). On the continental shelf, Viscido et al. (1997) studied small-scale spatial and seasonal distributions of sevenspine bay shrimp at three stations, a landward site at 12.5 m, a ridge-top site only slightly shallower, and a seaward site at about 14.5 m. The landward and seaward sites were separated by about 2 km, with the ridge top in between. Juveniles and adults of both sexes were found in all seasons. Minimal numbers of shrimp were found in June and July, with fairly high numbers in May and a temporal maximum in October. If there was a summer exit from nearby estuaries, it did not result in an obvious accumulation at these shallow shelf depths.

Only a little farther south, Price (1962) collected sevenspine bay shrimp by push net and seine at 11 shallow sites in the Delaware Bay region from October 1958 to February 1960, with the majority of samples taken in 1959. Sampling depths were less than 2 m. The major egg-carrying period lasted from March to October, but some berried females were found year-round, except for December in the shallowest waters. Haefner (1976) conducted monthly fish trawl surveys of sevenspine bay shrimp populations in the central channel of the York River out to the mouth of the Chesapeake Bay from January 1971 to April 1972. The trawl mesh could retain no juveniles. Berried females were most abundant from December through June and largely absent from July through October, probably due to low oxygen levels in the bottom water (Haefner 1976).

Haefner's (1976) and Price's (1962) results are difficult to compare because they sampled neither the same water depths nor the same sizes of animals. Sandifer (1973), however, took monthly plankton samples in the same region of Chesapeake Bay studied by Haefner (1976). He found larvae of sevenspine bay shrimp in every month of the year, but most abundantly from January through June, paralleling Haefner's (1976) phenology of berried females. Deevey's (1960) larval results also support a midwinter minimum in larvae for Delaware Bay, consonant with Price's (1962) observations of berried females. Although no special migration season is suggested for mid-Atlantic estuaries, and some larvae nearing competence may by chance enter the deep estuarine circulation before settlement, the majority of sevenspine bay shrimp larvae are likely exported from estuaries, with individuals migrating back as juveniles or adults after development on the shelf (Sandifer 1975). A single, long, reproductive period argues for an even longer period of immigration. The Chesapeake Bay results are compatible with migration of adults to either shallow water inside the estuary or to the shelf in summer. Price's (1962:252) words remain apt, however: "These data stress the importance of conducting shore zone and deep water surveys simultaneously in order to predict the reproductive behavior of this shrimp." A systematic sampling of estuary-shelf migrations is clearly in order for both sevenspine bay shrimp and *N. americana*.

Apparent Anomalies

Sevenspine bay shrimp occupies virtually the entire shelf both in the GoM and farther south and is occasionally found much deeper on the slope (Wigley and Theroux 1981; Theroux and Wigley 1998). It also is the only one of the three species observed routinely in tide pools. How can the species that is intermediate in size be so broadly distributed in shallow water, including clear water? A likely mechanism underlying this difference in distribution is a difference in behavior. Sevenspine bay shrimp achieves added optical protection through partial burial, which may be easier in its preferred sand substratum than in the mud substrata over which *N. americana* more often operates. Sevenspine bay shrimp still takes advantage of optical and sedimentary microenvironments; on the shallow New Jersey continental shelf, it is found more often in local depressions that contain finer sands and temporarily store wave-generated turbidty than on coarser ridge tops (Viscido et al. 1997). *Neomysis americana*, by contrast, remains at the sediment surface and stays active enough to maintain its heading into the current (H. Abello and S. Shellito, Darling Marine Center, personal communication of diving observations in the Damariscotta River estuary). Although analogous diel gut-fullness studies do not appear to have been done with *N. americana* that would demonstrate any relative advantage in this regard, sevenspine bay shrimp, in a round-the-clock study in Rhode Island, showed clear periodicity in gut fullness, with peaks at midnight and near dawn and a majority of specimens with empty guts in mid-afternoon (Wilcox and Jeffries 1974, Figure 1). Partial burial would appear to restrict daytime feeding opportunities, partially offsetting benefits of reduced predation risk.

Distributions of sevenspine bay shrimp in two barrier-bar estuaries also pose some conundrums. Able et al. (2002) reported that in the classic barrier-bar estuarine system of the Nauset River that empties eastward through Cape Cod, sevenspine bay shrimp is resident in the marsh system year-round as both juveniles and adults 15–50 mm long. They studied fish and decapod community structure and so did not provide further information on population dynamics. Locke et al. (2005) sampled sevenspine bay shrimp in the Kouchibouguac River, New Brunswick, that opens on Northumberland Strait in the Gulf of St. Lawrence. They reported that in this region, commercial smelt nets set under ice routinely catch egg-bearing sevenspine bay shrimp, and they collected berried females in their earliest seasonal trawls as soon as ice left the estuary (May 15–23). The commercial smelting season here closes by March. They also sampled for larvae as early as April 7, 1998 (through the ice) and April

30, 1998 (in open water) but found none until May 28, 1998. Sampling started later in 1997 (May 20), but larvae were not found until June 3. Berried females were never absent in samples, but larval abundance showed a unimodal peak in July (both years). Spring females were unusually large, not just with respect to females at other times in this estuary but also with respect to lengths reported from Newfoundland to North Carolina. Large prey size in an optical protection context suggests high turbidity (cf. Aksnes et al. 2004). This complement of unusually large females was gone by July, but whether through mortality or emigration is unknown. The single, broad reproductive peak for sevenspine bay shrimp is much more characteristic of southern coastal-plains estuaries than it is of other GoM estuaries (Corey 1987).

Conclusions, Open Questions, and Needed Approaches

Obvious but unanswered questions are whether population dynamics in coastal-plains, barrier-bar estuaries are materially different than elsewhere and, if so, how the difference is linked to optical properties. Although framed implicitly for sevenspine bay shrimp, the question is equally apt for *N. americana*. According to Johnson and Allen (2005), *N. americana* is most common in the deepest portions of estuaries. In the classic barrier-bar estuaries where it has been studied most, the deepest portions are inlets where water is generally clearer than upstream, and so deeper distribution follows from the need for optical protection. In the physiographically more variable, glacially sculpted estuaries that empty into the GoM, sills, narrows and holes are common, and depth distributions of mysids are not well known. In admittedly limited spatial coverage in Passamaquoddy Bay (Corey 1981), *N. americana* was most abundant year-round at the shallowest station farthest inshore. This pattern of habitat use is difficult to reconcile with optical protection as a major driver and leaves open questions of proximate and indirect causes, but a wealth of data supporting modulation of vertical habitat use in *N. americana* by optical properties includes its nocturnal entry into the pelagic habitat and the distinct differences in its vertical and horizontal distribution during daylight in waters differing in turbidity (e.g., Hulbert 1957; Carter and Dadswell 1983; Schiariti et al. 2006).

Northern shrimp has long been associated with fine-grained habitats (e.g., Warren and Sheldon 1968). It would be interesting to know whether the association with high values of equation (1) is even stronger, that is, whether optical properties play a role in habitat use and migration routes. Northern shrimp is a particularly highly valued food of Atlantic cod *Gadus morhua*—for which there is good evidence in population-level effects (Worm and Myers 2003). Despite amelioration by complementary use of chemosensing to aid vision in prey location, turbidity appears to have negative energetic consequences for juvenile Atlantic cod feeding on mysids (Meager et al. 2005; Meager and Batty 2007) and potentially even greater effects on larger Atlantic cod feeding on northern shrimp.

Timing and frequency of migrations between shelf and estuarine environments remain somewhat obscure. Fall emigration of *N. americana* from estuaries onto GoM shelves appears common but not universal. Evidence is suggestive but weaker for spring migration in the opposite direction. In the mid-Atlantic, there is little published evidence of emigration by *N. americana* from estuaries, but size-frequency data and data on leech infestation suggest net immigration in November. It would be particularly interesting to get migration data for *N. americana* in far northern barrier-bar estuaries on Cape Cod and in the Gulf of St. Lawrence.

A prolonged larval development complicates the picture for sevenspine bay shrimp, with the bulk of larvae being carried out of estuaries (Sandifer 1975). Corey (1987) suggested a transition just south of Connecticut, with two generations per year from distinct spring and fall reproduction events and a single but longer reproduction event to the south. In some GoM estuarine and shallow-water settings, there is evidence of an adult migration offshore in late fall or winter. A spring migration of large individuals into estuaries is also suggested by some size-frequency data and is clearly necessary to explain the spring hatch (March–June) where one has been demonstrated (Corey 1987). Sources of females that participate in the summer hatch within estuaries (mid-July–August) are generally not as clear because they may have matured to varying degrees either within or outside the estuary. Perhaps the clearest example of two generations and two seasonal peaks of immigration is for the Mystic River estuary (Modlin 1980). Embich's (1973) suggestions for migrations of Penobscot River estuary individuals are quite similar in terms of ages of individuals involved. Estuarine patterns are complicated by timing offsets of reproduction (due to food and temperature differences) from shelf populations (e.g., Bowdoin 2008). On the shelf, larvae are known earliest in the year from the shelf near New Hampshire and do not appear in the Bay of Fundy until after March. In the mid-Atlantic, the single, protracted summer breeding season results from some berried individuals in all seasons except perhaps January–February, and immigration pulses into estuaries are far less peaked. Larvae of sevenspine bay shrimp are still abundant in winter over the shelf, where they are the only common winter decapod larvae (Johnson and Allen 2005).

For both *N. americana* and sevenspine bay shrimp, simultaneous sampling in estuaries and the surrounding shelf would add substantially to the understanding of migration. Mysid statoliths present obvious problems and opportunities as tracers of provenance. A closed statocyst is found in the endopod of each uropod (Schram 1986). Just as in fish otoliths, the source of these extracellular precipitates is body fluid, not seawater. Otoliths have been used to document provenance (e.g., Thorrold et al. 1998). Unlike gradually accreted fish otoliths, new mysid statoliths are secreted rapidly after being shed with the exuvium (Morgan and Beeton 1978). *Neomysis americana*'s statoliths are made primarily of fluorite (Ariani et al. 1993), and no published data exist on their retention of trace elements. Rapid secretion may be required because statocysts appear to be essential to navigation by mysids (Neil 1975). The potential tracer, then, is useful only for the duration of the intermolt. Similar data on number of molts do not appear to be available for *N. americana*, but the larger *N. integer* matures after 15–16 molts in 110 d at 10°C (Winkler and Greve 2004). Large preadults (1.2 mm) molt every 19 d at 10°C (Winkler and Greve 2004). We suspect winter intermolt periods to be about 30 d in *N. americana* at ambient temperatures of 0–5°C (Pezzack and Corey 1979), but clearly this suspicion should be replaced with measurements from culture experiments. Mysid statoliths are remarkably large, with diameters of 1–3% of body length (Ariani et al. 1993). Relationship to body length is sufficiently precise that statolith measurements have been used to assign mysids to instars (Morgan and Beeton 1978).

It is difficult to find 10-cm-long estuarine or shelf fishes that do not eat mysids, and many fish species, both larger and smaller, do so as well. Rainbow smelt *Osmerus mordax* migrations are not limited to GoM habitats, but their prevalent winter feeding in GoM estuaries may be a factor in the apparent disappearance of *N. americana* and sevenspine bay shrimp from some GoM estuaries in winter. Larval rainbow smelt have intimate relations with turbidity (Dodson et al. 1989; Dauvin and Dodson 1990), and adults of a freshwater congener have been shown feeding sensitivity to turbidity (Horppila et al. 2004). Shrimp migrations and population dynamics need to be set in the context of the optical properties of the habitats that they occupy. For the two decapod shrimp, the problem of optical protection extends to larvae as well (e.g., Forward 1987; Morgan 1990). Studying both sevenspine bay shrimp and *N. americana* in the same places and times might, therefore, be very revealing as to critical life stages in habitat choice.

Last, an important question is how to remove study of these emergers from the weak methods of ad hoc, tentative explanation and place it in the more interesting and scientifically productive context of prediction and its testing. Given the long history of related research on diel vertical migrations by holoplankton and the many added complexities of crossing habitat boundaries, prospects for rapid advance to predictive understanding of emergence by reductionist testing of an analogous collection of hypotheses tailored to mysids or other emergers appear bleak. One may ask instead whether very different approaches have succeeded elsewhere in ecological problems of similar kinds and complexities and whether they show promise for application to seasonal emergence and migration. Individual-based modeling is an obvious answer (Grimm and Railsback 2005; Grimm et al. 2005).

Acknowledgments

I thank an anonymous reviewer and Jeff Runge for useful suggestions that improved the readability of this paper.

References

Abello, H. U., S. M. Shellito, L. H. Taylor, and P. A. Jumars. 2005. Light-cued emergence and re-entry events in a strongly tidal estuary. Estuaries 28:487–499.

Able, K. W., M. P. Fahay, K. L. Heck, Jr., C. T. Roman, M. A. Lazzari, and S. C. Kaiser. 2002. Seasonal distribution and abundance of fishes and decapod crustaceans in a Cape Cod estuary. Northeastern Naturalist 9:285–302.

Aksnes, D. L. 2007. Evidence for visual constraints in large marine fish stocks. Limnology and Oceanography 52:198–203.

Aksnes, D. L., J. Nejstgaard, E. Scedberg, and T. Sørnes. 2004. Optical control of fish and zooplankton populations. Limnology and Oceanography 49:233–238.

Allen, D. M., and W. B. Allen. 1981. Seasonal dynamics of a leech-mysid shrimp interaction in a temperate salt marsh. Biological Bulletin (Woods Hole) 160:1–10.

Ariani, A. P., K. J. Wittmann, and E. Franco. 1993. A comparative study of static bodies in mysid crustaceans: evolutionary implications of crystallographic characteristics. Biological Bulletin (Woods Hole) 185:393–404.

Bainbridge, R., and T. H. Waterman. 1957. Polarized light and the orientation of two marine Crustacea. Journal of Experimental Biology 34:342–364.

Bainbridge, R., and T. H. Waterman. 1958. Turbidity and polarized light orientation of the crustacean Mysidium. Journal of Experimental Biology 35:487–493.

Barr, L. 1970. Diel vertical migration of *Pandalus borealis* in Kachemak Bay, Alaska. Journal of the Fisheries Research Board of Canada 27:669–676.

Bates, M. J. 2007. Modeling physical controls on northern shrimp (*Pandalus borealis*) dispersal, retention and settlement success in the Gulf of Maine. Master's thesis. University of New Hampshire, Durham.

Beck, M. W., K. L. Heck, Jr., K. W. Able, D. L. Childers, D. B. Eggleston, B. M. Gillanders, B. Halpern, C. G. Hays, K. Hoshino, T. J. Minello, R. J. Orth, P. F. Sheridan, and M. P. Weinstein. 2001. The identification, conservation, and management of estuarine and marine nurseries for fish and invertebrates. BioScience 51:633–641.

Benfield, M. C., and T. J. Minello. 1996. Relative effects of turbidity and light intensity on reactive distance and feeding of an estuarine fish. Environmental Biology of Fishes 46:211–216.

Bergström, B. I. 2000. The biology of *Pandalus*. Advances in Marine Biology 38:55–256.

Bigelow, H. B., and M. Sears. 1939. Studies of the waters of the continental shelf, Cape Cod to Chesapeake Bay III. A volumetric study of zooplankton Memoirs of the Museum of Comparative Zoology 54:181–378.

Bisagni, J. J., K. W. Seeman, and T. P. Mavor. 2001. High-resolution satellite-derived sea surface temperature variability over the Gulf of Maine and Georges Bank region, 1993–1996. Deep-Sea Research Part II 48:71–94.

Blackwell, B. F., W. B. Krohn, and R. B. Allen. 1995. Foods of nestling double-crested cormorants in Penobscot Bay, Maine, USA: temporal and spatial comparisons Colonial Waterbirds 18:199–208.

Boehlert, G. W., and J. B. Morgan. 1985. Turbidity enhances feeding abilities of larval Pacific herring, *Clupea harengus pallasi*. Hydrobiologia 123:161–170.

Bowdoin, J. A. 2008. Life-history characteristics of *Crangon septemspinosa* and management implications. Master's thesis. University of Maine, Orono.

Bowman, T. E. 1984. Stalking the wild crustacean: The significance of sessile and stalked eyes in phylogeny. Journal of Crustacean Biology 4:7–11.

Brown, H., S. M. Bollens, L. P. Madin, and E. F. Horgan. 2005. Effects of warm water intrusions on populations of macrozooplankton on Georges Bank, northwest Atlantic. Continental Shelf Research 25:143–156.

Burreson, E. M., and D. M. Allen. 1978. Morphology and biology of *Mysidobdella borealis* (Johansson) *comb. n.* (Hirudinea: Piscicolidae) from mysids in the western North Atlantic Journal of Parasitology 64:1082–1091.

Carter, J. H. C., and M. J. Dadswell. 1983. Seasonal and spatial distribution of planktonic Crustacea in the lower Saint John River, a multibasin estuary in New Brunswick, Canada. Chesapeake Science 6:142–153.

Chiou, T.-H., S. Kleinlogel, T. Cronin, R. Caldwell, B. Loeffler, A. Siddiqi, A. Goldizen, and J. Marshall. 2008. Circular polarization vision in a stomatopod crustacean. Current Biology 18:429–434.

Clark, S. H., S. X. Cadrin, D. F. Schick, P. J. Diodati, M. P. Armstrong, and D. McCarron. 2000. The Gulf of Maine northern shrimp (*Pandalus borealis*) fishery: a review of the record. Journal of Northwest Atlantic Fisheries Science 27:193–226.

Coma, R., M. Carola, T. Riera, and M. Zabala. 1997. Horizontal transfer of matter by a cave-dwelling mysid. Pubblicazioni della Stazione Zoologica di Napoli I: Marine Ecology 18:211–226.

Corey, S. 1981. The life history of *Crangon septemspinosa* Say (Decapoda, Caridea) in the shallow sublittoral area of Passamaquoddy Bay, New Brunswick, Canada. Crustaceana 41:21–28.

Corey, S. 1987. Reproductive strategies and comparative fecundity of *Crangon septemspinosa* Say (Decapoda, Caridea). Crustaceana 52:25–28.

Corey, S. 1988. Quantitative distributional patterns and aspects of the biology of the Mysidacea (Crustacea: Peracarida) in the zooplankton of the Bay of Fundy region. Canadian Journal of Zoology 66:1545–1552.

Dadswell, M. J., G. D. Melvin, and P. J. Williams. 1983. Effect of turbidity on the temporal and spatial utilization of the inner Bay of Fundy by American shad (*Alosa sapidissima*) (Pisces: Clupeidae) and its relationship to local fisheries. Canadian Journal of Fisheries and Aquatic Sciences 40(S1):s322–s330.

Dauvin, J.-C., and J. J. Dodson. 1990. Relationship between feeding incidence and vertical and longitudinal distribution of rainbow smelt larvae (*Osmerus mordax*) in a turbid well-mixed estuary. Marine Ecology Progress Series 60:1–12.

Deevey, G. G. 1960. The zooplankton of the surface waters of the Delaware Bay region. Bulletin of the Bingham Oceanographic Collection Yale University 17:5–53.

De Robertis, A., J. S. Jaffe, and M. D. Ohman. 2000. Size-dependent visual predation risk and the timing of vertical migration in zooplankton. Limnology and Oceanography 45:1838–1844.

De Robertis, A., C. H. Ryer, A. Veloza, and R. D. Brodeur. 2003. Differential effects of turbidity on prey consumption of piscivorous and planktivorous fish. Canadian Journal of Fisheries and Aquatic Sciences 60:1517–1526.

Dodson, J. J., J.-C. Dauvin, R. G. Ingram, and B. d'Anglejan. 1989. Abundance of larval rainbow smelt (*Osmerus mordax*) in relation to the maximum turbidity zone and associate macroplanktonic fauna of the middle St. Lawrence Estuary. Estuaries 12:68–81.

Embich, T. R. 1973. Ecology of the sand shrimp, *Crangon septemspinosa* Say, 1818, in the Penobscot River estuary, Maine. Master's thesis. University of Maine, Orono.

Faulk, C. K., and G. J. Holt. 2005. Advances in rearing cobia *Rachycentron canadum* larvae in recirculating aquaculture systems: live prey enrichment and greenwater culture. Aquaculture 249:231–243.

Fiksen, Ø., D. L. Aksnes, M. H. Flyum, and J. Giske. 2002. The influence of turbidity on growth and survival of fish larvae: a numerical analysis. Hydrobiologia 484:49–59.

Fish, C. J. 1925. Seasonal distribution of the plankton of the Woods Hole region. Bulletin of the Bureau of Fisheries 41:91–179.

Forward, Jr., R. B. 1987. Larval release rhythms of decapod crustaceans: an overview. Bulletin of Marine Science 41:165–176.

Fosså, J. H. 1986. Aquarium observations on vertical zonation and bottom relationships of some deep-living hyperbenthic mysids (Crustacea:Mysidacea). Ophelia 25:107–117.

Friedland, K. D., G. C. Garman, A. J. Bejda, A. L. Studholme, and B. Olla. 1988. Interannual variation in diet and condition in juvenile bluefish during estuarine residency. Transactions of the American Fisheries Society 117:474–479.

Fulton, R. S., III. 1982. Predatory feeding of two marine mysids. Marine Biology 72:183–191.

Gilmurray, M. C., and G. R. Daborn. 1981. Feeding relations of the Atlantic silverside *Menidia menidia* in the Minas basin, Bay of Fundy. Maine Ecology Progress Series 6:231–235.

Goddard, S. M., and R. B. Forward. 1991. The role of the underwater polarized light pattern, in the sun compass navigation of the grass shrimp, *Paleomonetes vulgaris*. Journal of Comparative Physiology A 169:479–491.

Gordon, D. C., P. C. Cranford, and C. Desplanque. 1985. Observations on the ecological importance of salt marshes in the Cumberland basin, a macrotidal estuary in the Bay of Fundy. Estuarine Coastal and Shelf Science 20:205–227.

Grabe, S. A. 1996. Composition and seasonality of nocturnal peracarid zooplankton from coastal New Hampshire (USA) waters, 1978–1980. Journal of Plankton Research 18:881–894.

Grabe, S. A. 2003. Seasonal periodicity of decapod larvae and population dynamics of selected taxa in New Hampshire (USA) coastal waters. Journal of Plankton Research 25:417–428.

Gran, H. H., and T. Braarud. 1935. A quantitative study of the phytoplankton in the Bay of Fundy and the Gulf of Maine (including observations on hydrography, chemistry and turbidity). Journal of the Biological Board of Canada 1:279–467.

Grimm, V., and S. F. Railsback. 2005. Individual-based modeling and ecology. Princeton University Press, Princeton, New Jersey

Grimm, V., E. Revilla, U. Berger, F. Jeltsch, W. M. Mooj, S. F. Railsback, H.-H. Thulke, J. Weiner, T. Wiegand, and D. L. DeAngelis. 2005. Pattern-oriented modeling of agent-based complex systems: lessons from ecology. Science 310:987–991.

Grippo, M., J. W. Fleeger, R. Condrey, and K. R. Carman. 2009. High benthic microalgal biomass found on Ship Shoal, north-central Gulf of Mexico. Bulletin of Marine Science 84:237–256.

Haefner, P. A. 1969. Temperature and salinity tolerance of the sand shrimp, *Crangon septemspinosa* Say. Physiological Zoology 42:388–397.

Haefner, P. A. 1972. The biology of the sand shrimp, *Crangon semptemspinosa*, at Lamoine, Maine. Journal of the Elisa Mitchell Science Society 88:36–42.

Haefner, P. A. 1976. Seasonal distribution and abundance of sand shrimp *Crangon septemspinosa* in the York River-Chesapeake Bay estuary. Chesapeake Science 17:131–134.

Haynes, E. B., and R. L. Wigley. 1969. Biology of the northern shrimp, *Pandalus borealis*, in the Gulf of Maine. Transactions of the American Fisheries Society 98:60–76.

Herman, S. S. 1963. Vertical migration of the opossum shrimp, *Neomysis americana* Smith. Limnology and Oceanography 8:228–238.

Hitchcock, G. L., and T. J. Smayda. 1977. The importance of light in the initiation of the 1972–1973 winter-spring diatom bloom in Narragansett Bay. Limnology and Oceanography 22:126–131.

Hoffman, J. C., K. E. Limburg, D. A. Bronk, and J. E. Olney. 2008. Overwintering habits of migratory juvenile shad in Chesapeake Bay. Environmental Biology of Fishes 81:329–345.

Hopkins, T. L. 1965. Mysid shrimp abundance in surface waters of Indian River Inlet, Delaware. Chesapeake Science 6:86–91.

Horppila, J., A. Liljendahl-Nurminen, and T. Malinen. 2004. Effects of clay turbidity and light on the predator–prey interaction between smelts and chaoborids. Canadian Journal of Fisheries and Aquatic Sciences 61:1862–1870.

Horsted, S. A., and E. Smidt. 1956. The deep-sea prawn (*Pandalus borealis* Kr.) in Greenland waters. Meddelelser fra Danmarks Fiskeri-og Havundersogelser 1:1–118.

Hulburt, E. M. 1957. The distribution of Neomysis americana in the estuary of the Delaware River. Limnology and Oceanography 2:1–11.

Jahnke, R. A., J. R. Nelson, M. E. Richards, C. Y. Robertson, A. M. F. Rao, and D. B. Jahnke. 2008. Benthic primary productivity on the Georgia midcontinental shelf: benthic flux measurements and high-resolution, continuous in situ PAR records. Journal of Geophysical Research 113:C08022. DOI:10.1029/2008JC004745.

Johnson, C. L., and J. A. Hare. 2012. Zooplankton monitoring in the Gulf of Maine: past, present, and future. Pages 205–218 *in* R. L. Stephenson, J. H. Annala, J. A. Runge, and M. Hall-Arber, editors. Advancing an ecosystem approach in the Gulf of Maine. American Fisheries Society, Symposium 79, Bethesda, Maryland.

Johnson, W. S., and D. M. Allen. 2005. Zooplankton of the Atlantic and Gulf coasts: a guide to their identification and ecology. Johns Hopkins University Press, Baltimore, Maryland.

Jumars, P. A. 2007. Habitat coupling by mid-latitude, subtidal, marine mysids: import-subsidized omnivores. Oceanography and Marine Biology an Annual Review 45:89–138.

Koeller, P., C. Fuentes-Yaco, T. Platt, S. Sathyendranath, A. Richards, P. Ouellet, D. Orr, U. Skúladóttir, K. Wieland, L. Savard, and M. Aschan. 2009. Basin-scale coherence in phenology of shrimps and phytoplankton in the North Atlantic Ocean. Science 324:791–793.

Kühl, M., and B. B. Jørgensen. 1994. The light field of microbenthic communities: radiance distribution and microscale optics of sandy coastal sediments. Limnology and Oceanography 39:1368–1398.

Land, M. F. 1980. Eye movements and the mechanism of vertical steering in euphausiid Crustacea. Journal of Comparative Physiology 137:255–265.

Lankford, T. E., and T. E. Targett. 1997. Selective predation by juvenile weakfish: post-consumptive constraints on energy maximization and growth. Ecology 78:1049–1061.

Link, J. S., and L. P. Garrison. 2002. Trophic ecology

of Atlantic cod *Gadus morhua* on the northeast U.S. Continental Shelf. Marine Ecology Progress Series 227:109–123.

Locke, A., and S. Corey. 1988. Taxonomic composition and distribution of Euphausiacea and Decapoda (Crustacea) in the neuston of the Bay of Fundy, Canada. Journal of Plankton Research 10:195–198.

Locke, A., G. J. Klassen, R. Bernier, and V. Joseph. 2005. Life history of the sand shrimp, *Crangon septemspinosa* Say, in a southern Gulf of St Lawrence estuary. Journal of Shellfish Research 25:603–613.

Macia, A., K. G. S. Abrantes, and J. Paula. 2003. Thorn fish *Terapon jarbua* (Forskål) predation on juvenile white shrimp *Penaeus indicus* H. Milne Edwards and brown shrimp *Metapenaeus monoceros* (Fabricius): the effect of turbidity, prey density, substrate type and pneumatophore density. Journal of Experimental Marine Biology and Ecology 291:29–56.

Macquart-Moulin, C., and R. Maycas, E. 1995. Inshore and offshore diel migrations in European benthopelagic mysids, genera *Gastrosaccus, Anchialina* and *Haplostylus* (Crustacea, Mysidacea). Journal of Plankton Research 17:531–555.

Maurer, D., and R. L. Wigley. 1982. Distribution and ecology of mysids in Cape Cod Bay, Massachusetts, USA. Biological Bulletin (Woods Hole) 163:405–530.

McGillicuddy, D. J., Jr., D. R. Lynch, A. M. Moore, W. C. Gentleman, C. S. Davis, and C. J. Meise. 2002. An adjoint data assimilation approach to diagnosis of physical and biological controls on *Pseudocalanus* spp. in the Gulf of Maine-Georges Bank region. Fisheries Oceanography 7:205–218.

Meager, J. J., and R. S. Batty. 2007. Effects of turbidity on the spontaneous and prey-searching activity of juvenile Atlantic cod (*Gadus morhua*). Philosophical Transactions of the Royal Society (London) B 362:2123–2130.

Meager, J. J., T. Solbakken, A. C. Utne-Palm, and T. Oen. 2005. Effects of turbidity on the reactive distance, search time, and foraging success of juvenile Atlantic cod (*Gadus morhua*). Canadian Journal of Fisheries and Aquatic Sciences 62:1978–1984.

Modlin, R. F. 1980. The life cycle and recruitment of the sand shrimp, *Crangon septemspinosa*, in the Mystic River estuary, Connecticut. Estuaries 3:1–10.

Morgan, M. D., and A. M. Beeton. 1978. Life history and abundance of *Mysis relicta* in Lake Michigan. Journal of the Fisheries Research Board of Canada 35:1165–1170.

Morgan, S. G. 1990. Impact of planktivorous fishes on dispersal, hatching and morphology of estuarine crab larvae. Ecology 71:1639–1652.

Neil, D. M. 1975. The control of eyestalk movements in the mysid shrimp *Praunus flexuosus*. Journal of Experimental Biology 62:487–504.

Parsons, D. G. 2005. Predators of northern shrimp, *Pandalus borealis* (Pandalidae), throughout the North Atlantic. Marine Biology Research 1:48–58.

Pettigrew, N. A., J. H. Churchill, C. D. Janzen, L. J. Mangum, R. P. Signell, A. C. Thomas, D. W. Townsend, J. P. Wallingaa, and H. Xue. 2005. The kinematic and hydrographic structure of the Gulf of Maine Coastal Current. Deep-Sea Research Part II 52:2369–2391.

Pettigrew, N. R., D. W. Townsend, H. Xue, J. P. Wallinga, and P. Brickley. 1998. Observations of the eastern Maine Coastal Current and its offshore extensions in 1994. Journal of Geophysical Research 103:30,623–30:639.

Pezzack, D. S., and S. Corey. 1979. The life history and distribution of *Neomysis americana* (Smith) (Crustacea, Mysidacea) in Passamaquoddy Bay. Canadian Journal of Zoology 57:785–793.

Pihl, L., S. P. Baden, R. J. Diaz, and L. C. Shaffner. 1992. Hypoxia-induced structural changes in the diet of bottom-feeding fish and Crustacea. Marine Biology 112:349–361.

Price, Jr., K. S. 1962. Biology of the sand shrimp, *Crangon septemspinosa*, in the shore zone of the Delaware Bay region. Chesapeake Science 3:244–255.

Richards, S. W., and G. A. Riley. 1967. The benthic epifauna of Long Island Sound. Bulletin of the Bingham Oceanographic Collection Yale University 19:89–135.

Riley, G. A. 1942. The relationship of vertical turbulence and spring diatom flowerings. Journal of Marine Research 5:67–87.

Riley, G. A., H. Stommel, and D. F Bumpus. 1949. Quantitative ecology of the plankton of the western North Atlantic. Bulletin of the Bingham Oceanographic Collection Yale University 12:1–169.

Roman, C. T., N. Jaworski, F. T. Short, S. Findlay, and R. S. Warren. 2000. Estuaries of the northeastern United States: habitat and land use signatures. Estuaries 23:743–764.

Ruiz, G. M., A. H. Hines, and M. H. Posey. 1993. Shallow water as a refuge habitat for fish and crustaceans in non-vegetated estuaries: an example from Chesapeake Bay. Marine Ecology Progress Series 99:1–16.

Sandifer, P. A. 1973. Distribution and abundance of decapod crustacean larvae in the York River estuary and adjacent lower Chesapeake Bay, Virginia, 1968–1969. Chesapeake Science 14:235–257.

Sandifer, P. A. 1975. The role of pelagic larvae in recruitment to populations of adult decapod crustaceans in the York River estuary and adjacent lower Chesapeake Bay, Virginia. Estuarine and Coastal Marine Science 3:269–279.

Sato, M., and P. A. Jumars. 2008. Periods and phases of emergence rhythms dependent on season and depth. Limnology and Oceanography 53:1665–1677.

Schechner, Y. Y., and N. Karpel. 2005. Recovery of underwater visibility and structure by polarization analysis. IEEE (Institute of Electrical and Electronics Engineers) Journal of Oceanic Engineering 30:570–587.

Schiariti, A., A. D. Berasategui, D. A. Giberto, R. A. Guerrero, E. M. Acha, and H. W. Mianzan. 2006. Living in the front: *Neomysis americana* (Mysidacea) in the Rio de la Plata estuary, Argentina-Uruguay. Marine Biology 149:483–489.

Schneider, D. C., and B. A. Harrington. 1981. Timing of shorebird migration in relation to prey depletion. The Auk 98:801–811.

Schöne, H. 1954. Statozystenfunktion und statische Lageorientierung bei dekapoden Krebsen. [Statocyst function and gravitational orientation by decapod crabs]. Zeitschrift für vergleichende Physiologie 36:241–260.

Schöne, H. 1957. Control of course with statocysts. (Measurements on Crustacea). Zeitschrift für vergleichende Physiologie 39:235–240.

Schram, F. R. 1986. Crustacea. Oxford University Press, Oxford, UK.

Shumway, S. E., H. C. Perkins, D. F. Schick, and A. P. Stickney. 1985. Synopsis of biological data on the pink shrimp, *Pandalus borealis* Krøyer, 1838. NOAA Technical Report NMFS 30 (cross listed as FAO Fisheries Synopsis 144).

Simard, Y., and L. Savard. 1990. Variability, spatial patterns and scales of similarity in size-frequency distributions of the northern shrimp (*Pandalus borealis*) and its migrations in the Gulf of St. Lawrence. Canadian Journal of Fisheries and Aquatic Sciences 47:794–804.

Smith, S. I. 1879. The stalk-eyed crustaceans of the Atlantic coast of North America north of Cape Cod. Transactions of the Connecticut Academy of Arts and Sciences 5:27–138.

Spinrad, R. W. 1986. An optical study of the water masses of the Gulf of Maine. Journal of Geophysical Research 91:1007–1018.

Sverdrup, H. U. 1953. On conditions for the vernal blooming of phytoplankton. Journal du Conseil International pour l'Exploration de la Mer 18:287–295.

Taylor, D. L. 2005. Predation on post-settlement winter flounder *Pseudopleuronectes americanus* by sand shrimp *Crangon septemspinosa* in NW Atlantic estuaries. Marine Ecology Progress Series 289:245–262.

Theroux, R. B., and R. L. Wigley. 1998. Quantitative composition and distribution of the macroinvertebrate fauna of the continental shelf ecosystems of the northeastern United States. NOAA Technical Report NMFS 140:1–240.

Thiel, M. 1997. Extended parental care in estuarine amphipods. Doctoral dissertation. University of Maine, Orono.

Thomas, A. C., D. W. Townsend, and R. Weatherbee. 2003. Satellite-measured phytoplankton variability in the Gulf of Maine. Continental Shelf Research 23:971–989.

Thorrold, S. R., C. M. Jones, S. E. Campana, J. W. McLaren, and J. W. H. Lam. 1998. Trace element signatures in otoliths record natal river of juvenile American shad (*Alosa sapidissima*). Limnology and Oceanography 43:1826–1835.

Townsend, D. W. 1984. Comparison of inshore zooplankton and ichthyoplankton populations in the Gulf of Maine Marine Ecology Progress Series 15:79–90.

Townsend, D. W., L. M. Mayer, Q. Dortch, and R. W. Spinrad. 1992. Vertical structure and biological activity in the bottom nepheloid layer of the Gulf of Maine. Continental Shelf Research 12:367–387.

Townsend, D. W., and R. W. Spinrad. 1986. Early spring phytoplankton blooms in the Gulf of Maine. Continental Shelf Research 6:515–529.

Townsend, D. W., A. C. Thomas, L. M. Mayer, M. A. Thomas, and J. A. Quinlan. 2006. Chapter 5. Oceanography of the northwest Atlantic continental shelf. Pages 119–168 *in* A. R. Robinson and K. H. Brink, editors. The global coastal ocean: interdisciplinary regional studies and syntheses. The sea, volume 14A. Harvard University Press, Cambridge, Massachusetts.

Trites, M., I. Kaczmarska, J.M. Ehrman, P.W. Hicklin, and J. Ollerhead. 2005. Diatoms from two macro-tidal mudflats in Chignecto Bay, Upper Bay of Fundy, New Brunswick, Canada. Hydrobiologia 544:299–319.

Twining, B. S., J. J. Gilbert, and N. S. Fisher. 2000. Evidence of homing behavior in the coral reef mysid *Mysidium gracile*. Limnology and Oceanography 45:1845–1849.

Utne-Palm, A. C., A. G. V. Salvanes, B. Currie, S. Kaartvedt, G. E. Nilsson, V. A. Braithwaite, J. A. W. Stecyk, M. Hundt, M. van der Bank, B. Flynn, G. K. Sandvik, T. A. Klevjer, A. K. Sweetman, V. Brüchert, K. Pittman, K. R. Peard, I. G. Lunde, R. A. U. Strandabø, and M. J. Gibbons. 2010. Trophic structure and community stability in an overfished ecosystem. Science 329:333–336.

Viscido, S. V., D. E. Stearns, and K. W. Able. 1997. Seasonal and spatial patterns of an epibenthic decapod crustacean assemblage in north-west Atlantic continental shelf waters. Estuarine Coastal and Shelf Science 45:377–392.

Walter, J. F., III, and J. E. Olney. 2003. Feeding behavior of American shad during spawning migration in the York River, Virginia. Pages 201–209 *in* K. E. Limburg and J. R. Waldman, editors. Biodiversity, status, and conservation of the world's shads. American Fisheries Society, Symposium 35, Bethesda, Maryland.

Warren, P. J., and R. W. Sheldon. 1968. Association between *Pandalus borealis* and fine-grained sediment off Northumberland. Nature (London) 217:579–580.

Waterman, T. H. 2005. Reviving a neglected celestial underwater polarization compass for aquatic animals. Biological Reviews 81:1–5.

Weinberg, R. 1981. On the food and feeding habits of *Pandalus borealis* Krøyer 1838. Archiv für Fischereiwissenschaft 31:123–137.

Wenzhöfer, F., and R. N. Glud. 2004. Small-scale spatial and temporal variability in coastal benthic O_2 dynamics: effects of fauna activity. Limnology and Oceanography 49:1471–1481.

Whiteley Jr., G. C. 1948. The distribution of larger planktonic Crustacea on Georges Bank. Ecological Monographs 18:233–264.

Wigley, R. L., and B. R. Burns. 1971. Distribution and biology of mysids (Crustacea, Mysidacea) from the Atlantic coast of the United States in the NMFS Woods Hole collection. Fishery Bulletin (Washington, DC) 69:717–746.

Wigley, R. I., and R. B. Theroux. 1981. Macrobenthic invertebrate fauna from the Middle Atlantic Bight region: faunal composition and quantitative distribution. U.S. Department of the Interior, U.S. Geological Survey, Professional Paper 529 N, Washington, D.C.

Wilcox, J. R., and H. P. Jeffries. 1974. Feeding habits of the sand shrimp *Crangon septemspinosa*. Biological Bulletin (Woods Hole) 146:424–434.

Williams, A. B. 1972. A ten-year study of meroplankton in North Carolina estuaries: Mysid shrimps. Chesapeake Science 13:254–262.

Williams, A.B. 1984. Shrimps, lobsters and crabs of the Atlantic coast of the eastern United States, Maine to Florida. Smithsonian Institution Press, Washington, D.C.

Winkler, G., J. J. Dodson, N. Bertrand, D. Thivierge, and W. F. Vincent. 2003. Trophic coupling across the St. Lawrence River estuarine transition zone. Marine Ecology Progress Series 251:59–73.

Winkler, G., and W. Greve. 2004. Trophodynamics of two interacting species of estuarine mysids, *Praunus flexuosus* and *Neomysis integer*, and their predation on the calanoid copepod *Eurytemora affinis*. Journal of Experimental Marine Biology and Ecology 308:127–146.

Worm, B., and R. A. Myers. 2003. Meta-analysis of cod-shrimp interactions reveals top-down control in oceanic food webs. Ecology 84:162–173.

Zagursky, G., and R. J. Feller. 1985. Macrophyte detritus in the winter diet of the estuarine mysid *Neomysis americana*. Estuaries 8:355–362.

American Fisheries Society Symposium 79:345–359, 2012

Results of a Collaborative Project to Observe Coastal Zooplankton and Ichthyoplankton Abundance and Diversity in the Western Gulf of Maine: 2003–2008

Jeffrey A. Runge*
School of Marine Sciences, University of Maine and Gulf of Maine Research Institute 350 Commercial Street, Portland, Maine 04101, USA

Rebecca J. Jones
School of Marine Sciences, University of Maine and Gulf of Maine Research Institute 350 Commercial Street, Portland, Maine 04101, USA

Abstract.—In a collaborative project with a number of New England commercial fishermen, zooplankton was sampled two to three times a month between 2003 and 2005 at the GoMOOS (Gulf of Maine Ocean Observing System) Buoy "B" and between 2003 and 2008 at a station on Jeffreys Ledge in the western Gulf of Maine. Additionally, during 2007 and 2008 zooplankton and ichthyoplankton were sampled semimonthly at stations located in Massachusetts Bay and Ipswich Bay, New Hampshire. The authors report here on seasonal and interannual patterns in biomass, diversity, and abundance in the zooplankton at the Jeffreys Ledge station and in the ichthyoplankton at the Massachusetts and Ipswich Bay stations. Notable is the dominance of *Calanus finmarchicus* on Jeffreys Ledge and the dramatic decline in summer abundance of this species between 2003 and 2005, perhaps related to a shift to lower salinity water during this same period. Interannual differences in timing of peak abundance, and in species dominance of ichthyoplankton, were observed between 2007 and 2008. While these time series provide information and insight about change in the coastal planktonic communities in the western Gulf of Maine, currently there are no observing programs that sample coastal communities at frequency sufficient to show seasonal and interannual change in this region.

Introduction

A number of observing time series of pelagic ecosystem variables have been collected in the Gulf of Maine, some of which are still ongoing (Runge et al. 2010; Johnson et al. 2011; Johnson and Hare 2012, this volume). While these programs address various components of the Gulf of Maine ecosystem, very few have observed seasonal and interannual change in zooplankton abundance and diversity in the coastal zone (defined here as coastal waters inshore of the 100-m isobath). Sherman (1965, 1966, 1968, 1970) studied the seasonal variation of zooplankton in coastal waters of the Gulf of Maine between 1963 and 1969. The National Marine Fisheries Service (NMFS) collects and analyzes zooplankton on the EcoMon surveys (formerly MARMAP: Sherman et al. 2002) program, which includes four to six surveys a year since 1977, presently at about 30 stations sited randomly across the Gulf of Maine (Kane 2007; Johnson and Hare 2012). NMFS also supports a Continuous Plankton Recorder (CPR) transect that measures relative abundance of zooplankton at the surface in the deep, central Gulf of Maine at monthly intervals (e.g., Pershing et al. 2005). Seasonal to annual ichthyoplankton surveys have been conducted in the inshore coastal Gulf of Maine (Chenoweth 1973), as well as several Gulf

* Corresponding author: jeffrey.runge@maine.edu

of Maine estuaries and bays (Townsend 1984; Lazzari 2001), and at more offshore stations by the MARMAP and EcoMon surveys (e.g., Sherman et al. 1984; Pennington 1988).

From the perspective of an ecosystem approach to management, the relevance of observing zooplankton and ichthyoplankton in the coastal zone is associated with the feeding and spawning activity of many commercially harvested fish and invertebrate species, as well as the foraging activities of large whales. For example, Jeffreys Ledge and adjacent waters off the coast of New Hampshire are primary spawning and feeding areas for Atlantic herring *Clupea harengus* (Stevenson and Scott 2005). Ipswich Bay, west of Jeffreys Ledge, is a primary spawning area for Atlantic cod, and recruitment success of the Gulf of Maine cod populations depends in part on availability of zooplankton prey as larvae drift to nearshore juvenile nursery areas in Ipswich and Massachusetts bays (Runge et al. 2010). Northern shrimp *Pandalus borealis* spawn in the coastal zone along the Maine coast in winter, and larval stages feed on zooplankton, as well as well as phytoplankton, during the period of the spring bloom (Gordon 2008). The prominent anadromous fishes (alewife *Alosa pseudoharengus*, blueback herring *A. aestivalis*, and American shad *A. sapidissima*) return to the coastal ocean typically in June–July (Saunders et al. 2006), with depleted energy reserves postspawn; feeding on zooplankton in nearshore and coastal waters at this stage is essential to reconditioning and may be a factor in determining capacity to repeat spawning in the future. There is also a need for data on the abundance and life cycle of marine planktonic copepods, notably *Calanus finmarchicus*, in nearshore waters in order to understand the potential risk of entanglement of North Atlantic right whales *Eubalaena glacialis* that search for high concentrations of these copepods in coastal Gulf of Maine waters (Singer and Ludwig 2005).

Summarized here are results of a cooperative program with the fishing industry to observe zooplankton and ichthyoplankton abundance and diversity at fixed stations in the coastal zone of the western Gulf of Maine. The underlying conceptual approach is that sampling at fixed stations at intervals less than the normal generation time (i.e., sampling frequency >1 per month) of zooplankton provides indicators of seasonal and interannual change in fundamental ecosystem properties related to zooplankton abundance, diversity, and productivity. The fishing industry provided vessels and participated in collection of samples at weekly to semimonthly intervals. The sampling involved relatively inexpensive techniques and equipment that can be readily learned and routinely deployed, taking advantage of fishermen's skills and experience. The sampling frequency and method of the fixed station component of the Atlantic Zonal Monitoring Program (AZMP) conducted by Fisheries and Oceans Canada in coastal Canadian Atlantic waters (Therriault et al. 1998) were followed, allowing the possibility for comparison of change in zooplankton biodiversity across latitudes in the northwestern Atlantic coastal ocean. The primary objectives here were to (1) demonstrate the feasibility of a cooperative partnership to conduct plankton observing at reasonable cost, and (2) provide a baseline of observations of seasonal and interannual change of zooplankton and ichthyoplankton abundance and biodiversity in the western Gulf of Maine coastal waters. The fixed stations are presumed to be representative of large areas of the coastal, advectively coupled western Gulf of Maine.

Methods

Sampling was conducted at two stations in Massachusetts Bay and four stations off the coast of New Hampshire and southern Maine (Figure 1). The Massachusetts Bay stations were located just south of Gloucester, adjacent to GoMOOS (Gulf of Maine Ocean Observing System, www.gomoos.org) Buoy "A" (GA: 65 m depth), and at the northern tip of Stellwagen Bank (GSB: 82 m depth). Three New Hampshire stations were located along a transect, starting with a nearshore station south of the Isles of Shoals (P1: 70 m depth), a station in Scantum basin (P2: 125 m depth) and a station located on New Scantum on Jeffreys Ledge (S: 50 m depth). The fourth station was located adjacent to GoMOOS Buoy "B," six nautical miles east of York, Maine.

The nominal sampling interval was three times monthly, with some variation caused by weather, between 2003 and 2005 at Stations B and S, and semimonthly between 2007 and 2008

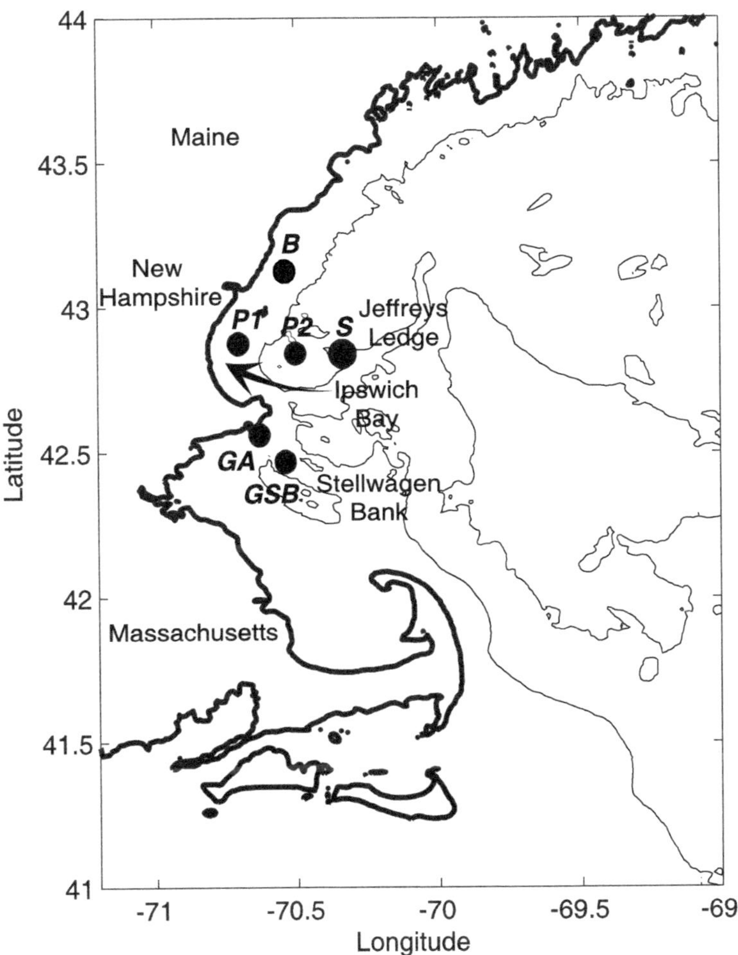

FIGURE 1. Locations of stations sampled during the cooperative partnership with the local fishing industry. 100-m isobath shown. Station S at the base of Jeffreys Ledge was sampled between 2003 and 2008 (except for 2006, when no sampling was conducted). Station B was active between 2003 and 2005. The other stations in Ipswich Bay (P1, P2) and in Massachusetts Bay (GA and GS) were sampled between 2007 and 2008.

at Stations S, P1, P2, GA, and GSB. No sampling was conducted in 2006 due to a gap in funding. The samples were collected aboard commercial fishing vessels equipped with a winch and gantry capable of towing bongo and ring zooplankton nets. Fishermen involved in this collaborative program were trained in the deployment of the equipment and handling of the samples, after which they carried out the sampling routine on their own. Regular communication and site visits were conducted to ensure that equipment was functioning properly and that protocols were being followed.

The sampling protocols relied on robust or low technology equipment that could be operated under variable sea conditions without need for specialized technical training. To standardize comparison among years and across regions, the same protocols used in the Canadian AZMP (Mitchell et al. 2002) were adopted here.

Zooplankton samples were collected at all stations with a 0.75-m-diameter, 200-μm mesh ring net in the mouth of which a General Oceanics flowmeter was suspended. The ring net was deployed in duplicate to within 5 m off the bottom and then towed vertically to the surface at a rate of 40 m/min. All samples were preserved in a 4% seawater-buffered formaldehyde solution. In the laboratory, all zooplankton samples were split in half using a Folsom plankton splitter. Half of the sample was archived for identification and enumeration of zooplankton species, and the other half was used to estimate dry weight biomass.

To estimate zooplankton diversity and abundance, the archived split was diluted and a 5- to 10-mL subsample taken to obtain a target number of 200 organisms. All copepods were identified to species and staged; all other plankton were identified to family or to species when possible. To minimize variability, an additional subsample of 75–200 individuals was analyzed to enumerate stages of *Calanus finmarchicus*, the typically dominant copepod. The counts were normalized to number/m^3 or number/m^2, taking into account the subsample dilution factor and split factors, and volume sampled by the net from the flowmeter reading.

The remaining half of the sample was analyzed for dry weight biomass. The sample was removed from preservative by sieving through a nitex mesh screen with mesh size less than or equal to the mesh size of the collection net. Samples were rinsed with 100 mL of freshwater to remove salts, and the screen and sieved sample were placed in a clean, preweighed Petri dish. Mesh filters and dishes were preweighed for subtraction from sample material weight after drying. The samples were dried in a Precision Econotherm laboratory oven at 65°C for 48 h. Dried samples were then weighed to the nearest 0.001 g on a PG403-S Delta Range Mettler Toledo precision scale. The dry biomass was calculated, taking into account the split factor and volume filtered by the plankton net.

Ichthyoplankton samples were collected at stations P1, GA, and GSB with a standard 0.65-m-diameter MARMAP bongo net (Sea-Gear Inc.) fitted with 500-μm mesh nitex plankton nets on both sides. Volume of water filtered was estimated by suspending a General Oceanics flowmeter in the mouth of each net. Bongo tows were standardized by limiting the tow duration to a total of 10 min, with the upcast and downcast divided evenly. Nominal maximum tow depth was within 5 m off the bottom. The collection protocol specified that tow speed not exceed 2 knots and that the angle of the tow cable be maintained at a 45-degree angle. Ichthyoplankton were sorted from the preserved bongo samples using a light table and transparent tray. All larvae and eggs were preserved in smaller vials for taxonomic identification and length measurements (not reported here), after which they were archived.

A SeaBird 19Plus conductivity-temperature-depth meter (CTD) with a WetLabs fluorometer was deployed at every station. Each port (Gloucester and Portsmouth) had its own CTD for convenience of sampling. Data were downloaded in Gloucester by the Massachusetts Fishermen's Cooperative and emailed to the Gulf of Maine Research Institute (GMRI) for analysis. Data from the Portsmouth CTD was downloaded by R. Jones and analyzed at GMRI. All CTD data were processed with the SeaBird Data Processing software and then binned to 1-m increments.

Between 2003 and 2005, Niskin water bottle samples were collected at six depths (0, 10, 20, 30, 40, and 50 m) at Stations B and S for measurement of chlorophyll-a concentrations. At sea, as quickly as possible after collection, duplicate 100-mL subsamples from each depth were filtered onto a Gelman GF/F glass fiber filter. The filters were folded into aluminum foil, frozen in liquid nitrogen, and stored at −80°C in the laboratory. At the time of analysis, each filter was placed in 10 mL of 90% acetone in a centrifuge, vortexed for several seconds, then cold extracted in the dark for 24 h. Extracted pigments were measured with a Turner Designs AU-10 fluorometer, and chlorophyll-a concentration was calculated following Welschmeyer (1994) initially (through mid-2004), then following Strickland and Parsons (1972) after acidification with 10% HCl, in order to conform with University of New Hampshire Coastal Observing Center protocols. Depth-integrated concentrations (mg Chlorophyll *a* per m^2) were calculated as the sum to 50 m of the values of the mean Chlorophyll-a concentration within each 10-m depth interval multiplied by 10 m.

Results

Highlighted here are results from the Jeffreys Ledge Station S, for which there is the longest time series. Data from all stations is served on the BCO-DMO (Biological and Chemical Oceanography Data Management Office) database located at the Woods Hole Oceanographic Institution (www.bco-dmo.org). A full data report is in preparation (R. Jones and J. Runge, University of Maine, unpublished manuscript), pending completion of all analyses.

The seasonal temperature minimum on Jeffreys Ledge occurred in March, when the water column was well mixed and temperatures ranged from 3.0°C to 4.5°C (Figure 2). Mean water temperature increased nearly linearly thereafter, reaching a maximum of 10–13°C in August–September. Maximum annual temperature attained 18–20°C in the surface 0–10 m and 7–9°C at 40 m. The warmest water at depth was recorded in 2004 and 2008, contributing to the highest late summer mean temperatures shown in Figure 2.

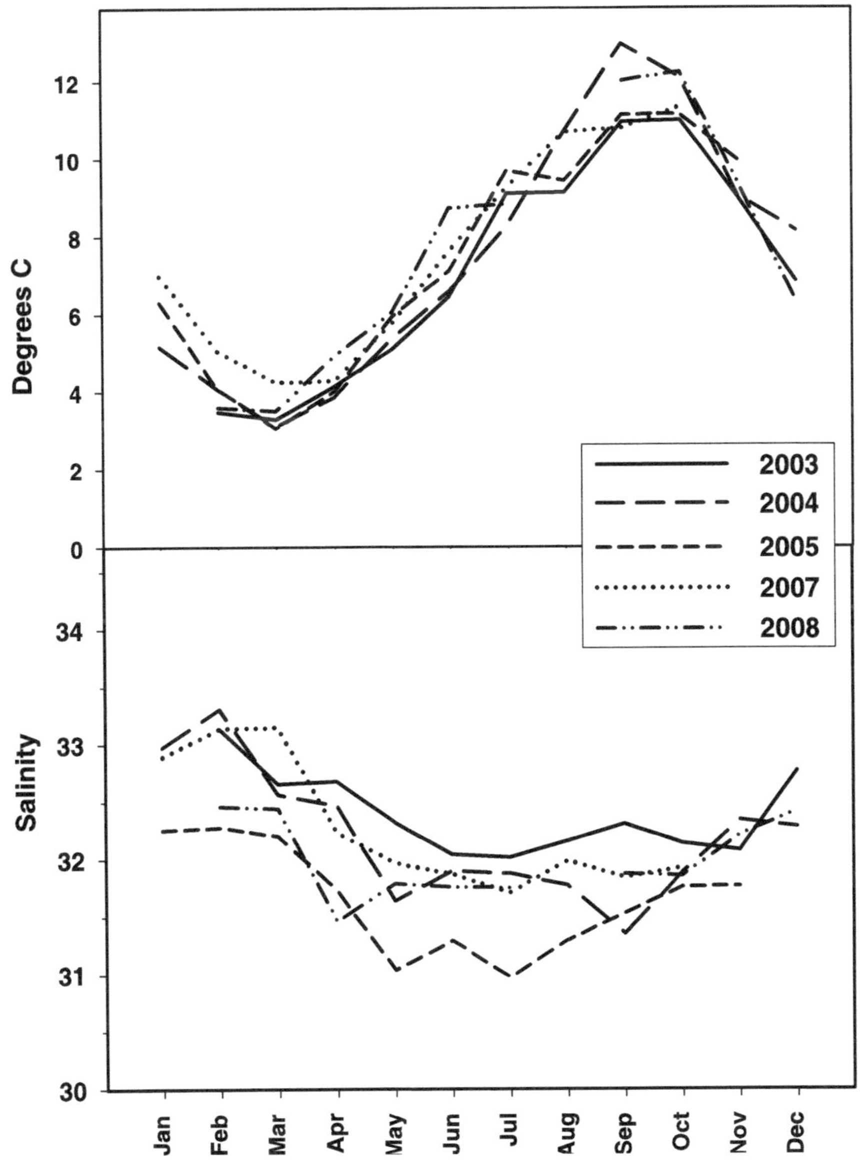

FIGURE 2. Monthly water column average (1–40 m at 1-m intervals) at Station S between 2003 and 2008 (no data in 2006). Upper panel: temperature (°C). Lower panel: salinity (PSU [practical salinity units])

There was considerable interannual variation in salinity on Jeffreys Ledge between 2003 and 2008 (Figures 2B and 3). Winter salinities greater than 33 were recorded in 2003, 2004, and 2007. Starting in summer 2004, and throughout 2005, salinities were markedly lower throughout the water column, by 0.5–1.0 units, than in 2003. These results are consistent with observations of lower surface salinities in 2005 in the western Gulf of Maine, measured as part of GNATS (the Gulf of Maine North Atlantic Time Series) along a transect between Portland, Maine and Yarmouth, Nova Scotia (Balch et al. 2008). Salinities were higher in 2007–2008 but still fresher than 2003.

The timing and magnitude of the seasonal cycle in phytoplankton biomass at Station S, as estimated by chlorophyll-a standing stocks, is shown in Figure 4. In 2003, the spring phytoplankton bloom peaked at the end of April. The period during which integrated concentrations were greater than 90 mg/m², the level above which production of the planktonic copepod *Calanus finmarchicus* is estimated to be unlimited by food (Runge et al. 2006), was approximately 50 d. In 2004, the spring bloom peaked near mid-April, and nonfood-limiting conditions were also approximately 50 d. In 2005, several peaks of high chlorophyll of lower magnitude were observed between mid-April and mid-June (keeping in mind the average interval between sampling was about 10 d). The total duration of unlimited food conditions for copepod production, as defined above, was about 40 d.

The zooplankton dry weight biomass at GoMOOS Buoy B and Jeffreys Ledge Station S showed pronounced seasonal and interannual variation (Figure 5). In 2003–2004, highest biomass levels ranged between 3.5 and 13.0 mg/m² during the summer months, falling off in early fall and winter to less than 3 mg/m². At both stations in 2005, the biomass peak in summer was smaller and of shorter duration than in 2003–2004. At Station S, the period of very low zooplankton biomass began in mid-August 2004. During 2005, summer biomass levels attained just 5 mg/m², and only for a short period in late July to early August. Summer zooplankton biomass at the Jeffreys Ledge station increased in 2007–2008 but only to intermediate levels compared to 2003.

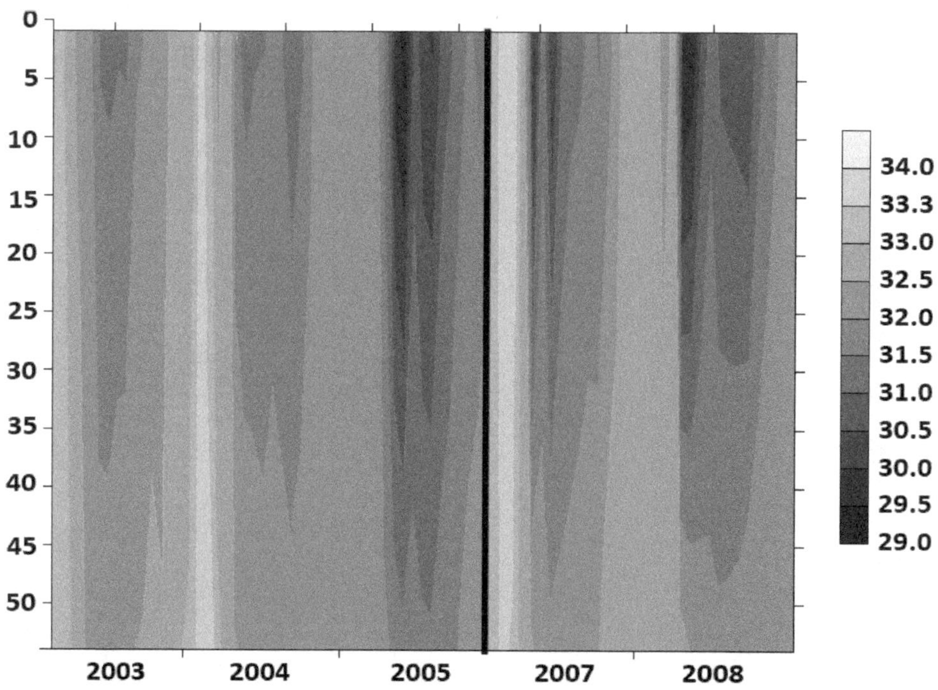

FIGURE 3. Salinity at Station S contoured over depth and time for the period 2003–2008; no data in 2006.

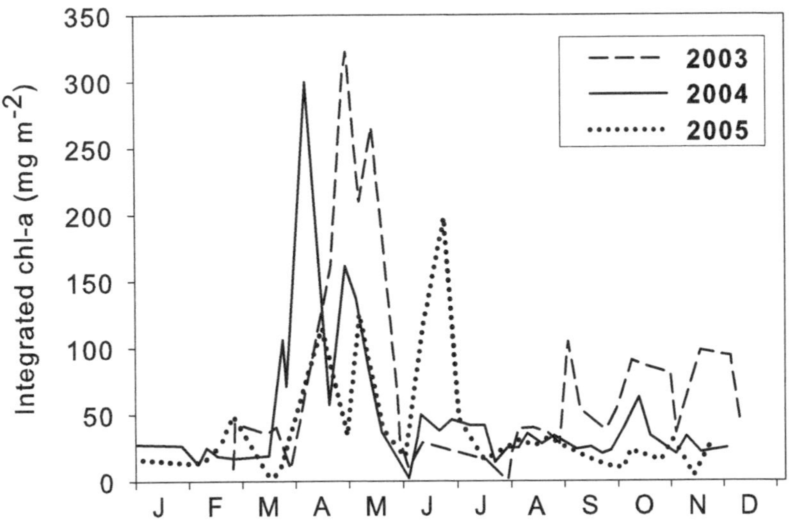

FIGURE 4. Seasonal variation in depth integrated chlorophyll *a* (0–50 m) at Station S.

The zooplankton at Station S were dominated by a relatively small number of species of planktonic copepods (Figure 6). *Calanus finmarchicus* was predominant in spring and early summer, when it constituted up to 80% of the total number of zooplankton collected with the 200-µm mesh net. Small cyclopoid copepods in the genus *Oithona* (primarily *O. similis*) were prominent on Jeffreys Ledge in all seasons and throughout the time series. *Oithona* spp. usually constituted

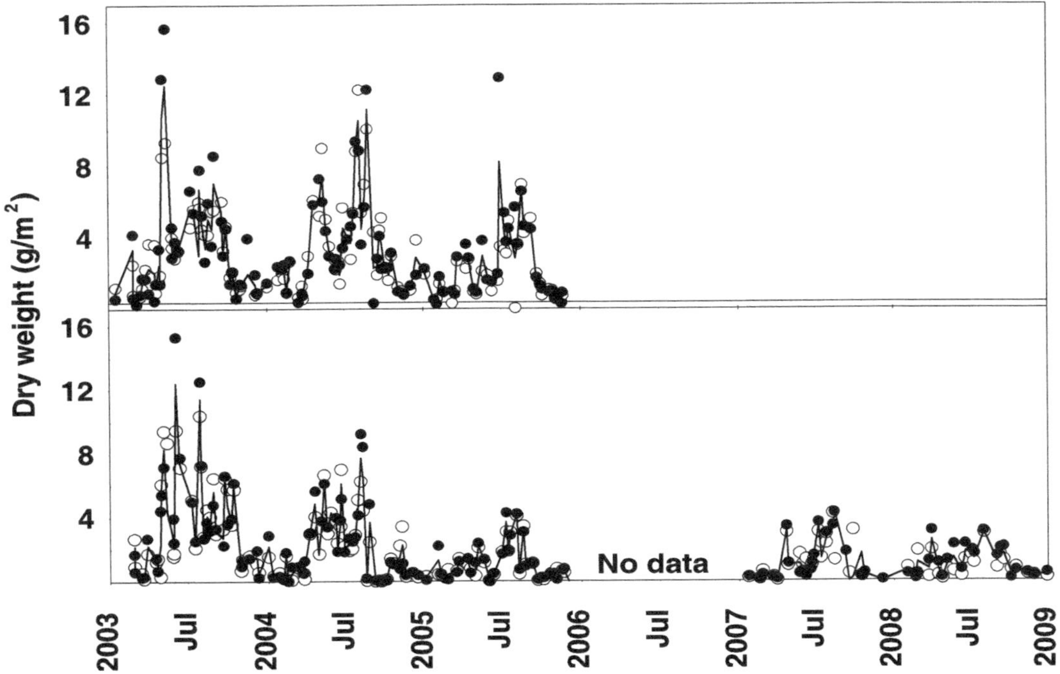

FIGURE 5. Vertically integrated dry weight biomass (mg/m²) of zooplankton collected with a 200-µm mesh ring net. Top panel: Station B (2003–2005). Bottom panel: Station S (2003–2008).

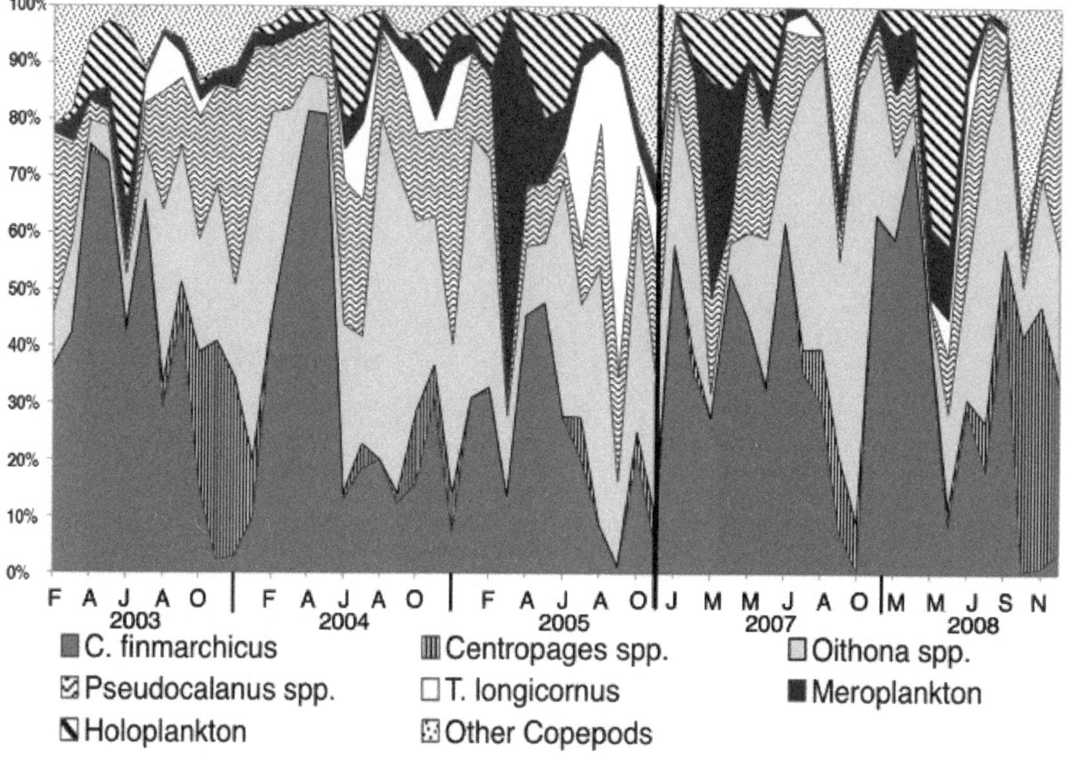

FIGURE 6. Species composition (% by number) of zooplankton collected with a 200-μm mesh ring net on Jeffreys Ledge at Station S between 2003 and 2008 (no data in 2006). The most dominant taxa are the planktonic copepods, *Calanus finmarchicus*, *Oithona* spp., *Pseudocalanus* spp., Centropages (>90% of which is *C. typicus*), and *Temora longicornis*.

20–80% (in late 2007) of the total catch, which does not include the early life stages, too small to be efficiently captured by the mesh size of the net. In fall and winter, the presence of *C. finmarchicus* diminished, succeeded in prominence by smaller copepod species such as *Pseudocalanus* spp. and *Centropages typicus*, as well as *Oithona*. Notably, *C. typicus* was relatively rare in fall of 2004 and 2005, replaced in 2005 by *Temora longicornis*. Meroplankton (bivalve, gastropod, and polychaete larvae) were relatively abundant in late summer 2004 and in early spring of 2005 and 2007 (primarily barnacle larvae). Dominant holoplankton were larvaceans (*Oikopleura* sp.) and euphausids in winter–spring, euphausids in summer 2003, and cladoceran species (*Podon* and *Evadne* spp.) in summers 2004–2008.

The interannual variation in abundance of *C. finmarchicus* reflects the decline in total zooplankton biomass between 2003 and 2005 (Figure 7). *Calanus* abundance was highest in spring and summer of 2003 and lowest in 2005. The highest late-summer *Calanus* abundances, when the population was predominantly in the lipid-rich stage CV, occurred in 2003. In that year, the late-summer *Calanus* CV concentration was estimated to be 1–10 *Calanus* per liter, depending upon whether they were uniformly distributed throughout the water column or assumed to be in a layer 10 m thick. These concentrations are sufficient to attract foraging North Atlantic right whales (Mayo and Marx 1989), which feed primarily on *Calanus*, and presumably the higher concentrations would also be attractive to Atlantic herring and other planktivores as well. Notable is the seasonal decrease in magnitude of *Calanus* abundance on Jeffreys Ledge in early fall in 2003 and in late summer between 2004 and 2008. The abundance of *C. finmarchicus* on Jeffreys Ledge in late summer (August–September)

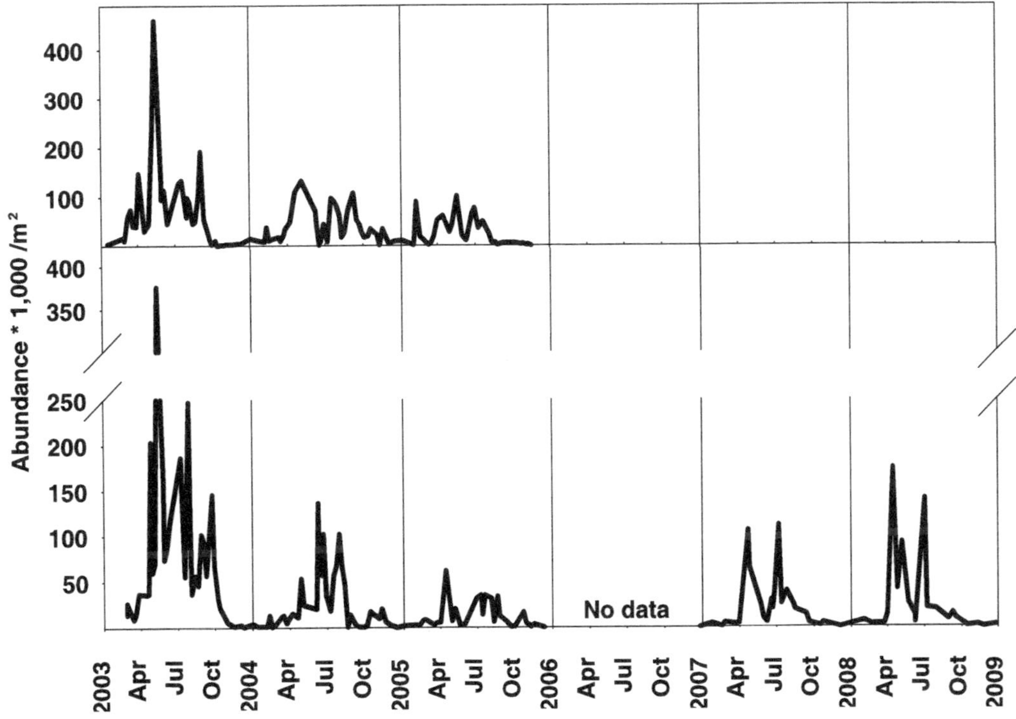

FIGURE 7. Interannual variation in vertically integrated total abundance (nauplius and copepodid stages) of the planktonic copepod *Calanus finmarchicus* at Station B (2003–2005) and Station S (2003–2008; no data in 2006).

in 2005 was lower than 2003 levels by a factor of 5–10.

Analysis of the catch from the bongo net tows revealed the seasonal and interannual variability in abundance and diversity of the larval stages of the western Gulf of Maine fish community (Figures 8 and 9). A fall–winter (October–February) and summer (May–August) peak in total ichthyoplankton abundance are evident in Massachusetts Bay. Only a smaller summer (July–August) peak is apparent in Ipswich Bay; however, Station P1 was not well sampled in fall 2007. Atlantic herring *Clupea harengus*, pollock *Pollachius virens*, and sand lance *Ammodytes* spp. were most abundant in winter months. Cunner *Tautogolabrus adspersus*, silver hake *Merluccius bilinearis*, red hake *Urophycis chuss*, American plaice *Hippoglossoides platessoides*, daubed shanny *Leptoclinus* (formerly *Lumpenus*) *maculatus*, and Acadian redfish (*Sebastes fasciatus*) were most abundant in summer. While 33 species were found over the 2-year period, these nine species constituted 77% of the total number captured. Notable also in the record is the presence of Atlantic cod *Gadus morhua* and Atlantic menhaden *Brevoortia tyrannus* larvae. Atlantic cod larvae were observed in concentrations up to 17 per 100 m^3, primarily at stations GA or GSB, in December–February and in mid-April to July. Higher concentrations of gadoid eggs were also observed during the same time periods but could not be positively identified as Atlantic cod. Atlantic menhaden larvae were observed only in 2008, mainly at Stations GA or GSB during July–August in concentrations of up to 10 per 100 m^3. Mean concentrations of these two species are lower, hence their absence in the legend for Figure 9.

Discussion

This project has demonstrated a successful partnership between academic research and local fishermen for a contribution to monitoring of the coastal pelagic ecosystem in the western Gulf of

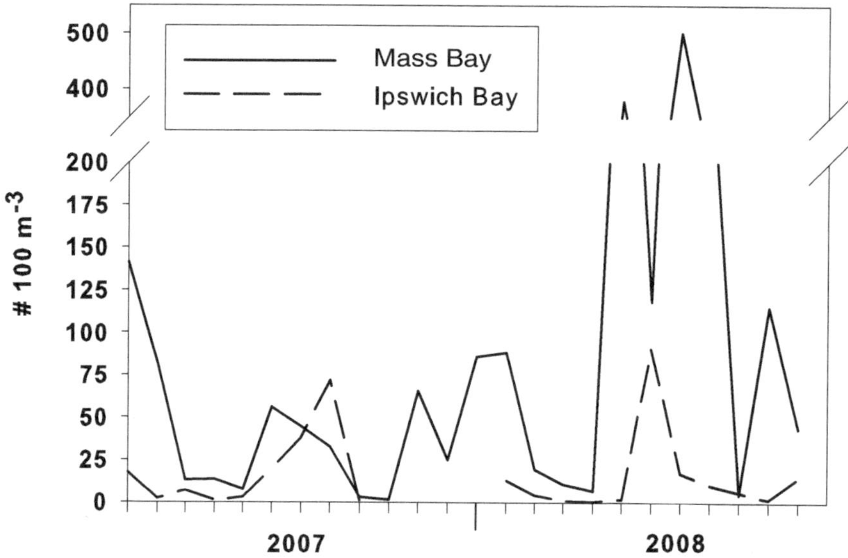

FIGURE 8. Seasonal and interannual variation in total abundance of fish larvae (fish eggs excluded) captured in Ipswich Bay (Station P1) and Massachusetts Bay (Stations GA and GSB) from semimonthly bongo samples.

Maine. The sampling frequency (two to three station visits per month throughout the year) has not been achieved by other Gulf of Maine zooplankton sampling programs. The results provide new information and insights on seasonal cycles of diversity and abundance of the coastal zooplankton and ichthyoplankton community in this area. The analysis here focuses on possible causes and implications of zooplankton variability at Station S on Jeffreys Ledge and an assessment of the value of the ichthyoplankton results collected in 2007–2008.

Jeffreys Ledge is a narrow and relatively shallow (50 m) glacial deposit that juts out alongside the deep (300 m) central Gulf of Maine water of Wilkinson basin to the east. It is separated from the nearshore shelf by the relatively deep (120 m) waters of Bigelow Bight and Scantum basin (Figure 1). Jeffreys Ledge lies in the path of the western Maine Coastal Current (WMCC), part of the Gulf of Maine Coastal Current, which flows south–southeast off the New Hampshire and Massachusetts coasts. The strength of the WMCC is determined, in part, by the extent to which it is supplied by the eastern Maine Coastal Current (EMCC) (Churchill et al. 2005; Pettigrew et al. 2005; Manning et al. 2009). The EMCC is driven by inflow of Scotian Shelf water around the south coast of Nova Scotia, as well as freshwater river runoff (Balch et al. 2008). The circulation and geophysical attributes of Jeffreys Ledge undoubtedly play a role in determining its zooplankton diversity and abundance patterns.

The zooplankton diversity on Station S reflects nearshore coastal and deep Gulf of Maine influence over the 5 years of sampling. Comparison with other studies of Gulf of Maine zooplankton composition provides insight into the sources of the zooplankton community at station S, despite the differences in sampling gear and mesh sizes among studies. In 2003, the four seasonally dominant copepod taxa, *Calanus finmarchicus*, *Oithona* spp., *Centropages typicus*, and *Pseudocalanus* spp., are also prominent in the zooplankton composition in offshore waters and banks of the Gulf of Maine (Sherman et al. 1983; Davis 1987; Jossi and Kane 2000; Durbin et al. 2003; Pershing et al. 2005; Durbin and Casas 2006). Starting in mid-2004 and continuing through 2005, coupled with the decline in *Calanus finmarchicus* abundance (Figure 7), there was an increased presence of the neritic copepod *Temora longicornis* in spring and summer at Station S (Figure 6). In a study in 2002 (using a 150-μm mesh net), *T. longicornis* dominated the

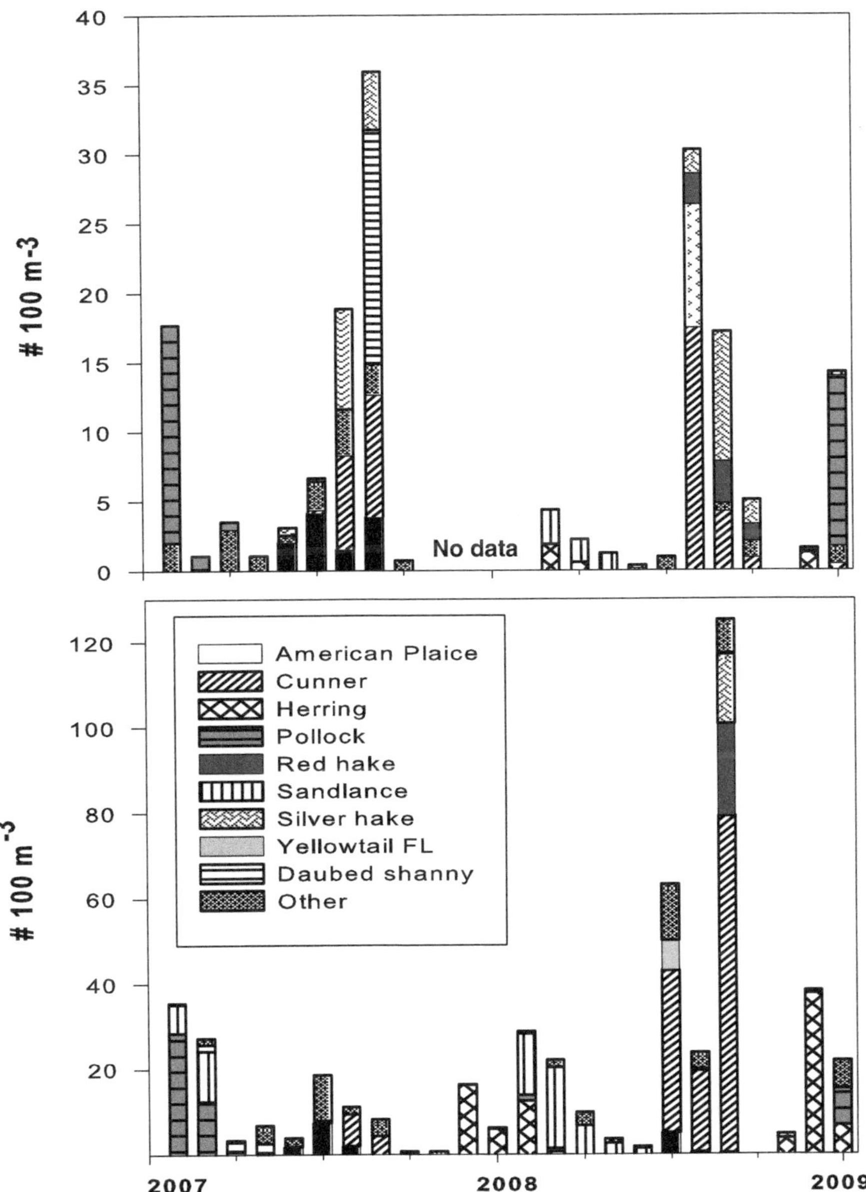

FIGURE 9. The 2-year (2007–2008) seasonal and interannual variation in diversity of fish larvae captured in Ipswich Bay (Station P1: top panel) and Massachusetts Bay (bottom panel; mean of Stations GA and GSB). A total of 33 species were observed ($N = 5427$).

adult zooplankton biomass in summer at nearshore stations to the west of Jeffreys Ledge off the mouth of Great Bay, New Hampshire, suggesting a nearshore coastal influence at Station S (Manning and Bucklin 2005). An increase in relative abundance of meroplankton and cladocera in 2005 is also consistent with greater nearshore influence. A moderate return of *C. finmarchicus* in spring and early summer 2007 and 2008 suggests a more oceanic influence on Jeffreys Ledge, although meroplankton were also prominent in spring of 2007.

The seasonal decline of *C. finmarchicus* in late fall and winter at Station S is likely related to the shallowness of the station, which prevents the overwintering stage CV of this species to undertake its normal ontogenetic migration to deeper in the water column. Studies of *C. finmarchicus* vertical distribution in fall and winter in the Gulf of St. Lawrence and Gulf of Maine (Plourde et al. 2001; Zakardjian et al. 2003; P. Wiebe, Woods Hole Oceanographic Institution, unpublished data) show that *Calanus* overwinters at mid-depth or deeper in the coastal ocean, well below the depth of Jeffreys Ledge, and observations of *Calanus* abundance in fall and winter 2004–2008, in the adjacent Wilkinson basin, indicate that *Calanus* was abundant there during the Station S time series (Maps et al. 2012). It is hypothesized that late stage CVs residing on Jeffreys Ledge in fall were either eaten or swept off, and *Calanus* was not replenished on the ledge until late winter, when females were again in the surface layer.

Three alternate hypotheses to explain the interannual decline in *C. finmarchicus* at Station S are (1) an increase in summer abundance of planktivorous predators (including Atlantic herring and Atlantic mackerel *Scomber scombrus*) affecting late stage mortality rates on Jeffreys Ledge; (2) a change in timing and or quality of the spring phytoplankton bloom affecting egg production and early life stage mortality of *Calanus* in spring and early summer; or (3) a change in physical forcing associated with the appearance of fresher water on Jeffreys Ledge, affecting advection of *Calanus* onto, or out of, the coastal shelf. At present, there are no data to assess the relative role of interannual variation in predation mortality on *C. finmarchicus* abundance on Jeffreys Ledge. Consistent with hypothesis (2), the magnitude and duration of the spring phytoplankton bloom was lower in 2005 (Figure 4), possibly leading to lower egg production and higher daily mortality rate of early life stages, through either increased cannibalism, predation from other suspension-feeding copepods and zooplankton, or starvation in the first feeding naupliar stages. Under this hypothesis, the reduced spring and summer recruitment would carry over to much lower stage CIV/CV abundance in late summer and fall.

Consistent with hypothesis (3), analysis of current data from a mooring in the Northeast Channel (Smith et al. 2012, this volume) indicates of a shift in the flow of water in and out of the Gulf of Maine sometime between 2000, when the previous mooring was removed, and 2004, when a new mooring was installed. The shift, especially noticeable in 2004–2005, suggests greater inflow of Scotian Shelf water into the Gulf of Maine. In addition, and perhaps related to this shift in current flow in the Northeast Channel, precipitation in southern New Hampshire and Maine was unusually high in fall 2004 and spring 2005. The wettest year in the 104-year U.S. Geological Survey precipitation record for the Penobscot River occurred in 2005, and 2006 was the eighth wettest year on record (Balch et al. 2008). Balch et al. (2008) concluded that freshwater supplied by Maine rivers caused lower surface salinities in 2005 and 2006 in the western Maine Coastal Current. The appearance of lower salinity water on Jeffreys Ledge (Figures 2B and 3) may therefore indicate a greater influx of Scotian Shelf water into the WMCC and/or a local source derived from the increased freshwater river discharge.

The fresher water evident on Jeffreys Ledge would represent either transport of fewer Calanus in the WMCC, if a dominant Scotian Shelf influence, or a consequence of dilution of coastal waters with local river runoff. Sherman (1968) observed a qualitative correlation between salinity and *C. finmarchicus* abundance in Gulf of Maine coastal waters, which he attributed to increased advection of the GMCC, transporting *Calanus* off the coastal shelf. Increased freshwater runoff may also inhibit advection in winter–spring of female *Calanus* onto the inner coastal shelf from the deep Gulf of Maine basins. Kane and Prezioso (2008) argued that increased inflow of Scotian Shelf water during the 1990s brought with it high abundances of *T. longicornis*, resulting in observations of much higher *Temora* abundance during this period. They also observed substantially higher abundance of *T. longicornis* in the Gulf of Maine CPR record in 2005, relative to 2003 and 2004. Alternatively, increases in relative meroplankton abundance, as well as *T. longicornis* on Jeffreys Ledge in 2005, implicate expansion of the nearshore environment that may be related to increased river discharge along the Maine and NH coasts. The magnitude and duration of the spring bloom were also lower in 2005 relative to 2003 (Figure 3; see also Balch et al. 2008), which may have inhibited reproduction by the *Calanus* females that did make their way onto Jeffreys Ledge, thereby adding to the observations of low total abundance.

Whatever the cause of the decline in *C. finmarchicus* abundance, planktivorous predators on Jeffreys Ledge would likely have been affected. Atlantic herring and North Atlantic right whales that feed in summer and fall on Jeffreys Ledge, likely in response to high concentrations of lipid-rich, late-stage *Calanus*, may have either suffered reduced energy intake after mid-2004 or altered their distribution and dispersal patterns in search of higher prey levels. Feeding conditions for Atlantic cod larvae spawned in spring (e.g., J. Runge et al. 2010) may also have been affected. These implications suggest that physical forcing due to climate change will have significant influences on distribution and productivity of marine resources in the western Gulf of Maine. It is anticipated that new, coupled *Calanus*-circulation modeling (e.g., Pershing et al. 2009; J. Runge et al. 2010) will provide insight into the underlying controls on population dynamics and distribution of *Calanus* in the coastal Gulf of Maine. This information could be used to compare with distribution and migrations of feeding aggregations of North Atlantic right whales and Atlantic herring in this region.

The time series of ichthyoplankton diversity and abundance shows a community composition in the coastal western Gulf of Maine dominated by cunner, pollock, Atlantic herring, sand lance, red hake, and silver hake. It resembles, but is nevertheless different from, the offshore Gulf of Maine MARMAP stations, where sand lance, Atlantic herring, Atlantic mackerel, cunner, Acadian redfish, and silver hake constituted 76% of the ichthyoplankton community in combined spring and fall trawl surveys (Pennington 1988). The western Gulf of Maine community is very different from the diversity of ichthyoplankton in mid-coast Gulf of Maine estuaries, dominated by rock gunnel *Pholis gunnellus* and sculpin larvae (e.g., Townsend 1984). One value of a long-term, fully annual coastal time series would be the capacity to identify shifts in ichthyoplankton diversity related to climate forcing.

The results of this study indicate that acquisition of fixed-station time series data on coastal zooplankton and ichthyoplankton diversity and abundance, through cooperative partnerships, would provide information support for understanding climate impacts on the Gulf of Maine ecosystem within the framework of an ecosystem approach to management. The present observing system for the northeast region (NERACOOS) does not presently observe change in coastal Gulf of Maine zooplankton and ichthyoplankton communities. The potential benefit to fisheries and coastal management is informed decision making, based on understanding of changes to biodiversity and abundance of key trophic species, determining distribution of, and recruitment into, fish stocks in the coastal western Gulf of Maine.

Acknowledgments

We gratefully acknowledge Erich Anderson, Peter Kendall, George Littlefield, Craig Maverikis, Dennis Robillard, and Alan Vangile, commercial fishermen operating out of the Portsmouth or Yankee Fisherman's Coops, New Hampshire, and B.G. Brown, Peter Marshall, and especially Daniel Murphy, commercial fisherman operating out of Gloucester and members of the Massachusetts Fishermen's Partnership. We thank Olivia Rugo-Free and Angela Sanfilipino of the Massachusetts Fishermen's Partnership for their administrative support. This research was supported by awards from the Northeast Consortium, the University of New Hampshire Coastal Observing Center, and the National Science Foundation through award numbers OCE-0733910, OCE-0733907, and OCE 0815336.

References

Balch, W. M., D. T. Drapeau, B. C. Bowler, E. S. Booth, L. A. Windecker, and A. Ashe. 2008. Space-time variability of carbon standing stocks and fixation rates in the Gulf of Maine along the GNATS transect between Portland, ME, USA and Yarmouth, Nova Scotia, Canada. Journal of Plankton Research 30:119–139.

Chenoweth, S. B. 1973. Fish larvae of the estuaries and coast of central Maine. Fishery Bulletin 71:105–113.

Churchill, J., J. Runge, and C. Chen. 2011. Processes controlling retention of *spring-spawned Atlantic cod (Gadus morhua)* in the western Gulf of Maine and their relationship to an index of recruitment success. Fisheries Oceanography 20:32–46.

Churchill, J. H., N. R. Pettigrew, and R. P. Signell.

2005. Structure and variability of the western Maine Coastal Current. Deep-Sea Research Part II 52:2392–2410.

Davis, C. S. 1987. Zooplankton life cycles. Pages 256–267 *in* R. H. Backus, editor. Georges Bank. MIT Press, Cambridge, Massachusetts.

Durbin, E., and M. Casas. 2006. Abundance and spatial distribution of copepods on Georges Bank during the winter/spring period. Deep-Sea Research Part II 52:2537–2569.

Durbin, E. G., R. G. Campbell, M. C. Casas, M. D. Ohman, B. Niehoff, J. Runge, and M. Wagner. 2003. Interannual variation in phytoplankton blooms and zooplankton productivity and abundance in the Gulf of Maine during winter. Marine Ecology Progress Series 254:81–100.

Gordon, E. 2008. Distribution and feeding behavior of early life stages of the northern shrimp, *Pandalus borealis*, in relation to the spring phytoplankton bloom in the western Gulf of Maine. Master's thesis. University of New Hampshire, Durham.

Johnson, C. L., and J. A. Hare. 2012. Zooplankton monitoring in the Gulf of Maine: past, present, and future. Pages 205–218 *in* R. L. Stephenson, J. H. Annala, J. A. Runge, and M. Hall-Arber, editors. Advancing an ecosystem approach in the Gulf of Maine. American Fisheries Society, Symposium 79, Bethesda, Maryland.

Johnson, C., J. Runge, A. Bucklin, K. A. Curtis, E. Durbin, J. A. Hare, L. S. Incze, J. Link,

G. Melvin, T. O'Brien, and L. Van Guelpen. 2011. Biodiversity and ecosystem function in the Gulf of Maine: pattern and role of zooplankton and pelagic nekton. PLoS (Public Library of Science) One [online serial] 6:e16491. DOI: 10.1371/journal.pone.0016491.

Jossi, J. W., and J. Kane. 2000. An atlas of seasonal mean abundances of the common zooplankton of the United States northeast continental shelf ecosystem. Bulletin of the Sea Fisheries Institute 3:1–21.

Kane, J. 2007. Zooplankton abundance trends on Georges Bank, 1977–2004. ICES Journal of Marine Science 64:909–919.

Kane, J., and J. Prezioso. 2008. Distribution and multi-annual abundance trends of the copepod *Temora longicornis* in the US northeast shelf ecosystem. Journal of Plankton Research 30:619–632.

Lazzari, M. A. 2001. Dynamics of larval fish abundance in Penobscot Bay, Maine. Fishery Bulletin 99:81–93.

Manning, C. A., and A. Bucklin. 2005. Multivariate analysis of the copepod community of nearshore water in the western Gulf of Maine. Marine Ecology Progress Series 292:233–249.

Manning, J. P., D. J. McGillicuddy, N. R. Pettigrew, J. H. Churchill, and L. S. Incze. 2009. Drifter observations of the Gulf of Maine Coastal Current. Continental Shelf Research 29:835–845.

Maps, F., J. Runge, A. Leising, A. Pershing, N. Record, S. Plourde, and J. Pierson. 2012. Modeling the timing and duration of dormancy in populations of *Calanus finmarchicus* on the northwest Atlantic shelf. Journal of Plankton Research 34:36–54.

Mayo, C. A., and M. K. Marx. 1989. Surface foraging behaviour of the North Atlantic right whale, *Eubalaena glacialis*, and associated zooplankton characteristics. Canadian Journal of Zoology 68:2214–2220.

Mitchell, M. R., G. Harrison, K. Pauley, A. Gagné, G. Maillet, and P. Strain. 2002. Atlantic zonal monitoring program sampling protocol. Canadian Technical Report of Hydrography and Ocean Sciences 223.

Pennington, M. 1988. Absolute abundance estimates using trawl and ichthyoplankton survey data. Pages 112–119 *in* W. G. Smith, editor. An analysis and evaluation of ichthyoplankton survey data from Northeast Continental Shelf Ecosystem. National Oceanic and Atmospheric Administration, National Marine Fisheries Service, NOAA Technical Memorandum NMFS-F/NEC-57, Silver Spring, Maryland.

Pershing, A. J., C. H. Greene, J. W. Jossi, L. O'Brien, J. K. T. Brodziak, and B. A. Bailey. 2005. Interdecadal variability in the Gulf of Maine zooplankton community with potential impacts on fish recruitment. ICES Journal of Marine Science 62:1511.

Pershing, A. J., N. R. Record, B. C. Monger, D. E. Pendleton, and L. A. Woodard. 2009. Model-based estimates of *Calanus finmarchicus* abundance in the Gulf of Maine. Marine Ecology Progress Series 378:227–243.

Pettigrew, N. R., J. H. Churchill, C. D. Janzen, L. J. Mangum, R. P. Signell, A. C. Thomas, D. W. Townsend, J. P. Wallinga, and H. Xue. 2005. The kinematics and hydrographic structure of the Gulf of Maine Coastal Current. Deep-Sea Research Part II 52:2369–2391.

Plourde, S., P. Joly, J. A. Runge , B. Zakardjian, and J. Dodson. 2001. Life cycle of *Calanus finmarchi-*

cus in the lower St. Lawrence estuary: the imprint of circulation and late timing of the spring phytoplankton bloom. Canadian Journal of Fisheries and Aquatic Sciences 58:647–658.

Runge, J. A., A. Kovach, J. Churchill, L. Kerr, J. R. Morrison, R. Beardsley, D. Berlinsky, C. Chen, S. Cadrin, C. Davis, K. Ford, J. H. Grabowski, W. H. Howell, R. Ji, R. Jones, A. Pershing, N. Record, A. Thomas, G. Sherwood, S. Tallack, and D. Townsend. 2010. Understanding climate impacts on recruitment and spatial dynamics of Atlantic cod in the Gulf of Maine: integration of observations and modeling. Progress in Oceanography 87:251–263.

Runge, J. A., S. Plourde, P. Joly, E. Durbin, and B. Niehoff. 2006. Characteristics of egg production of the planktonic copepod, *Calanus finmarchicus*, on Georges Bank: 1994–1999. Deep-Sea Research Part II 53:3618–2631.

Saunders, R., M. A. Hachey, and C. W. Fay. 2006. Maine's diadromous fish community: past, present, and implications for Atlantic salmon recovery. Fisheries 31:537–547.

Sherman, K. 1965. Seasonal and areal distribution of Gulf of Maine coastal zooplankton, 1963. International Commission for the Northwest Atlantic Fisheries Special Publication 6:611–623.

Sherman, K. 1966. Seasonal and areal distribution of zooplankton in the coastal waters of the Gulf of Maine, 1964. U.S. Fish and Wildlife Service Special Scientific Report Fisheries 530:1–11.

Sherman, K. 1968. Seasonal and areal distribution of zooplankton in the coastal waters of the Gulf of Maine, 1965 and 1966. U.S. Fish and Wildlife Service Special Scientific Report Fisheries 562:1–11.

Sherman, K. 1970. Seasonal and areal distribution of zooplankton in coastal waters of the Gulf of Maine, 1967 and 1968. U.S. Fish and Wildlife Service Special Scientific Report Fisheries 594:1–8.

Sherman, K., J. R. Green, J. R. Goulet, and L. Ejsymont. 1983. Coherence in zooplankton of a large northwest Atlantic ecosystem. Fishery Bulletin 81:855–862.

Sherman, K., J. Kane, S. Murawski, W. Overholtz, and A. Solow. 2002. The U.S. northeast shelf large marine ecosystem: zooplankton trends in fish biomass recovery. Pages 195–215 *in* K. Sherman and H.-R. Skjoldal, editors. Large marine ecosystems of the North Atlantic. Elsevier, Amsterdam.

Sherman, K., W. Smith, W. Morse, M. Berman, J. Green, and L. Ejsymont. 1984. Spawning strategies of fishes in relation to circulation, phytoplankton production, and pulses in zooplankton off the northeast United States. Marine Ecology Progress Series 18:1–19.

Singer, L., and L. Ludwig. 2005. Right whale foraging in nearshore waters of the northern Gulf of Maine. Proceedings of a workshop, April 15, 2005. Gulf of Maine Research Institute, Portland.

Smith, P. C., N. R. Pettigrew, P. Yeats, D. W. Townsend, and G. Han. 2012. Regime shift in the Gulf of Maine. Pages 185–203 *in* R. L. Stephenson, J. H. Annala, J. A. Runge, and M. Hall-Arber, editors. Advancing an ecosystem approach in the Gulf of Maine. American Fisheries Society, Symposium 79, Bethesda, Maryland.

Strickland, J. D. H., and T. R. Parsons. 1972. A practical handbook of seawater analysis. Bulletin of the Fisheries Research Board of Canada 167:201.

Stevenson, D. K., and M. L. Scott. 2005. Atlantic herring, *Clupea harengus*, life history and habitat characteristics. National Oceanic and Atmospheric Administration, National Marine Fisheries Service, NOAA Technical Memorandum NMFS-NE-192, Silver Spring, Maryland.

Therriault, J-C. B. Petrie, P. Pepin, J. Gagnon, D. Gregory, J. Helbig, A. Herman, D. Lefaivre, M. Mitchell, B. Pelchat, J. Runge, and D. Sameoto. 1998. Proposal for a northwest Atlantic zonal monitoring program. Canadian Technical Report of Hydrography and Ocean Sciences 194:1–57.

Townsend, D. 1984. Comparison of inshore zooplankton and ichthyoplankton populations of the Gulf of Maine. Marine Ecology Progress Series 15:79–90.

Welschmeyer, N. 1994. Fluorometric analysis of chlorophyll a in the presence of chlorophyll b and pheopigments. Limnology and Oceanography 39:1985–1992.

Zakardjian, B. A., J. Sheng, J. A. Runge, K. R. Thompson, Y. Gratton, I. A. McLaren, and S. Plourde. 2003. Effects of temperature and circulation on the population dynamics of *Calanus finmarchicus* in the Gulf of St. Lawrence and Scotian Shelf: study with a coupled three-dimensional hydrodynamic, stage-based life-history model. Journal of Geophysical Research 108(C11):8016.

American Fisheries Society Symposium 79:361–373, 2012

Organismal Biology in the Age of Ecosystem-Based Management

Jacob P. Kritzer*
Oceans Program, Environmental Defense Fund
18 Tremont Street, Suite 850, Boston, Massachusetts 02108, USA

Jamie M. Cournane
Ocean Process Analysis Laboratory, University of New Hampshire
8 College Road, Durham, New Hampshire 03824, USA

Abstract.—Ecosystem-based management rests upon a scientific foundation involving greater structural complexity and larger spatial and temporal scales of predictive models. Increases in complexity and scale impose constraints upon the extent to which complexity at the level of an organism (life history, behavior, and physiology) can be modeled. Although the earliest ecosystem models, notably Ecopath and its successors, eliminated much organismal detail and instead adopted a biomass dynamic approach, newer models such as Atlantis and multispecies virtual population analysis retain greater detail at the organismal level. However, effective ecosystem-based management will require appreciation of the limitations of complex models and reliance upon parallel insights on key processes gained through empirical research that are not captured by models. These insights should be used to evaluate differences between model predictions and management outcomes, modify models where possible, and adjust management measures suggested by models otherwise. Important empirical research at the organismal level includes effects of habitat, contaminants, and physical and chemical properties of water on organismal traits. It is then important to understand reciprocal effects of those changes on the ecosystem (e.g., temperature-driven changes in growth that alter trophic interactions). Intrapopulation variability in life history traits, behavior, and physiology can increase resilience to ecosystem change, and those benefits and processes that erode that variability need to be understood. Ultimately, the continued evolution toward ecosystem-based science and management should strive for a balance between reality and tractability in underlying models and incorporate empirical research in addition to model outputs to best shape management measures.

Introduction

Many marine scientists are calling for a shift in fisheries management from a species-by-species approach to an ecosystem-based approach that considers interactions among species, effects of fishing on habitat and of habitat on stock productivity, and effects of other physical and chemical factors such as pollution and climate change (Pikitch et al. 2004; Worm et al. 2009; Thrush and Dayton 2010). This transition within fisheries management is occurring alongside policy directives for a broader change in management of coastal and ocean resources that will integrate multiple, and often competing, uses within a single governance framework or at least within a forum for coordination among governing institutions (Pew Ocean Commission 2003; U.S. Commission on Ocean Policy 2004; Council on Environ-

* Corresponding author: jkritzer@edf.org

mental Quality 2010). However, ecosystem-based fisheries management (EBFM) in most cases will precede more comprehensive ecosystem-based management (EBM) because EBFM can occur largely within existing governance structures but with different goals and expectations and a fundamentally different scientific foundation (Mangel and Levin 2005; Levin et al. 2009).

The science underlying traditional single-species fisheries management has undergone a considerable evolution over the past few decades, and that evolution will continue with the shift toward EBFM. Simpler biomass dynamic models that use gross measures of population size coupled with logistic population growth functions to predict stock dynamics have given way to more complex size-, stage- and age-structured models that incorporate more detailed estimates of somatic growth, mortality, maturity, and other life history traits (per-recruit models, delay difference models, virtual population analysis, and statistical catch-at-age or catch-at-length models; Hilborn and Walters 1992; Quinn and Deriso 1999). Single-species models continue to evolve, particularly by incorporating spatial structure within the larger stock being modeled (Cadrin and Secor 2009). In parallel, an evolution has been underway toward building and refining marine ecosystem models needed to inform EBFM (Arkema and Samhouri 2012, this volume).

Adding spatial complexity to single-species models aims to improve accuracy but at the cost of both greater data requirements and potentially greater uncertainty due to the number of parameters estimated. These include estimates of life history parameters specific to different locations, subpopulations or substocks, and the processes of dispersal and migration that describe connections among those spatial units (see Kritzer and Sale 2006 and chapters therein). Similarly, ecosystem models aim to better capture reality by incorporating interspecific interactions (e.g., predation, competition), habitat effects on stock productivity, and abiotic (physical and/or chemical) factors, also at the cost of greater data needs and possibly greater uncertainty.

The first marine ecosystem model to see widespread use among marine ecologists and fisheries scientists was Ecopath, which is essentially a steady-state multispecies biomass dynamic model. Temporal and spatial heterogeneity were later added to Ecopath in the development of Ecosim and Ecospace, respectively (the evolution of all three models is reviewed by Pauly et al. 2000). All of these models are based upon principles of energy flow and mass balance and include relatively little parameterization of individual species' biology. The structure of these models and their increasingly widespread use since first created in the mid-1980s might have signaled an implicit decision among the scientific community to forego model complexity at the level of an individual organism for the sake of tractability at community and ecosystem scales. However, other ecosystem models have emerged that retain more aspects of organismal biology. To the extent that models with greater organismal complexity are used in management, continued attention to processes at the organismal level will be needed. However, regardless of the type of model chosen for a given purpose, effective EBM will require more than just ecosystem modeling. Empirical understanding of biological and ecological processes not captured in models will be needed to understand the limits of models, refine models accordingly, and adjust model outcomes in setting management measures.

We examine the role of science at the level of an organism within larger fisheries scientific and management frameworks at the scale of ecosystems. We begin by briefly reviewing selected marine ecosystem models that retain detailed organismal biology, even while addressing more complex interactions and larger-scale processes to inform EBM. However, we do not evaluate the relative merits of different models because our goal is to show where and how aspects of organismal biology are being used in EBFM rather than to draw universal conclusions on the utility of particular models. Given that insights attainable from any model are necessarily limited by structural decisions about scale and complexity, we next look beyond ecosystem models at empirical research revealing both important effects of ecosystem change on species' biology (e.g., temperature-driven changes in growth) and reciprocal effects of those changes on the surrounding ecosystem (e.g., trophic shifts due to size-dependent predation patterns). We conclude by reviewing the role of organismal biology in determining resilience and adaptability at the scale of both populations and ecosystems and offering some summary thoughts on the future of organismal biology in an age of ecosystem-based management.

Throughout this review, we draw upon a range of examples from the Gulf of Maine and from other locations focused on species found in the Gulf of Maine, as well as examples focused on foreign species and locales that are particularly illustrative. Also, although we briefly consider applications to lower trophic levels when reviewing models that incorporate organismal biology, our review focuses primarily on fishes and macroinvertebrates, since those species are the targets of fisheries management. However, we recognize that the processes we examine are certainly relevant at planktonic and microbial levels as well and in fact might be stronger at those scales in many cases.

Ecosystem Models Incorporating Organismal Biology

One of the most common stock assessment tools within contemporary fisheries science, at least in data-rich settings, is virtual population analysis (VPA). In the late 1980s, the VPA approach was expanded into multispecies VPA (MSVPA), one of the first ecosystem-based assessment models to retain organismal detail despite increasing structural complexity (Hilborn and Walters 1992). In fact, MSVPA requires even greater complexity at the organismal level since not only are somatic growth and mortality parameters required, but also parameters for age- or size-specific predation rates on particular age- or size-classes of different prey species for each predator–prey combination in the model. Total mortality of prey species then becomes a function of the abundance and population structure of different predators, fishing mortality, and a "background" mortality estimate accounting for species not included in the model. The limited number of species included in a given MSVPA can be a strength if the system is simple in terms of number of interacting species or if the interactions among the modeled species are strong, but the tool might become more limiting if the trophic structure is more complex.

MSVPA has been applied to a range of systems, including the major commercial groundfish and forage fish species of Georges Bank at the southern extent of the Gulf of Maine (Tsou and Collie 2001) and many of these same species in the North Sea (Rice and Gislason 1996). Garrison et al. (2010) used MSVPA to examine trophic interactions of key commercial species in the estuaries of the mid-Atlantic, the location of the primary spawning grounds of striped bass *Morone saxatilis* that migrate seasonally to the Gulf of Maine. MSVPA has also been used to model interactions with seabird and mammal predators, in addition to fish in the North Sea (Furness and Tasker 1997) and the Bering Sea (Livingston and Jurado-Molina 2000), but, to date, has not been applied to species other than fish in the Gulf of Maine.

Focused on the earliest life stage of fishes and macroinvertebrates, as well as a very different set of purely planktonic species and different processes than traditional fisheries models, coupled biological-physical models of plankton dynamics represent another arena of considerable progress toward linking organismal biology with broader ecosystem processes. Practitioners have recognized the seemingly overwhelming, and especially multiscalar, complexity of oceanographic processes that seemed to represent an impediment to understanding important dynamics (Denman and Gargett 1995). However, considerable progress has been made toward understanding and modeling organismal processes such as the behavior and life history of both phytoplankton and zooplankton within the oceanographic system (Hare and Kane 2012, this volume), which has greatly increased explanatory power (Mann and Lazier 2006; e.g., deYoung et al. 2004; Kristiansen et al. 2009; Churchill et al. 2011). For example, Ji et al. (2008) recently examined nitrogen limitation and primary productivity on George's Bank and in the Gulf of Maine by incorporating growth, grazing, respiration, and photosynthesis parameters for zooplankton and phytoplankton.

Organismal biology also plays a significant role in emerging ecosystem models that include many more species and interactions than most MSVPAs. Prominent among these is Atlantis, which incorporates oceanographic processes and plankton dynamics and represents a step toward unifying coupled biological-physical models with models focused on higher trophic levels (Fulton et al. 2004a). Atlantis is a spatially explicit ecosystem model of water quality and food webs, which can incorporate fishing behavior and economics (Fulton et al. 2004a; Smith et al. 2007). Food webs are simulated using growth, mortality, and recruitment parameters of individual species and through the predator–prey relationships between those species. Therefore, like MSVPA, Atlantis incorporates predation parameters but adds feeding efficiency and other metabolic parameters (e.g., oxygen limitation parameters in the bacterial submodel; Fulton et al. 2004a). However, a high level of physiological detail might not be necessary to capture system dynamics well (Fulton et al. 2004b). Applications of Atlantis often combine species into larger functional group units based on similar life history characteristics, which means some species-specific detail can be lost.

Perhaps because it awaits more rigorous testing and has much greater data requirements than models currently in use, or perhaps because the policy frameworks for such comprehensive ecosystem-level decision making are not yet in place, Atlantis has not yet seen widespread application in management. In the Gulf of Maine, Atlantis has been used to explore the likely impacts of major ecosystem changes and management decisions (Overholtz and Link 2009), although Ecopath with Ecosim has also been used for similar exploration in the region (Worm et al. 2009). However, neither has yet been used as the basis for management decisions. Whether those explorations will evolve into application, or whether the evolution toward EBFM will rely on models that are more focused on specific interactions of interest rather than the entirety of the ecosystem (e.g., Overholtz and Link 2007), remains to be seen. In all likelihood, a variety of modeling tools will be used for different applications.

Although the evolution of marine ecosystem models from Ecopath and its variants to tools like MSVPA and Atlantis has meant reintroducing and even expanding the level of detail at the organismal level, interactions between organisms and the larger ecosystem are far more complex than any model to date can (and perhaps should) fully capture. In the following sections, we review empirical research on the interactions between organismal biology and ecosystem processes, independent of whether those interactions are captured in contemporary models. These types of research should be used to gauge the limitations of model outcomes and adjust management decisions accordingly and, where warranted and possible, modify model structure going forward.

Ecosystem Effects on Organismal Biology

There have been numerous studies documenting changes in life history traits of marine fishes along environmental gradients, particularly latitudinal and temperature gradients. The most extensively studied species that perhaps best illustrates these patterns and processes is the Atlantic silverside *Menidia menidia*, which is distributed throughout the Gulf of Maine coast and the eastern seaboard of North America. Conover, Munch, and their collaborators have done extensive field and laboratory investigations of this small forage fish, illustrating a variety of adaptive responses to temperature and seasonality. Conover and Present (1990) first documented the pattern of countergradient variation in silversides, whereby fish from cold high-latitude waters compensate for a shorter growing season by increasing growth rate. Conover (1990) then demonstrated the same pattern in several other temperate northwestern Atlantic species. However, this increased growth comes at a cost, as the required physiological and behavioral investments increase predation risk and therefore result in higher mortality (Munch and Conover 2003; Arnott et al. 2006; Chiba et al. 2007).

The silverside model and similar studies focused on other species such as mummichog *Fundulus heteroclitus* (Schultz et al. 1996) and Atlantic cod *Gadus morhua* (Björnsson et al. 2001) from the Gulf of Maine, and turbot *Scophthalmus maximus* from the northeast Atlantic (Imsland et al. 2000), lend insights into how climate-driven changes in temperature profiles might drive adaptive responses in life history, behavior, and physiology. These types of changes are often stronger at larval and early juvenile life stages, and can drive recruitment patterns (e.g., Atlantic cod in the Gulf of Maine and elsewhere; Drinkwater 2005). Climate-driven effects on recruitment and postrecruitment growth and mortality have important implications for stock productivity and fisheries management strategies, even in a single-species assessment and management context. Populations of silversides and other species that exhibit slower growth and lower mortality in warmer, low-latitude areas might see those traits in higher latitudes in response to a warming trend. If so, those changes would have opposing effects, with lower mortality increasing productivity and slower growth decreasing productivity, all else being equal. However, there is evidence for a genetic basis to these latitudinal adaptations, operating in conjunction with phenotypic responses (Conover 1990; Schultz et al. 1996). Whether, to what extent, and at what pace, adaptations will track ecosystem changes is also unclear, so any opposing effects of slower growth and lower mortality that do eventuate might not be equal.

Of course, temperature is not the only climate-driven ecosystem influence on the organismal traits that drive stock dynamics. Changes in water chemistry due to increased CO_2 concentration have especially important implications for life histories and physiology of organisms that produce calcareous skeletal structures. Among commercially harvested species, bivalve shellfish will likely be affected most significantly by changes in CO_2 concentration (Talmage and Gobler 2010), but a wide range of crustacean, foraminiferan, and other species will also be affected (Fabry et al. 2008). Calcification that is necessary for shell growth is greatly reduced under high CO_2 conditions, which increases mortality and compromises feeding and lipid storage. Importantly, different shellfish species exhibit differing degrees of susceptibility to these effects (Cooley and Doney 2009). Understanding those differences will be critical in predicting future sustainable harvest strategies in the Gulf of Maine, where several shellfish species support commercial fishing and aquaculture industries.

Climate-driven changes are not the only ecosystem-level changes that can affect organismal biology of fishes and invertebrates. High contaminant loads can cause mortality directly by disrupting vital cellular and physiological processes. Such direct mortality is more common in smaller and simpler organisms (i.e., microorganisms and larval stages; Boening 2000). However, dramatic and disastrous mortality events can also be caused by indirect effects of contaminant levels in larger and more complex organisms. One of the best known examples of such an event was the 1999 die-off of American lobster *Homarus americanus* in Long Island Sound, which occurred when stress induced by the combination of high water temperatures and contaminants agitated from bottom sediments by severe weather events made lobsters physiologically incapable of fending off the disease paramoebiasis (reviewed by Pearce and Balcom 2005). High temperatures not only stressed lobsters directly, but also caused them to migrate to deeper areas within Long Island Sound. There, they were exposed more directly to toxic bottom sediments in more crowded conditions, which allowed more rapid spread of disease. This illustrates how physiological and behavioral effects can interact synergistically to increase stress and mortality. Of course, lobsters in the Gulf of Maine occupy a very different physical and chemical environment than those in southern New England and Long Island Sound, so analogous processes may be of less concern in the region.

More often, effects of contaminants will be sublethal, resulting in compromised growth, condition, and reproduction, but not immediate mortality of the affected individual. For example, catadromous eels undertake extensive migrations for a one-time spawning event at sea, including the American eel *Anguilla rostrata*, common in freshwater tributaries and nearshore marine and estuarine areas across the Gulf of Maine. To complete this extensive migration, catadromous eels rely extensively on stored lipids, and transport of lipophilic toxins to the gonads when lipids are mobilized during migration might severely reduce fertility and partly explain the North Atlantic basin-wide decline of catadromous eels (Robinet and Feunteun 2002; also see Kime 1995 for a broader review of contaminant effects on reproduction in fishes). Fortunately, some species exhibit capacity to adapt to high contaminant loads, such as mummichog, which is common across the Gulf of Maine (Nacci et al. 1999). However, such adaptive responses might have implications for both genetic and life history diversity within the population.

Another important ecosystem change that can affect organismal biology is loss or degradation of benthic habitat, including both sessile biota and abiotic structure. The most severe, visible, and widespread habitat impacts in the marine realm are likely to be those affecting biotic habitat features such as corals, shellfish beds and reefs, submerged aquatic vegetation, and tidal marshes. However, soft sediments provide important and often surprisingly complex, yet underappreciated, habitats that can be affected by a range of fishing and nonfishing impacts (Thrush et al. 2001). Whether biotic or abiotic, benthic habitat attributes influence growth, survival, reproduction, and behavior in ways that are generally poorly understood. Progress has been made toward linking growth and survival with habitat metrics in freshwater and nearshore reef systems that are easier to observe and experimentally manipulate than deeper areas of the ocean. Studies along these lines have focused on Gulf of Maine species such as the anadromous Atlantic salmon *Salmo salar* during its freshwater phase (Finstad et al. 2007) and the reef-dwelling cunner *Tautogolabrus adspersus* (Tupper and Boutilier 1997). These studies often show strong effects, especially during early life stages when individuals are more site-attached and more susceptible to predation. Unlike temperature, water chemistry, and contaminant effects, habitat effects on life history traits are less likely to be tied directly to cellular or physiological effects, and instead are determined by effects on behavior, feeding success, and exposure.

The arrangement of different habitat types in mosaics can also have pronounced effects on the biology of constituent species. For example, Irlandi and Crawford (1997) found that pinfish *Lagodon rhomboides* living in tidal marshes adjacent to sea grass beds grew nearly twice as large as those alongside unvegetated sediments. Pinfish can be found in the Gulf of Maine but are uncommon. However, other species more common in the region that use nearshore vegetated habitats might exhibit similar responses to the arrangement of habitat types (e.g., mummichog; Halpin 2000). Effects of habitat mosaics on life history, behavior, and physiology are even less well understood than effects of individual habitat types, but the consequences of any actions that either reduce or create areas of a given habitat will depend upon how a given patch is situated relative to others.

Many traditional single species stock assessment models rely on some combination of growth, mortality, and reproductive parameters to predict stock dynamics. However, in most models, the parameters are static, and rarely are effects of temperature, water chemistry, contaminants, habitat, or other ecosystem attributes considered. These effects can have important consequences, even when the variation is random and nondirectional (Freon et al. 1993), and are more important when changes are directional, due to broader trends in ecosystem alteration. For example, during the 2005 assessments of U.S. groundfish stocks, it was noted that there was a trend of declining mean length and weight at age for many stocks (Mayo and Terceiro 2005). In response, the subsequent 2008 assessments used only the most recent size-at-age data, rather than the entire time series in order to capture the traits most likely driving recent and upcoming stock dynamics, and this change in part contributed to changed estimates of stock productivity and reference points (Northeast Fisheries Science Center 2008). The 2008 assessment also began a more in-depth exploration of causes of the changes, including density-dependence and oceanographic variables (O'Boyle et al. 2007). Of course, fishing can affect patterns of growth and other life history and behavioral traits (Hutchings and Fraser 2008; Uusi-Heikkilä et al. 2008) and is a contributing factor that should be considered when analyzing other ecosystem-driven changes.

Reciprocal Effects of Changes in Organism Biology on the Surrounding Ecosystem

When any combination of temperature, water chemistry, contaminant load, habitat, fishing, and other effects results in changes to organism-level traits and processes in constituent species, those changes can in turn effect important changes upon the surrounding ecosystem. Many of these changes will be trophic in nature. Most stock assessment and other population models attempt to describe changes in abundance or biomass of a given species due to the combined effects of its growth, mortality, fecundity, and maturity traits. Physiology and behavior also determine abundance, even if they are not modeled. On a macroscale, abundance determines the extent to which a given population feeds upon and is fed upon by other species and therefore determines abundance of those species as well as system-wide fluxes in carbon, nutrients, and energy. To the extent that changes in any given life history, behavioral, or physiological traits affect population dynamics of a species (e.g., size-dependent effects on reproductive success of Atlantic cod; Murawski et al. 2001), they will in turn affect ecosystem structure and dynamics.

Ecosystem-driven changes in organismal traits need not, however, fundamentally change abundance of species in order to have ecosystem-level consequences. Many trophic interactions are size-dependent, and populations of a given species with similar abundance or biomass but with different size structure due to differences in growth, mortality, or condition can also have very different ecosystem effects. For example, striped bass exhibit pronounced differences in diet among size-classes, especially a major shift away from small invertebrates toward a predominantly fish diet, as well as differences in size and type of fish prey among larger size-classes (reviewed by Walter et al. 2003). Any changes in growth or mortality that shift the size structure of striped bass will therefore not only affect the aggregate volume of prey consumed, but also the type. This could lead to decreases in abundance of certain prey species and increases in others, with further ecosystem implications of those shifts likely due to the differing ecosystem functions of the prey species. However, the nature and strength of size-dependent predator–prey relationships varies considerably among different types of both predators and prey (Scharf et al. 2000), so the ecosystem consequences of changes in organismal traits that determine body size will depend upon exactly which species are affected.

Changes in life history, behavior, and physiology of a given species can not only have ecosystem implications through their roles in determining abundance and population structure, but also in how they determine distribution of a species. In the review of dietary information for striped bass by Walter et al. (2003), predation patterns differ not only among size-classes, but also among locations along the coast. As a migratory species, a single individual can feed in multiple locations, and ecosystem-driven changes in these migratory patterns can affect both local and regional structure and processes. Ecosystem determinants of changes in migration and distribution of marine species can include point impacts such as dams (Drinkwater and Frank 1994), a widespread concern across the Gulf of Maine watershed. Distribution patterns can also be driven by much larger scale shifts in temperature profiles, which have already been documented for many fish species in the Gulf of Maine (Nye et al. 2009). The result of these physiological- and behavior-driven changes in distributions is that the number and type of species interacting in a given area can change dramatically. The consequences of these distributional changes have been comparatively easy to understand in rivers and lakes but will be much more difficult to predict and monitor in the open ocean. Ecosystem models will have important roles to play in trying to predict those consequences and understand their implications.

Resilience and Adaptability

Our review so far has focused on ecosystem-driven changes in the population mean of pertinent organismal traits. However, intrapopulation variability in life history and behavioral attributes in particular is a common phenomenon and one that has important implications for the capacity of species to adapt to large-scale changes in the ecosystem. Intrapopulation variation in organismal traits can simply represent a contiguous distribution of values around the mean, with individual differences generated by a combination of inherent genetic inequalities and effects of random interactions with the environment that might be positive or negative. Even this "normal" variance that is often ignored in population models can have important implications for population dynamics (Bjørnstad and Hansen 1994; but see Pfister and Stevens 2003 for situations when individual variation can be ignored). For example, Rice et al. (1993) used a freshwater predator–prey system to show that populations with greater variation in growth rate realize both greater survival and a higher final mean growth rate after exposure to high predation pressure. In other words, populations with greater variability in key traits will likely be more resilient to stressors they might confront as ecosystems change.

Not all intrapopulation variation is contiguous or random, however. Mulligan and Leaman (1992) demonstrate that individual variation in size at age might actually represent fundamentally different growth types within a population.

These growth types represent trade-offs between growth and mortality, with faster growth requiring greater metabolic investment and more behavioral risks and therefore incurring greater mortality (as predicted by life history theory; Roff 1992). The result is a range of growth types from those with larger body size and shorter longevity to those with smaller body size and greater longevity. The former will be advantageous when years with high recruitment success are more frequent, whereas the latter will be advantageous during prolonged periods of poor recruitment. As conditions shift from one state to the other through time, the relative frequency of different growth types in the population should shift accordingly, but the presence of both allows the overall population to be more productive when conditions are good and more resilient when conditions are poor. Such responsiveness to changing conditions will be especially important as ecosystems continue to change, as long as the population response can keep pace with the rate of ecosystem change.

Growth is not the only organismal trait that can exhibit fundamentally different types within a population. Kerr et al. (2010) illustrate the implications of different behavioral strategies within a population, specifically the relative proportion of resident versus migratory fish. They show that a greater proportion of migratory fish increases both productivity and resilience but decreases stability. There are benefits to population attributes that promote resilience, productivity, and stability, and Kerr et al. urge development of assessment and management strategies that both account for and preserve different behavioral types within a population, especially spatial strategies that preserve a diversity of habitats. In a review of the interactive effects of fishing and climate change on marine populations, Planque et al. (2010) make a similar case for ensuring that a diversity of life history traits, behavioral traits, and spatial refuges are available to increase capacity of species to respond to ecosystem change.

Summary and Conclusions

As fisheries management, as well as broader management of coasts and oceans, continues to evolve toward more complex and larger scale ecosystem-based approaches, research and monitoring of life history, behavioral, and physiological traits will remain an important scientific need. Multiple ecosystem models still require parameterization at the organismal level like their single-species predecessors and, in fact, often require greater detail due to the inclusion of species interactions and effects of physical and chemical processes. The growing importance of these models, and especially their use in approaches such as management strategy evaluation and integrated ecosystem assessment (Levin et al. 2009), should guide empirical research on organismal biology to provide and improve needed parameter estimates, especially those revealed by sensitivity analyses to have the greatest influence on model outcomes (e.g., Fulton et al. 2004b).

Application of a model introduces a risk of constraining our focus and understanding only to the elements of the model. Empirical research should therefore not be entirely constrained by modeling needs. Indeed, the components and structure of any ecological model originate in empirical research, and models will not evolve if research does not explore new avenues. Continued attention to biological and ecological processes not captured by models should motivate refinements of those models or development of new ones. This has happened with respect to interspecies interactions, which are a core component of Ecopath and its variants, MSVPA, Atlantis, and other tools. In contrast, effects of benthic habitat features and physical and chemical properties of water have arguably seen too little incorporation into ecosystem models to date.

Of course, any model faces limitations of exactly how much detail can be incorporated without introducing unrealistic computational demand and generating excessive uncertainty in outputs. However, information on organismal biology of priority species, and other aspects of the ecosystem for that matter, can be used as information for decision making alongside model outputs in generating management strategies. Schnute and Richards (2001) envision a process wherein a multidisciplinary team, including fishermen and other nonscientists, collectively determine management measures, starting with model outputs but modifying those based on information on oceanography, habitat, and interacting species. Their hypothetical quota-setting process takes place within a single-species assessment and management context but represents an interim step toward EBFM in that additional ecosystem components are factored into management. The same adjustments can also be made when using more complex ecosystem models. Adjustments might be based on observed changes in growth, behavior, or other organismal traits of the target species, predators, prey, or competitors not fully captured by the model but known to have important effects. This is essentially an application of the traffic light approach to determining catch limits (Caddy 1999) to include a broader range of ecosystem considerations.

Data on organismal traits are not only useful in constructing, parameterizing, and adjusting the outputs of single species or ecosystem models. Those data can also be used to set and monitor new management targets. Fisheries are typically assessed and managed on the basis of aggregate biomass. However, a given biomass can represent very different degrees of stability, productivity, resilience, and ecosystem function depending upon how it is composed in terms of age-classes, size-classes, life history types, behavioral types, and other characteristics. Given the importance of these features, a large biomass with a very homogenous makeup might actually be less likely to achieve management goals than a smaller biomass with more diverse structure. Therefore, management should consider multiple indices in determining status and developing strategies. For example, Atlantic cod in the Gulf of Maine exhibit different color morphs associated with different migratory behaviors (Sherwood and Grabowski 2010), and the relative proportion of these morphs would be relatively easy to monitor. Since life history and behavioral diversity carries important advantages (Kerr et al. 2010; Planque et al. 2010), a management strategy could then be developed that adjusts our perception of stock status by the diversity of morphs in the population. Similar adjustments could be made using demographic information, even if size-specific fertility is not modeled, given its importance in stock productivity (Murawski et al. 2001).

In conclusion, organismal biology will remain a critical research and management focus, even as we move toward EBFM and more comprehensive EBM. Although these new directions promise more complete understanding of ecosystems and, as a result, more effective management, important challenges will remain, including those related to understanding and accounting for organismal detail. Research agendas will therefore need to be more strategic and efficient to best navigate the trade-offs between complexity and detail. Models should help guide research and monitoring by identifying those components with the strongest influences or uncertainties but should not be the sole determinant. Collaboration will be critical in extracting emergent lessons from separate data sets and most efficiently planning new research to chart the way forward in organismal biology. After all, it is individual organisms that ultimately live, die, feed, grow, move, and reproduce to determine the states and benefits of nature.

Acknowledgments

We thank the speakers, panelists, and other attendees in the technical session "Life Histories of Fishes and Invertebrates in the Gulf of Maine" at the 2009 Gulf of Maine Symposium for stimulating many of the ideas in this paper. We also thank the symposium organizers for encouraging both the technical session and this paper. Comments from an anonymous reviewer and the editorial committee of this volume significantly improved the quality of the manuscript. Roger Pocock of the Biotech Research and Innovation Center at the University of Copenhagen provided logistical support during the final preparation of the manuscript. The authors were supported by funding from the Gordon and Betty Moore Foundation and the National Fish and Wildlife Foundation.

References

Arkema, K. K., and J. F. Samhouri. 2012. Linking ecosystem health and services to inform marine ecosystem-based management. Pages 9–25 *in* R. L. Stephenson, J. H. Annala, J. A. Runge, and M. Hall-Arber, editors. Advancing an ecosystem approach in the Gulf of Maine. American Fisheries Society, Symposium 79, Bethesda, Maryland.

Arnott, S. A., S. Chiba, and D. O. Conover. 2006. Evolution of intrinsic growth rate: metabolic costs drive trade-offs between growth and swimming performance in *Menidia menidia*. Evolution 60:1269–1278.

Bjørnstad, O. N., and T. F. Hansen. 1994. Individual variation and population dynamics. Oikos 69:167–171.

Björnsson, B., A. Steinarsson, and M. Oddgeirsson. 2001. Optimal temperature for growth and feed conversion of immature cod (*Gadus morhua* L.). ICES Journal of Marine Science 58:29–38.

Boening, D. W. 2000. Ecological effects, transport, and fate of mercury: a general review. Chemosphere 40:1335–1351.

Caddy, J. F. 1999. Deciding on precautionary management measures for a stock based on a suite of limit reference points (LRPs) as a basis for a multi-LRP harvest law. NAFO (Northwest Atlantic Fisheries Organization) Scientific Council Studies 32:55–68.

Cadrin, S. X., and D. H. Secor. 2009. Accounting for spatial population structure in stock assessment: past present, and future. Pages 405–426 *in* R. J. Beamish and B. J. Rothschild, editors. The future of fisheries science in North America. Springer, New York.

Chiba, S., S. A. Arnott, and D. O. Conover. 2007. Coevolution of foraging behavior with intrinsic growth rate: risk-taking in naturally and artificially selected growth genotypes of *Menidia menidia*. Oecologia 154:237–246.

Churchill, J., J. Runge, and C. Chen. 2011. Processes controlling retention of spring-spawned Atlantic cod (*Gadus morhua*) in the western Gulf of Maine and their relationship to an index of recruitment success. Fisheries Oceanography 20:32–46.

Conover, D. O. 1990. The relation between capacity for growth and length of growing season: evidence for and implications of countergradient variation. Transactions of the American Fisheries Society 119:416–430.

Conover, D. O., and T. M. C. Present. 1990. Countergradient variation in growth rate: compensation for length of the growing season among Atlantic silversides from different latitudes. Oecologia 83:316–324.

Cooley, S. R., and S. C. Doney. 2009. Anticipating ocean acidification's economic consequences for commercial fisheries. Environmental Research Letters 4:1–8.

Council on Environmental Quality. 2010. Final recommendations of the interagency ocean policy task force. Council on Environmental Quality, Washington, D.C.

Denman, K. L., and A. E. Gargett. 1995. Biological-physical interactions in the upper ocean: the role of vertical and small-scale transport processes. Annual Review of Fluid Mechanics 27:225–255.

deYoung, B., F. Heath, F. E. Werner, B. Chai, B. A. Megrey, and P. Monfray. 2004. Challenges of modeling decadal variability in ocean basin ecosystems. Science 304:1463–1466.

Drinkwater, K. F. 2005. The response of Atlantic cod (*Gadus morhua*) to future climate change. ICES Journal of Marine Science 62:1327–1337.

Drinkwater, K. F., and K. T. Frank. 1994. Effects of river regulation and diversion on marine fish and invertebrates. Aquatic Conservation 4:135–151.

Fabry, V. J., B. A. Seibel, R. A. Feely, and J. C. Orr. 2008. Impacts of ocean acidification on marine fauna and ecosystem processes. ICES Journal of Marine Science 65:414–432.

Finstad, A. G., S. Einum, T. Forseth, and O. Ugedal. 2007. Shelter availability affects behaviour, size-dependent and mean growth of juvenile Atlantic salmon. Freshwater Biology 52:1710–1718.

Freon, P., F. Gerlotto, and O. A. Misund. 1993. Consequences of fish behaviour for stock assessment. ICES Marine Science Symposium 196:190–195.

Fulton, E.A., M. Fuller, A.D.M. Smith, and A. Punt. 2004a. Ecological indicators of the ecosystem effects of fishing: final report. Australian Fisheries Management Authority, Report Number R99/1546, Canberra.

Fulton, E. A., J. S. Parslow, A. D. M. Smith, and C. R. Johnson. 2004b. Biogeochemical marine ecosystem models, II: the effect of physiological detail on model performance. Ecological Modelling 173:371–406.

Furness, R. W., and M. L. Tasker. 1997. Seabird consumption in sand lance MSVPA models for the North Sea, and the impact of industrial fishing on seabird population dynamics. Pages 147–169 *in* Forage fishes in marine ecosystems. Alaska Sea Grant College Program, University of Alaska, AK-SG-97-01, Fairbanks.

Garrison, L. P., J. S. Link, D.P. Kilduff, M.D. Cieri, B. Muffley, D.S. Vaughan, A. Sharov, B. Mahmoudi, and R. J. Latour. 2010. An expansion of the MSVPA approach for quantifying predator–prey interactions in exploited fish communities. ICES Journal of Marine Science 67:856–870.

Halpin, P. M. 2000. Habitat use by an intertidal salt-marsh fish: trade-offs between predation and growth. Marine Ecology Progress Series 198:203–214.

Hare, J. A., and J. Kane. 2012. Zooplankton of the Gulf of Maine—a changing perspective. Pages 115–137 *in* R. L. Stephenson, J. H. Annala, J. A. Runge, and M. Hall-Arber, editors. Advancing an ecosystem approach in the Gulf of Maine. American Fisheries Society, Symposium 79, Bethesda, Maryland.

Hilborn, R., and C.J. Walters. 1992. Quantitative fisheries stock assessment: choice, dynamics and uncertainty. Kluwer Academic Publishers, Norwell, Massachusetts.

Hutchings, J. A., and D. J. Fraser. 2008. The nature of fisheries- and farming-induced evolution. Molecular Ecology 17:294–313.

Imsland, A. K., A. Foss, G. Naevdal, T. Cross, S.W. Bonga, E.V. Ham, and S. O. Stefansson. 2000. Countergradient variation in growth and food conversion efficiency of juvenile turbot. Journal of Fish Biology 57:1213–1226.

Irlandi, E. A., and M. K. Crawford. 1997. Habitat linkages: the effect of intertidal saltmarshes and adjacent subtidal habitats on abundance, movement, and growth of an estuarine fish. Oecologia 110:222–230.

Ji, R., C. Davis, C. Chen, and R. Beardsley. 2008. Influence of local and external processes on the annual nitrogen cycle and primary productivity on Georges Bank: a 3-D biological-physical modeling study. Journal of Marine Systems 73:31–47.

Kerr, L. A., S. X. Cadrin, and D. H. Secor. 2010. The role of spatial dynamics in the stability, resilience, and productivity of an estuarine fish population. Ecological Applications 20:497–507.

Kime, D. E. 1995. The effects of pollution on reproduction in fish. Reviews in Fish Biology and Fisheries 5:52–95.

Kristiansen, T., R. G. Lough, F. E. Werner, E. A. Broughton, and L. J. Buckley. 2009. Individual-based modeling of feeding ecology and prey selection of larval cod on Georges Bank. Marine Ecology Progress Series 376:227–243.

Kritzer, J. P., and P. F. Sale, editors. 2006. Marine metapopulations. Academic Press, San Diego, California.

Levin, P. S., M. J. Fogarty, S. A. Murawski, and D. Fluharty. 2009. Integrated ecosystem assessments: developing the scientific basis for ecosystem-based management of the ocean. PLoS (Public Library of Science) Biology [online serial] 7(1):e1000014. DOI: 10.1371/journal.pbio.1000014.

Livingston, P. A., and J. Jurado-Molina. 2000. A multispecies virtual population analysis of the eastern Bering Sea. ICES Journal of Marine Science 57:294–299.

Mangel, M., and P. S. Levin. 2005. Regime, phase and paradigm shifts: making community ecology the basic science for fisheries. Philosophical Transactions of the Royal Society 360:95–105.

Mann, K. H., and J. R. N. Lazier, editors. 2006. Dynamics of marine ecosystems: biological-physical interactions in the oceans. Blackwell Publishing, Malden, Massachusetts.

Mayo, R. K., and M. Terceiro, editors. 2005. Assessment of 19 northeast groundfish stocks through 2004. National Oceanic and Atmospheric Administration, National Marine Fisheries Service, Northeast Fisheries Science Center, Reference Document 05–13, Woods Hole, Massachusetts.

Mulligan, T. J., and B. M. Leaman. 1992. Length-at-age analysis: can you get what you see? Canadian Journal of Fisheries and Aquatic Sciences 49:632–643.

Munch, S. B., and D. O. Conover. 2003. Rapid growth results in increased susceptibility to predation in Menidia menidia. Evolution 57:2119–2127.

Murawski, S. A., P. J. Rago, and E. A. Trippel. 2001. Impacts of demographic variation in spawning characteristics on reference points for fishery management. ICES Journal of Marine Science 58:1002–1014.

Nacci, D., L. Coiro, D. Champlin, S. Jayaraman, R. McKinney, T. R. Gleason, W. R. Munns, Jr., J. L. Specker, and K. R. Cooper. 1999. Adaptations of wild populations of the estuarine fish Fundulus heteroclitus to persistent environmental contaminants. Marine Biology 134:9–17.

Northeast Fisheries Science Center. 2008. Assessment of 19 northeast groundfish stocks through 2007. National Oceanic and Atmospheric Administration, National Marine Fisheries Service, Northeast Fisheries Science Center, Reference Document 08–15, Woods Hole, Massachusetts.

Nye, J. A., J. S. Link, J. A. Hare, and W. J. Overholtz. 2009. Changing spatial distribution of fish stocks in relation to climate and population size on the northeast United States Continental Shelf. Marine Ecology Progress Series 393:111–129.

O'Boyle, R., V. Crecco, L. Van-Eeckhaute, D. Kahn, C. Needle, B. Rothschild, S. Smith, and J. H. Vølstad. 2007. Report of the groundfish assessment review meeting (GARM III), part 1. data methods. National Oceanic and Atmospheric

Administration, National Marine Fisheries Service, Northeast Fisheries Science Center, Woods Hole, Massachusetts.

Overholtz, W. J., and J. S. Link. 2007. Consumption impacts by marine mammals, fish, and seabirds on the Gulf of Maine–Georges Bank Atlantic herring (*Clupea harengus*) complex during the years 1977–2002. ICEs Journal of Marine Science 64:83–96.

Overholtz, W., and J. Link. 2009. A simulation model to explore the response of the Gulf of Maine food web to large-scale environmental and ecological changes. Ecological Modelling 220:2491–2502.

Pauly, D., V. Christensen, and C. Walters. 2000. Ecopath, Ecosim, and Ecospace as tools for evaluating ecosystem impact of fisheries ICES Journal of Marine Science 57:697–706.

Pearce, J., and N. Balcom. 2005. The 1999 Long Island Sound lobster mortality event: findings of the comprehensive research initiative. Journal of Shellfish Research 24:691–697.

Pew Ocean Commission. 2003. America's living oceans: charting a course for sea change. Pew Ocean Commission, Final Report, Washington, D.C.

Pfister, C. A., and F. R. Stevens. 2003. Individual variation and environmental stochasticity: implications for matrix model predictions. Ecology 84:496–510.

Pikitch, E. K., C. Santora, E.A. Babcock, A. Bakun, R. Bonfil, D. O. Conover, P. Dayton, P. Doukakis, D. Fluharty, B. Heneman, E. D. Houde, J. Link, P. A. Livingston, M. Mangel, M. K. McAllister, J. Pope, and K. J. Sainsbury. 2004. Ecosystem-based fishery management Science 305:346–347.

Planque, B., J-M. Fromentin, P. Cury, K. F. Drinkwater, S. Jennings, R. I. Perry, and S. Kifani. 2010. How does fishing alter marine populations and ecosystems sensitivity to climate? Journal of Marine Systems 79:403–417.

Quinn, T. J., and R. B. Deriso. 1999. Quantitative fish dynamics. Oxford University Press, New York.

Rice, J., and H. Gislason. 1996. Patterns of change in the size spectra of numbers and diversity of the North Sea fish assemblage, as reflected in surveys and models. ICES Journal of Marine Science 53:1214–1225.

Rice, J. A., T. J. Miller, K. A. Rose, L. B. Crowder, E. A. Marschall, A. S. Trebitz, and D. L. DeAngelis. 1993. Growth rate variation and larval survival: inferences from an individual-based size-dependent predation model. Canadian Journal of Fisheries and Aquatic Sciences 58:133–142.

Robinet, T. T., and E. E. Feunteun. 2002. Sublethal effects of exposure to chemical compounds: a cause for the decline in Atlantic eels? Ecotoxicology 11:265–277.

Roff, D.A. 1992. The evolution of life histories. Chapman and Hall, New York.

Scharf, F. S., F. Juanes, and R. A. Rountree. 2000. Predator size–prey size relationships of marine fish predators: interspecific variation and effects of ontogeny and body size on trophic-niche breadth. Marine Ecology Progress Series 208:229–248.

Schnute, J. T., and L. J. Richards. 2001. Use and abuse of fishery models. Canadian Journal of Fisheries and Aquatic Sciences 58:10–17.

Schultz, E. T., K. E. Reynolds, and D. O. Conover. 1996. Countergradient variation in growth among newly hatched *Fundulus heteroclitus:* geographic differences revealed by common-environment experiments. Functional Ecology 10:366–374.

Sherwood, G. D., and J. H. Grabowski. 2010. Exploring the life-history implications of colour variation in offshore Gulf of Maine cod (*Gadus morhua*). ICES Journal of Marine Science 67:1640–1649.

Smith, A. D. M., E. J. Fulton, A. J. Hobday, D. C. Smith, and P. Shoulder. 2007. Scientific tools to support the practical implementation of ecosystem-based fisheries management. ICES Journal of Marine Science 64:633–639.

Talmage, S. C., and C. J. Gobler. 2010. Effects of past, present, and future ocean carbon dioxide concentrations on the growth and survival of larval shellfish. Proceedings of the National Academy of Sciences 107:17246–17251.

Thrush, S. F., and P. K. Dayton. 2010. What can ecology contribute to ecosystem-based management? Annual Review of Marine Science 2:419–441.

Thrush, S. F., J. E. Hewitt, G.A. Funnell, V.J. Cummings, J. Ellis, D. Schultz, D. Talley, and A. Norkko. 2001. Fishing disturbance and marine biodiversity: the role of habitat structure in simple soft-sediment systems. Marine Ecology Progress Series 223:277–286.

Tsou, T. S., and J. S. Collie. 2001. Predation-mediated recruitment in the Georges Bank fish community. ICES Journal of Marine Science 58:994–1001.

Tupper, M., and R. G. Boutilier. 1997. Effects of habitat on settlement, growth, predation risk and survival of a temperate reef fish. Marine Ecology Progress Series 151:225–236.

U.S. Commission on Ocean Policy. 2004. An ocean blueprint for the 21st century. U.S. Commission on Ocean Policy, Final Report, Washington, D.C.

Uusi-Heikkilä, S., C. Wolter, T. Klefoth, R. Arlinghaus. 2008. A behavioral perspective on fishing-induced evolution. Trends in Ecology and Evolution 23:419–421.

Walter, J. F., A. S. Overton, K. H. Ferry, and M. E. Mather. 2003. Atlantic coast feeding habits of striped bass: a synthesis supporting a coast-wide understanding of trophic biology. Fisheries Management and Ecology 10:349–360.

Worm, B., R. Hilborn, J. K. Baum, T. A. Branch, J. S. Collie, C. Costello, M. J. Fogarty, E. A. Fulton, J. A. Hutchings, S. Jennings, O. P. Jensen, H. K. Lotze, P. M. Mace, T. R. McClanahan, C. Minto, S. R. Palumbi, A. M. Parma, D. Ricard, A. A. Rosenberg, R. Watson, and D. Zeller. 2009. Rebuilding global fisheries. Science 325:578–585.

American Fisheries Society Symposium 79:375–391, 2012

Representation of Biodiversity Knowledge within Ecosystem-Based Management Approaches for the Gulf of Maine

PETER LAWTON*
Fisheries and Oceans Canada, Biological Station,
531 Brandy Cove Road, St. Andrews, New Brunswick E5B 2L9, Canada

LEWIS S. INCZE[1] AND SARA L. ELLIS[2]
Aquatic Systems Group, University of Southern Maine
Portland, Maine 04101, USA

Abstract.—Conceptual organization of biodiversity knowledge and explicit acknowledgement of its current limits can help identify priority scientific questions and inform the evolving partnership between science and management in evaluating approaches and priorities for implementing ecosystem-based management. We outline a four-part approach that describes the *composition and size spectrum of known diversity*; provides estimates of the presently *unknown diversity*; outlines the *compositional, structural, and functional elements* that mold diversity patterns at different hierarchical levels (from genes to ecoregions); and identifies *spatial domains* relative to both scientific discovery and management application. This overarching representation of how biodiversity knowledge can be organized builds on earlier considerations for biodiversity monitoring and is informed by recent work under the Gulf of Maine Area Program, the regional ecosystem pilot project of the international Census of Marine Life. Understanding the specific set of biodiversity elements that particular ocean policy objectives and management strategies are directed at can help to target biodiversity research, conservation tactics, and monitoring approaches. A framework for representing biodiversity knowledge increases the capacity to visualize ramifications of management actions in a complex physical-biological system and also provides the essential two-directional conduit for communicating ideas, priorities, and findings between the regional science and management communities.

Introduction

Acknowledgment of the need to incorporate ecosystem considerations in management of human interactions with ocean ecosystems has increased significantly in the past decade (Sinclair and Valdimarsson 2003; Millennium Ecosystem Assessment 2005; Rosenberg and McLeod 2005; Murawski 2007; Ruckelshaus et al. 2008; McLeod and Leslie 2009a; Palumbi et al. 2009; Curtin and Prellezo 2010; Lester et al. 2010). "Ecosystem approaches to management" (EAM; Murawski 2007) and "ecosystem-based management" (EBM; McLeod et al. 2005) have been broadly advocated as terms to reflect this new approach, and each promotes conserving biodiversity to maintain ecosystem function and adaptation over long periods of time (Yachi and Loreau 1999; Levin and Lubchenco 2008; Palumbi et al. 2009). Distinctions between the two largely reflect differing perceptions of how far along we are, or can eventually progress, towards managing for

* Corresponding author: peter.lawton@dfo-mpo.gc.ca
[1] Current address: School of Marine Sciences, University of Maine, Orono, Maine 04469, USA
[2] Current address: 98 Old Pine HIll Road, Berwick, Maine 03901, USA

long-term resilience based on knowledge of ecosystem structure and function. Here, EBM is adopted, as the authors cover a broader range of considerations of regional biodiversity than have been typically considered under EAM discussions.

Ecosystem-based management is creating unprecedented interdisciplinary connections (Shackeroff et al. 2009; Lester et al. 2010; see also Delauer et al. 2012; Feurt 2012; Fogarty et al. 2012; Hale and Westhead 2012; all this volume). Regional interfacing of science, humanities, and management is thus considerably more complex than when the last Gulf of Maine science symposium was held in 1996 (Wallace and Braasch 1997). As a consequence, it is important that concepts, terminologies, and technical approaches be clearly articulated for researchers, EBM practitioners, and stakeholders. Development and implementation of national-level EBM systems by five governments, including the United States and Canada, were recently reviewed by Rosenberg et al. (2009). Their review highlighted common issues related to the jurisdictional complexity of implementing EBM and the need for clear political directives with long-term support for EBM (see also Murawski 2007). Rosenberg et al. (2009) also concluded that successful future implementation of EBM requires a participatory, proactive, and transparent learning process that can adapt to changing knowledge, environmental conditions, and improved technologies.

The Census of Marine Life, a 10-year scientific initiative that was completed in 2010, has sought to assess and explain the diversity, distribution, and abundance of life in the oceans: past, present and future (Yarincik and O'Dor 2005; McIntyre 2010; Snelgrove 2010). The majority of its field programs focused on discovery of new species and communities in underexplored areas. In contrast, the Gulf of Maine was chosen as the census' ecosystem pilot program because it is a well-studied, relatively data-rich body of water with a long history of human use and associated management needs. The Gulf of Maine Area (GoMA) program of the census was designed to identify and assemble biological knowledge relevant to EBM across the full range of trophic levels (from microbes to mammals) and habitats (from shallow inshore to deep offshore) (O'Dor and Gallardo 2005). Incze et al. (2010) summarize GoMA's development, implementation, and synthesis phases while Ellis et al. (2011) compare the design and program experiences of GoMA with three other ecosystem-level studies of marine biodiversity. Through its consideration of available information for various scales of marine biodiversity organization, GoMA identified the need for a more comprehensive biodiversity knowledge framework that would help to guide both the further development of biodiversity science and the use of that science to support regional EBM applications.

As Cogan et al. (2009) recently did for marine habitat mapping, considered by them as key support for EBM, we deconstruct biodiversity organization into its principal constituent elements and identify some of the opportunities and constraints to incorporating biodiversity knowledge into EBM objectives within the Gulf of Maine. We identify new ways in which the full spectrum of biodiversity research on a regional ecosystem such as the Gulf of Maine can be visualized. Specifically, we assert that an evolving conceptual framework is needed to efficiently connect knowledge from different disciplines and stimulate new research and analysis. Our intent is to provide a broad canvas upon which different interest groups should be able to draw their own domain of interest and influence and also understand that of other stakeholders and enablers of EBM.

Biodiversity is often considered simply in terms of species richness or the number of species in a given area. However, biodiversity is organized across multiple temporal and spatial scales, from genetic diversity within populations to species diversity within communities to community diversity across landscapes and ecosystems (Wilson 1992; Sala et al. 2000). Biodiversity operates in both direct and indirect ways to affect ecosystem functioning (Loreau 2000; Gamfeldt et al. 2008). The number of species and potential interactions is generally large, so that the effects of altering species composition and patterns of abundance are difficult to predict (Naeem and Wright 2003). As Guichard and Peterson (2009) have recently argued, the potential for cross-scale interactions within the natural system increases uncertainty in terms of system response when specific management actions are undertaken, and in common with the natural system, management and conservation strategies themselves are multiscale processes (McLeod and Leslie 2009b; Lester et al. 2010). This potentially limits the controllability of ecological processes by management action under EBM. Indeed, there are some things that from a practical standpoint could be considered unknowable—that is, we will have to proceed without that information for the foreseeable future (Palumbi et al. 2009). Faced with these limits to knowledge, how well prepared are we to use biodiversity information as part of the implementation of EBM in the Gulf of Maine? We outline an overarching biodiversity framework that we have found helpful for focusing on this question across biological, technological, and management application dimensions.

Biodiversity Representation within Ecosystem-Based Management

Recent experience shows what may emerge as a common occurrence in the development of EBM approaches. Initial formulation of the ways in which the natural system is structured, and how human use interacts with ecosystem properties, results in a multilayered hierarchical representation (Jamieson et al. 2001) that attempts to encompass the full breadth of conceptual foundations and multiple issues. Subsequent implementation of EBM within regional ocean management areas, while recognizing this system complexity, tactically focuses in on a reduced set of system properties and management issues requiring inputs from science and management (O'Boyle and Jamieson 2006; O'Boyle and Worcester 2009). We appreciate the practical need to do this: an operational EBM system must contend with specific qualitative and quantitative metrics for measuring status and trends against formal objectives, strategies, tactics, and reference points within the relevant societal and jurisdictional system. Nonetheless, one consequence of focus on management and retrenchment from full representations of biocomplexity has been to disconnect academic and applied science perspectives. Considerations for monitoring programs developed through a focus on applications tend to gravitate towards pragmatic approaches around components of the system that are relatively well studied and/or directly under management control. These are not necessarily the components of the natural system that are important structuring elements. Conversely, academic scientists working on cutting-edge approaches on less-understood processes or parts of the system are unlikely to posit their research as policy- and management-relevant. Indeed, there is only minimal incentive to do so and less than adequate understanding of how the information could be used. Retaining a comprehensive biodiversity framework, within which the present approaches to implementing EBM remain nested, should provide for progressive improvements in system understanding that would be obtained from academic science, as well as more effective use of data (perhaps in the form of information and insights) by management. A framework provides the essential two-directional conduit for communicating ideas, priorities, and findings.

In undertaking the GoMA program, we assembled a four-part approach for representing biodiversity knowledge. The first approach deals with organizing our inventory of the marine species that are known from the Gulf of Maine. The known marine biodiversity within the region has now been compiled within a Gulf of Maine Register of Marine Species (GoMRMS). The register includes a vetted, taxonomically ordered list of species known to occur in the Gulf of Maine, as well as a list of provisional entries requiring validation (Incze et al. 2010). An "up-to-date" and "fully developed" register (in reality, a constant work in progress) provides a valuable tool for examining the recorded biodiversity of a region, seeing if an organism has been reported there, and obtaining relevant references and electronic links to other sources of information on the species. It facilitates research and contributes to answering the question "What kind of system is this?"

The second approach was to compare this known biodiversity with an estimate of the presently unknown biodiversity, principally through a preliminary size-spectrum analysis (Figure 1). This emphasized that almost all of our well-known biodiversity and virtually all of our "managed" species are at the large end of the spectrum and that an equal or greater portion of regional marine biodiversity (much more so if the bacteria and archaea are considered) remains poorly described or completely unknown at the smaller end of the size spectrum. This is not surprising, but illustrating this fact in a semiquantitative way

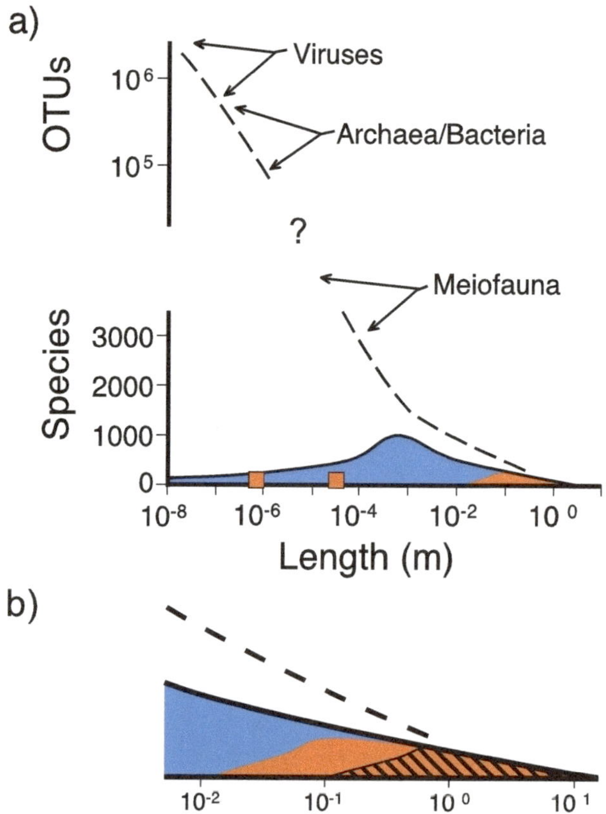

FIGURE 1. Biodiversity size spectrum within the Gulf of Maine ecosystem, adapted from Incze et al. (2010). Length-based schematic shows the approximate size distribution of adult stages of named species (solid line; blue fill) and a suggestion of the possible extent of the unknown biodiversity (broken line). The line for known (named) taxa approximates species numbers for groups of organisms contained within size groupings of $10^x \pm 10^{0.5x}$ m, where x is a whole number from –8 to 1. Note that these numbers are from November 2009. For the prokarya and viruses, diversity is expressed as "types" (operational taxonomic units, OTUs), since there is no agreement on what makes a species in these groups. The shape of the curve of unknown biodiversity from meiofauna to viruses, and the maximum number of types cannot be projected with any certainty. Lower figure enlarges the lower right portion of the size-diversity curve to illustrate where most "monitored" (orange fill) and "managed" (diagonal stripes) species occur. Examples of monitored species are unmanaged species caught in fisheries assessment surveys and seabird abundances at long-term study sites; managed species are those with management plans such as commercial fish, crustaceans and mollusks, cetaceans, and threatened or endangered species. Solid squares (orange fill) on the x-axis in the main portion of the figure represent harmful algal species and coliform bacteria for which there are also monitoring programs in the Gulf of Maine.

makes clear that the greatest ignorance exists in that portion of the size spectrum that contains most of the biomass and supports the majority of ecosystem processes (Whitman et al. 1998; Amaral-Zettler et al. 2010).

The third approach was to adopt existing hierarchical classification schemas (Noss 1990; Cogan and Noji 2007; Cogan et al. 2009) that partition the various factors and processes that influence the *expression* of biodiversity within the system into compositional, structural, and functional elements (Table 1). A clarification needed here is that in these classification approaches, functional elements operate "internally" within the system to affect or modulate biodiversity characteristics; thus, they are not directly analogous to the "func-

TABLE 1. Hierarchy of biodiversity elements with contextual examples of biodiversity research, biodiversity information resources, and ecosystem-based management (EBM) considerations. For comprehensive accounts of the classification approach, and further examples of research and monitoring applications, see Noss (1990), Cogan and Noji (2007), Cogan et al. (2009), Incze et al. (2010), and Ellis et al. (2011). Further discussion of international and national policy context, modes of coordination, and challenges on implementing EBM in Canada and the United States is provided by Rosenberg et al. (2009).

Hierarchy	Biodiversity research and information resources	EBM considerations
Compositional elements		Convention on Biological Diversity[a]
Ecoregion	Cold temperate northwest Atlantic[b]; Acadian Atlantic[c]	
Communities	Definition of major (typical) groups of co-occurring and interacting species often grouped by dominant organisms and/or habitats	
Species	Gulf of Maine Register of Marine Species (GoMRMS); Ocean Biogeographic Information System (OBIS); species discovery	
Genes	Extent of variation within a population; cryptic species	
Structural elements		Convention on Biological Diversity[a]
Ecoregion heterogeneity and complexity	Characterization of physiographic regions, landscapes	Ecologically and biologically significant areas[d]
Habitat structure, physiognomy	Biotic and abiotic determinants of pattern (surrogates)	
Population structure	Influence on population structure, adaptation, persistence; age, size, frequencies, affecting populations	Ecosystem approach to fisheries[e]
Genetic structure	How is the genetic variability distributed?	
Functional elements		Convention on Biological Diversity[a]
Ecological processes and disturbances; resource use	Effects of ocean warming and acidification; consequences of habitat simplification due to trawling	Ecosystem status and trends[f]
Interspecific interactions, ecosystem processes	Hinging effect of keystones (climate–copepods; humans–fish), roles of the rare	Forage species[g]
Demographic processes	Migration, recruitment, survivorship	
Genetic processes	Metapopulation and genetic diversity analyses; analyses; gene flow	Endangered Species/ Species at Risk[h]

[a] Convention on Biological Diversity obligations span compositional, structural, and functional biodiversity elements (relevance to Canada and USA EBM systems discussed by Rosenberg et al. 2009).
[b] Spalding et al. (2007).
[c] Wilkinson et al. (2009).
[d] DFO (2004).
[e] Christensen et al. (2007).
[f] Ecosystem Assessment Program (2009).
[g] DFO (2006).
[h] Species at Risk (Canada): Findlay et al. (2009); Endangered Species (USA): Schwartz (2008).

tion" of regional marine biodiversity in supporting various ecosystem services. Such benefits to human populations are best considered separately. However, human uses of the marine environment, including disturbances to the environment and alteration of populations (through targeted removals and other effects), are considered within the set of functional elements because they can affect biodiversity composition, structure, and function. Originally proposed as a means to organize strategies and technical approaches for biodiversity monitoring in terrestrial systems (Noss 1990), the classification approach was subsequently adapted by Cogan and Noji (2007) and Cogan et al. (2009) as a means to focus debate on how marine biodiversity analysis and marine habitat mapping applications might be used to implement EBM. Here, we contend that this classification approach can serve an even broader purpose by helping narrow the current disconnect between basic biodiversity science and the management of ocean space and resources.

Our fourth approach was to consider how to organize biodiversity information and ecological concepts to deal with issues of scale, detail, and environmental heterogeneity when trying to inform EBM decisions that must integrate across multiple scales, from an area as small as a marsh or estuary up to the Gulf of Maine ecosystem. This is perhaps the most difficult challenge because it will not be possible, certainly not in the foreseeable future, to know everything we would like to know, everywhere. While this problem is generally recognized, a strategy for dealing with it remains elusive. Addressing it can help focus opportunities and benefits for both science and management.

An overarching framework for representing biodiversity information and knowledge should provide for progress without being too proscriptive. In particular, a framework should stimulate discussion and innovation at both the basic research and the pragmatic application end of the science-management continuum. The size-spectrum approach developed by the authors (for known and unknown biodiversity) can be used as a lens through which to assess the understanding of composition, structure, and function for different components of regional biodiversity while a complementary conceptual model for spatial domains and ecological interactions, relative to EBM interventions, can lead more easily into marine spatial planning applications. Results of such an overarching framework, which would evolve over time, should lead to a clearer definition and understanding by managers and scientists of how research results can inform EBM. A framework provides a conceptual model for communication: a way for managers to distill, prioritize, and express needs for information that meet their criteria and a way for scientists to identify opportunities for research that are both fundamentally interesting and translatable to management needs, either "as is" or as essential increments to the knowledge needed to implement EBM. This explicitly recognizes that EBM will be an evolving and long-term undertaking; a framework provides an evolving tool about which the interests and strategies of scientists, managers, funding agencies, and public stakeholders can coalesce. Ecosystem-based management is complex by its nature, and by systematizing its components and schematizing a strategy, we hope to move EBM forward. We caution, however, that EBM will never be simple in practice, and to reduce it to overly simple concepts is likely to increase risk of errors, miss opportunities to strengthen the underlying science, and fail to incorporate fundamental understanding of biodiversity processes into an integrated process of well-informed decision making.

Most knowledge of biodiversity in marine systems has accrued to the larger end of the size spectrum (Figure 1), owing to the obvious and often charismatic nature of larger organisms (Nee 2004), the relative ease of sampling and identifying them, and/or commercial interests in using them in some way. Marine mammals, fishes, mollusks, large crustaceans, echinoderms, and seabirds are in this group (but see Diamond 2012, this volume, for a discussion of large migratory species that are typically excluded from management plans). Marine ecologists also focused for many years on intertidal and shallow subtidal environments because their accessibility provided affordable, abundant, and diverse venues for examining and testing fundamental ecological processes such as recruitment, competition, predation, succession, and the role of stochastic biological and physical disturbances. These studies extend the number of comparatively well-studied organisms toward the smaller size range. Appreciably, there has been little overlap between these nearshore autecological or community studies and the population-level focus of studies on the larger biota. Other knowledge about smaller organisms centers on phytoplankton, some conspicuous protozoans, dinoflagellates, crustacean zooplankton, and some bacteria. However, regular monitoring of this part of the size spectrum is for only a small subset of the known species (Johnson et al. 2011; Li et al. 2011). Some of the data sets are now being considered for the ecological context they provide, particularly with respect to detecting climate change effects that might influence upper trophic levels (Pershing et al. 2005). New technologies and analyses have revealed a tremendous diversity among the microbes, estimated as "types" in Figure 1 (Sogin et al. 2006; Incze et al. 2010; Li et al. 2011).

Sampling of smaller organisms and community structure in deeper waters has been hampered historically by high costs, the great expanse of deep water, and perhaps some ambiguity about the usefulness of the information, at least in the short term (further explored by Johnson and Hare 2012; Runge and Jones 2012; both this volume). Today, the desire for a more comprehensive view of the oceans reveals large gaps in the knowledge of which organisms live in the Gulf of Maine, their quantitative contributions to ecosystem function, the spatial and temporal dynamics of their populations, and their responses to disturbance. The shape and height of the "unknown biodiversity" curve is speculative in detail but grossly correct in its depiction of the wide gap in compositional, let alone structural and functional, knowledge of biota in the Gulf of Maine area. Although the number of unknown species is impossible to estimate accurately, the more essential point is to illustrate where knowledge is most deficient. From this perspective, we can consider how these deficiencies affect our understanding of local communities and marine ecosystem processes and what strategies might be used to understand the distribution, function, and conservation requirements of these poorly known and unknown parts of the ecosystem.

How do we address this gap? We cannot know everything everywhere all the time, so what is an efficient and efficacious strategy for comprehending composition, structure, and function across such a wide-sized spectrum, mapped onto the heterogeneous and dynamic expanse of the Gulf of Maine? Relevant questions, strategies, and applications will vary as one moves up the size range from microbes to whales. A comparison of relative knowledge on microbes and upper trophic level predators discloses how research program evolution might differ. For microbes, the species concept itself is not applicable, and even in terms of types (Figure 1), we are unlikely to completely document compositional diversity any time soon. What may be of more interest to the research community is the functional diversity of this group, and the application of this information to EBM might be as an indicator of regional and subregional function and change. Conversely, the compositional diversity of upper trophic level predators in the Gulf of Maine area is completely known, at least at the species level (273 extant species), and for some species, considerable information exists on compositional diversity at population and genetic levels (S. Kraus and coworkers, New England Aquarium, unpublished manuscript). Ecosystem-based management interests are likely to focus on these species as targets of management intervention and indicators of environmental change or management performance. Between the bacteria and the upper trophic level predators, and aside from the fishes and a few abundant and commercially valuable "shellfish" (bivalves; decapods crustaceans), which are also the focus of management attention, there remains a huge diversity of eukaryotic organisms and a wide range of organism sizes, required sampling methods, and life histories for which EBM seeks relevant information.

Here, we present a capsule overview of a hierarchical scheme for depicting biodiversity organization, originally developed by Noss (1990) for monitoring in terrestrial environments and later applied by Cogan and Noji (2007), Cogan et al. (2009), and Ellis et al. (2011) to marine systems. Detailed catalogs of biodiversity descriptors, indicators, and metrics have been elaborated in these earlier papers. Within this hierarchical scheme (Table 1) three principal elements, each of which is further represented through a hierarchical (nominally spatially referenced) structure ranging from the ecoregion to genetic levels, are used to describe how biodiversity is organized and its expression modulated within the physical-biological system.

Compositional marine biodiversity represents the identity and variety of biodiversity elements within the system, from genes to the complement of species occurring within particular ecoregions and biogeographic provinces. Structural marine biodiversity is concerned with the physical organization or pattern of biodiversity occurring within the system, most commonly from community to ecoregion levels, and includes both biotic and abiotic variables that modulate patterns. At the ecoregion level, structural diversity analysis can be linked to the arrangement of physiographic regions and landscapes that contribute to the internal makeup of the ecoregion, and this knowledge is a composite of findings derived at fine scale (habitats) as well as larger scales. Functional marine biodiversity elements are processes that operate at various spatial, ecological, and evolutionary scales to mold biodiversity composition and structure. The functional elements range from genetic processes to regional forcing variables, both natural and anthropogenic (Cogan and Noji 2007). Within this approach to deconstructing biodiversity, functional diversity elements are the factors that need to be considered as influential in driving biodiversity patterns and may also be significant as potential constraints on management actions. In this sense, these functional elements operate "internally" within the physical-biological system and so are distinguished from the function of regional marine biodiversity in supporting various ecosystem services to human society (e.g., provisioning, regulating, supporting, and cultural services; see Levin and Lubchenco 2008 for a recent overview). The utility of deconstructing biodiversity into these three principal elements comes from being able to identify specific aspects of biodiversity organization that may require management attention and thus better scientific understanding. The hierarchical systems reviewed here are not intended as explanations of how an ecosystem "works," but rather as a way to visualize the various structuring factors and forces that influence the expression of biodiversity within the system over time.

Accepting this important distinction, consideration of the biological size spectrum reveals the enormity of the task of filling in the outstanding gaps in our knowledge of the full biodiversity within the Gulf of Maine system, meaning not just compositional and structural biodiversity elements, but many aspects of function as well. For compositional diversity, the highest level of the hierarchy aligns with a biogeographic or ecoregion perspective and represents, in the aggregate, the overall reservoir of biodiversity represented within the system (for example, the type of broadscale assessment called for under the Convention on Biological Diversity; see Table 1). It would also place the Gulf of Maine in a global context, wherein experience and insights from other systems might be transferable. Such knowledge, even for just the composition, would take many years to acquire, and some "shortcuts" will almost certainly prove necessary and practical to make some headway in the quest to build understanding and improve the basis for management. An expedited strategy for filling the large gaps in knowledge might include detailed studies of species composition and abundance in different substrates in a select number of habitats. In a subset of these, functional studies might be conducted on topics such as trophic role, geochemical role, life history, population variance, functional diversity and redundancy among community members, and responses to different types and levels of disturbance. Genetic tools (Savolainen et al. 2005) may provide indices of diversity before most species have been identified by traditional means, and community-level functional properties and rates may be determined before the roles of individual species have been learned. As preliminary as these approaches will be in their first years, they initiate a process that gains depth and breadth over time. We anticipate that the interests and needs of both the scientific and management communities will be enhanced by a framework that requires and uses insights from both ends.

Structural diversity elements (Table 1) represent the component of regional biodiversity for which we have gained the most significant advances in documentation and broadscale understanding, albeit largely focused on partial segments of the biodiversity size spectrum (Figure 1). For both demersal and benthic components of the Gulf of Maine system, we have shelf-scale surveys that provide some representation of ecosystem structure and, in certain cases, capability to evaluate trends. Fisheries trawl surveys (National Marine Fisheries Service in the United States;

Department of Fisheries and Oceans in Canada) dating back to the 1960s form the longest shelf-wide time series on fish (largely demersal) and some epibenthic biota. Although there are certain historical broadscale investigations of benthos in the Gulf of Maine (e.g., Theroux and Wigley 1998), there is at present no comprehensive benthic sampling program analogous to the fisheries trawl surveys. Following advances in seabed mapping technologies and in response to the need for predictive tools for understanding seabed diversity (McArthur et al. 2010), marine benthic habitat mapping programs are undergoing a significant expansion (see Cogan et al. 2009 for a recent review and Grabowski and Hart 2012, this volume).

There remain major limitations in our current understanding of how ecosystem function and adaptability are linked to biodiversity, and functional diversity as a *property* of the natural system is a current focus of research. In theory, more complex ecosystem services are expected from more complex (diverse) systems, and the simplification of ecosystems (through reductions in diversity at various levels) results in a reduction in functional aspects such as services or resilience or both (Palumbi et al. 2009). Attempts to demonstrate this experimentally involve simplifications or other artifacts themselves, but the majority of studies support this general theoretical prediction (Cardinale et al. 2006), and Duffy (2009) argues that experiments likely underestimate the importance of biodiversity to ecosystem functioning. Most studies have focused on abundant and/or highly interactive species, but the majority of species are comparatively rare, and many ecological interactions take place among lesser players in diverse landscapes. Their roles in supporting ecosystem function are less well known in the present and impossible to predict for the future (e.g., Bellwood et al. 2006). The conservation of biodiversity at all levels, including what we are calling structural diversity (after Cogan and Noji 2007), is thus one of the precautionary objectives of EBM.

Given this inability to fully understand how functional diversity supports valued ecosystem services, a pragmatic approach to EBM is to attempt to conserve function through the proxy of composition and structure, particularly by taking advantage of various spatial management approaches to reduce pressures on particular species, biological communities, and the habitats they occupy (Foley et al. 2010; Fogarty et al. 2012). Both of these principal elements of biodiversity organization are also more amenable to initial assessment and routine monitoring, but such an approach does carry risks and considerable uncertainty. What are the relationships between structure and function? What connections to management would this information support? What are the linkages such that one objective can be managed and, in turn, help to realize other objectives? Continuing efforts to identify priority areas for marine conservation planning consideration, such as the definition of ecologically and biologically significant areas within Canadian marine waters (DFO 2004), and other regional planning efforts to build networks of marine protected areas (e.g., Crawford et al. 2006), will afford some protection to particular geographical areas within the Gulf of Maine. These are typically locations for which there is reasonably comprehensive inventory, with high local species richness (e.g., Buzeta and Singh 2008), and which by proxy will contribute to supporting biodiversity.

The set of biodiversity elements at ecoregion, community, population, species, and genetic levels of organization (Table 1) can result in implied spatial hierarchies that are not necessarily correct. Genes, as entities, are obviously much smaller in size than other biodiversity organizational elements, but they are not necessarily spatially restricted within the system. We thus advocate developing a separate facet for the biodiversity framework, at present a preliminary conceptual model for spatial organization of the system (Figure 2) within which domains of activity, interest, and capacities of curiosity-driven science, applied (use-directed) science, monitoring approaches, and management actions can be better visualized. This graphic is not fundamentally different from arguments in Cogan et al. (2009) for marine habitat and biodiversity mapping (their usage) organized within an overall research-management framework (their Figure 3). However, developing a conceptual model for spatial organization that complements the composition, structure, and function hierarchies (Table 1) should help to identify promising research avenues for spatially explicit modeling of biodiversity patterns and processes and the inter-

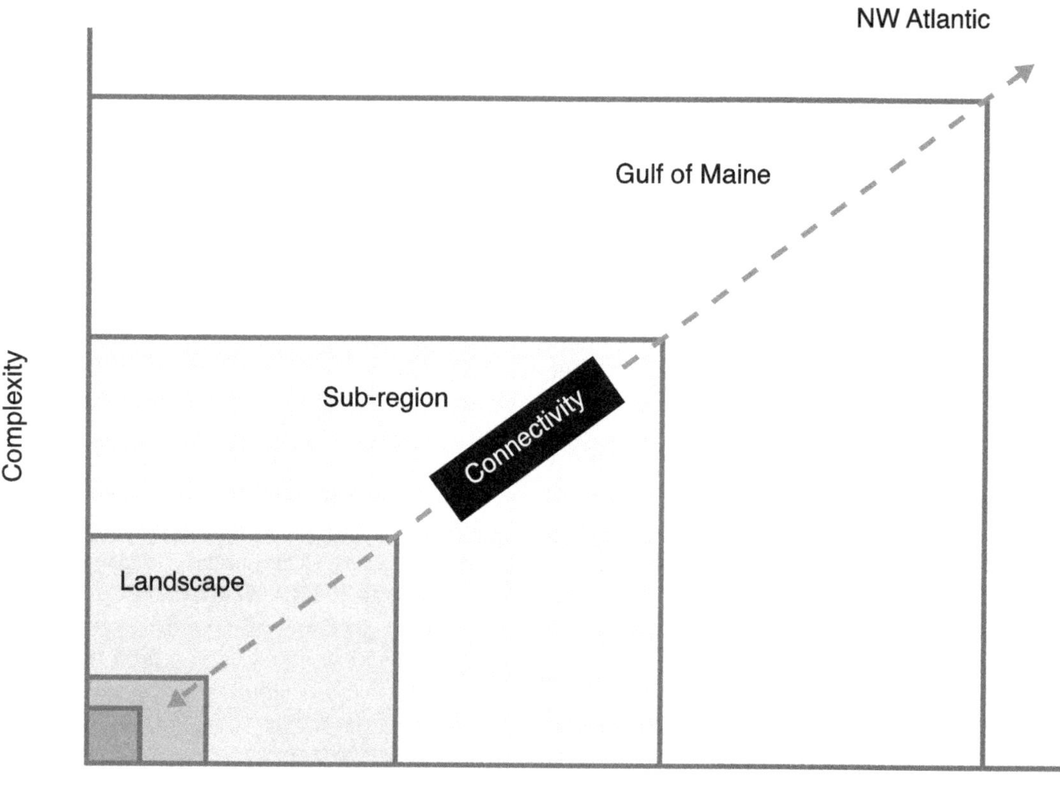

Figure 2. A schematic showing the nested spatial domains of the Gulf of Maine ecosystem and connectivity (dashed line) between them resulting from both biological processes (such as larval dispersal, adult migrations, and benthic-pelagic coupling) and ocean processes (such as transport and mixing). The smallest box represents habitat types, and the slightly larger one represents species interactions that extend beyond habitat boundaries. Landscapes are extensive areas of a general habitat type or mixtures of habitats that are considered together for geographical, ecological, research, or management purposes (for example, Stellwagen Bank National Marine Sanctuary). A subregion may consist of many habitats and landscape types that are grouped together because of geography, depth range, ocean conditions, or other geophysical reasons. The Bay of Fundy, the Massachusetts Bay/Cape Cod Bay system, and the Maine coastal shelf (from shore to 100 m) are examples. This type of representation of spatial domains can help identify needs and opportunities for linking fundamental research on biodiversity with the application of that knowledge in ecosystem-based management.

actions between them, both scale-dependent and cross-scale (Guichard and Peterson 2009).

The Gulf of Maine ecosystem can be considered as a nested set of spatial domains ranging from fine-scale habitat features, through interconnected landscapes, and several subregions to a coastal shelf ecoregion (Gulf of Maine) nested within, and influenced by, the northwest Atlantic (Figure 2). Ecoregions in this context, although they are defined biogeographically and with reference to their physical oceanographic setting, are principally conservation planning units, constructed to facilitate broad assessments of continental and global biodiversity patterns (Spalding et al. 2007; Wilkinson et al. 2009). Technically, GoMA combines two ecoregions (Gulf of Maine/Bay of Fundy and Scotian Shelf) because of their essential connectedness. Oceanographically modulated processes, such as bentho-pelagic coupling, transport and mixing, and animal migrations, promote connectivity across all of these spatial domains and should represent important considerations for spatial management under EBM. Temporal variability, including year-to-year variations (natural variability and human-use activity) and secular change (climate and community trajectories), also influence ocean management considerations. Although discovery-based research may be directed across the full range of spatial domains, and may consider temporal variability and secular change, programs for routine monitoring and spatial planning largely remain focused on periodic assessments and decision making above a certain minimum domain size and typically do not incorporate detailed considerations of system complexity. An understanding of this complexity, however, provides the necessary insights to understand and predict change and to help define precautionary limits to human activities that influence ecosystem function. In this regard, in particular, an understanding of how ecosystem function relates to ecosystem services (supporting, regulating, provisioning, cultural) will be an important part of implementing EBM.

Promising research developments underway within the Gulf of Maine research community bring the prospect of mapping organism distribution and community structure (whether by acoustics, optical imagery, or combined systems) across a broad range of scales, from tens to hundreds of kilometers in extent, preserving resolution (fine ecological grain), and thus permitting spatially explicit structural diversity measures (for example, as demonstrated by Makris et al. 2006, 2009; York 2009). As new types of exploration, inventory, and assessment tools become more widely available (Hofmann and Gaines 2008; Grabowski and Hart 2012), there will be a significant surge in the demand for and availability of spatially explicit biodiversity metrics from which to assess different marine spatial planning options, and this is best organized within a spatial context.

Support for developing a more complete conceptual model of the spatial scaling of regional biodiversity stems from emerging considerations of cross-scale interactions in ecological systems. Guichard and Peterson (2009) explored spatial modeling approaches that recognize the scale-dependence of both ecological and governance processes, and they argue for recognition of cross-scale dynamics phenomena as fundamental characteristics of nonequilibrium ecological systems. Due to feedbacks between local ecological and governance processes and large-scale ecosystem fluctuations, they caution that it may be difficult to achieve management success strictly by managing for spatial pattern (such as with networks of protected areas) or just by managing processes (such as by changes in fishing effort). Rather, cross-scale interactions between pattern and process will require the development of EBM approaches that integrate multiple scales and recognize that small perturbations can potentially have long-term and disproportionate impacts on social-ecological systems (Guichard and Peterson 2009; see also Levin and Lubchenco 2008 for other considerations of resilience and robustness in marine ecosystems). There remains a need for further development of a comprehensive, conceptual structure for spatial analysis that can adequately capture multiscale attributes of biodiversity organization and potential cross-scale interaction.

Discussion

Biodiversity operates dynamically across multiple scales. Understanding this and managing for the conservation of this dynamic process, is extremely complex, yet cannot be delayed. Despite the long history of scientific research and management of human usage within the Gulf of Maine area, we are still a considerable way from fully implementing biodiversity considerations within EBM. A deeper understanding of biodiversity organization within this regional ecosystem and an understanding of the limits to which we can characterize biodiversity, monitor for change, and manage human use to conserve key elements should substantially help to enable EBM. Conserving biodiversity to conserve ecosystem function and adaptability requires compelling arguments because it involves trade-offs against other human uses.

A barrier to moving forward has been a conceptual disconnect or lack of a common meeting ground between the broader academic science community and applied science/ocean management. Marine scientists are interested in furthering their understanding of the composition, structure, and function of marine biodiversity within marine ecosystems across all spatial domains.

To responsibly inform ocean management decision making, applied scientists must necessarily deal with development of specific policy-relevant science advice that incorporates risk and uncertainty projections. Such advice, when articulated towards tactical measures, progresses our strategies to achieve EBM objectives (O'Boyle and Jamieson 2006; O'Boyle and Worcester 2009) but largely restricts the scope of enquiry to development of tangible metrics for certain monitored or managed species (Figure 1) over a partial set of spatial domains that ultimately are referenced to ocean management administrative boundaries (Figure 2). The additional constraint to move incrementally toward EBM while servicing existing management approaches adds pressure on the time scale for response and may reinforce this partial or skewed representation of marine ecosystem properties. Rather than circumscribing the bounds for EBM consideration, these more technical approaches for articulating science advice for important societal goals, such as sustainable fisheries, should remain more clearly nested within an overarching conceptual framework.

This disconnect might simply stem from differing interpretations of two ubiquitous words, "research" and "monitoring," within both the scientific and the management/stakeholder communities involved with development of EBM. When saying "biodiversity research to support EBM," it is quite easy for some to conclude that the studies are directed only towards a long-term aspiration to understand how the system works and therefore unlikely to yield near-term benefits. Conversely, if "biodiversity monitoring to support EBM" is used, then part of the community will understand the need but will not consider it likely to lead to system understanding because it is focused on observable properties and not on drivers. Importantly, while current monitoring approaches may be restricted to a specific range of interest, new science insights and innovations may yield new applied approaches. Fully inclusive frameworks are therefore of value to the overall community in defining current limits for the application of basic research and can thus foster new targeted research to guide or provide solutions.

The discovery imperative "the closer we look, the more we see" fuels curiosity-driven research, and we can expect continuing advances in ecological understanding, driven both from new scientific approaches and integrative concepts as well as enabling technology. As Noss (1990) and Cogan and Noji (2007) noted for biodiversity monitoring frameworks in general, and Cogan et al. (2009) articulated for marine habitat mapping, conceptual frameworks should be robust enough to adapt to technology innovation, rather than the new technology driving changes in core classification approaches. A fully inclusive conceptual framework can also address Rosenberg et al.'s (2009) recommendation for a participatory, proactive, and transparent EBM learning process that can adapt to changing knowledge, environmental conditions, and improved technologies.

While not exhaustive, our list of policy and ocean management linkages (Table 1) is intended to show that biodiversity conservation is often targeted at specific levels within the overall complex that forms regional biodiversity. Options for management (and system responses) may be constrained by higher hierarchical levels as noted by Noss (1990) or have either direct or indirect consequences at different levels through cross-scale interactions (Guichard and Peterson 2009). A recent comprehensive treatment of current international and national EBM programs also underscores the multiscale nature of management systems (Rosenberg et al. 2009).

The simple term "biodiversity conservation" is often invoked as an essential consideration in managing marine ecosystems, but its vagueness or generality has hampered full consideration of biodiversity-related conservation objectives within science-management forums or when communicating with stakeholders and the public. Deconstructing biodiversity into three principal categories, compositional, structural, and functional, each containing a hierarchical set of elements relevant to biodiversity organization within the system, can help to conceptualize how different projects in different systems have addressed biodiversity and to assess broad areas of knowledge and gaps. In our recent comparison of four regional ecosystem research projects (Ellis et al. 2011), we used a tabular representation of these biodiversity hierarchies to assess relative emphasis within each science project. We did not attempt a formal comparison of relative scientific investments across projects and systems, but this could be undertaken in future research program design and assessment. The biodiversity framework presented here could also be developed more fully to address social and economic dimensions of ecosystem science program evolution. Similarly, other overviews could be constructed of the relative breadth and depth of knowledge across composition, structure, and function for different ecosystem "compartments," both within and across ecoregions.

Building a general approach to capture how various studies have focused on different aspects of biodiversity organization will contribute to ongoing meta-analyses of worldwide experience in EBM-related research. This will also help guide interdisciplinary integration of different communities of practice, identify management objectives for the ecosystem, including natural and human goals, and ensure that the governance structure matches with the scale over which ecosystem elements are measured and managed (identified as promising next steps for EBM by Ruckelshaus et al. 2008). In presenting this expanded framework of biodiversity considerations, our approach echoes Cogan et al. (2009) in their own framing of marine habitat mapping within an EBM context in which they called for clarification on the context, opportunities, constraints, and contributions of specific components of EBM. It is seldom possible, beyond general principles of ecology and evolution, to show how conserving all biodiversity relates to human well-being outside of cultural, aesthetic, and ethical arguments. Whenever this can be articulated in practical terms from research results, it will benefit stakeholder understanding and public support for research and the implementation of EBM.

Conclusions

1. Hierarchical or partly hierarchical ways of representing our knowledge and investigations about biodiversity, ecosystem structure, and dynamics contribute to a shared understanding of basic principles and options. We advocate building on them. Feedback loops, possible cross-scale interactions, and semantics mean that allocating particular attributes or concepts to a specific level of a hierarchy will likely always represent a compromise. Nonetheless, the overarching biodiversity framework we have presented can explicitly accommodate biological size spectra, including both known and presently unknown biodiversity, along with compositional, structural, and functional biodiversity elements that influence the expression of biodiversity with the system, within an organized spatial context. The framework is thus quite adaptable to evolving needs. Recent efforts under the GoMA program have shown that even our knowledge of compositional elements of biodiversity for the Gulf of Maine ecosystem remains incomplete (the biological size spectrum being illustrative of this), despite many excellent but isolated pieces of work (the focused way that science generally operates). As we remain in the discovery phase for biodiversity characterization within the Gulf of Maine system, ongoing support is warranted for taxonomy, field work, modeling, analysis, and open database construction and maintenance so that data are available to all potential users.
2. To gain more of an understanding of how ecosystem function and adaptability may be linked to biodiversity, we need the means to conduct experimental investigations at a range of scales. Consolidating some regional research capacities within "defined ocean spaces" where ecological structure and function can be assessed across different temporal and spatial scales, along with evaluation of comprehensive data integration and modeling techniques, could represent a key step towards testing and implementation of EBM approaches. These defined ocean spaces could benefit from being implemented within legally defined conservation areas, such as marine protected areas and sanctuaries (e.g., Stellwagen Bank, USA), as well as across circumscribed regions within the Gulf of Maine that cover a representative range of physically and biologically defined seascapes and levels of human use.
3. Science progresses through a curiosity-driven, creative process (often individually conceived) with an aspirational goal to build more comprehensive understanding of how the natural world is structured and/or the various ways in which humans intervene and

manipulate it. We feel that the pace of progress towards EBM will be increased by further developing a comprehensive adaptive biodiversity framework that describes how fundamental science and management application come together, that is, a framework that creates new opportunities.

Acknowledgments

We thank the co-leaders of six expert groups convened by GoMA to synthesize current biodiversity knowledge on different ecosystem compartments for stimulating academic discussion and insights on regional biodiversity: Catherine Johnson, Noreen Kelly, Scott Kraus, Peter Larsen, Bill Li, Jeff Runge, Mike Sieracki, and Kent Smedbol. We also recognize the contributions of GoMA team members: Chelsie Archibald, Michelle Greenlaw, Ashley Holmes, Suzy Ryan, and Nick Wolff. For data and discussions, we thank Lou Van Guelpen, Stephen Hale, Mary Kennedy, Tom Noji, Gerhard Pohle, and Tom Trott. Financial support was provided by the Alfred P. Sloan Foundation, Fisheries and Oceans Canada (DFO), and the University of Southern Maine. We also acknowledge reviews by Andrew Cooper, Craig Brown, and two anonymous reviewers that helped improve the manuscript.

References

Amaral-Zettler, L., L. F. Artigas, J. Baross, P. A. Loka Bharathi, A. Boetius, D. Chandramohan, G. Herndl, K. Kogure, P. Neal, C. Pedrós-Alió, A. Ramette, S. Schouten, L. Stal, A. Thessen, J. de Leeuw, and M. Sogin. 2010. A global census of marine microbes. Pages 223–245 *in* A. D. McIntyre, editor. Life in the world's oceans: diversity, distribution and abundance. Wiley-Blackwell Publishing, Oxford, UK.

Bellwood, D. R., T. P. Hughes, and A. S. Hoey. 2006. Sleeping functional group drives coral reef recovery. Current Biology 16:2434–2439.

Buzeta, M-I., and R. Singh. 2008. Identification of ecologically and biologically significant areas in the Bay of Fundy, Gulf of Maine. Volume 1: areas identified for review, and assessment of the Quoddy region. Canadian Technical Report of Fisheries and Aquatic Sciences 2788.

Cardinale, B. J., D. S. Srivastava, J. E. Duffy, J. P. Wright, A. L. Downing, M. Sankaran, and C. Jouseau. 2006. Effects of biodiversity on the functioning of trophic groups and ecosystems. Nature (London) 443:989–992.

Christensen, V., K. A. Aiken, and M. C. Villanueva. 2007. Threats to the ocean: on the role of ecosystem approaches to fisheries. Social Science Information 46:67–86.

Cogan, C. B., and T. T. Noji. 2007. Marine classification, mapping, and biodiversity analysis. Pages 129–139 *in* B. Todd and H. G. Greene, editors. Mapping the seafloor for habitat characterization. Geological Association of Canada, St. John's, Newfoundland, Canada.

Cogan, C. B., B. J. Todd, P. Lawton, T. T. Noji. 2009. The role of marine habitat mapping in ecosystem-based management. ICES Journal of Marine Science 66:2033–2042.

Crawford, J. D., R. Rangeley, J. Smith, S. Clarke Stuart, K. Larade, H. Alidina, M. King, R. Cook, P. Brooks, J. Laughren, and J. C. Roff. 2006. Marine ecosystem conservation for New England and Maritime Canada: a science-based approach to identifying priority areas for conservation. Conservation Law Foundation, Boston and World Wildlife Fund-Canada, Toronto.

Curtin, R., and R. Prellezo. 2010. Understanding marine ecosystem based management: a literature review. Marine Policy 34:821–830.

DeLauer, V., S. Ryan, I. Babb, P. Taylor, and P. DiBona. 2012. Linking science to management and policy through strategic communication. Pages 89–101 *in* R. L. Stephenson, J. H. Annala, J. A. Runge, and M. Hall-Arber, editors. Advancing an ecosystem approach in the Gulf of Maine. American Fisheries Society, Symposium 79, Bethesda, Maryland.

DFO (Fisheries and Oceans Canada). 2004. Identification of ecologically and biologically significant areas. DFO, Canadian Science Advisory Secretariat, Ecosystem Status Report 2004/06, Ottawa.

DFO (Fisheries and Oceans Canada). 2006. National science workshop: development of criteria to identify ecologically and biologically significant species (EBSS). DFO, Canadian Scientific Advisory Secretariat, Proceedings Series 2006/028, Ottawa.

Diamond, A. W. 2012. Managing for migrants: the Gulf of Maine as a global "hotspot" for long-distance migrants. Pages 311–320 *in* R. L. Stephenson, J. H. Annala, J. A. Runge, and M. Hall-Arber, editors. Advancing an ecosystem approach in the Gulf of Maine. American Fisheries Society, Symposium 79, Bethesda, Maryland.

Duffy, J. E. 2009. Why biodiversity is important to the functioning of real-world ecosystems. Frontiers in Ecology and the Environment 7:437–444.

Ecosystem Assessment Program. 2009. Ecosystem status report for the northeast United States continental shelf large marine ecosystem. National Oceanic and Atmospheric Administration, National Marine Fisheries Service, Northeast Fisheries Science Center, Reference Document 09–11, Woods Hole, Massachusetts.

Ellis, S. L., L. S. Incze, P. Lawton, B. R. MacKenzie, H. Ojaveer, C. R. Pitcher, T. C. Shirley, M. Eero, J. W. Tunnell, Jr., P. J . Doherty, and B. M. Zeller. 2011. Four regional marine biodiversity studies: approaches and contributions to ecosystem-based management. PLoS (Public Library of Science) ONE [online serial] 6(4):e18997. DOI: 10.1371/journal.pone.0018997.

Feurt, C. B. 2012. Collaborative learning strategies to overcome barriers to ecosystem management in coastal watersheds of the Gulf of Maine. Pages 73–88 *in* R. L. Stephenson, J. H. Annala, J. A. Runge, and M. Hall-Arber, editors. Advancing an ecosystem approach in the Gulf of Maine. American Fisheries Society, Symposium 79, Bethesda, Maryland.

Findlay, C. S., S. Elgie, B. Giles, and L. Burr. 2009. Species listing under Canada's Species at Risk Act. Conservation Biology 23:1609–1617.

Fogarty, M. J., D. Townsend, and E. Klein. 2012. Advances in understanding ecosystem structure and function in the Gulf of Maine. Pages 261–272 *in* R. L. Stephenson, J. H. Annala, J. A. Runge, and M. Hall-Arber, editors. Advancing an ecosystem approach in the Gulf of Maine. American Fisheries Society, Symposium 79, Bethesda, Maryland.

Foley, M. M., B. S. Halpern, F. Micheli, M. H. Armsby, M. R. Caldwell, C. M. Crain, E. Prahler, N. Rohr, D. Sivas, M. W. Beck, M. H. Carr, L. B. Crowder, J. E. Duffy, S. D. Hacker, K. L. McLeod, S. R. Palumbi, C. H. Peterson, H. M. Regan, M. H. Ruckelshaus, P. A. Sandifer, and R. S. Steneck. 2010. Guiding ecological principles for marine spatial planning. Marine Policy 34:955–966.

Gamfeldt, L., H. Hillebrand, and P. R. Jonsson. 2008. Multiple functions increase the importance of biodiversity for overall ecosystem functioning. Ecology 89:1223–1231.

Grabowski, J. H., and T. Hart. 2012. Seafloor mapping for ecosystem management in the Gulf of Maine. Pages 273–280 *in* R. L. Stephenson, J. H. Annala, J. A. Runge, and M. Hall-Arber, editors. Advancing an ecosystem approach in the Gulf of Maine. American Fisheries Society, Symposium 79, Bethesda, Maryland.

Guichard, F., and G. Peterson. 2009. Ecological cross-scale interactions. Pages 74–91 *in* K. L. McLeod and H. M. Leslie, editors. Ecosystem-based management for the oceans. Island Press, Washington, D.C.

Hale, S. S., and M. Westhead. 2012. Ecosystem services in the Gulf of Maine. Pages 1–8 *in* R. L. Stephenson, J. H. Annala, J. A. Runge, and M. Hall-Arber, editors. Advancing an ecosystem approach in the Gulf of Maine. American Fisheries Society, Symposium 79, Bethesda, Maryland.

Hofmann, G. E., and S. D. Gaines. 2008. New tools to meet new challenges: emerging technologies for managing marine ecosystems for resilience. Bioscience 58:43–51.

Incze, L. S., P. Lawton, S. L. Ellis, and N. H. Wolff. 2010. Biodiversity knowledge and its application in the Gulf of Maine Area. Pages 43–63 *in* A. D. McIntyre, editor. Life in the world's oceans: diversity, distribution and abundance. Blackwell Publishing, Oxford, UK.

Jamieson, G., R. O'Boyle, J. Arbour, D. Cobb, S. Courtenay, R. Gregory, C. Levings, J. Munro, I. Perry, and H. Vandermeulen. 2001. Proceedings of the national workshop on objectives and indicators for ecosystem-based management. Sidney, British Columbia, 27 February-2 March 2001. Fisheries and Oceans Canada, Canadian Science Advisory Secretariat, Proceedings Series 2001/09, Ottawa.

Johnson, C. L., J. A. Runge, K. A. Curtis, E. G. Durbin, J. A. Hare, L. S. Incze, J. S. Link, G. D. Melvin, T. D. O'Brien, and L. Van Guelpen. 2011. Biodiversity and ecosystem function in the Gulf of Maine: pattern and role of zooplankton and pelagic nekton. PLoS (Public Library of Science) ONE [online serial] 6(1):e16491. DOI: 10.1371/journal.pone/0016491.

Johnson, C. L., and J. A. Hare. 2012. Zooplankton monitoring in the Gulf of Maine: past, present, and future. Pages 205–218 *in* R. L. Stephenson, J. H. Annala, J. A. Runge, and M. Hall-Arber, editors. Advancing an ecosystem approach in the Gulf of Maine. American Fisheries Society, Symposium 79, Bethesda, Maryland.

Lester, S. E., K. L. McLeod, H. Tallis, M. Ruckelshaus, B. S. Halpern, P. S. Levin, F. P. Chavez, C. Pomeroy, B. J. McCay, C. Costello, S. D. Gaines, A. J. Mace, J. A. Barth, D. L. Fluharty, and J. K. Parrish. 2010. Science in support of ecosystem-based management for the US west coast and beyond. Biological Conservation 143:576–587.

Levin, S. A., and J. Lubchenco. 2008. Resilience, robustness, and marine ecosystem-based management. Bioscience 58:27–32.

Li, W. K., R. A. Anderson, D. J. Gifford, L. S. Incze, J. L. Martin, C. H. Pilskaln, J. N. Rooney-Varga, M. E. Sieracki, W. H. Wilson, and N. H. Wolff. 2011. Planktonic microbes in the Gulf of Maine area. PLoS (Public Library of Science) ONE [online serial] 6(6):e20981. DOI: 10.1371/journal.pone.0020981.

Loreau, M. 2000. Biodiversity and ecosystem functioning: recent theoretical advances. Oikos 91:3–17.

Makris, N. C., P. Ratilal, S. Jagannathan, Z. Gong, M. Andrews, I. Bertsatos, O. R. Godø, R. W. Nero, and J. M. Jech. 2009. Critical population density triggers rapid formation of vast oceanic fish shoals. Science 323:1734–1737.

Makris, N. C., P. Ratilal, D. T. Symonds, S. Jagannathan, S. Lee, and R. W. Nero. 2006. Fish population and behavior revealed by instantaneous continental shelf-scale imaging. Science 311:660–663.

McArthur, M., B. Brooke, R. Przeslawski, D. Ryan, V. Lucieer, S. Nichol, A. McCallum, C. Mellin, I. Cresswell, and L. Radke. 2010. On the use of abiotic surrogates to describe marine benthic biodiversity. Estuarine Coastal and Shelf Science 88:21–32.

McIntyre, A. D., editor. 2010. Life in the world's oceans: diversity, distribution, and abundance. Blackwell Publishing, Oxford, UK.

McLeod, K. L., and H. M. Leslie, editors. 2009a. Ecosystem-based management for the oceans. Island Press, Washington, D.C.

McLeod, K., and H. M. Leslie. 2009b. State of practice. Pages 314-321 *in* K. L. McLeod and H. M. Leslie, editors. Ecosystem-based management for the oceans. Island Press, Washington, D.C.

McLeod K. L., J. Lubchenco, S. R. Palumbi, and A. A. Rosenberg. 2005. Scientific consensus statement on marine ecosystem-based management. Communication Partnership for Science and the Sea. Available: www.compassonline.org/sites/all/files/document_files/EBM_Consensus_Statement_v12.pdf (February 2012).

Millennium Ecosystem Assessment. 2005. Millennium ecosystem assessment: synthesis report. Island Press, Washington, D.C.

Murawski, S. A. 2007. Ten myths concerning ecosystem approaches to marine resource management. Marine Policy 31:681–690.

Naeem, S., and J. P. Wright. 2003. Disentangling biodiversity effects on ecosystem functioning: Deriving solutions to a seemingly insurmountable problem. Ecology Letters 6:567–579.

Nee, S. 2004. More than meets the eye: Earth's real biodiversity is invisible, whether we like it or not. Nature (London) 429:804–805.

Noss, R. F. 1990. Indicators for monitoring biodiversity: a hierarchical approach. Conservation Biology 4:355–364.

O'Boyle, R., and G. Jamieson. 2006. Observations on implementation of ecosystem-based management: experiences on Canada's east and west coasts. Fisheries Research 79:1–12.

O'Boyle, R., and T. Worcester. 2009. Eastern Scotian Shelf, Canada. Pages 253–267 *in* K. L. McLeod and H. M. Leslie, editors. Ecosystem-based management for the oceans. Island Press, Washington, D.C.

O'Dor, R. K., and V. A. Gallardo. 2005. How to cen sus marine life: ocean realm field projects. Scientia Marina 69(Supplement 1):181–199.

Palumbi, S. R., P. A. Sandifer, J. D Allan, M. W. Beck, D. G. Fautin, M. J. Fogarty, B. S. Halpern, L. S. Incze, J. Leong, E. Norse, J. J. Stachowicz, and D. H. Wall. 2009. Managing for ocean biodiversity to sustain marine ecosystem services. Frontiers in Ecology and the Environment 7:204–211.

Pershing, A. J., C. H. Greene, J. W. Jossi, L. O'Brien, J. K.T. Brodziak, and B. A. Bailey. 2005. Interdecadal variability in the Gulf of Maine zooplankton community with potential impacts on fish recruitment. ICES Journal of Marine Science 62:1511–1523.

Rosenberg, A. A., and K. L. McLeod. 2005. Implementing ecosystem-based approaches to management for the conservation of ecosystem services. Marine Ecology Progress Series 300:270–274.

Rosenberg, A. A., M. L. Mooney-Seus, I. Kiessling, C. B. Morgensen, R. O'Boyle, and J. Peacey. 2009. Lessons from national-level implementation across the world. Pages 294–313 *in* K. L. McLeod and H. M. Leslie, editors. Ecosystem-based management for the oceans. Island Press, Washington, D.C.

Ruckelshaus, M., T. Klinger, N. Knowlton, and D. P. DeMaster. 2008. Marine ecosystem-based management in practice: scientific and governance challenges. Bioscience 58:53–63.

Runge, J. A., and R. J. Jones. 2012. Results of a collaborative project to observe coastal zooplankton and ichthyoplankton abundance and diversity in the western Gulf of Maine: 2003–2008. Pages 345–359 *in* R. L. Stephenson, J. H. Annala, J. A. Runge, and M. Hall-Arber, editors. Advancing an ecosystem approach in the Gulf of Maine. American Fisheries Society, Symposium 79, Bethesda, Maryland.

Sala, O. E., F. S. Chapin, J. J. Armesto, E. Berlow, J. Bloomfield, R. Dirzo, E. Huber-Sanwald, L. F. Huenneke, R. B. Jackson, A. Kinzig, R. Leemans, D. M. Lodge, H. A. Mooney, M. Oesterheld, N. LeRoy Poff, M. T. Sykes, B. H. Walker, M. Walker, and D. H. Wall. 2000. Global biodiversity scenarios for the year 2100. Science 287:1770–1774.

Savolainen, V., R. S. Cowan, A. P. Vogler, G. K. Roderick, and R. Lane. 2005. Towards writing the encyclopaedia of life: an introduction to DNA barcoding. Philosophical Transactions of the Royal Society B 360:1805–1811.

Schwartz, M. W. 2008. The performance of the Endangered Species Act. Annual Review of Ecology Evolution, and Systematics 39:279–299.

Shackeroff, J. M., E. Hazen, and L. B. Crowder. 2009. The oceans as peopled landscapes. Pages 33–54 *in* K. L. McLeod and H. M. Leslie, editors. Ecosystem-based management for the oceans. Island Press, Washington, D.C.

Sinclair, M., and G. Valdimarsson, editors. 2003. Responsible fisheries in the marine ecosystem. Food and Agriculture Organization of the United Nations, Rome and CABI Publishing, Cambridge, Massachusetts.

Snelgrove, P. V. R. 2010. Discoveries of the Census of Marine Life: making ocean life count. Cambridge University Press, Cambridge, UK.

Sogin, M. L., H. G. Morrison, J. A. Huber, D. M. Welch, S. M. Huse, P. R. Neal, J. M. Arrieta, G. J. Herndl. 2006. Microbial diversity in the deep sea and the underexplored "rare biosphere." Proceedings of the National Academy of Sciences of the United States of America 103:12115–12120.

Spalding, M. D., H. E. Fox, G. R. Allen, N. Davidson, Z. A. Ferdaña, M. Finlayson, B. S. Halpern, M. A. Jorge, A. Lombana, S. A. Lourie, K. D. Martin, E. McManus, J. Molnar, C. A. Recchia, and J. Robertson. 2007. Marine ecoregions of the world: a bioregionalization of coastal and shelf areas. Bioscience 57:573–582.

Theroux, R. B., and R. L. Wigley. 1998. Quantitative composition and distribution of the macrobenthic invertebrate fauna of the continental shelf ecosystems of the northeastern United States. NOAA Technical Report NMFS 140.

Wallace, G. T., and E. F. Braasch, editors. 1997. Proceedings of the Gulf of Maine ecosystem dynamics scientific symposium and workshop. Regional Association for Research on the Gulf of Maine, RARGOM Report 97-1, Hanover, New Hampshire.

Whitman, W. B., D. C. Coleman, and W. J. Wiebe. 1998. Prokaryotes: the unseen majority. Proceedings of the National Academy of Sciences of the United States of America 95:6578–6583.

Wilkinson, T., E. Wiken, J. Bezaury-Creel, T. Hourigan, T. Agardy, H. Herrmann, L. Janishevski, C. Madden, L. Morgan, and M. Padilla. 2009. Marine ecoregions of North America. Commission for Environmental Cooperation, Montreal.

Wilson E. O. 1992. The diversity of life. Harvard University Press, Cambridge, Massachusetts.

Yachi, S., and M. Loreau. 1999. Biodiversity and ecosystem productivity in a fluctuating environment: the insurance hypothesis. Proceedings of the National Academy of Sciences of the United States of America 96:1463–1468.

Yarincik, K., and R. K. O'Dor. 2005. The census of marine life: goals, scope and strategy. Scientia Marina 69(Supplement 1):201–208.

York, A. D. 2009. HabCam sheds light on invading tunicate: habitat mapping camera system monitors ecosystem change over time and space. Sea Technology 50(8):41–46.

American Fisheries Society Symposium 79:393–407, 2012

Thermal Phenological Factors Affecting the Survival of Atlantic Salmon in the Gulf of Maine

Kevin D. Friedland*
National Marine Fisheries Service
28 Tarzwell Drive, Narragansett, Rhode Island 02882, USA

James P. Manning and Jason S. Link
Northeast Fisheries Science Center
166 Water Street, Wood Hole, Massachusetts 02543, USA.

Abstract.—The marine survival of Atlantic salmon *Salmo salar* stocks in the Gulf of Maine area has declined as sea surface temperature in the coastal ocean has increased; in part, this change in recruitment can be attributed to a phenological mismatch between the timing of smolt migrations and initial conditions postsmolts find during their early marine phase. Salmon juvenile migrations to the ocean are released by photoperiod and spring transitional freshwater temperatures, neither of which have changed significantly in recent decades, thus actuating the migrations at nearly the same time each year. However, ocean water temperatures have increased during the spring transition period, suggesting that smolts have entered the ocean under varying physical and biological conditions. The phenological effect observed in the Gulf of Maine is consistent with the relationship observed for the North American stock complex. In light of recent findings related to the growth of postsmolts, the contrast in recruitment for North American salmon, and Gulf of Maine stocks in particular, appears to be the result of mortality that occurs during the first months at sea. This mortality covaries with the thermal changes in the coastal ocean, which we suspect is associated with variation in the predator field.

Introduction

The recruitment of Atlantic salmon *Salmo salar* appears to be governed by climatological drivers affecting both physical and biological pressures on postsmolt salmon during their first year at sea. Remarkably, these basin-scale drivers are having a differential effect on the stocks from Europe and North America. The recruitment of salmon in Europe appears to be controlled by factors that affect the growth of postsmolts during their first summer at sea. There has been a shift in thermal conditions in the northeast Atlantic (Richardson and Schoeman 2004), which has actuated a change in the structure of the food web (Beaugrand and Reid 2003) in the area of the postsmolt nursery (Holm et al. 2000). A consequence of this shift has been a concomitant shift in postsmolt growth, which correlates with the resultant survival of salmon cohorts (Friedland et al. 2009a). It is hypothesized that postsmolt growth mediates survival by the functional relationship between postsmolts and their predators. Increased growth rate, specifically growth during the summer months of the postsmolt year, is associated with increased survivorship (Peyronnet et al. 2007; McCarthy et al. 2008; Friedland et al. 2009b). Central to this hypothesis is the notion that growth rate controls the time salmon postsmolts remain at sizes vulnerable to predators (Sogard 1997).

Recruitment of North American salmon stocks is differentiated from European stocks by the role of postsmolt growth and climate. First,

* Corresponding author: kevin.friedland@noaa.gov

postsmolt growth and survival in European Atlantic salmon is positively correlated, whereas there is a lack of such a relationship for North American salmon. The data to support this assertion comes from hatchery-dependent and wild populations. An analysis of postsmolt growth for one and two seawinter returns to the Penobscot River, a hatchery-dependent stock from Maine, did not detect any significant relationships between return rate and seasonal indices of postsmolt growth (Friedland and Haas 1996). In a complementary analysis, Friedland et al. (1996) compared the postgrowth for the hatchery-dependent Penobscot and Connecticut River stocks, which have had much lower marine survival rates. The postsmolt growth of the Penobscot fish tended to be higher than for the Connecticut fish of the same smolt year-class, suggesting that the growth differences were responsible for the differences in survival rates. Aside from examining the comparable performance of the two stocks, the data also showed that during a period of decreasing marine survival, postsmolt growth had increased in both stocks. The evidence from wild populations comes from river systems in both Canada and the United States. Postsmolt growth for the Margaree stock from Nova Scotia increased during a period that the North American stock complex was declining in abundance and experiencing lower marine survival rates (Friedland et al. 2005). The Margaree time series of postsmolt growth data were relatively short; hence, it does not cover the full excursion of marine survival conditions seen in North America. A longer time series of the postsmolt growth was analyzed for returns to the Miramichi River, the largest salmon-producing river in North America. In this stock, postsmolt growth was negatively correlated with survival during the second month the fish were at sea (Friedland et al. 2009b). This correlation is an obverse pattern to that for European stocks, but also suggests that density-dependent growth occurs that is a reflection of the rate of early marine mortality that defines the size of the year-class. The data for two coastal rivers from Maine, the Machias and the Narraguagus, provides growth data back to 1946. The postsmolt growth was found to be uncorrelated to local indicators of marine survival as well as abundance measures for the North American stock complex (Hogan and Friedland 2010). The data from both wild and hatchery stocks are consistent with findings relating climate variability during the months after smolts migrate to sea and recruitment variability, suggesting that the mortality is a short-duration event independent of growth conditions.

The role of climate variation on the recruitment of North American salmon is differentiated from the pattern seen for European salmon by the seasonality of the effect and the added issue of the migration timing of smolts. A negative correlation between spring sea surface temperatures (SST) and recruitment emerges using two study approaches. The first employs long-term data sets based on integrated catch data and the second employs contemporary high resolution data that differentiates the effects on specific cohorts (Friedland et al. 2003a, 2003b). Despite these differences in approach, the results were the same, showing that warming conditions in the coastal ocean during spring are associated with poor survival; however, they both suffered from the limitation that the analyses were conditioned on trends for the North American stock complex and thus emphasize the effect of climate in the Gulf of St. Lawrence. Furthermore, Friedland et al. (2003b) raised the issue of a phenological mismatch between smolt migration timing, which is triggered by photoperiod and temperature in freshwater habitats (Whalen et al. 1999), and the conditions postsmolt experience during the first weeks to months at sea.

In this paper, we seek to examine the applicability of recruitment hypotheses developed for the North American stock complex to stocks local to the Gulf of Maine. We start by re-examining the relationship between sea surface temperature and stock complex recruitment in finer spatial and temporal resolution than has been done before. To test effects specific to the Gulf of Maine, we developed a recruitment indicator more reflective of local conditions and evaluated the relationship of this indicator to spring SST conditions and the effect of smolt migration timing in relation to the timing of spring SST transitions. Finally, we examined basin-wide trends in SST range as an indicator of seasonal temperature dynamics and relate these findings to the conditions in the Gulf of Maine.

Methods

Recruitment Benchmarks for North America and the Gulf of Maine

The recruitment patterns for Atlantic salmon populations in North America and the Gulf of Maine were represented by complementary indices. The North American stock complex was represented by a catch-based index that has been used in previous studies of North American salmon recruitment (Friedland et al. 2003b) and was recently updated for contemporary years (Friedland et al. 2009b). It is based on salmon landings in North America and in Greenland of fish that can be attributed to North American origin. Because the catch index is not age-disaggregated, the data are relevant to recruitment at 1- and 2-year lags to match the smolt years for both one seawinter (1SW) and two seawinter (2SW) salmon. We analyzed the 2-year lag data because the recruitment contrast of the complex is associated with changes in the 2SW component.

The recruitment pattern in the Gulf of Maine was represented by time series of return rates for a hatchery-dependent stock that captures the regional variation in survivorship. The Penobscot River has a hatchery-dependent population of Atlantic salmon that provides an indication of marine survival conditions for salmon populations located in the Gulf of Maine area. The Penobscot River is located on the Maine coast and empties into the Penobscot Bay, a large coastal embayment that provides multiple migration corridors for postsmolts to enter the Gulf of Maine. Penobscot smolts were released during the month of May, which matches the migration period of local stocks (Whalen et al. 1999). Returns rates of the Penobscot fish were based on the recapture of adults as either 1SW or 2SW fish from 1- and 2-year smolt releases in the Penobscot River during the years 1970–2004. Returns are number of returning fish to the river partitioned by sea age and primarily recaptured at the trapping facility at the Veazie Dam located within ~70 km of Penobscot Bay. There are multiple factors that affect the interpretation of return rate data, including procedures related to the rearing and release of smolts, variation in fishery removals at sea, and variation in trapping conditions when the fish return. The return rate data for the Penobscot has been scrutinized in past studies, and two important points are evident. First, when compared to return rates for hatchery and wild populations to the north and south, Penobscot return rates were coherent with the return rates of other systems (Friedland et al. 1993, 1996). This indicates that the return rate reflects broadscale forcing affecting all the stocks. Second, the Penobscot return rate data are highly correlated with the estimates of prefishery abundance for the North American stock complex, which explicitly takes into account fishing mortality (Chaput et al. 2005). The Pearson product-moment correlations between Penobscot 1SW return rate and 1SW prefishery abundance, 2SW return rate and 2SW prefishery abundance, and combined 1SW and 2SW return rate and combined prefishery abundance are 0.55, 0.84, and 0.78, respectively. All three correlations were significant ($p < 0.01$), with the probability adjusted for time series autocorrelation using the adjustment to the test procedure where an "effective" number of degrees of freedom are applied (Pyper and Peterman 1998). All of the probabilities of correlations in this study using time series data are adjusted in this fashion.

Sea Surface Temperature Trends

Trends in sea surface temperature in the northwest Atlantic and Gulf of Maine areas were characterized using the extended reconstructed SST data set (ERSST [Extended Reconstructed Sea Surface Temperature], version 3). This data set is based on the SST compilation of ICOADS (the International Comprehensive Ocean-Atmosphere Data Set) and represents interpolation procedures that reconstruct SST fields in regions with sparse data (Smith et al. 2008). The spatial resolution of the data is 2° longitude by 2° latitude bins and the temporal resolution is monthly; we extracted two subsets of the data. To characterize SST conditions of the northwest Atlantic relevant to the stock complex, we examined data from the area 39–61°N and 49–71°W over the period 1908–2003, thus matching the spatial distribution of all stocks in the stock complex. We restricted the annual temporal coverage of the data to the period of May through December to represent the duration of the postsmolt year. The monthly SST fields were correlated to the catch index of

North American stock complex to yield correlation response surfaces for the months of May to December. To characterize SST conditions of the Gulf of Maine relevant to the Penobscot survival index, we examined data from the area 41–45°N and 67–71°W over the period 1970–2004. We restricted this analysis to the period of May through July to match the temporal period that Penobscot juveniles are in the Gulf of Maine.

Interaction of Air and Sea Temperatures

To test the hypothesis that smolt migrations are synchronized to deliver postsmolts into specific ocean conditions in the Gulf of Maine, we examined the synchronization of the date of arrival of target temperatures in the vicinity of the freshwater rearing areas to the date of arrival for target temperatures in the coastal ocean. The annual arrival dates of target temperatures in the rearing areas were based on surface air temperatures from the NCEP (National Centers for Environmental Protection)/NCAR (National Center for Atmospheric Research) reanalysis project data set (Kalnay et al. 1996), assuming a relationship between air and stream temperatures in the region (Mohseni and Stefan 1999; Morrill et al. 2005). Using monthly mean surface air temperatures during March through June for the area bounded approximately by 44–46°N and 67–71°W, we computed a regression and estimated arrival date (Julian day) for a range of target temperatures. The range of target temperatures was designed to bracket the temperature commonly observed to initiate smolt migration for stocks in the region, that being 10°C (Whalen et al. 1999). A similar procedure was used to estimate the annual arrival date of coastal SSTs; an annual regression was estimated using the ERSST data over April through July for the area bounded by 43–45°N and 67–71°W. The date of arrival of target surface air temperature in the vicinity of the rearing areas minus the date of arrival of SST was used as an index of migration synchronization. The difference in dates was correlated to the return rate of 2SW fish and plotted as a response surface.

Temperature Range in the North Atlantic

The dynamic nature of ocean climate conditions for regions of the North Atlantic was characterized with sea surface temperature data. Using the ERSST data, spring and annual temperature range was characterized. Spring temperature range was determined for each grid location by differencing the mean SST for May and July. The annual range was determined by differencing the data for the months of February and August.

Results

Recruitment Benchmarks for North America and the Gulf of Maine

The catch index of the North American stock complex reveals a pattern of multidecadal recruitment variability. Recruitment to the complex was at localized maxima during the 1930s and 1970s, preceded and followed by minima resulting in a declining trend over the past three decades (Figure 1a). The periodicity of the oscillating pattern of recruitment is approximately 48 years, but because the pattern sampling is below the Nyquist rate, that estimate of periodicity must be used with caution. There is also a conservative pattern of decline in recruitment over the past century that is associated with multiple factors, including significant insult to salmon habitat.

The adult returns from smolt releases in the Penobscot River are dominated by the return of 2SW salmon. The return rate for 1SW salmon averaged 0.67 fish per 1,000 smolts released over the study period compared to an average return rate of 3.88 fish for 2SW salmon (Figure 1b). The return rate for the two sea ages shows distinct patterns of change over time. The return of 1SW salmon was virtually unchanged during the study period, whereas the return of 2SW fish showed approximately a fourfold decline over the past three decades. The time series of 1SW and 2SW return rates are uncorrelated. The return rates for 1SW and 2SW salmon were used in subsequent analyses to test the potential effect of environmental and biological factors on return rate; because the Penobscot population is dominated by the change in return of 2SW salmon, some tests were limited to correlation to the 2SW return rate only.

Sea Surface Temperature Trends

Monthly correlation response surfaces representative of the postsmolt year reveal a pattern of

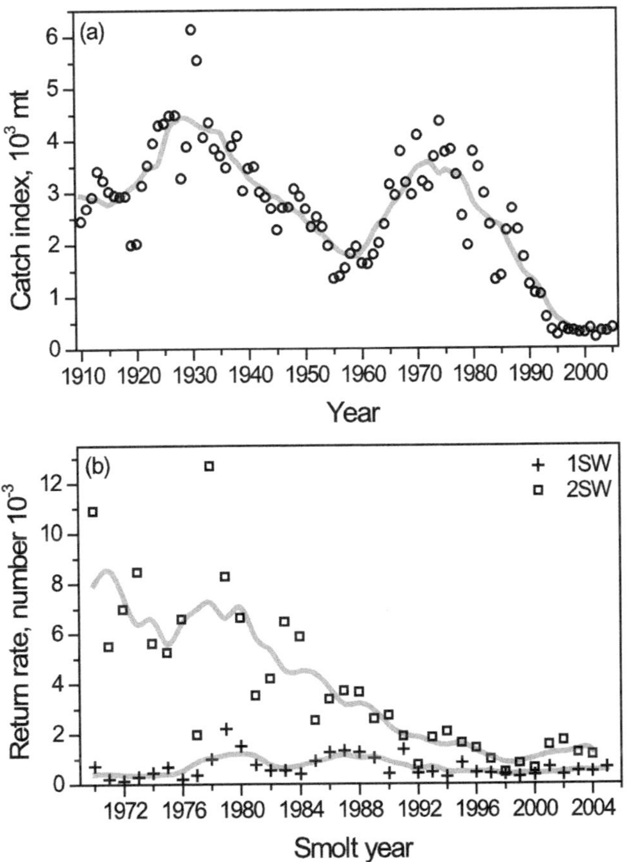

FIGURE 1. Time series of a catch index of North American Atlantic salmon abundance (a) versus the year corresponding to the smolt year for the two seawinter (2SW) component of the catch. Smoothed line is 10-point adjacent averaging. Return rates of one seawinter (1SW) and 2SW returns versus smolt year to the Penobscot River (b). Smoothed lines are 5-point adjacent averaging.

high negative correlative density during the first months at sea followed by diminished correlation thereafter. The correlation response surfaces for May through July show significant negative correlation between SST and the catch index in the coastal ocean waters of the Gulf of Maine, of the Gulf of St. Lawrence, and, to a lesser extent, off Labrador (Figure 2a–c). Oceanic waters of the Labrador Sea, Grand Banks, and Scotian shelf do not show any significant correlation pattern. As the postsmolt year progresses, as shown for the months of August through October, this correlation pattern dissipates (Figure 2d–f). By late fall into early winter, there were remnant patterns of negative correlation in the Gulf of St. Lawrence only (Figure 2g, 2h), but it should be noted that by this time of the year, postsmolts are known to have migrated out of the gulf in response to rapidly cooling conditions (Dutil and Coutu 1988). Of greater importance is the continued absence of any correlative density associated with offshore waters.

The Gulf of Maine has undergone a warming of spring ocean temperatures over the past three decades that is correlated with the return rates of Penobscot salmon. May, June, and July SSTs have increased on the order of 1.1°C, 0.8°C, and 0.5°C, respectively, based on a linear trend over the study period. Several correlates have emerged. The return rate of 2SW fish and the combined sea-age return rate were found to be negatively correlated with May SST in three of the four SST data bins (Table 1). There were no significant correlations associated with June SSTs, but there was nega-

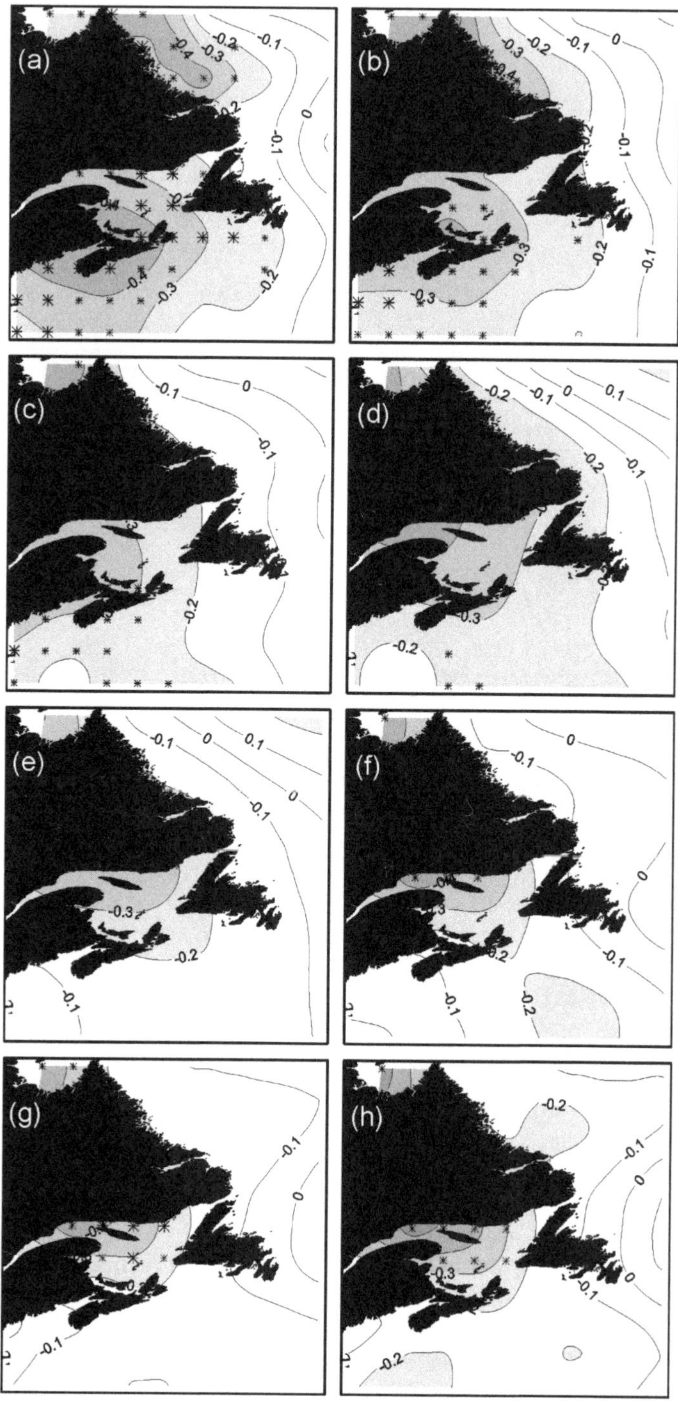

FIGURE 2. Correlation between the catch index of North American Atlantic salmon abundance and smolt-year sea surface temperatures for the months May through December, panels (a) through (h), respectively. Light grey shading marks approximate regions with uncorrected correlations significant at $p = 0.05$, and dark grey represents approximate regions significant at $p = 0.01$. * denotes discrete location where correlation is significance at $p = 0.05$ level corrected for time series autocorrelation, ✱ denotes significance at $p = 0.01$.

TABLE 1. Correlation between sea-age return rates and the sea surface temperature over three months for four 2° square longitude by latitude areas. * denotes significance at $p = 0.05$ level corrected for time series autocorrelation.

Month	Longitude	Latitude	One seawinter R	Two seawinter R
May	–68	42	0.193	–0.416*
May	–68	44	0.207	–0.350
May	–70	42	0.246	–0.426*
May	–70	44	0.214	–0.453*
June	–68	42	–0.115	–0.231
June	–68	44	–0.107	–0.207
June	–70	42	–0.036	–0.153
June	–70	44	–0.032	–0.266
July	–68	42	–0.371*	–0.175
July	–68	44	–0.368*	–0.123
July	–70	42	–0.246	–0.247
July	–70	44	–0.244	–0.255

tive correlation between July SST in two bins and the return rate for 1SW fish. The mean May and July SSTs for the study area show wide interannual variability in the data, but the trends reflect the twofold difference in temperature increase over the 3-month period (Figure 3). For that multi-month period that postsmolts are migrating through the Gulf of Maine, the most demonstrative changes in thermal conditions occurred during May.

Interaction of Air and Sea Temperatures

Smolt migrations appear to be synchronized to deliver postsmolts into specific ocean conditions in the Gulf of Maine. The correlation between the synchronization index (the difference in date of arrival of target air temperatures in the vicinity of the freshwater rearing areas and date of arrival for target sea temperatures in the coastal ocean) and return rate of 2SW salmon was greatest around target temperatures of 10°C (Figure 4). The negative correlation indicates that when the synchrony was poor due to large differences in the two arrival dates, return rate was low; when synchrony was good due to small differences in the arrival dates, return rate was high. Though the strongest correlations had uncorrected probabilities at $p < 0.01$, after the corrections for autocorrelation was applied, only the correlates for target air temperature of 10°C and higher and SST of 11°C and higher remained significant at $p = 0.05$. The dates of arrival of 10°C air should be considered unchanged because the slope of the dates over time was not significant; sea surface temperatures were 6.4 d earlier, respectively, based on linear trends over the study period (Figure 5). This means that the principal cue for the smolt migration appears to have occurred 1–2 d later in the year and the arrival of warmer ocean conditions were advanced by nearly a week.

Temperature Range in the North Atlantic

The spring and annual ranges of SST differ between the eastern and western sides of the Atlantic basin. The eastern portion of the basin has a spring SST range of 4°C or less, with the exception of the North Sea (Figure 6a). The western portion of the basin tends to have a spring range in excess of 8°C from the Middle Atlantic Bight extending north and east to southern Newfoundland, including the Gulf St. Lawrence. This range difference suggests that salmon stocks through most of Europe are not faced with a phenology issue between the timing of smolt migrations and marine conditions postsmolts encounter when they enter the ocean. In contrast, North American salmon stocks must time their smolt migrations to place postsmolts into a rapidly changing set of oceans conditions because spring transitional warming will occur at a faster rate and over a wider range

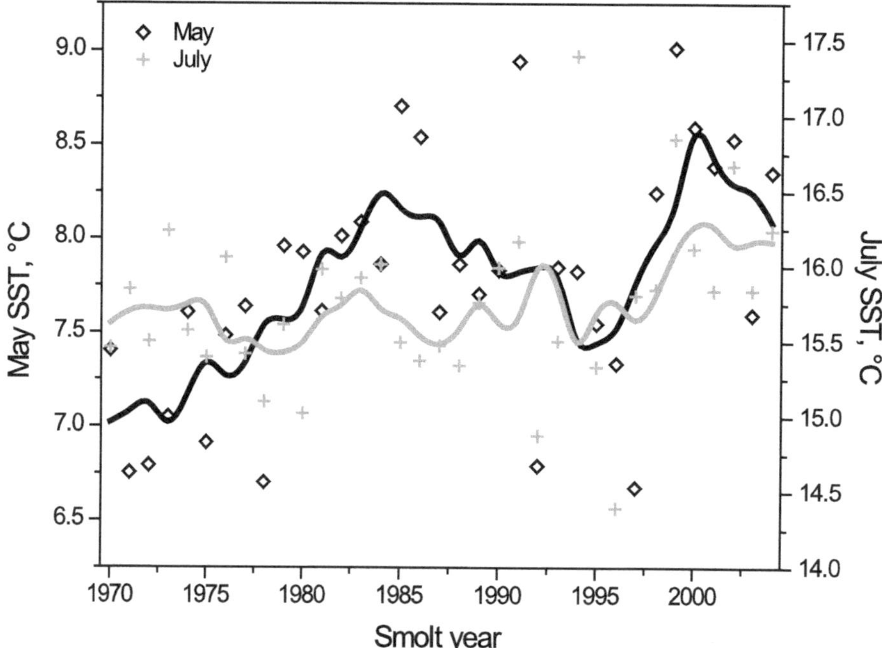

FIGURE 3. Sea surface temperature (SST) in May (area 69–71°W, 41–45°N) and July (area 67–68°W, 41–45°N) versus smolt year. Smoothed lines are 5-point adjacent averaging.

of temperatures. The same trends are evident in the annual SST ranges, showing that the eastern portion of the basin tends to have ranges of 6°C or less and the western portion in excess of 12°C. In addition to the issue presented for migration timing of smolts, the annual data suggest that much of the eastern portion of the basin remains salmon habitat for much of the year, whereas areas like the Gulf of Maine warm to temperatures considered to be outside the thermal habitat of salmon.

Discussion

Rapid spring warming of the coastal ocean in North America is associated with a decline in Atlantic salmon recruitment, both throughout the greater portion of the species range and, in the context of this study, in the Gulf of Maine as well. For analyses of the stock complex, the pivotal region is the Gulf of St. Lawrence because the most productive salmon rivers are in that area. The Gulf of St. Lawrence has shown a pattern of rapid spring warming in recent years, and with recruitment benchmarks conditioned primarily by data for Gulf of St. Lawrence stocks, the added dimension of a phenology mismatch between smolt migration timing and ocean conditions has been observed (Friedland et al. 2003b). This same mechanism appears to be at work in the Gulf of Maine, suggesting that smolts enter a Gulf of Maine ecosystem that has shifted over recent decades from one that supported successful postsmolt migration and survival to one that now causes mortality rates that may not be sustainable. The shift in thermal conditions could be associated with a number of mortality vectors, but a general class of issues would seem illogical in the context of the patterns of postsmolt growth observed for resident populations.

Individuals from hatchery and wild populations of Atlantic salmon in the Gulf of Maine have experienced generally improved postsmolt growth during recent decades of declining survivorship, suggesting that mortality vectors associated with feeding and growth are not at the center of the survival issue. Until we see evidence to the contrary, we suspect that shifting ecosystem productivity is not negatively affecting postsmolts and, instead, may have improved to support the increased growth. Gulf of Maine salmon have

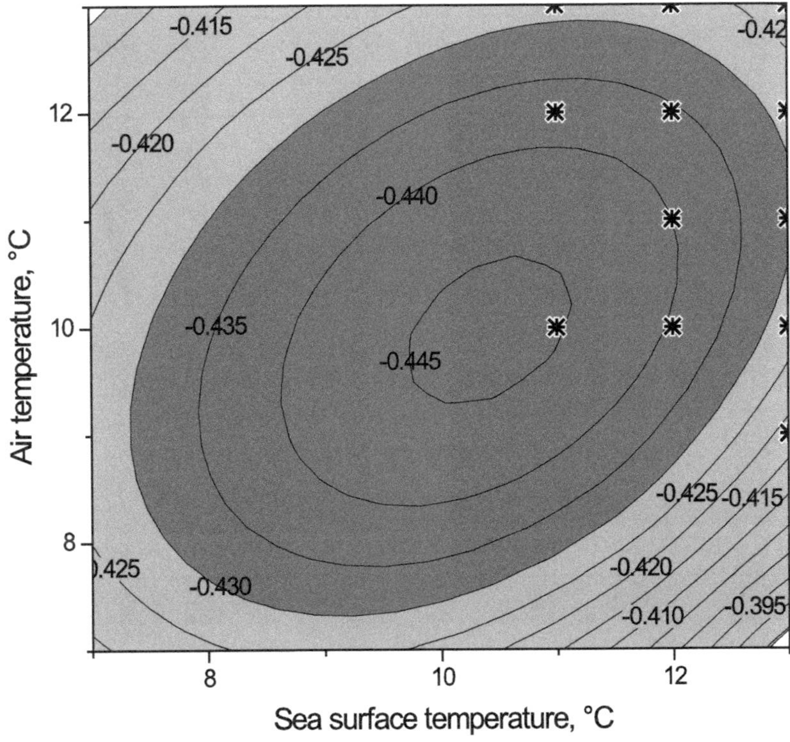

FIGURE 4. Response surface of correlation between Penobscot two seawinter return rate and the difference in arrival dates, in number of days, of air and sea surface temperatures in the Penobscot freshwater rearing area and the coastal Maine area, respectively. * denotes significance at $p = 0.05$ level corrected for time series autocorrelation.

been migrating into an increasingly warmer ocean and thus could be affected by the effects of warmer ocean conditions or by the associated changes in biological communities associated with this climate shift. The latter is straight forward, regardless of their growth response; there simply is a more aggressive predator field present in the gulf associated with the warmer conditions. The former is more complex because it represents a host of possible mortality vectors actuated by the warmer conditions or by some combination of effects, including those associated with freshwater.

Warmer ocean conditions could negatively affect the physiology of postsmolts, making them more susceptible to predation or stimulate a mortality factor independent of predation either associated with the entry into the marine environment or a factor carried with them from freshwater. Increased temperature conditions will increase metabolic demands in juvenile salmon and specifically increase the metabolic costs of swimming (Enders et al. 2005). In theory, postsmolts may have a swimming deficit and suffer higher predation mortality after being rendered less fit to escape predators. It is not known whether postsmolts are maladapted in this way; confirmation would require experimental observations and innovative field observations. Tracking experiments with postsmolts in the region do not suggest that migration rates and swimming speeds were impaired (Lacroix and Knox 2005; Hedger et al. 2008), but there exists no practical standard of comparison, so the point remains speculative. McCormick et al. (2009) highlighted the concern that stressors in freshwater may impair seawater performance and thus increase predation rate because of reduced growth and swimming performance. It is also speculated that these stressors may increase mortality due to disease because stressed fish are more likely to develop infections (Loge et al. 2005). It is not known if freshwater stressors have increased in recent decades, and at least one stressor identi-

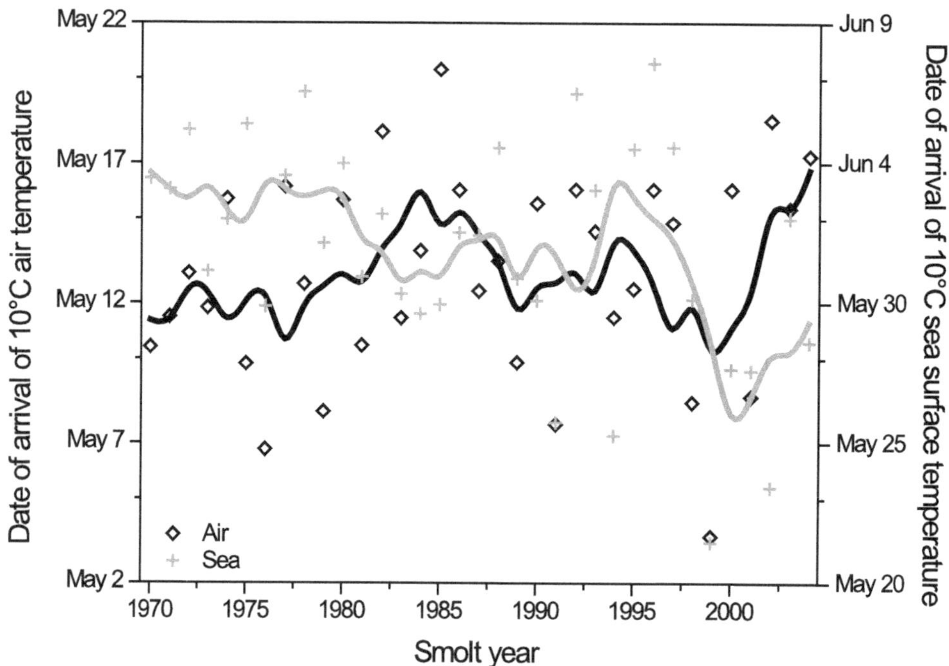

FIGURE 5. Date of arrival of 12°C air and sea surface temperatures in the Penobscot freshwater rearing area and the coastal Maine area, respectively. Smoothed lines are 5-point adjacent averaging.

fied by McCormick et al. (2009), acid deposition, has declined (Kahl et al. 2004); nor is it known whether these mortality vectors are affected by elevated seawater temperatures.

There is evidence that the biological community in the Gulf of Maine has changed in recent decades, suggesting the species composition and abundance of potential predators on postsmolts have also changed. A multidecadal trend of warming has resulted in a number of distribution shifts of pelagic piscivores affecting the nature of the predator field in the migration corridors for postsmolts in the Gulf of Maine (Nye et al. 2009). However, some candidate predators have taken on diminished ecological roles due to changes in their abundance caused by overexploitation (Auster and Link 2009; Link et al. 2009). The predation problem can be partitioned into the subareas of the marine ecosystem where the recruitment patterning mortality might be occurring. Predators like striped bass *Morone saxatilis*, which are associated with estuaries and have increased in abundance as salmon survival has declined (Grout 2006), may be an important factor along the Maine coast as their abundance and hence distribution continues to expand. Salmon postsmolts are a food item of striped bass in the special case of predation associated with fishways (Blackwell and Juanes 1998), but they also appear to be a minor component of the diet when considered over a range of habitats (Walter et al. 2003). There is also evidence that the postsmolt transit through estuaries is sufficiently fast that predation there has only a minor effect (Lacroix et al. 2005; Serrano et al. 2009). In contrast, transit through the Gulf of Maine would introduce a broader spectrum of potential predators and take markedly longer. Though piscivorous fish species such as pollock (also known as saithe) *Pollachius virens* and Atlantic cod *Gadus morhua* have been documented to feed upon postsmolts in other areas (Hvidsten and Møkkelgjerd 1987; Hvidsten and Lund 1988; Jepsen et al. 2006), predation by the same or similar species in the Gulf of Maine is minimal (Link and Almeida 2000).

There are generally two schools of thought on the mechanism of fish predation on postsmolts. One view is that the relationship between predators and postsmolts is formed in terms of the functional predator–prey relationship between

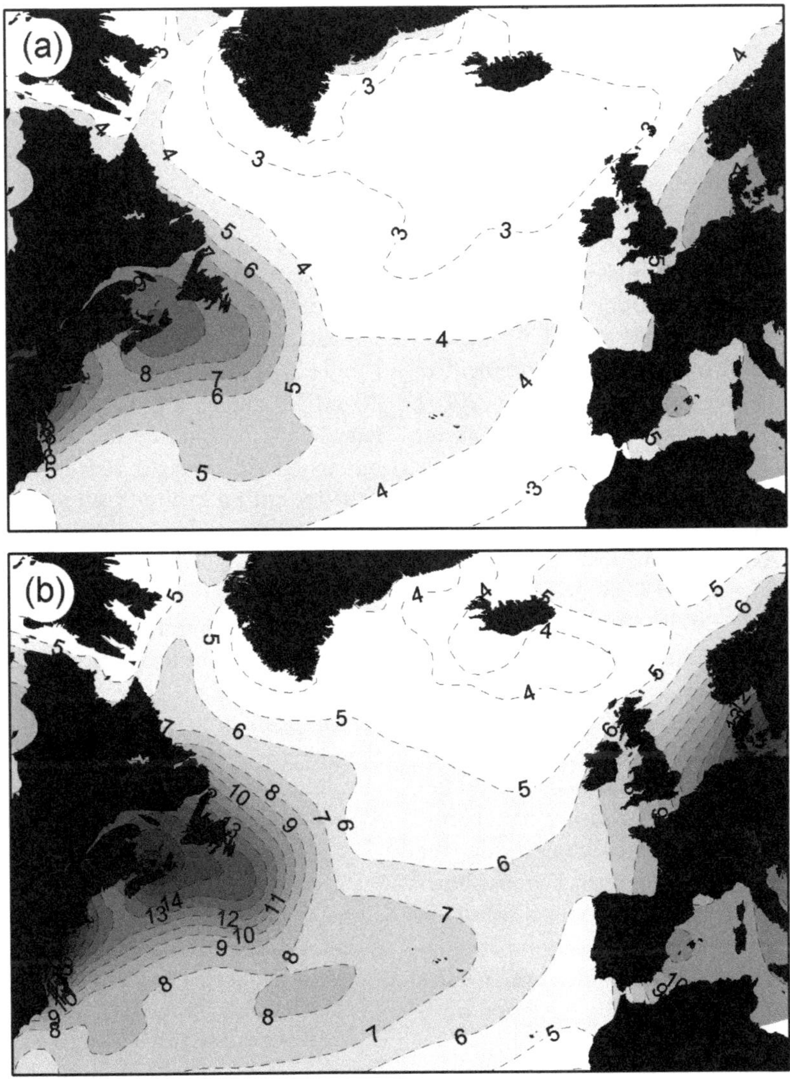

FIGURE 6. Spring (a) and annual (b) temperature range in the North Atlantic area in degrees centigrade.

them (Emmett and Sampson 2007; Sturdevant et al. 2009). The other school takes a more complex perspective relating the predation rate on postsmolts to the abundance of alternative prey abundances (Cooney et al. 2001; Svenning et al. 2005; Kline et al. 2008). The alternate prey hypothesis may be less viable relative to Atlantic salmon in the Gulf Maine during the estuarine portion of the postsmolt migration because other groups of diadromous fishes presenting similarly to predators, such as blueback herring *Alosa aestivalis*, alewife *A. pseudoharengus*, and American shad *A. sapidissima* juveniles, tend to migrate much later in the year compared to salmon smolts (Iafrate and Oliveira 2008), but that does not consider the ambient prey field. The Gulf of Maine proper has many species that would offer alternative targets to salmon postsmolt; the role of alternative prey may emerge from data collections documenting predation on postsmolts (Link and Almeida 2000).

In addition to potential predation by finfish, there may be significant predation by marine mammals and seabirds on postsmolts during their migrations from natal rivers and through the Gulf of Maine. A number of bird species have been demonstrated to prey upon salmon postsmolts, including cormorants (Blackwell et al. 1997), herring gulls (Dieperink et al. 2002), and gannets (Montevecchi and Cairns 2007). The predation effects of bird predators will vary in time and space (Overholtz and Link 2007) as affected by foraging range. Like fish predators, birds also migrate in response to environmental cues; thus, their arrival in feeding areas may shift annually in response to cues related to the movement of other predator or prey groups (Jonzén et al. 2007). There has been a demonstrated shift in marine mammal abundance in the Gulf of Maine during recent decades; in particular, the abundance of harbor seals Phoca vitulina concolor has increased (Gilbert et al. 2005). Harbors seals are of particular interest because their distribution and feeding ecology would suggest that they are capable of taking salmon postsmolts; a recent regional study of harbor seal diet does not show predation on salmon postsmolts, a result perhaps related to the low of abundance of salmon making them below detection limits in the diet (Wood 2001). Middlemas et al. (2007) did not find evidence that marine mammals were the predation factor causing the decline in salmonid abundance in Europe; however, this was attributed, in part, to the difficulty of assessing marine mammal predation. Until predation can be documented and consumption rates estimated, hypotheses concerning predation mortality vectors will remain speculative.

The persistent range of Atlantic salmon in North America may be defined by the seasonal dynamic of ocean thermal conditions. Prognosis for Atlantic salmon populations under climate change scenarios has primarily focused on the anticipated change of conditions in the freshwater rearing areas (Mather et al. 2008; Lassalle and Rochard 2009). Though not an issue now, terrestrial temperature forecasts raise the concern that summer temperature may exceed tolerable limits for juvenile salmon and that the extant usable rearing habitat will shrink. However, marine thermal conditions are already a problem for southern tier salmon stocks in North America and Europe for varying reasons. It would stand to reason that the concomitant increase in northern tier stocks is a result of these respective thermal effects (Niemelä et al. 2004; Chaput et al. 2005). The phenological mismatch we identified as critical to North American stocks is thought to be associated with the part of the western Atlantic with an annual temperature range exceeding 12°C, yet in the Gulf of Maine alone, a temperature increase on the order of 2°C is expected over the next century (Fogarty et al. 2008). Considering that increased temperatures in the middle-latitude, large marine ecosystem of the northeast U.S. continental shelf has resulted in increasing annual range (Friedland and Hare 2007), phenological issues with smolt migration timing will only become more problematic. It is recognized that the nuisances of the mechanistic linkages between recruitment and climate are critical to any climate change assessment (Cury et al. 2008; Hollowed et al. 2009); we would thus argue that a core issue for modeling Atlantic salmon response to future climate change in North America is the control of smolt migration in regard to ocean conditions for postsmolts. Because annual thermal range of the coastal ocean associated with salmon rivers in most of Europe is less than 6°C, climate-change driven change to species range and recruitment will reflect shifting ocean productivity during summer in the northeast Atlantic.

References

Auster, P. J., and J. S. Link. 2009. Compensation and recovery of feeding guilds in a northwest Atlantic shelf fish community. Marine Ecology Progress Series 382:163–172.

Beaugrand, G., and P. C. Reid. 2003. Long-term changes in phytoplankton, zooplankton and salmon linked to climate. Global Change Biology 9:801–817.

Blackwell, B. F., and F. Juanes. 1998. Predation on Atlantic salmon smolts by striped bass after dam passage. North American Journal of Fisheries Management 18:936–939.

Blackwell, B. F., W. B. Krohn, N. R. Dube, and A. J. Godin. 1997. Spring prey use by double-crested cormorants on the Penobscot River, Maine, USA. Colonial Waterbirds 20:77–86.

Chaput, G., C. M. Legault, D. G. Reddin, F. Caron, and P. G. Amiro. 2005. Provision of catch advice taking account of non-stationary in productivity of Atlantic salmon (*Salmo salar* L.) in the northwest Atlantic. ICES Journal of Marine Science 62:131–143.

Cooney, T. T., J. R. Allen, M. A. Bishop, D. L. Eslinger, T. Kline, B. L. Norcross, C. P. McRoy, J. Milton, J. Olsen, V. Patrick, A. J. Paul, D. Salmon, D. Scheel, G. L. Thomas, S. L. Vaughan, and T. M. Willette. 2001. Ecosystem controls of juvenile pink salmon (*Oncorhynchus gorbuscha*) and Pacific herring (*Clupea pallasi*) populations in Prince William Sound, Alaska. Fisheries Oceanography 10(Supplement 1):1–13.

Cury, P. M., Y. J. Shin, B. Planque, J. M. Durant, J. M. Fromentin, S. Kramer-Schadt, N. C. Stenseth, M. Travers, and V. Grimm. 2008. Ecosystem oceanography for global change in fisheries. Trends in Ecology and Evolution 23:338–346.

Dieperink, C., B. D. Bak, L. F. Pedersen, M. I. Pedersen, and S. Pedersen. 2002. Predation on Atlantic salmon and sea trout during their first days as postsmolts. Journal of Fish Biology 61:848–852.

Dutil, J. D., and J. M. Coutu. 1988. Early marine life of Atlantic salmon, *Salmo salar*, postsmolts in the northern Gulf of St. Lawrence. Fishery Bulletin 86:197–211.

Emmett, R. L., and D. B. Sampson. 2007. The relationships between predatory fish, forage fishes, and juvenile salmonid marine survival off the Columbia River: a simple trophic model analysis. California Cooperative Oceanic Fisheries Investigations Reports 48:92–105.

Enders, E. C., D. Boisclair, and A. G. Roy. 2005. A model of total swimming costs in turbulent flow for juvenile Atlantic salmon (*Salmo salar*). Canadian Journal of Fisheries and Aquatic Sciences 62:1079–1089.

Fogarty, M., L. Incze, K. Hayhoe, D. Mountain, and J. Manning. 2008. Potential climate change impacts on Atlantic cod (*Gadus morhua*) off the northeastern USA. Mitigation and Adaptation Strategies for Global Change 13:453–466.

Friedland, K., G. Chaput, and J. MacLean. 2005. The rmerging tole of climate in post-smolt growth of Atlantic salmon. ICES Journal of Marine Science 62:1338–1349.

Friedland, K. D., and R. E. Haas. 1996. Marine post-smolt growth and age at maturity of Atlantic salmon (*Salmo salar*). Journal of Fish Biology 48:1–15.

Friedland, K. D., R. E. Haas, and T. F. Sheehan. 1996. Post-smolt growth, maturation, and survival of two stocks of Atlantic salmon. Fishery Bulletin 94:654–663.

Friedland, K. D., and J. A. Hare. 2007. Long-term trends and regime shifts in sea surface temperature on the continental shelf of the northeast United States. Continental Shelf Research 27:2313–2328.

Friedland, K. D., J. C. MacLean, L. P. Hansen, A. J. Peyronnet, L. Karlsson, D. G. Reddin, N. Ó. Maoiléidigh, and J. L. McCarthy. 2009a. The recruitment of Atlantic salmon in Europe. ICES Journal of Marine Science 66:289–304.

Friedland, K. D., D. Moore, and F. Hogan. 2009b. Retrospective growth analysis of Atlantic salmon from the Miramichi River, Canada. Canadian Journal of Fisheries and Aquatic Sciences 66:1294–1308.

Friedland, K. D., D. G. Reddin, and M. Castonguay. 2003a. Ocean thermal conditions in the post-smolt nursery of North American Atlantic salmon. ICES Journal of Marine Science 60:343–355.

Friedland, K. D., D. G. Reddin, and J. F. Kocik. 1993. Marine survival of North American and European Atlantic salmon: effects of growth and environment. ICES Journal of Marine Science 50:481–492.

Friedland, K. D., D. G. Reddin, J. R. McMenemy, and K. F. Drinkwater. 2003b. Multidecadal trends in North American Atlantic salmon (*Salmo salar*) stocks and climate trends relevant to juvenile survival. Canadian Journal of Fisheries and Aquatic Sciences 60:563–583.

Gilbert, J. R., G. T. Waring, K. M. Wynne, and N. Guldager. 2005. Changes in abundance of harbor seals in Maine, 1981–2001. Marine Mammal Science 21:519–535.

Grout, D. E. 2006. Interactions between striped bass (*Morone saxatilis*) rebuilding programmes and the conservation of Atlantic salmon (*Salmo salar*) and other anadromous fish species in the USA. ICES Journal of Marine Science 63:1346–1352.

Hedger, R. D., F. Martin, D. Hatin, F. Caron, F. G. Whoriskey, and J. J. Dodson. 2008. Active migration of wild Atlantic salmon *Salmo salar* smolt through a coastal embayment. Marine Ecology Progress Series 355:235–246.

Hogan, F., and K. D. Friedland. 2010. Retrospective growth analysis of Atlantic salmon *Salmo salar* and implications for abundance trends. Journal of Fish Biology 76:2502–2520.

Hollowed, A. B., N. A. Bond, T. K. Wilderbuer, W. T. Stockhausen, Z. T. A'mar, R. J. Beamish, J. E. Overland, and M. J. Schirripa. 2009. A framework for modelling fish and shellfish responses to future climate change. ICES Journal of Marine Science 66:1584–1594.

Holm, M., J. C. Holst, and L. P. Hansen. 2000. Spatial and temporal distribution of post-smolts of Atlantic salmon (*Salmo salar* L.) in the Norwegian Sea and adjacent areas. ICES Journal of Marine Science 57:955–964.

Hvidsten, N. A., and R. A. Lund. 1988. Predation on hatchery reared and wild smolts of Atlantic salmon in the estuary of River Orkla, Norway. Journal of Fish Biology 33:121–126.

Hvidsten, N. A., and P. I. Møkkelgjerd. 1987. Predation on salmon smolts in the estuary of the River Surna, Norway. Journal of Fish Biology 30:273–280.

Iafrate, J., and K. Oliveira. 2008. Factors affecting migration patterns of juvenile river herring in a coastal Massachusetts stream. Environmental Biology of Fishes 81:101–110.

Jepsen, N., E. Holthe, and F. Okland. 2006. Observations of predation on salmon and trout smolts in a river mouth. Fisheries Management and Ecology 13:341–343.

Jonzén, N., A. Hedenström, and P. Lundberg. 2007. Climate change and the optimal arrival of migratory birds. Proceedings of the Royal Society B 274:269–274.

Kahl, J. S., J. L. Stoddard, R. Haeuber, S. G. Paulsen, R. Birnbaum, F. A. Deviney, J. R. Webb, D. R. DeWalle, W. Sharpe, C. T. Driscoll, A. T. Herlihy, J. H. Kellogg, P. S. Murdoch, K. Roy, K. E. Webster, and N. S. Urquhart. 2004. Have U.S. surface waters responded to the 1990 Clean Air Act amendments? Environmental Science and Technology 38:484A–490A.

Kalnay, E., M. Kanamitsu, R. Kistler, W. Collins, D. Deaven, L. Gandin, M. Iredell, S. Saha, G. White, J. Woollen, Y. Zhu, M. Chelliah, W. Ebisuzaki, W. Higgins, J. Janowiak, K. C. Mo, C. Ropelewski, J. Wang, A. Leetmaa, R. Reynolds, R. Jenne, and D. Joseph. 1996. The NCEP/NCAR 40-year reanalysis project. Bulletin of the American Meteorological Society 77:437–470.

Kline, T. C., J. L. Boldt, E. V. Farley, L. J. Haldorson, and J. H. Helle. 2008. Pink salmon (*Oncorhynchus gorbuscha*) marine survival rates reflect early marine carbon source dependency. Progress in Oceanography 77:194–202.

Lacroix, G. L., and D. Knox. 2005. Distribution of Atlantic salmon (*Salmo salar*) postsmolts of different origins in the Bay of Fundy and Gulf of Maine and evaluation of factors affecting migration, growth, and survival. Canadian Journal of Fisheries and Aquatic Sciences 62:1363–1376.

Lacroix, G. L., D. Knox, and M. J. W. Stokesbury. 2005. Survival and behaviour of post-smolt Atlantic salmon in coastal habitat with extreme tides. Journal of Fish Biology 66:485–498.

Lassalle, G., and E. Rochard. 2009. Impact of twenty-first century climate change on diadromous fish spread over Europe, North Africa and the Middle East. Global Change Biology 15:1072–1089.

Link, J. S., and F. P. Almeida. 2000. An overview and history of the food web dynamics program of the Northeast Fisheries Science Center, Woods Hole, Massachusetts. National Oceanic and Atmospheric Administration, National Marine Fisheries Service, Northeast Fisheries Science Center, Technical Memorandum NMFS-NE-159, Woods Hold, Massachusetts.

Link, J. S., B. Bogstad, H. Sparholt, and G. R. Lilly. 2009. Trophic role of Atlantic cod in the ecosystem. Fish and Fisheries 10:58–87.

Loge, F. J., M. R. Arkoosh, T. R. Ginn, L. L. Johnson, and T. K. Collier. 2005. Impact of environmental stressors on the dynamics of disease transmission. Environmental Science and Technology 39:7329–7336.

Mather, M. E., D. L. Parrish, C. A. Campbell, J. R. McMenemy, and J. M. Smith. 2008. Summer temperature variation and implications for juvenile Atlantic salmon. Hydrobiologia 603:183–196.

McCarthy, J. L., K. D. Friedland, and L. P. Hansen. 2008. Monthly indices of the post-smolt growth of Atlantic salmon from the Drammen River, Norway. Journal of Fish Biology 72:1572–1588.

McCormick, S. D., D. T. Lerner, M. Y. Monette, K. Nieves-Puigdoller, J. T. Kelly, and B. T. Björnsson. 2009. Taking it with you when you go: how perturbations to the freshwater environment, including temperature, dams, and contaminants, affect marine survival of salmon. Pages 195–214 *in* A. Haro, K. L. Smith, R. A. Rulifson, C. M. Moffitt, R. J. Klauda, M. J. Dadswell, R. A. Cunjak, J. E. Cooper, K. L. Beal, and T. S. Avery, editors. Challenges for diadromous fishes in a dynamic global environment. American Fisheries Society, Symposium 69, Bethesda, Maryland.

Middlemas, S. J., J. D. Armstrong, and P. M. Thompson. 2007. The significance of marine mammal predation on salmon and sea trout. Pages 41–60 *in* D. Mills, editor. Salmon at the edge, Blackwell Scientific Publications, Oxford, UK.

Mohseni, O., and H. G. Stefan. 1999. Stream temperature/air temperature relationship: a physical interpretation. Journal of Hydrology 218:128–141.

Montevecchi, W. A., and D. K. Cairns. 2007. Predation on post-smolt Atlantic salmon by gannets: research implications and opportunities. Pages 61–77 *in* D. Mills, editor. Salmon at the edge, Blackwell Scientific Publications, Oxford, UK.

Morrill, J. C., R. C. Bales, and M. H. Conklin. 2005. Estimating stream temperature from air temperature: implications for future water quality. Journal of Environmental Engineering 131:139–146.

Niemelä, E., J. Erkinaro, J. B. Dempson, M. Julkunen, A. Zubchenko, S. Prusov, M. A. Svenning, R. Ingvaldsen, M. Holm, and E. Hassinen. 2004. Temporal synchrony and variation in abundance of Atlantic salmon (*Salmo salar*) in two subarctic Barents Sea rivers: influence of oceanic conditions. Canadian Journal of Fisheries and Aquatic Sciences 61:2384–2391.

Nye, J. A., J. S. Link, J. A. Hare, and W. J. Overholtz. 2009. Changing spatial distribution fish stocks in relation to climate and population size within the northeast US continental shelf. Marine Ecology Progress Series 393:111–129.

Overholtz, W. J., and J. S. Link. 2007. Consumption impacts by marine mammals, fish, and seabirds on the Gulf of Maine–Georges Bank Atlantic herring (*Clupea harengus*) complex during the years 1977–2002. ICES Journal of Marine Science 64:83–96.

Peyronnet, A., K. D. Friedland, N. Ó Maoiléidigh, M. Manning, and W. R. Poole. 2007. Links between patterns of marine growth and survival of Atlantic salmon *Salmo salar*, L. Journal of Fish Biology 71:684–700.

Pyper, B. J., and R. M. Peterman. 1998. Comparison of methods to account for autocorrelation in correlation analyses of fish data. Canadian Journal of Fisheries and Aquatic Sciences 55:2127–2140.

Richardson, A. J., and D. S. Schoeman. 2004. Climate impact on plankton ecosystems in the northeast Atlantic. Science 305:1609–1612.

Serrano, I., P. Rivinoja, L. Karlsson, and S. Larsson. 2009. Riverine and early marine survival of stocked salmon smolts, *Salmo salar* L., descending the Testebo River, Sweden. Fisheries Management and Ecology 2009 16:386–394.

Smith, T. M., R. W. Reynolds, T. C. Peterson, and J. Lawrimore. 2008. Improvements to NOAA's historical merged land-ocean surface temperature analysis (1880–2006). Journal of Climate 21:2283–2296.

Sogard, S. M. 1997. Size-selective mortality in the juvenile stage of teleost fishes: a review. Bulletin of Marine Science 60:1129–1157.

Sturdevant, M. V., M. F. Sigler, and J. A. Orsi. 2009. Sablefish predation on juvenile Pacific salmon in the coastal marine waters of Southeast Alaska in 1999. Transactions of the American Fisheries Society 138:675–691.

Svenning, M.-A., R. Borgstrøm, T. O. Dehli, G. Moen, R. T. Barrett, T. Pedersen, and W. Vader. 2005. The impact of marine fish predation on Atlantic salmon smolts (*Salmo salar*) in the Tana estuary, north Norway, in the presence of an alternative prey, lesser sandeel (*Ammodytes marinus*). Fisheries Research 76:466–474.

Walter, J. F., A. S. Overton, K. H. Ferry, and M. E. Mather. 2003. Atlantic coast feeding habits of striped bass: a synthesis supporting a coast-wide understanding of trophic biology. Fisheries Management and Ecology 10:349–360.

Whalen, K. G., D. L. Parrish, and S. D. McCormick. 1999. Migration timing of Atlantic salmon smolts relative to environmental and physiological factors. Transactions of the American Fisheries Society 128:289–301.

Wood, S. A. 2001. Summary of harbor seal (*Phoca vitulina concolor*) food habits in mid-coast Maine: summer 2000. Marine Environmental Research Institute, Blue Hill, Maine.

American Fisheries Society Symposium 79:409–415, 2012

A Perspective on Advancing Ecosystem Research for the Gulf of Maine

Robert L. Stephenson*
St. Andrews Biological Station
531 Brandy Cove Road, St. Andrews, New Brunswick E5B 2L9, Canada

Abstract.—There have been significant advances in our understanding of the Gulf of Maine, and federal, provincial, and state jurisdictions in Canada and the United States have moved forward with ecosystem approaches to management. The international context for an ecosystem based approach has evolved, and there have been legislative changes in both the United States and Canada. There has also been an evolution of the public perspective with respect to the need for enhanced stewardship in the face of ecological change. There is increasing awareness of the need for development of an integrated approach to the management of multiple human activities in relation to a more diverse set of objectives that include a higher standard of ecological integrity and diverse aspects of sustainability, in a changing environment.

The priorities for implementation of an integrated or ecosystem approach appear to be (1) consensus in developing (and leadership in implementing) a common vision, (2) common treatment of all activities, (3) management of cumulative impacts of multiple overlapping activities, (4) an expanded set of objectives with higher standards of implementation, (5) an interdisciplinary approach, and (6) participatory governance. The resulting research priorities include (a) articulation of objectives and desired attributes/services (with relevant performance indicators), (b) decision-support methods, and (c) methods for evaluating cumulative impacts.

Introduction

The Gulf of Maine (GOM) is arguably one of the best studied marine ecosystems in the world. The laboratories at Woods Hole, Massachusetts and St. Andrews, New Brunswick are among the oldest marine research facilities in the world, and interest in the Gulf of Maine's physical environment (including Bay of Fundy), fisheries (including Georges Bank), and Canada/USA boundary have resulted in considerable research attention in the Gulf of Maine for more than a century. The Gulf of Maine is also highly managed by two advanced nations with the recent commitment to implementing an ecosystem approach to management (EAM) or ecosystem-based management (EBM). The Gulf of Maine is, therefore, a logical case study in which to investigate the evolution of the ecosystem approach to management and the supporting research.

This perspective paper is written from the view of both a researcher with more than 25 years of experience in fisheries and marine research in the Gulf of Maine and a manager who has been involved with shaping the research of a major research institution (the St. Andrews Biological Station) to meet the needs of future EBM. I suggest at the outset that the need is for development of an integrated approach to the management of multiple activities, in relation to a more diverse set of objectives with a higher standard of ecological integrity, in a changing environment.

* Corresponding author: robert.stephenson@dfo-mpo.gc.ca

The Evolving Landscape of Management

The landscape of management has evolved, both in the Gulf of Maine and internationally, in recent years. Prompted by international agreements and conventions, now referred to in national legislation (Canada's Oceans Act and Species at Risk Act; USA Ocean Policy), the management of fisheries and other activities is evolving rapidly towards more holistic ecosystem-based and integrated approaches. While not yet fully articulated or operational, these approaches will inevitably include more diverse objectives with respect to the impacts of fishing and other activities on the ecosystem, as well as increased awareness of the impact of the ecosystem (including ecosystem change) on management and will demand integration of conservation (including maintenance of productivity at the organism and trophic levels, conservation of biodiversity of populations and communities, and protection of habitat), as well as social, economic, and institutional aspects.

Accompanying, and in fact helping to stir, this management change has been a recent increase in public awareness of the diverse suite of ecosystem services that humankind derives from the coastal environment and an increase in appreciation of and respect for the concept of sustainability. This has been manifest, for example, in marketplace pressure for products that are seen as being sustainable and a blossoming of programs to certify products as being sustainable or harvested responsibly. Increasingly, there is the need for fisheries to consider views of the general public (especially consumers) and to maintain a social license through certification or other forms of social acceptance.

These developments are causing profound change in evaluation and management of fishing and other human activities in the coastal zone. Fisheries and other marine activities are facing a number of challenges, including the combined issues of sustainability, viability, and competitiveness, as well as the uncertainties of environmental and economic variability of evolving EBM approaches.

The changing landscape of management is illustrated in Figure 1. Individual activities (fishing, aquaculture, transportation, etc.) are being managed with an expanding suite of objectives that include conservation, social, economic, and institutional elements. At the same time, these activities are being forced to conform to external marketplace pressures for certification of sustainability or responsibility. Further, there is increasing need to resolve space use conflicts among these activities that are currently managed separately (in many cases by different levels of government using different processes) and pressure to consider the cumulative effects of multiple activities in relation to overall sustainability of a region or within an ecosystem. There has been an increase in the need to make decisions between competing uses and among conflicting objectives. Increased public awareness of, and concern for, the coastal zone has resulted in demand for a more participatory governance and management structure and more transparency with respect to decisions. It is also anticipated that the management system will soon evolve to review or audit the cumulative impact of all activities of an area in terms of the combined impact (an ecosystem assessment).

Having watched recent events, including the evolution of traditional management of fishing and aquaculture, the early development of EAM, and experiments such as the Southwest New Brunswick Marine Resource Planning Initiative (http://bofmrp.ca/home), I suggest that there are five major challenges of management:

1. There is no common vision for an ecosystem approach or for an operational system of in tegrated management (IM). I have been asked by industry, "How much do I need to change to comply with an ecosystem approach?" But there is no minimum standard, and I suggest that it is fair to say that we would have a hard time determining when we have reached an ecosystem approach.
2. There is insufficient consideration of the cumulative impacts of single activities (for example, the sum of all fisheries or of all aquaculture sites) and certainly no consideration of the cumulative impacts of multiple activities (aquaculture plus fishing, etc.). Our current structures of evaluation and of management are not yet dealing with this issue.
3. We face the complication of managing multiple activities by different groups using different tools, standards, and approaches. In the lower Bay of Fundy, for example, fisher-

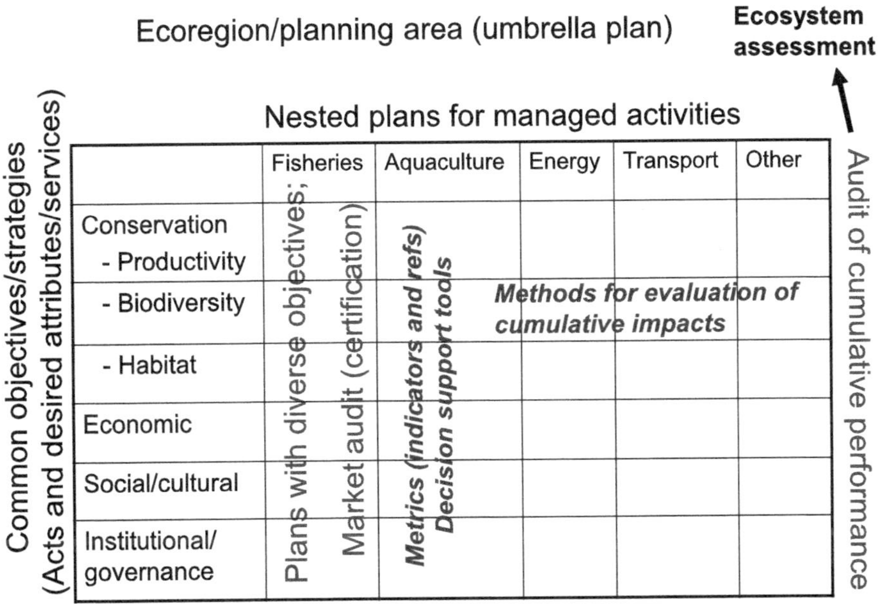

FIGURE 1. The evolving landscape of management. An ecosystem approach may be achieved with nested plans within an ecoregion or planning area that conform to a common set of objectives (based on regulations and desired attributes or ecosystem services). Activities are likely to be evaluated (audited) in three ways: (1) individual management plans, (2) marketplace certification, and (3) the evaluation of cumulative performance (=ecosystem assessment). This evolution will require a participatory governance regime. Key research needs include the metrics associated with an expanded list of objectives of ecosystem approach to management (including indicators and reference points), decision support tools, and methods for evaluation the cumulative impact of multiple activities.

ies are regulated federally and aquaculture is regulated provincially with very different approaches.

4. There seems to be insufficient public appreciation of the treatment of trade-offs among conflicting objectives. There is insufficient feedback regarding decisions and insufficient documentation of any consideration of how allowing one activity impacts others.
5. There is a perception of a lack of transparency of governance and a lack of participation in management. While we have a considerable amount of consultation, it is claimed by some that there is insufficient feedback following deliberation and on decisions and that those with interest in activities want to be part of the governance process.

While it could be argued that points 4 and 5 could be fixed within current management structures, I suggest at least the first three (the common vision for EAM/IM, the consideration of cumulative effects, and the consistent management of multiple activities) require major changes in management approach, governance, and methods.

Perspectives on Ecosystem Approach to Management/Ecosystem-Based Management

The presentations and papers arising from the 2009 GOM symposium demonstrate the diversity of current thinking regarding the ecosystem approach to management.

The recently published book edited by McLeod and Leslie (2009), *Ecosystem-Based Management for the Oceans*, adopts a broad definition of EBM based on the 2005 "Scientific Consensus Statement on Marine Ecosystem-Based Management" (McLeod et al. 2005):

> Ecosystem-based management is an integrated approach to management that considers the entire ecosystem, including

> humans. The goal of ecosystem-based management is to maintain an ecosystem in a healthy, productive and resilient condition so that it can provide the services humans want and need. Ecosystem-based management differs from current approaches that usually focus on single species, sector, activity, or concern; it considers the cumulative impacts of different sectors.

This definition is general (as are most others), and within it there is a great deal of scope for interpretation. Most papers in the symposium referred explicitly to an ecosystem approach, yet the papers differed greatly in scope and concept.

Several papers are rooted in the notion that an ecosystem approach requires better understanding of the ecosystem and that we need to understand more about the ecosystem in order to maintain a healthy, productive, and resilient system (Fogarty et al. 2012b; Hale 2012; Johnson and Hare; Lawton et al. 2012; Walmsley 2012; all this volume). Other papers focus on how management of human activities must change, recognizing that we do not manage the ecosystem, but that the scope for action and improvement is in how we manage human activities (McCay 2012; Arkema and Samhouri 2012; d'Entrement 2012; McGee 2012; all this volume).

Our understanding of the ecosystem is certainly incomplete. Further our understanding is also biased towards large organisms and processes that have been known to be important to past activities. We can predict that an ecosystem approach will require new and different information. Yet I fear that we do not have the research capacity to focus on understanding the ecosystem. I favor an applied approach whereby the needs for ecosystem understanding are identified through (and prioritized by) the needs of management—and that research into ecosystem understanding is tied explicitly to the use of that information in management decisions. In this approach, EAM or EBM is really an approach to integrated management.

The papers presented at the symposium, and represented in this volume, provide a useful time stamp on the development of the applied ecosystem-based approach. The paper by Fogarty et al. (2012a, this volume), for example, outlines an indicator-based approach to tracking the state of the ecosystem in a driver-pressure-state-impact-response (DPSIR) framework being piloted at the National Marine Fisheries Service. The symposium presentation by Gavaris (published as Gavaris 2009) outlined a pragmatic approach to the progressive evolution of fisheries management through expanded set of objectives, strategies, and tactics that has been adopted by Maritimes Region of Fisheries and Oceans Canada as a practical implementation of EAM. Walmsley (2012), for example, outlines a Web-based approach to state-of-the-environment reporting within the DPSIR framework being proposed by the Gulf of Maine Council on the Marine Environment.

While there may be consensus on the general direction of EAM, EBM, or IM, there has been insufficient experience to allow consensus on practical application, and there is currently a range of approaches in the Gulf of Maine.

The Challenges of Implementing an Ecosystem Approach

The challenge of an ecosystem approach is to implement an integrated approach to the management of multiple activities in relation to a more diverse set of objectives with a higher standard of ecological integrity in a changing environment (Figure 2). This is about managing people and activities in an integrated governance regime. I suggest that there are a number of requirements:

1. *Common vision for EAM/IM (and leader ship in implementation).*—There is need for consensus on a vision of what EAM is and how it will be implemented. At present, there is progress under the flag of EAM in several areas of the Gulf of Maine, but no vision for how the various pieces will fit together. It is as if we are sewing squares of cloth for a quilt with no design for the finished quilt and no idea who is going to sew the pieces together. It seems obvious that we need to (a) define a common strategy or a complimentary set of strategies (the quilt of EAM/IM), and (b) move management of existing activities to ward EAM, recognizing that management of a single activity is not in itself sufficient, but

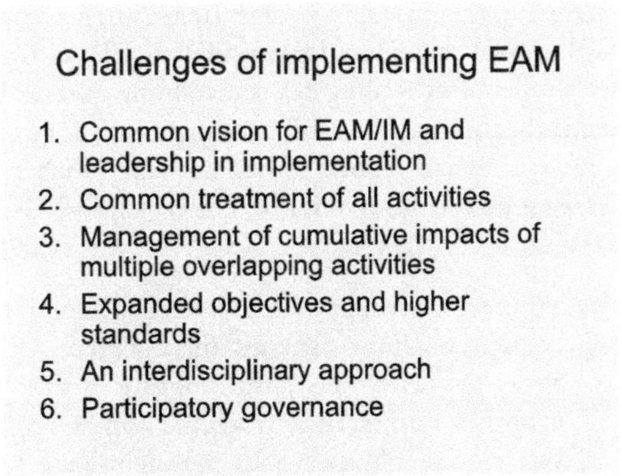

Challenges of implementing EAM

1. Common vision for EAM/IM and leadership in implementation
2. Common treatment of all activities
3. Management of cumulative impacts of multiple overlapping activities
4. Expanded objectives and higher standards
5. An interdisciplinary approach
6. Participatory governance

FIGURE 2. The proposed challenges or requirements of an integrated or ecosystem approach to management.

that all activities need to be managed in a manner consistent with EAM.

2. *Common treatment of all activities.*—We currently have separate management structures for different activities, and there is insufficient integration across activities. We need some form of integration across fisheries, as well as among fisheries, aquaculture, transportation, energy, and other activities. This is a prerequisite to consideration of the cumulative effects of multiple activities.
3. *Emphasis on management and governance.*—The ecosystem approach to management is primarily about management rather than the ecosystem. The challenge of EAM is the management of a diverse set of people undertaking multiple activities in a complex ecological landscape. We need policy implementation in spite of limited knowledge of ecosystem structure and function. To do this, we need to employ techniques of management science, as suggested by Stephenson and Lane (e.g., Stephenson and Lane 2010) in an appeal for fisheries management science.
4. *More diverse objectives with higher standards.*—EAM requires a more diverse set of objectives (conservation, social, economic, and institutional) and higher standards. Ocean use has been undertaken without an appreciation of, and due regard for, the overall sustainability of activities. Recent international conventions demand higher standards, and consumers are increasingly demanding assurance of sustainability. Consideration of multiple activities requires consideration of cumulative impacts of multiple activities and a look at longer term performance with the view toward true sustainability.
5. *An interdisciplinary approach.*—The objectives of EAM (conservation, social, economic, and institutional) are inherently interdisciplinary. Further, the challenge of management is interdisciplinary in that it demands governance of human activity interacting with a natural system. Collaboration is key. EAM/IM requires the coordinated expertise of diverse disciplines that, for the most part, exist in separate worlds (stovepipes) with different languages and methodologies and have no tradition of working together.
6. *A participatory governance process.*—We have learned in fisheries management that we cannot manage (or at least it is very difficult to manage) an unwilling constituency. Lack of participation is one of the aspects of current management that is most criticized. There is a wealth of literature on the benefits of a participatory management regime.

The Scientific Requirements for Ecosystem Research

The rapidly evolving management landscape, which includes implementation of an ecosystem approach and the challenge of sustainability in a climate of economic and environmental variability, poses a number of research challenges and important research requirements. What are the research requirements of EAM/EBM?

First, there is an obvious need for research in support of the expanded set of objectives of EAM. While the choice of objectives requires a broader discussion, the scientific community can certainly predict the need for development of performance indicators and reference points and of monitoring related to the conservation objectives associated with productivity, biodiversity, and habitat. Implicit in this is the question of the impact of environmental change on reference points. Further, we can predict the need for articulation of performance indicators and reference points for a suite of yet unspecified objectives related to social, economic, and institutional goals. There is consensus on the high-level conservation objectives related to productivity, biodiversity, and habitat (Sinclair et al. 2002; Gavaris 2009) but no agreement to date on an analogous set of social, economic, and institutional objectives. Research into the objectives themselves and the performance indicators and reference points is required.

A second important line of research is related to the development of decision-support systems that will help inform management in relation to the trade-offs among potentially conflicting objectives. An expanded set of objectives, required by EAM, will inevitably lead to the need to quantify the relative performance of different potential management scenarios and to trade-offs among objectives in management decisions (including overall cost–benefit-type analyses). This requires development or adoption of methods of management science (Stephenson and Lane 2010).

A third, important, major line of research is on the cumulative impact of multiple activities. While there has long been recognition of the need to consider the sum of activities (the total impact of an activity such as fishing and the combined impact of fishing plus aquaculture plus others) in light of carrying capacity of an area, this has not been achieved. There is urgent need for both methodological research and evaluation of governance options that would allow evaluation and management of cumulative impacts.

These are all applied areas of research and are directed explicitly to improved decision making and management rather than to understanding of the ecosystem structure and function. There is a need to enhance knowledge of aquatic ecosystems health and productivity, and of ecosystem services, but in relation to the management of human activities. This includes increased knowledge of the human impacts on vulnerable habitats and aquatic ecosystems to mitigate these impacts through sustainable operations and improved management decision making and knowledge that will fill critical gaps to allow evaluation of the effectiveness of management strategies on activities.

In my opinion, the priority is for targeted research that is directly relevant to emerging decision making and management. Increasingly there is the need to undertake research that links the four pillars (bio-ecological, social, economic, and institutional) of sustainability, as expressed, for example, in recent reports of the Canadian Fisheries Resource Conservation Council (e.g., FRCC 2007:9):

> While the concept of "sustainability" has been implicit in fisheries management from the early days, the concept has evolved more recently from the conservation of single species to the inclusion of both ecosystem and human factors, with a balance between resource conservation and human concerns. In the context of sustainable utilization, the modern concept of sustainability is seen as having ecological, social, economic and institutional components.

The research required to meet the challenge of sustainability of activities in this rapidly changing environment demands collaboration of industry, government (science and management), and the academic community in an approach that is interdisciplinary and strategic.

References

Arkema, K. K., and J. F. Samhouri. 2012. Linking ecosystem health and services to inform marine ecosystem-based management. Pages 9–25 *in* R. L. Stephenson, J. H. Annala, J. A. Runge, and M. Hall-Arber, editors. Advancing an ecosystem approach in the Gulf of Maine. American Fisheries Society, Symposium 79, Bethesda, Maryland.

d'Entrement, J. G. 2012. Fishermen without borders. Pages 41–45 *in* R. L. Stephenson, J. H. Annala, J. A. Runge, and M. Hall-Arber, editors. Advancing an ecosystem approach in the Gulf of Maine. American Fisheries Society, Symposium 79, Bethesda, Maryland.

Fogarty, M. J., K. D. Friedland, L. Col, R. Gamble, J. Hare, K. Hyde, J. S. Link, S. Lucey, H. Liu, J. Nye, W. J. Overholtz, D. Richardson, B. Rountree, and M. Taylor. 2012a. Status of the northeast U.S. Continental Shelf large marine ecosystem: an indicator-based approach. Pages 139–165 *in* R. L. Stephenson, J. H. Annala, J. A. Runge, and M. Hall-Arber, editors. Advancing an ecosystem approach in the Gulf of Maine. American Fisheries Society, Symposium 79, Bethesda, Maryland.

Fogarty, M. J., D. Townsend, and E. Klein. 2012b. Advances in understanding ecosystem structure and function in the Gulf of Maine. Pages 261–272 *in* R. L. Stephenson, J. H. Annala, J. A. Runge, and M. Hall-Arber, editors. Advancing an ecosystem approach in the Gulf of Maine. American Fisheries Society, Symposium 79, Bethesda, Maryland.

FRCC (Fisheries Resource Conservation Council). 2007. Sustainable framework for Atlantic Lobster 2007. FRCC, FRCC.07.R1, Ottawa.

Gavaris, S. 2009. Fisheries management planning and support for strategic and tactical decisions in an ecosystem approach context. Fisheries Research 100:6–14.

Hale, S. S. 2012. Spatial patterns of subtidal benthic invertebrates and environmental factors in the nearshore Gulf of Maine. Pages 167–183 *in* R. L. Stephenson, J. H. Annala, J. A. Runge, and M. Hall-Arber, editors. Advancing an ecosystem approach in the Gulf of Maine. American Fisheries Society, Symposium 79, Bethesda, Maryland.

Johnson, C. L., and J. A. Hare. 2012. Zooplankton monitoring in the Gulf of Maine: past, present, and future. Pages 205–218 *in* R. L. Stephenson, J. H. Annala, J. A. Runge, and M. Hall-Arber, editors. Advancing an ecosystem approach in the Gulf of Maine. American Fisheries Society, Symposium 79, Bethesda, Maryland.

Lawton, P., L. S. Incze, and S. L. Ellis. 2012. Representation of biodiversity knowledge within ecosystem-based management approaches for the Gulf of Maine. Pages 375–391 *in* R. L. Stephenson, J. H. Annala, J. A. Runge, and M. Hall-Arber, editors. Advancing an ecosystem approach in the Gulf of Maine. American Fisheries Society, Symposium 79, Bethesda, Maryland.

McCay, B. J. 2012. Peopling the marine ecosystem. Pages 27–33 *in* R. L. Stephenson, J. H. Annala, J. A. Runge, and M. Hall-Arber, editors. Advancing an ecosystem approach in the Gulf of Maine. American Fisheries Society, Symposium 79, Bethesda, Maryland.

McGee, S. 2012. We really can get there from here: important steps toward ecosystem-based fishery management. Pages 35–40 *in* R. L. Stephenson, J. H. Annala, J. A. Runge, and M. Hall-Arber, editors. Advancing an ecosystem approach in the Gulf of Maine. American Fisheries Society, Symposium 79, Bethesda, Maryland.

McLeod, K. L., and H. Leslie. 2009. Ecosystem-based management for the oceans. Island Press, Washington, D.C.

McLeod, K. L., J. Lubchenco, S. R. Palumbi, and A. A. Rosenberg. 2005. Scientific consensus statement on marine ecosystem-based management. Signed by 221 academic scientists and policy experts with relevant expertise and published by the Communication Partnership for Science and the Sea. Available: www.compassonline.org/science/EBM_CMSP/EBMconsensus. (January 2012).

Sinclair, M., Arnason, R., Csirke, J., Karnicki, Z., Sigurjonsson, J., Rune Skjoldal, H., and G. Valdimarsson. 2002. Responsible fisheries in the marine ecosystem. Fisheries Research 58:255–265.

Stephenson, R. L., and D. E. Lane. 2010. Fisheries management science. Pages 630–637 *in* R. Q. Grafton. R. Hilborn, D. Squires, M. Tait, and M. Williams, editors. Handbook of marine fisheries conservation and management. Oxford University Press, Oxford, UK.

Walmsley, J. 2012. State-of-the-environment reporting for the Gulf of Maine. Pages 61–72 *in* R. L. Stephenson, J. H. Annala, J. A. Runge, and M. Hall-Arber, editors. Advancing an ecosystem approach in the Gulf of Maine. American Fisheries Society, Symposium 79, Bethesda, Maryland.